Parasitic Protozoa of Farm Animals and Pets

Monica Florin-Christensen
Leonhard Schnittger
Editors

Parasitic Protozoa of Farm Animals and Pets

Springer

Editors
Monica Florin-Christensen
Center for Research on Veterinary and Agronomic Sciences
National Institute of Agricultural Technology
Hurlingham, Argentina

National Council of Scientific and Technological Research (CONICET)
Buenos Aires, Argentina

School of Exact, Chemical and Natural Sciences (FCEQN), University of Morón
Morón, Province of Buenos Aires
Argentina

Leonhard Schnittger
Center for Research on Veterinary and Agronomic Sciences
National Institute of Agricultural Technology
Hurlingham, Argentina

National Council of Scientific and Technological Research (CONICET)
Buenos Aires, Argentina

School of Exact, Chemical and Natural Sciences (FCEQN), University of Morón
Morón, Province of Buenos Aires
Argentina

ISBN 978-3-030-09936-7 ISBN 978-3-319-70132-5 (eBook)
https://doi.org/10.1007/978-3-319-70132-5

Printed on acid-free paper

This Springer imprint is published by the registered company Springer International Publishing AG part of Springer Nature
The registered company address is: Gewerbestrasse 11, 6330 Cham, Switzerland

Dedicatory

To our children, Nicolas, Ana, Julieta, and Uriel

Contents

About the Editors

Monica Florin-Christensen and Leonhard Schnittger are biologists and researchers of the National Council of Scientific and Technological Research of Argentina (CONICET). They lead a research group that investigates on molecular parasitology, focusing on pathogenic protists of veterinary importance, at the Institute of Pathobiology, National Institute of Agricultural Technology (INTA, Argentina). Both are Professors of molecular biology and population genetics at the School of Exact, Chemical and Natural Sciencies, University of Moron, Argentina.

Introduction into Parasitic Protozoa

1

Leonhard Schnittger and Monica Florin-Christensen

Abstract

Parasitic protozoans are protists—i.e., unicellular eukaryotes. Besides, they share the common features of being heterotrophic, motile in at least one of their stages, and dependent on a host for survival. Protists are a highly diverse evolutionary unrelated grouping that is at present taxonomically divided into at least five supergroups or eukaryotic kingdoms: SAR, Excavata, Amoebozoa, Archaeplastida, and Ophistokonta. The protozoan parasites that infect farm animals and pets, and are the subject of this book, belong to the first three supergroups. These organisms can cause significant morbidity and mortality in domestic animals, thus provoking substantial economic losses associated with livestock production or major concern for pet owners. Additionally, some are zoonotic and of public health relevance. Human domestication has led to a huge world population of a few selected farm and pet animals, usually raised in close proximity. This has provided an ideal and spacious ecological niche for parasitic protozoans that invade and thrive in these hosts. In addition, international travel has promoted global distribution of most parasitic protozoa. A long history of coadaptation has existed between parasites and their wild ancestor hosts, whereas adaptation to domestic animals has been much shorter from an evolutionary perspective. While the first scenario resulted in unapparent infections that ensured

L. Schnittger (✉) • M. Florin-Christensen
Center for Research on Veterinary and Agronomic Sciences
National Institute of Agricultural Technology
Hurlingham, Argentina

National Council of Scientific and Technological Research (CONICET), Buenos Aires, Argentina

School of Exact, Chemical and Natural Sciences (FCEQN), University of Morón, Morón, Province of Buenos Aires, Argentina
e-mail: schnittger.leonhard@inta.gob.ar

M. Florin-Christensen, L. Schnittger (eds.), *Parasitic Protozoa of Farm Animals and Pets*, https://doi.org/10.1007/978-3-319-70132-5_1

parasite perpetuation, the second represents an unbalanced situation resulting in clinical signs and mortality. This calls for continuous efforts in the development and application of control measures.

1.1 Protozoa as a Subgroup of Protists

Protozoa represent a subgroup of the protists. Then, what are protists? Originally, the name referred to all eukaryotic unicellular organisms that were joined into the taxonomic kingdom Protista in the year 1866 by Ernst Haeckel. The group of protists was further subdivided into (1) animal-like protozoans, i.e., heterotrophic/parasitic motile eukaryotes; (2) plantlike protophyta or algae, i.e., autotrophic unicellular eukaryotes; and (3) fungus-like saprophytes, i.e., unicellular eukaryotic decomposers with extracellular digestion and spore production. However, with the advancement of science, it became clear that these subdivisions are artificial, as they do not reflect valid taxonomic groups based on evolutionary relatedness. Nevertheless, all protists that are non-photosynthetic and non-fungus-like are, for convenience, often referred to as protozoans.

The simplest definition of a protist is exclusive: a eukaryote that is neither an animal nor a plant or a fungus. However, based on genetic similarities or differences, protists represent a paraphyletic assembly, i.e., an artificial collection of diverse organisms—possibly only connected to the last common eukaryotic ancestor—that spans the entire eukaryotic tree and consists of at least five monophyletic supergroups. Each of these supergroups represents a taxonomic kingdom that is more diverse than and different from the traditional taxonomic kingdoms of animals, plants, or fungi. Possibly, protists are the least understood group of organisms as compared to prokaryotes, animals, fungi, and plants. This is likely because they have undergone by far the most extensive, continuous, and ongoing changes in their classification and naming over the last decades, making it difficult to approach this group by anybody than the expert. Consequently, protists have been gradually omitted from most biology textbooks, as noticed by Adl et al. (2007). Protists display an immense morphological variety at a cellular level, reflecting the extensive capacity of the eukaryotic cell to evolve and adapt to an enormous diversity of ecological niches. This morphological variety is, in contrast to higher animals and plants, only accessible—i.e., can only be visualized and studied—by the use of light or electron microscopy. Importantly, morphological characteristics of protists that can be studied by light microscopy—e.g., an amoeboid cell form—are often counterintuitive with regard to their phylogenetic origin as they may represent convergent evolutionary developments. They are therefore of limited value to classify and scientifically name these organisms. In contrast, ultrastructural differences that can only be appreciated by electron microscope studies—e.g., the apical complex of apicomplexan protozoans—reveal character traits that often allow recognition and delineation of phylogenetically related groups.

1.2 Molecular Phylogeny of Protists

Based on molecular phylogenetic studies of the last 25 years, the classification of protists has been and continues to be extensively modified. Our understanding of eukaryotic phylogeny has been revolutionized in at least two major steps. In a first step, molecular phylogenetics using single sequences—e.g., the 18S RNA gene—has led to the Archezoa hypothesis, which proposes that certain taxa of protists, such as microsporidians, diplomonads, and parabasalids that do not contain a mitochondrion, represent the common ancestor of the eukaryote lineage from which all other protists, as well as animals, fungi, and plants, descend (Cavalier-Smith 1987, 1989; Sogin 1991). An improved methodology of tree building, as well as an increased availability of genome data from a wider variety of eukaryotic taxons coupled with the advent of phylogenomics—i.e., the concatenation of multiple sequences based on a large set of genes—allowed a complete reorganization and reclassification toward a further improved reflection of the phylogenetic history of protists (Stiller and Hall 1999; Philipe 2000; Delsuc et al. 2005). This approach allowed delineating at least five taxonomic kingdoms or supergroups within Eukaryota—referred to as (1) SAR (Stramenopiles, Alveolata, Rhizaria), (2) Archaeplastida, (3) Excavata, (4) Amoebozoa, and (5) Ophistokonta (Adl et al. 2012; Klinger et al. 2016). In this classification system, plants are phylogenetically connected—i.e., represent a crown group—to Archaeplastida, while animals and fungi represent each a crown group of Ophistokonta. Correspondingly, Archaeplastida and Ophistokonta ancestor lineages—stem groups—are formed by protists. The other three protist groups, on the other hand, did not lead to the development of multicellular organisms.

The protozoan parasite species—parasitic protists—covered in the present book belong to three of the five well-defined major eukaryotic lineages (Table 1.1):

- SAR, represented by Apicomplexa that can be further subdivided into Cryptogregarina (*Cryptosporidium*), Coccidia (*Eimeria*, *Neospora*, *Sarcocystis*, *Toxoplasma*, *Besnoitia*), and Piroplasmida (*Babesia* and *Theileria*)
- Excavata, that can be split into the subgroups: Discoba (*Tritrichomonas* and *Trichomonas*) and Metamonada (*Leishmania*, *Trypanosoma*, and *Naegleria*)
- Amoebozoa, represented by Conosa (*Entamoeba*) and Lobosa (*Neoparamoeba*)

Except Apicomplexa, which comprise almost exclusively parasitic protozoans, the sublineages Discoba and Metamonada of the Excavata, and Conosa and Lobosa of the Amoebozoa, jointly include free-living and parasitic protozoans. This observation demonstrates a polyphyletic origin of parasitic protozoan lineages—i.e., the parasitic lifestyle evolved independently in diverse and evolutionary unrelated protist lineages.

Table 1.1 Current taxonomy of protists covered in this book (Adl et al. 2012; Cavalier-Smith 2014)

Supergroup/Kingdom	Taxonomic unit 1	Taxonomic unit 2	Taxonomic unit 3	Taxonomic unit 4	Genus	Chapter
SAR (Alveolata, Rhizaria, and Stramenopiles)	Apicomplexa	Gregarine	Gregarinomorphea	Cryptogregaria	*Cryptosporidium*	2
		Conoidasida	Coccidia	Eimeriorina	*Eimeria*	3
					Sarcocystis	4
					Neospora	5
					Toxoplasma	6
					Besnoitia	7
		Aconoidasida	Haemosporidia	Piroplasmida	*Babesia*	9, 10, 11
					Theileria	8, 11
Excavata	Metamonada	Parabasalia		Trichomonadea	*Trichomonas*	14
				Tritrichomonadea	*Tritrichomonas*	14
	Discoba	Kinetoplastea	Metakinetoplastina	Trypanosomatida	*Trypanosoma* *Leishmania*	12, 13
		Discicristata	Heterolobosea	Tetramitia	*Heterolobosea*	15
Amoebozoa	Conosa	Archamoebae	Entamoebida	Entamoebidae	*Entamoeba*	15
	Lobosa	Discosea	Centramoebida	Balamuthiidae	*Balamuthia*	15
				Acanthamoebidae	*Acanthamoeba*	15
			Dactylopodinab	Paramoebidae	*Paramoeba*	15

1.3 Convergent Evolution of Parasitic Protozoans

A comparison of parasitic lineages with those of their free-living relatives allows identifying evolutionary changes in response to environmental pressures common to the parasitic lifestyle.

The evolutionary trends of mitochondria and mitochondria-related organelles observed in the diverse lineages of parasitic protozoans have recently been reviewed by Klinger et al. (2016). Evidence has been presented that a single ancestral endosymbiotic event gave rise to the currently observed metabolic diversity and variation of classical mitochondria and mitochondria-related organelles, such as hydrogenosomes and mitosomes (Embley and Martin 2006; Shiflett and Johnson 2010). Parasitic protozoans represent endoparasites that need to adapt to a microaerobic or even anaerobic—in particular gut protozoan endoparasites—yet nutrient-rich environment in the host. This environmental pressure has been shown to result in a convergent evolution toward biochemical adaptation, reduction, or even elimination of the classical mitochondrial metabolism of ATP generation by oxidative phosphorylation—i.e., aerobic respiration. Mitochondria evolved into hydrogenosomes that produce hydrogen and generate ATP in the organelle lumen by nitrate/fumarate reduction, i.e., anaerobic respiration, or substrate-level phosphorylation, i.e., fermentation, respectively. A complete loss of genome and oxidative phosphorylation took place in mitosomes; thus mitosome-containing organisms generate ATP in the cellular cytosol by fermentation. Importantly, mitosomes have never been found in free-living protozoans, substantiating the notion that this organelle represents an adaptation to an endoparasitic lifestyle. Such convergent evolutionary adaptations to parasitism are observed across all protist lineages and are accomplished by the development of analogous biochemical pathways. As an example, hydrogenosomes and mitosomes that evolved in different protist lineages used different structures and strategies to accomplish the same goal.

1.4 Protozoan Diseases of Major Veterinary Importance

In order to allow a fast overview and orientation, we present here a shortlist of the diseases caused by parasitic protozoans of farm animals and pets with high (1) economic or (2) zoonotic impact. Most of these diseases are widespread and are of continental, or even global, distribution. This shortlist is necessarily subjective, as estimates of the economic losses caused by these diseases are prone to errors and scarce.

1. Economic impact
   (a) Coccidiosis caused by *Eimeria* spp.-infection of poultry and to a lesser degree of cattle, sheep, goat, horse, and rabbit (see Chap. 3)
   (b) Neosporosis caused by *Neospora caninum*-infection of cattle (see Chap. 5)
   (c) Ovine theileriosis caused by *Theileria lestoquardi* in sheep and goats (see Chap. 8)

(d) East coast fever of cattle caused by *Theileria parva* (see Chap. 8)
(e) Tropical theileriosis of cattle cause by *Theileria annulata* (see Chap. 8)
(f) Bovine babesiosis of cattle caused by *Babesia bovis* and *Babesia bigemina* (see Chap. 9)
(g) Ovine babesiosis of sheep and goats cause by *Babesia ovis* and *Babesia motasi* (see Chap. 9)
(h) Equine piroplasmosis caused by *Theileria equi* and *Babesia caballi* in horses, donkeys, and mules (see Chap. 11)
(i) Trypanosomiasis caused by *Trypanosoma brucei*, *T. congolese*, and *T. vivax* in cattle but also in pigs, camels, goats, and sheep (see Chap. 12)
(j) Amoebiasis caused by *Neoparamoeba perurans* especially in salmon raised in fish farms (see Chap. 15)

2. Zoonotic transmission
   (a) Cryptosporidiosis caused by *Cryptosporidium parvum* and *C. hominis* with calves as reservoir host (see Chap. 2)
   (b) Toxoplasmosis caused by *Toxoplasma gondii* with cats as reservoir host (see Chap. 6)
   (c) Visceral leishmaniasis caused by *Leishmania donovani*, *L. infantum* in the Old World, and by *L. chagasi* in the New World with dogs and wild canines as reservoir host (see Chap. 13)

Pathogenic protozoans that can principally be well controlled, are of more local importance, or are caused by opportunistic infections are represented by *Besnoitia besnoiti* infecting cattle (see Chap. 7), *Trichomonas* spp. (*Tritrichomonas foetus* infecting cattle and cat and *Trichomonas gallinae* infecting chicken) (see Chap. 14), and Amoebae (*Acanthamoeba* spp. infecting bovines, horses, and dogs, and *Balamuthia mandrillaris* infecting horses and sheep) (see Chap. 15).

1.5 World Population of Protozoan Parasites and Farm and Pet Animals

When contemplating the large variety and diversity of parasitic protozoan species, one might wonder why relatively few parasite species are of major importance in the veterinary field (Adl et al. 2007). This is possibly for historical reasons: the current human population of about seven billion worldwide, corresponding to a biomass of 287 million tonnes, have selected a very restricted number of animals to cover their food demands, i.e., cattle, sheep, goat, pig, chicken, and turkey, or as pets, i.e., cat, dog, and horse (Walpole et al. 2012). Worldwide, the biomass of all productive animals—as determined based on the number of cattle, pigs, sheep, and chicken—has been estimated to amount to a biomass of about 700 million tonnes and that of the dog as a pet to about ten million tonnes. In comparison, all large wild animal vertebrates—whales, deer, lions, reptiles, etc.—have been estimated to

Cryptosporidium

2

Mariela L. Tomazic, Carlos Garro, and Leonhard Schnittger

Abstract

Cryptosporidium sp. infects the gastrointestinal tract of a wide range of vertebrates, including domestic and livestock animals as well as humans. Cryptosporidiosis of neonatal farm ruminants causes considerable economic losses as the disease is commonly associated with intense diarrhea leading to an impaired growth, a decreased performance and production, and often animal death. The highly infective oocyst stage is excreted with the feces and disseminated into the environment, contaminating water and food. Neonatal calves are a major reservoir of the zoonotic *C. parvum*, which causes, in addition to the anthroponotic *C. hominis*, human cryptosporidiosis and is of considerable public

M.L. Tomazic
Institute of Pathobiology, Center for Research on Veterinary and Agronomic Sciences (CICVyA), National Institute of Agricultural Technology (INTA-Argentina), Hurlingham, Provincia de Buenos Aires, Argentina

National Council of Scientific and Technological Research (CONICET), Godoy Cruz, Buenos Aires, Argentina

C. Garro
Institute of Pathobiology, Center for Research on Veterinary and Agronomic Sciences (CICVyA), National Institute of Agricultural Technology (INTA-Argentina), Hurlingham, Provincia de Buenos Aires, Argentina

L. Schnittger (✉)
Institute of Pathobiology, Center for Research on Veterinary and Agronomic Sciences (CICVyA), National Institute of Agricultural Technology (INTA-Argentina), Hurlingham, Provincia de Buenos Aires, Argentina

National Council of Scientific and Technological Research (CONICET), Godoy Cruz, Buenos Aires, Argentina

School of Exact, Chemical and Natural Sciences (FCEQN), University of Morón, Morón, Province of Buenos Aires, Argentina
e-mail: schnittger.leonhard@inta.gob.ar

M. Florin-Christensen, L. Schnittger (eds.), *Parasitic Protozoa of Farm Animals and Pets*, https://doi.org/10.1007/978-3-319-70132-5_2

health concern. Currently, no vaccine or efficient drug is available against the disease. From a veterinarian economical point of view, *C. parvum* is the most important species among the 30 recognized species infecting bovines, lamb, goats, pigs, horses, and dogs worldwide. After the discovery of *C. parvum* by Tyzzer in the year 1912, the taxon *Cryptosporidium* has been classified into coccidia. However, recent findings on the *Cryptosporidium* life cycle and molecular phylogenetic evidence resulted in the reclassification of *Cryptosporidium* into the gregarines. This novel placement appreciates previously underestimated and/or neglected features of *Cryptosporidium* that are common to gregarines such as a low host specificity and/or the possibility to survive without a host.

2.1 Morphology, Life Cycle, and Host-Pathogen Interaction

Among *Cryptosporidium* species, *C. parvum* is the most extensively studied because of its veterinarian economic importance in the veterinarian field, because it is a causative agent of human cryptosporidiosis, and because of its wide host range. For this reason, the life cycle and host-pathogen characteristic of *Cryptosporidium* is best illustrated based on the accumulated knowledge of this species. Recently, the parasite stages and life cycle of *C. parvum* have been restudied and reevaluated demonstrating that *Cryptosporidium* shares biological and morphological features with gregarines. In addition, molecular phylogenetic data exclude *Cryptosporidium* from coccidia and include this genus into a novel subclass *Cryptogregaria* into the class Gregarinomorphea within gregarines (reviewed by Cavalier-Smith 2014; Clode et al. 2015; Ryan and Hijjawi 2015; Aldeyarbi and Karanis 2016a; Ryan et al. 2016).

2.1.1 Morphology

2.1.1.1 Oocysts

Cryptosporidium sp. oocysts are usually spherical and, depending on the species, range in size from 3.2 to 8.3 μm (Table 2.1, Sect. 2.2.2). In the case of *C. parvum* the oocyst is bounded by a smooth cleft-including surface and has a spherical to ovoid shape of a width and length of 4.5 to 7.0 μm, respectively (Reduker et al. 1985; Fayer 2008; Borowski et al. 2010).

During the life cycle two types of oocysts are generated, (1) thick-walled oocysts (TKW) that are excreted and disseminated with the feces into the environment and (2) thin-walled (Tw) oocysts characterized by the lack of a thick multi-zoned inner layer that are autoinfective. Oocysts contain four infective sporozoites released during excystation, and a residual large lipid body that contains amylopectin granules, a crystalline protein inclusion, ribosomes, and cytomembranes (Fayer 2008).

Conventional thin-section electron microscopy of the TKW parasite stage has revealed that the wall is composed of three electron-dense and an intermediate

Table 2.1 *Cryptosporidium* species infecting farm animals

Host	Species	Site of infection	Oocyst size (μm)	Zoonotic	Public health significance	Reference
Cattle	*C. parvum*[a]	Intestinal	4.5 × 7.0	Yes	High	Wyatt et al. (2010) and Tomazic et al. (2013)
	C. ryanae	Intestinal	3.2 × 3.7	No	No	Fayer et al. (1998) and Santín et al. (2008)
	C. bovis	Unknown	4.9 × 4.6	No	No	Fayer et al. (2010a, b)
	C. andersoni	Gastric	7.4 × 5.5	Yes	Minor	Fayer et al. (2006)
Pig	*C. parvum*[a]	Intestinal	4.5 × 7.0	Yes	High	Ryan et al. (2004)
	C. suis	Intestinal	4.6 × 4.2	Yes	Minor	Xiao et al. (2006)
	C. muris	Gastric	7.0 × 5.0	Yes	Minor	Kvác et al. (2009)
Sheep	*C. parvum*[a]	Intestinal	4.5 × 7.0	Yes	High	Santin (2013)
	C. ubiquitum	Intestinal	5.0 × 4.7	Yes	Minor	Fayer and Santín (2009)
	C. xiaoi[b]	Intestinal	3.9 × 3.4	No	No	Imre et al. (2013)
Goat	*C. parvum*[a]	Intestinal	4.5 × 7.0	Yes	High	Noordeen et al. (2002) and Karanis et al. (2007)
	C. xiaoi[b]	Intestinal	3.9 × 3.4	No	No	Park et al. (2006) and Giles et al. (2009)
Horse	*C. parvum*[a]	Intestinal	4.5 × 7.0	Yes	High	Grinberg et al. (2008)
	C. hominis	Intestinal	5.2 × 4.9	Yes	Major	Deng et al. (2017)
Dog	*C. parvum*[a]	Intestinal	4.5 × 7.0	Yes	High	Hajdušek et al. (2004)
	C. canis	Intestinal	5.0 × 4.7	Yes	Minor	Giangaspero et al. (2006);
	C. meleagridis[c]	Intestinal	5.2 × 4.6	Yes	Moderate	Ellis et al. (2010)
	C. muris	Gastric	7.0 × 5.0	Yes	Minor	Fayer et al. (2001)
Cat	*C. felis*	Intestinal	5.0 × 4.5	Yes	Minor	Iseki (1979) and Santín et al. (2006)
	C. muris	Gastric	7.0 × 5.0	Yes	Minor	Pavlasek and Ryan (2007)
Chicken	*C. galli*	Gastric	8.3 × 6.3	No	No	Pavlasek (2001)
	C. baileyi	Intestinal, BF, respiratorial, conjunctive	6.2 × 4.6	No	No	Current et al. (1986)
	C. meleagridis	Intestinal, BF		Yes	Moderate	Ryan (2010)

(continued)

Table 2.1 (continued)

Host	Species	Site of infection	Oocyst size (µm)	Zoonotic	Public health significance	Reference
Fish	*C. scophthalmi*	Intestinal	4.4 × 3.9	No	No	Fayer (2008) and Ryan (2010)
	C. molnari	Gastric	4.7 × 4.5	No	No	Alvarez-Pellitero and Sitjà-Bobadilla (2002)
Reptiles	*C. serpentis*	Gastric	2.8 × 3.6	No	No	Fayer (2008) and Santin (2013)
	C. varanii	Intestinal, gastric	4.8 × 4.7	No	No	Pavlasek and Ryan (2008)

BF Bursa of fabricius
[a]syn. *C. pestis*
[b]syn. *C. agni*
[c]syn. *C. tyzzeri*

electron-translucent layer. Beneath the intermediate electron-translucent layer, structural proteins are located that are thought to provide much of the strength and flexibility characteristic for the oocyst wall. A suture structure, which opens to release the four sporozoites contained in the oocyst, is overlaid by the outer dense and the middle layer (Jenkins et al. 2010).

A model of the *C. parvum* oocyst wall has been proposed in which the external surface is composed by the glycocalyx, followed by a lipid hydrocarbon layer, proteins, and a structural polysaccharide layer. The oocyst-covering glycocalyx is composed of glycoproteins and acidic polysaccharides. Fatty acids contained in the lipid layer are thought to confer the acid-fast staining feature, widely used for diagnostic purposes (Jenkins et al. 2010). Additionally, cysteine-rich domains contained in oocyst wall proteins (COWP) are thought to be responsible for the wall rigidity since intermolecular disulfide bonds are formed (Spano et al. 1997; Jenkins et al. 2010).

Recently, the formation of novel TKW and Tw oocysts in a host cell-free culture after completion of the *C. parvum* life cycle has been described. In accordance with Jenkins et al. (2010), the formed TKWs have been reported to be composed of four layers, (1) an outer electron-dense layer, (2) a central electron-lucent layer which is interposed by (3) a dark osmophilic electro-dense layer, and (4) an inner thicker granular layer. In contrast to TKWs, the Tws generated in cell-free culture system exhibit layers (1), (2), and (3) but lack the innermost layer (4) (Aldeyarbi and Karanis 2016b).

2.1.1.2 Sporozoites

C. parvum sporozoites are long-shaped of about 5.0 × 0.5 µm in size and are surrounded by the pellicle below which the subpellicular microtubules are located. The anteriormost cellular structure is the apical complex, involved in the invasion and attachment to the host cell, and consisting of apical rings, the conoid, and the secretory organelles such as a rhoptry, multiple micronemes, and dense granules (Fig. 2.1). Interestingly, unlike other apicomplexans such as *Toxoplasma*,

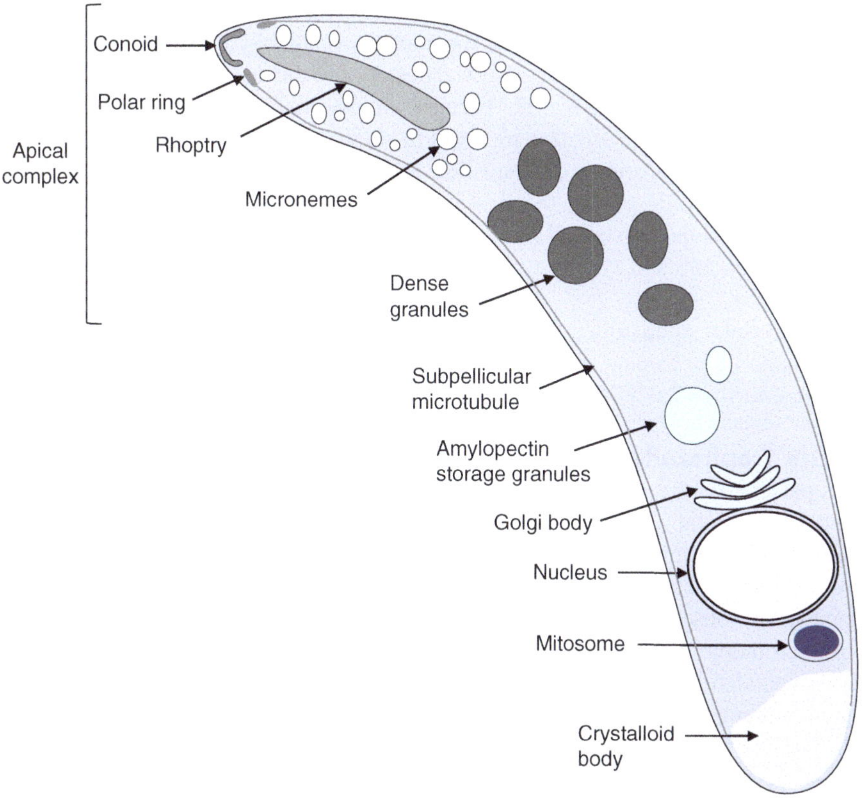

Fig. 2.1 *Cryptosporidium* sporozoite. In the anteriormost region, the apical complex consisting of apical rings, the conoid, and the secretory organelles consisting of a rhoptry, micronemes are situated. Morphologically, dense granules—or secretory bodies—do not belong to the apical complex, but they are functionally related as they are involved in the establishment of the parasite in the host cell after invasion. In the central region, amylopectin storage granules, the Golgi organelle, and the nucleus are located, whereas in the posterior region, the crystalloid body is depicted. Between the nucleus and the crystalloid body, there is a mitosome—a relict mitochondrion—located

Plasmodium, *Eimeria*, *Sarcocystis*, or *Babesia* that possess two or more rhoptries, *C. parvum* has a single rhoptry (Šlapeta and Keithly 2004; Snelling et al. 2007). Upon invasion, the rhoptry and micronemes release their content in the apical region of the host cell. Importantly, when invading, the parasite does not enter into the host cell cytosol but remains at its surface in a membrane-bound compartment called parasitophorous vacuole (PV). Consequently, after invasion this results in an epicellular location—an intracellular and extracytoplasmic parasite location with respect to the host cell (see Sect. 2.1.3.4). During the invasion process, micronemal proteins are involved in gliding motility, whereas the rhoptry is essential for the formation of the PV. Dense granules discharge their content on the zoite surface—i.e., sporozoites, trophozoites, or merozoites—and are responsible for host cell modification after invasion (see Sect. 2.1.3.4) (Tetley et al. 1998; Borowski et al. 2008).

Organelles and other cellular structures that are found in the central region of the parasite cell are the Golgi apparatus, the nucleus, small amylopectin granules, and ribosomes. A crystalloid body is observed posterior to the nucleus, but its origin and function are unknown. However, because of the juxtaposition of the enzyme pyruvate:NADP$^+$ oxidoreductase involved in energy metabolism with amylopectin granules in some apicomplexan parasites, it has been suggested that this structure may play a role in the energy metabolism (Lemgruber and Lupetti 2012). Between the nucleus and the crystalloid body, there is a mitosome—a relict mitochondrion lacking a genome and typical tubular cristae—an organelle not seen in coccidia (Fayer 2008).

Extracellular sporozoites have been also observed in cell-free cultures. They were found to be banana-shaped, measured 2.2 × 1.6 μm, had an apical complex, and were surrounded by a pellicle (Aldeyarbi and Karanis 2016b).

2.1.1.3 Trophozoite

This parasite stage resides within the PV and acquires different sizes depending on the developmental stage. In an in vitro system that uses the host cell line HCT-8 for parasite propagation, early trophozoite stages—corresponding to a time of cultivation 6 h post-inoculation—measured less than 1.0 μm. In contrast, late trophozoite stages—i.e., observed after 24 h post-inoculation—measured up to 2.5 μm. The observed mature or late stage trophozoites were attached to the host surface but were separated by a feeder organelle—a structure that is formed by the folding of the membrane of the PV and facilitates the uptake of nutrients from the host cell. The feeder organelle has a smooth surface, and the basal membrane forms a hood-like shape and cytoplasmic granulations (Borowski et al. 2010).

Extracellular trophozoites were also observed in axenic cultures—i.e., cultures lacking host cells—measured 2 μm in size, and had a rough surface (Hijjawi et al. 2004; Borowski et al. 2010).

2.1.1.4 Meronts I

This parasite stage has an epicellular location and contains six to eight merozoites I. Its size varies according to its developmental stage; early meronts I measure 1.5 μm, while mature meronts I measure 2.5 μm. In an in vitro system using the host cell line HCT-8, merozoites I measured 1.0 × 0.4 μm in size and looked like rods with a pointed apical region and a rough surface (Borowski et al. 2010).

Meronts I were also observed in cell-free cultures and released actively motile merozoites of a circular to oval shape and a small size of 1.2 × 1.0 μm (Hijjawi et al. 2004).

2.1.1.5 Meronts II

This parasite stage is situated epicellular and contains merozoites II. Meronts II measure 3.5 μm, are considerably larger than meronts I, and possess a thicker outer membrane. Merozoites II are round, measure between 1.0 to 0.5 μm in size, and have a rough surface.

In noninfected host cell cultures, extracellular meronts II have also been observed. Extracellular stages were rounded, found to be larger than epicellular meronts—8.0 μm—and contained merozoites of 2.0 μm that were densely packed (Borowski et al. 2010). In axenic cultures—i.e., cultures lacking host cells—pleomorphic forms of merozoites II, i.e., forms that change in cell size as well as shape, have been observed such as (1) bean, (2) short spindle, and (3) long spindle shapes (Hijjawi et al. 2004; Thompson et al. 2005; Karanis et al. 2008).

2.1.1.6 Microgamonts

This parasite stage is round and measures 2.0 × 2.0 μm, contains large number of microgametes, and is surrounded by a host cell-derived membrane and a feeder organelle. Some microgamonts have a stalklike structure that might be the product of host cell detachment, supporting the observation that they have little contact with the host cell in an in vitro system (Borowski et al. 2010). Within microgamonts, microgametes are densely packed and are of spherical shape, measuring 0.1 μm in size, and are non-flagellated (Borowski et al. 2010). Additionally, microgamonts that contain microgametes and that were surrounded by a double membrane have been observed also in a cell-free culture, but significant differences in size were reported. For example, Hijjawi et al. (2004) reported microgamonts of a size of 5.6 × 5.0 μm and microgametes of a size of 2.2 × 1.6 μm, while Aldeyarbi and Karanis (2016a) have observed microgamonts of 1.53 × 1.18 μm and microgametes of 0.34 × 0.26 μm in size (Hijjawi et al. 2004; Aldeyarbi and Karanis 2016b). Interestingly, remnants of the apical complex have been observed in the cytoplasm of microgamonts (Aldeyarbi and Karanis 2016b).

2.1.1.7 Macrogamonts

They have an ovular form sizing 5.0 × 4.0 μm and have a rough surface. As in the case of microgamonts, they own a stalklike structure supporting the notion that this parasite stage has little contact with the host cell and has an extracellular location (Borowski et al. 2010).

This parasite stage has been also observed in cell-free cultures, differing in size (Hijjawi et al. 2004; Aldeyarbi and Karanis 2016b). Recently, Aldeyarbi and Karanis (2016a) have reported that extracellular macrogamonts in a cell-free culture had a spherical to ovoid form, measured 1.6 × 1.1 μm, and were surrounded by two membranes (Aldeyarbi and Karanis 2016b). These authors also observed two maturation stages of macrogamonts. The early macrogamont stage contained lipid vacuoles and amylopectin granules in their cytoplasm, whereas the advanced stage had two types of wall-forming bodies (WFB I and II) and amylopectin granules in their cytoplasm. The two types of WFB are responsible for the posterior development of the oocyst wall (Aldeyarbi and Karanis 2016b).

2.1.1.8 Zygote

This stage results from the fertilization of the macrogamont by a microgamete. There is no morphological description of this stage reported from an in vitro system using host cells, but this stage has been recently reported in cell-free culture system. The zygote measures 2.9 × 1.5 μm and includes a large number of amylopectin bodies (Aldeyarbi and Karanis 2016b).

2.1.1.9 Gamont-like

Gigantic gamont-like stages have been exclusively observed in cell-free cultures. They appear 72 h after parasite cultivation in axenic media and initially exhibit a size of 5.3 × 2.3 μm. At a later time of cultivation, they grow up to a size of 16.6 × 7.6 μm (Hijjawi et al. 2004).

2.1.2 Life Cycle

Cryptosporidium sp. features a monoxenous life cycle that starts with the ingestion of the infective oocyst stage and its passage through the digestive system followed by its excystation resulting in sporozoite release (Fig. 2.2). Free sporozoites invade host cells in an intracellular but extracytoplasmic fashion referred to as epicellular location. However, extracellular stages of the parasite have been also observed in a cell-free system as well as in an in vitro culture without invasion of host cells, suggesting that the parasite is able to complete its life cycle also without its host representing a facultative epicellular Apicomplexa (Hijjawi et al. 2004; Borowski et al. 2010). Borowski et al. (2010) have suggested that extracellular stages form part of the life cycle or, alternatively, that they may coexist, resembling rudimentary stages of an ancestral life cycle (Borowski et al. 2010).

Before invasion, sporozoites attach to the host cell apical surface membrane, which embraces the sporozoite, initiating the formation of a PV. At the site of attachment to the host cell, an electron-dense region is found that forms the bottom part of the feeder organelle allowing the parasite to acquire nutrients and energy from the host cell. Subsequently, sporozoites develop into trophozoites. It has been demonstrated that in axenic cultures, trophozoites associate end-to-end, resembling syzygy—a cellular association observed in gregarines with end-to-end pairing without sexual fusion (Clode et al. 2015). Recent studies have shown the dominance of the trophozoite stage within the life cycle, which undergoes merogony—or schizogony—and forms meronts I (Hijjawi et al.

Fig. 2.2 Life cycle of *C. parvum*. While thin-walled oocysts are autoinfective and maintain a proliferation cycle in the intestine, thick-walled oocysts are excreted and disseminated with the feces in the environment. After their ingestion, excystation of oocysts takes place in the intestine resulting in the release of four sporozoites. The excystation process and the mode of multiplication—gametogony, sporogony, merogony, and binary fission—are highlighted in bold. The parasitophorous vacuole—represented in purple—and the feeder organelle are thought to be formed in all *C. parvum* stages, with the exception of gamont, gamont-like, and oocyst stages. The asexual and sexual parts of the life cycle are depicted in green and pink, respectively. Resemblances with gregarines are highlighted in red and refer to the occurrence of pleomorphism, extracellular stages, and syzygy. All parasite stages have been observed in an epicellular location as well as in the extracellular medium with the exception of gamont-like extracellular stages that were demonstrated only in host cell-free systems (Borowski et al. 2010; Clode et al. 2015; Ryan and Hijjawi 2015)

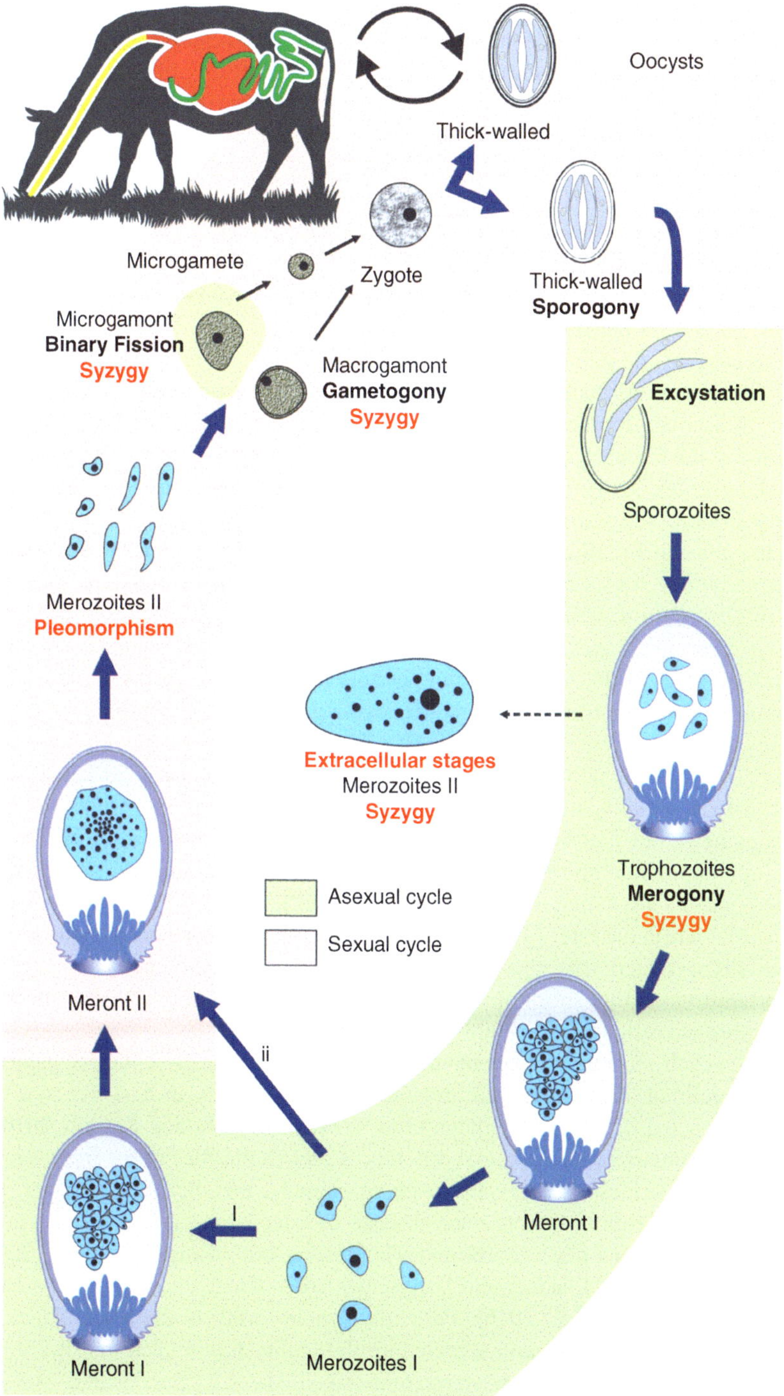
Oocysts
Thick-walled
Thick-walled
Sporogony
Zygote
Microgamete
Microgamont
Binary Fission
Syzygy
Macrogamont
Gametogony
Syzygy
Excystation
Sporozoites
Merozoites II
Pleomorphism
Extracellular stages
Merozoites II
Syzygy
Trophozoites
Merogony
Syzygy
Asexual cycle
Sexual cycle
Meront II
ii
Meront I
I
Meront I
Merozoites I

2004; Borowski et al. 2010). Each meront I contains six to eight merozoites. Merozoites I leave the meront I by rupture of the membrane that engulfs the parasite, invade further enterocytes, and initiate replication of either (1) an asexual propagation cycle resulting in the development of meronts I or (2) a sexual propagation cycle resulting in the formation of meronts II that contain merozoites II. Interestingly, earlier studies showed the pleomorphic feature of the merozoite stage—defined as the change and variation in cell size and shape—in host cell-free cultures, which has been confirmed more than a decade later and is characteristic for gregarines (Hijjawi et al. 2004; Karanis et al. 2008; Aldeyarbi and Karanis 2016a, b). Micro- and macrogamonts are formed after the invasion of enterocytes by merozoites II, initiating the sexual reproduction cycle, known as gametogony or gamogony. Microgamonts, which are considered the male form, become multinucleated, and each nucleus is incorporated into a microgamete. Subsequently, they are released from microgamonts and fertilize macrogamonts—the female forms—producing zygotes (Borowski et al. 2008; Lendner and Daugschies 2014). It has been shown that microgamonts are able to undergo binary fission, suggesting that in addition to merogony, asexual reproduction also happens during this stage (Borowski et al. 2010). The occurrence of syzygy in extracellular stages and gamonts has been demonstrated in axenic and host cell in vitro culture systems. Furthermore, it has been suggested that predominantly gamont stages—i.e., micro- and macrogamonts—employ syzygy (Borowski et al. 2008; Aldeyarbi and Karanis 2016a). Noteworthy, extracellular parasite stages seem to appear as the life cycle progresses, given that micro- and macrogamonts are observed in in vitro systems 4 days post-inoculation, after having little or even no interaction with host cells (Rosales et al. 2005; Borowski et al. 2010). Finally, the zygotes—or fertilized macrogamont, containing the microgamete and macrogamont nucleus—undergoes meiosis resulting in the development of four sporozoites each owning a single nucleus. Thin- and thick-walled oocysts are generated; the former (Tw) initiate an autoinfective propagation cycle, while the later (TKW)—representing the infective oocyst stage—are eventually excreted with the feces and disseminated in the environment. It is important to note that sporulation—sporogony—of TKW oocysts takes place in situ before excretion (Fayer 2008).

All life cycle stages were seen in host cell cultures with the exception of gamont-like extracellular stages that were demonstrated exclusively in host-free systems and are depicted in Fig. 2.2 (Hijjawi et al. 2004; Aldeyarbi and Karanis 2016a). Table 2.2 summarizes the principal differences and similarities of the life cycle of coccidian, gregarines, and *Cryptosporidium*, namely, the site of attachment, the mode of multiplication, the presence/absence of a feeder organelle, and the presence/lack of an apicoplast (Barta and Thompson 2006; Valigurová et al. 2008; Borowski et al. 2010; Lendner and Daugschies 2014; Clode et al. 2015; Ryan and Hijjawi 2015; Ryan et al. 2016). Particularly noteworthy is the resemblance of *Cryptosporidium* with gregarines including the ability to complete the life cycle without the need of a host, the existence of extracellular—i.e., gamont-like—stages, and the occurrence of syzygy.

Table 2.2 Main biological differences and similarities of the life cycle of coccidian, gregarines, and *Cryptosporidium* (Barta and Thompson 2006; Thompson et al. 2016)

Biological feature	Coccidian	Gregarines	*Cryptosporidium*
Interaction with host cell	Intracellular	Transmembrane and extracellular	Epicellular and extracellular
Feeder organelle	Not present	Present (epimerite)	Present (epimerite)
Extracellular development	No	Yes	Yes
Multiplication	Multiple fission, merogony, gametogony, sporogony, endodyogeny, and endopolygeny (in some)	Multiple fission, merogony (in some), gametogony, sporogony, binary fission, syzygy	Multiple fission, merogony, gametogony, sporogony, binary fission, syzygy
Apicoplast	Present	Not present in some gregarines	Not present

2.1.3 Host-Pathogen Interaction

Although the mechanistic interactions between the host cell and *Cryptosporidium* are still not very well understood, a significant progress has been made regarding the identification of involved molecules and proteins. An extensive overview of the molecules and processes implicated in host cell invasion has been recently done by Lendner and Daugschies (2014). The most relevant will be outlined in brief in the following. Upon ingestion of oocysts, their excystation takes place in the gastrointestinal tract, and host cells are subsequently infected by sporozoites in a two-step process, (1) attachment and gliding motility and (2) invasion. Once the parasite has invaded host cells, it has later to egress to complete its life cycle. Finally, the host cell initiates defense mechanisms to control the infection. All these processes are discussed in the following.

2.1.3.1 Excystation

Once infective oocysts are ingested, the passage through the digestive system triggers excystation, during which the oocyst cleft opens and sporozoites exit and invade epithelial cells of the small intestine of the host. Interestingly, very different levels of excystation of oocysts have been observed in vitro and in cell-free culture systems. It has been described that the excystation process in *Cryptosporidium* begins with a single stimulus, in contrast to other apicomplexa like *Toxoplasma* or *Eimeria* that need two or more stimuli for excystation (Fayer and Leek 1984). Among stimuli or factors that have been described to favor excystation are the temperature, the presence of bile salts, reducing conditions, pH, and digestive enzymes (Fayer and Leek 1984; King et al. 2012; Lendner and Daugschies 2014). It has been also suggested that there is no established order of triggers of excystation nor a synergism among the factors that favor this process (Aldeyarbi and Karanis 2016b). However, it has been observed that pretreatment of oocyst with sodium hypochlorite and parasite-derived proteases, like cysteine and serine proteases, enhances excystation in in vitro systems (Smith et al. 2005).

2.1.3.2 Attachment

Following excystation, receptors and other surface molecules are expressed on infective stages—i.e., sporozoites, trophozoites, merozoites, or gametocytes—mediating attachment to epithelial cells. Several of these surface molecules represent well-characterized lectins of which one, p30, is expressed in the apical region of sporozoites and binds exclusively galactose-N-acetylgalactosamine. p30 is known to associate with two other glycoproteins involved in invasion: gp40/15 and gp900 (Bhat et al. 2007). The former, gp40/15, is a complex of two fragments generated by proteolytic cleavage of a 60 kDa precursor glycoprotein, designated gp60. gp40 is non-covalently associated with gp15, which is anchored to the surface membrane by glycosylphosphatidylinositol (GPI). gp15 is located on the whole surface of sporozoites and merozoites, while gp40 is confined to the apical region (Cevallos et al. 2000a, b; Wanyiri et al. 2007; O'Connor et al. 2007). Importantly, the gp60 gene encodes a highly polymorphic polyserine domain situated within the gp40 fragment, which is used for subtyping of *C. parvum* isolates and strains (Alves et al. 2003; Xiao 2010). Another protein, gp900, is a mucin-like glycoprotein of 900 kDa containing cysteine-rich as well as polythreonine domains and a large membrane-proximal N-glycosylated core region. This glycoprotein is located in micronemes and has been proposed to define host specificity since the attachment site seems to differ between *Cryptosporidium* species (Petersen et al. 1992; Jakobi and Petry 2006).

2.1.3.3 Gliding Motility

Ensuing attachment, apical organelles of sporozoites such as the rhoptry, micronemes, and dense granules secrete substances facilitating and effecting gliding motility, an exclusive way of locomotion of apicomplexan parasites that is a prerequisite for host cell invasion. It has been proposed that gliding motility and invasion involves a set of molecules that bind to the host cells and couple the translocation of surface adhesins to an actin-myosin motor located beneath the parasite plasma membrane (Sibley 2004). Although the mechanism is well studied in apicomplexan model parasites like *Toxoplasma gondii* and *Plasmodium* spp., there is lesser information available regarding the gliding motility of *Cryptosporidium*. In *C. parvum* gliding motility has been observed on host cell microvilli in an in vitro system, as evidenced by trails of elongated microvilli between excysted oocyst and new trophozoites (Borowski et al. 2010). The authors have also noted that not all sporozoites caused these gliding trails but that some invaded directly at their origin of excystation. Wetzel et al. (2005) have observed that the parasite undergoes a more rapid circular and helical gliding as compared to the *Toxoplasma* sporozoite and trophozoite stages. They also confirmed that *Cryptosporidium* gliding motility is generated by an actin-myosin motor complex in accordance with the model described for other Apicomplexa, since motor inhibition blocked movement and parasite invasion but not attachment to the host cell (Chen et al. 2004; Wetzel et al. 2005). Some of the proteins released from *C. parvum*-micronemes and secretory vesicles do not have the typical conserved domains or motifs observed in *Toxoplasma* or *Plasmodium*. For instance, proteins involved in locomotion and invasion are commonly proteins that contain type I repeats of human platelet thrombospondin (TSP1) domains (Lendner and Daugschies 2014). Accordingly, also 12 proteins that contain a TSP1-like domain have been identified in *Cryptosporidium*. However, in contrast to *Toxoplasma* and *Plasmodium*, these proteins contain an additional Apple or Kringle domain while lacking a vWF domain suggesting different sites of host cell attachment (Naitza et al. 1998; Deng et al. 2002; Lendner and Daugschies 2014).

2.1.3.4 Invasion and Egress

As in other apicomplexans, host cell invasion of *Cryptosporidium* occurs after attachment and is dependent from gliding motility. Sporozoites bind to the apical membrane of enterocytes—i.e., at the anterior cellular end between microvilli—and reorient themselves before invasion takes place. It has been shown that calcium is a key messenger for invasion and that its level is regulated, at least partially, by bile salts (Chen et al. 2004; King et al. 2012). There are significant differences between *Cryptosporidium* vs. *Toxoplasma* and *Plasmodium* in the mode and mechanism of invasion (Huang et al. 2004; Wetzel et al. 2005). Most importantly, the parasite does not enter deep into the cytosol but rather remains at the cell surface in the PV (Wetzel et al. 2005). It has been reported that this epicellular invasion process induces considerable reorganizations within the host cell leading to the generation of epithelial protrusions that embrace the parasite (Huang et al. 2004). Although protrusions appear to be induced by extension of the host cell microvillus, they have been also observed in extracellular stages without host cell encapsulation, and the term of vacuole was proposed instead (Koh et al. 2014; Clode et al. 2015). The membrane of this vacuole folds to develop the feeder organelle termed epimerite that facilitates uptake of nutrients from the host cell. With the exception of the thick-walled and thin-walled oocysts, the feeder organelle is thought to be formed in all parasite stages (Barta and Thompson 2006; Borowski et al. 2010; Clode et al. 2015). It has been hypothesized that the feeder organelle does besides acquiring nutrients from the host cell also function to acquire nutrients directly from the environment (Koh et al. 2014; Clode et al. 2015; Aldeyarbi and Karanis 2016a).

Egress from the host is a key step to complete the life cycle. It has been hypothesized that the parasite may use the programmed cell death mechanism—i.e., apoptosis—to escape from the host cell. Induction of apoptosis by the parasite was first demonstrated in vitro (Chen et al. 1998). Also in vivo studies have shown that *Cryptosporidium* induces apoptosis, but it is not clear if the process is altered by the parasite or executed by the host, since apoptosis is a well-known defense mechanism (Sasahara et al. 2003). Recent ex vivo studies suggest that modulation of apoptosis is addressed by the host and not by *Cryptosporidium* (Foster et al. 2012).

2.1.3.5 Host Immune Response

The host defense mechanism against *Cryptosporidium* infection involves innate as well as adaptive immune responses. Innate immunity is a non-specific first acting response, implying cells and humoral factors. Among response cells are intestinal epithelial cells, which produce proinflammatory cytokines, like interleukin-18 (IL-18), and proteins such as defensins—also known as antimicrobial peptides—and the complement system (Tarver et al. 1998; Zaalouk et al. 2004). Other cells that are implied in the defense against *Cryptosporidium* are natural killer (NK) cells, phagocytes such as macrophages and neutrophils, and dendritic cells (DCs). It has been demonstrated that NK cells produce interferon-γ (IFN-γ) which has been shown to be a key player of the innate immune response against *C. parvum* because it prevents the initiation and extension of the infection (Ungar et al. 1991; Chen et al. 1993b; Tessema et al. 2009). Furthermore, it has been demonstrated that an increased production of free radicals of nitric oxide (NO) by phagocytes such as macrophages and neutrophils has a protective effect against *C. parvum* infection (Gookin et al. 2004). Although it has been shown that DCs contribute to the clearance of *C. parvum*, additional studies are needed to comprehend the DC mechanisms involved (Auray et al. 2007; Petry et al. 2010). It has been suggested that the intestinal flora in calves is an important competitor for sporozoite attachment (Harp 2003; Wyatt et al. 2010). A key player of the humoral innate immune response is the complement system, and it has been suggested that complement factors are produced in *C. parvum* infection (Wagner et al. 2003; Petry et al. 2010).

The adaptive immune system is the second acting response, which is more specific and effective than the innate response. It involves cells such as T and B lymphocytes and humoral factors, like antibodies. T cells are important for the recovery of mice from *Cryptosporidium* infection, specifically helper T cells ($CD4^+$) but not cytotoxic T cells ($CD8^+$) (Heine et al. 1984; Chen et al. 1993a). However, in calves it has been demonstrated that $CD8^+$ cells increased upon *C. parvum* infection (Abrahamsen et al. 1997). T cells recovered from villous lymphoid cells from calves 6 days post-infection have shown a large amount of activated $CD8^+$ cells compared to controls, demonstrating a local immune response against *C. parvum* (Wyatt et al. 1997). Moreover, a downregulation of helper T cells of type 1 (Th1) has been observed, consistent with the development of diarrhea and the progression of the infection (Wyatt et al. 2010). It has been shown that $CD4^+$ cells produce key cytokines such as IFN-γ, IL-4, IL-12, and IL-18, and antibody production of B cells is mediated by helper T cells of type 2 (Th2) (Petry et al. 2010). After a *C. parvum* infection titers of parasite-specific immunoglobulins (Ig) of class IgG and IgA are increased. In particular, mucosal IgA avoids attachment of the infective stages to epithelial host cells (Petry et al. 2010). It has been demonstrated that recovering of calves from the infection—i.e., from clinical disease—is associated with antibody production and Th1 response in the intestinal mucosa (Wyatt et al. 2001).

In an in vitro assay in which peripheral blood mononuclear cells from recovered calves was incubated with recombinant P23—an antigen that has demonstrated protection in vaccination trials—the proliferation of $CD4^+$ and production of INF-γ have been demonstrated, suggesting a Th1-like memory response (Wyatt et al. 2005). A role of other T cell subsets such as T cells of type 17 has not been yet investigated (Petry et al. 2010).

2.2 Diagnosis, Epidemiology, and Economic Impact

Molecular characterization has allowed identifying, defining, and differentiating species and subtypes that have been unknown so far. Molecular approaches are best suited to identify sources of infection, the origin of outbreaks, and the grade of public health significance of *Cryptosporidium* spp. Furthermore, molecular studies carried out on epidemiology, diagnosis, food, and water surveillance have provided insight into the routes of transmission of circulating species and subspecies. They have allowed establishing much improved methods of parasite species and subtype detection. These findings are of particular importance for the surveillance and control of pathogenic *Cryptosporidium* spp. as they are worldwide distributed; many species are zoonotic and/or oligoxenous—i.e., show a relatively wide host range as compared to the genus *Eimeria* that is homoxenous (see Chap. 3)—and thrive under a large range of climatic conditions. In addition, the resistance and abundance of its infective stage complicate surveillance and control. For pathogenic parasite species and subtypes further improved monitoring and control strategies are globally needed (Fayer et al. 2000; Xiao 2010; Cacciò and Chalmers 2016).

Molecular characterization and experimental data have validated 30 species that infect amphibians, reptiles, birds, fish, and mammals (Fayer 2010; Šlapeta 2013). Additionally, it has been estimated that every year a novel additional species is identified and named and more than 40 isolates—currently referred to as genotypes—have been reported and await their description (Fayer 2010; Šlapeta 2013). Noteworthy, at least 150 mammalian species have been described to host *Cryptosporidium* spp., and an increasing number of species and genotypes have been reported in fish, amphibian, reptile, bird, and vertebrates (Xiao et al. 2004). Among them is *C. parvum*—syn. *C. pestis*—which is commonly found in bovines up to 2 months of age and is the major zoonotic species. Previously, it has been shown that *C. parvum* consists of two genotypes, the genotype H or 1, exclusively infecting humans, and the bovine or zoonotic genotype C or 2, infecting bovines and humans (Morgan et al. 1995; Carraway et al. 1996; Peng et al. 1997). The former has been categorized as *C. hominis*, while the latter has been proposed to be named *C. pestis* (Šlapeta 2013). This proposal was justified based on the priority rule of the International Code of Zoological Nomenclature (ICZN). As *C. parvum* has been originally described by Tyzzer (1912) in mice—sometimes referred to as *C. parvum* mouse I genotype—the newly recognized species *C. parvum* bovine genotype, or genotype 2, should, in accordance with this rule, receive a novel species name (Šlapeta 2006; Šlapeta 2013). However, the name *C. parvum* has been used for cattle isolates ever since Tyzzer's discovery in 1912. Therefore, it has been also put forward to continue to use the name *C. parvum* for the *Cryptosporidium* bovine genotype in order to maintain the universality and stability of the scientific names, as also stated by the ICZN (Fayer 2010; Šlapeta 2012; Xiao et al. 2012). Furthermore, the name *C. tyzzeri* has been proposed for the *C. parvum* mouse I genotype (Šlapeta 2012; Xiao et al. 2012). For the reason of clarity, we will use the name *C. parvum* for the bovine genotype and the name *C. tyzzeri* for the mouse I genotype until this discussion has finally settled.

2.2.1 Diagnostic

Diagnosis of cryptosporidiosis can be challenging for the reason that clinical signs are not pathognomonic and because of the frequently intermittent and low shedding of the small-sized oocyst (Santin 2013). It is therefore desirable to increase the sensitivity of utilized diagnostic tools and test multiple stool specimens as the disease is commonly sub-diagnosed—e.g., horse cryptosporidiosis (Santín and Trout 2008a). Currently, there is no prescribed method for diagnosis, and the World Organization for Animal Health states that the demonstration of *Cryptosporidium* species oocysts or antigen in a properly collected and handled sample is sufficient for a positive diagnosis (Smith 2008a). The selection of an appropriate diagnostic method may depend on a number of criteria in accordance to the needs and possibilities of the researchers and diagnostic laboratories—i.e., the technical expertise, the available time, the required sensitivity and specificity, the need to identify concomitant enteropathogens, and the financial resource (Chalmers and Katzer 2013). As there is no universally accepted gold-standard detection test, comparison of sensitivity and specificity between different diagnostic tests can be done by appointing a method as gold-standard or by using proper statistical tests allowing test comparison (De Waele et al. 2011).

Diagnostic tools rely on stool sample, which is a complex matrix that needs to be properly transported and preserved, since it contains microorganisms that break down the sample, generating odors and interfering with the downstream isolation of oocysts or even DNA. Thus, samples have to be either processed immediately or frozen until further use. Alternatively, potassium dichromate as a preservative may be added in a final concentration of 2.5% and the sample stored at 4 °C, which allows to store contained oocyst for extended periods—6 to 12 months—and also neutralizes odors. Other compounds used for preservation are formaldehyde or saline formalin at 10%, respectively, which are additionally reducing the viability of *Cryptosporidium* oocysts allowing a safer manipulation of samples. However, their use depends on the final application since it may modify protein epitopes and/or DNA (Smith 2008; Chalmers and Katzer 2013).

Many techniques are available for oocyst detection (Table 2.3). Traditionally, diagnostics has been based on microscopic examination of oocysts in fecal smears. Specimens are usually treated with a modified acid-fast staining protocol such as Ziehl-Neelsen or Kinyoun. On microscopic examination, *Cryptosporidium* oocysts appear as red-stained spherules. This is still often considered the method of choice as it is simple, rapid, and of low cost (Silverlås et al. 2013). In order to improve oocyst detection, feces is often purified by passage through sieves and oocysts concentrated by centrifugation and/or oocyst flotation. The most common procedure for oocyst concentration is sucrose or sodium chloride flotation or—in case a high purity is desired—sucrose gradient centrifugation. To allow oocyst flotation, the solutions need to have a higher final specific density of 1.15–1.20 g/m^3 than that of intact oocysts of approximately 1.05 g/m^3 (Smith 2008b).

The most efficient method for oocyst concentration is immunomagnetic separation (IMS). This method consists of the use of paramagnetic beads coated with

Table 2.3 Advantages and disadvantages of diagnostic tools and methods (Chalmers and Katzer 2013)

Tools	Diagnostic method	Sample	Advantage	Disadvantage
Microscopic	AFM and FM	Stool and water	Simplicity, rapid, low cost	Technical expertise, low sensitivity and specificity, time consuming
	IFM	Stool and water	High sensitivity and specificity, less subjective than AF and FM	Technical expertise, work effort, cross-reaction Variability of species antigen
Molecular	PCR	DNA	High sensitivity and specificity, species differentiation	Preservation of samples may inhibit PCR, requires efficient DNA extraction
	rq-PCR, MT-PCR	Stool and water	Detection of other pathogens	No differentiation of *Cryptosporidium* sp.
Immunological	EIA	Stool	Low sensitivity and specificity, can be automated, detection of other pathogens	High cost, variability of species antigen, may compromise diagnosis
	ICFL	Stool	Low specificity, few skills, detection of other pathogens	Low sensitivity, high cost, variability of species antigen, may compromise diagnosis

AFM acid-fast staining and microscopy, *FM* fluorescence microscopy, *IFM* immunofluorescence microscopy, *rq-PCR* real-time PCR, *MT-PCR* multilocus PCR, *EIA* enzyme immunoassay, *ICFL* immunochromatographic lateral flow

anti-oocyst wall antigen-monoclonal antibodies (mAb). After binding of beads to the oocyst surface, a bead-mAb-oocyst complex is formed that is separated from the sample matrix by use of a magnet. After magnetic separation, the complex is later dissociated in an acidic solution (Bukhari et al. 1998). IMS accounts for highest detection sensitivities when coupled with appropriate detection methods. After acid-fast staining and microscopic examination, the crescentic shape of the residing four sporozoites can be occasionally observed in purified oocyst samples (Fig. 2.3).

Possibly the most specific and sensitive method and therefore by some laboratories considered to represent the current gold-standard of oocyst detection is the direct immunofluorescent antibody (DIA) assay. In this assay immunofluorescent stains such as Auramine O are used that enhance sensitivity and allow easier *Cryptosporidium* oocyst detection than acid-fast staining methods upon microscopic inspection (Chalmers and Katzer 2013). As the DIA assay is usually developed based on *C. parvum*, antigen variability between species may compromise recognition as has been observed in the case of *C. felis* (Agnamey et al. 2011).

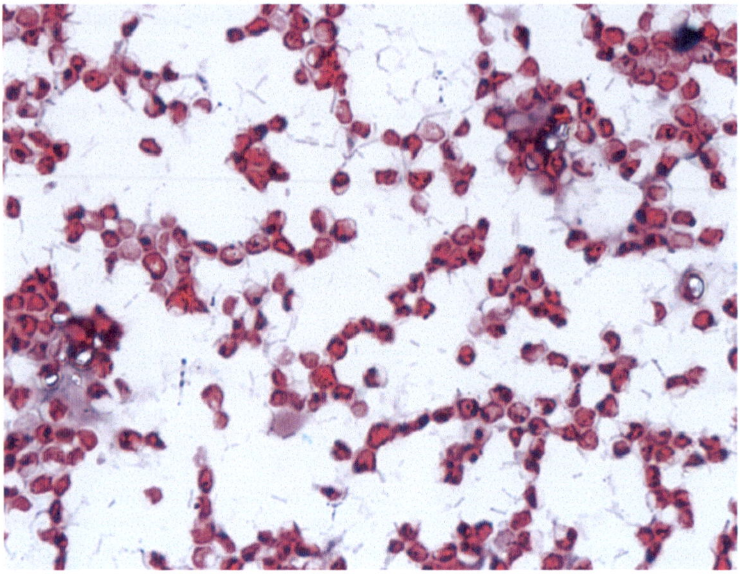

Fig. 2.3 *C. parvum* oocysts. After sucrose flotation purification, staining was done using the modified Ziehl-Neelsen protocol; final magnification of 1000×

In the last decade enzyme immunoassays (EIA) based on the detection of oocyst antigens are increasingly used (Savin et al. 2008). EIAs have the advantage that they can be automated and are more sensitive and specific than microscopy (Giadinis et al. 2012). Although exhibiting a lower sensitivity, the immunochromatographic lateral flow (ICLF) assay has the advantage that it can be performed without the need of specialized equipment. Both methods may be designed to detect also other pathogens such as *Giardia* and *Entamoeba histolytica* for humans (Budu-Amoako et al. 2012; Goñi et al. 2012). Unfortunately, none of the serological tests is able to distinguish between different *Cryptosporidium* species. However, a joined PCR-EIA format has been described that allows to distinguish between *C. parvum* and *C. hominis* (Savin et al. 2008).

Molecular PCR-based detection tools are currently the only available method for species identification and subtyping (Santín and Trout 2008b; Xiao 2010). Target genes used for PCR detection are the 18S RNA gene, the oocyst wall protein gene (COWP), the heat shock protein gene (HSP-70), and actin genes using either a direct or a nested PCR format. Species determination is commonly done by PCR of the 18S RNA gene followed by enzymatic digestion and gel electrophoresis (PCR-RFLP) and/or sequencing (Xiao et al. 1999; Coupe et al. 2005). Phylogenetic analysis of the amplified and sequenced 18S RNA gene fragment allows determining the species or genotype (Šlapeta 2013). Also highly sensitive species-specific real-time PCR assays for the detection of *C. parvum* and *C. hominis* and multiplex real-time PCR assays for the detection of *Cryptosporidium* spp. and other enteroprotozoans have been established (Bruijnesteijn van Coppenraet et al. 2009; De Waele et al. 2011; Elwin et al. 2012; Jex et al. 2012; Lalonde et al. 2013).

For subtyping of *C. parvum* and *C. hominis*, the highly polymorphic GP60-encoding gene is widely used (Xiao 2010). GP60 subtypes are defined by phylogenetic segregation of the GP60 gene and also the sequence pattern and period number of a microsatellite tandem repeat contained in its ORF. Importantly, zoonotic transmission and/or the severity of disease can often be associated with GP60 subtypes (Plutzer and Karanis 2009). As the identification of species subtyping is limited when using a single locus such as the GP60 gene, multilocus typing tools based on microsatellite tandem repeats or multilocus sequence typing (MLST) have been and are being developed and are increasingly applied to define zoonotic or pathogenic strains (Widmer and Lee 2010; Díaz et al. 2012). In a recent comparison of *C. parvum*-multilocus genotypes (MLGs) identified in humans, calves, sheep, and goat, a specific subpopulation could be defined in the latter host (Drumo et al. 2012).

2.2.2 Epidemiology

A wide range of *Cryptosporidium* sp. of which the most common are listed in Table 2.1 infects farm animals such as cattle, sheep, goat, pig, chicken, and fish of fish farms as well as companion animals such as horse, dog, and cat. Among the species that have been reported in farm animals, *C. parvum* and *C. hominis* are of high and major significance for human public health, respectively. Young bovines of under 2 months of age are considered the most important animal reservoir of the zoonotic *C. parvum* followed by goats. In addition, *C. parvum* infects the companion animals horse and dog that are in close contact with humans (Santin 2013). Transmission of the infective oocyst stage of *C. parvum* happens via the fecal-oral route, either anthroponotic, by direct person to person, or zoonotic, animal to person contact. Additionally, indirect transmission occurs through contaminated food, drinking, or recreational water. In contrast, transmission of *C. hominis* is commonly observed by direct anthroponotic transmission or by indirect transmission via food, drinking, and recreational water (Fig. 2.4). However, in a recent study this parasite has been identified as the predominant species in horses in China suggesting that zoonotic transmission may also occur (Deng et al. 2017). In humans, *Cryptosporidium* infects the gastrointestinal tract typically causing a self-limiting diarrhea that lasts about 2–3 weeks, but in individuals with a suboptimal immune system such as malnourished children, elderly, or immunocompromised subjects, the infection may fulminate and develop into a life-threatening condition and death (Thompson et al. 2005; Mor and Tzipori 2009; Checkley et al. 2015).

The infective oocyst exhibits an outstanding resistance against chlorine-based chemical disinfectants commonly used in water treatment systems—e.g., 6% sodium hypochlorite—as well as against 70% ethanol and other commercial preparations commonly used domestically or in animal husbandry (Weir et al. 2002). It has been shown that the oocyst can withstand up to 6 months in the environment under stressful conditions such as low and high temperatures (5–40 °C), freezing, and desiccation (Robertson et al. 1992). Importantly, the ability for survival, persistence, and propagation in the environment independent from its vertebrate host has

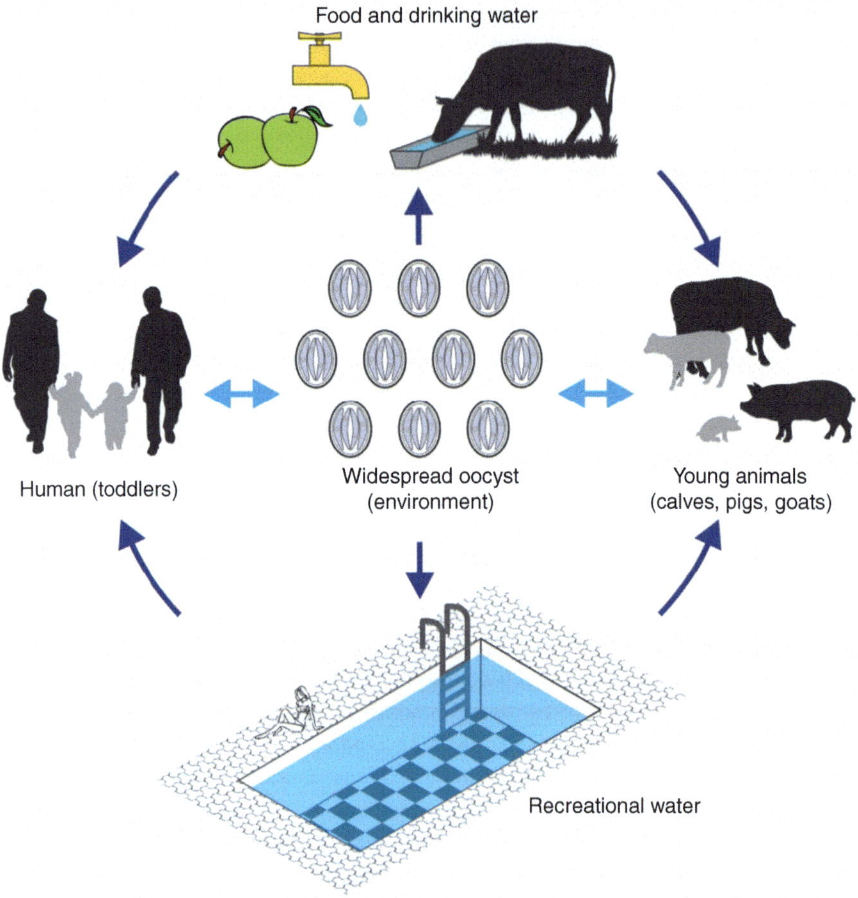

Fig. 2.4 Transmission route of infective *Cryptosporidium* oocysts. Blue light arrows represent direct transmission through person to person, anthroponotic, mainly *C. hominis*; animal to animal, non-zoonotic *Cryptosporidium* spp.; or animal to person, zoonotic, mainly *C. parvum*. Indirect transmission occurs through contaminated food, drinking, or recreational water and is designated by *dark blue arrows*

been recently demonstrated for *Cryptosporidium* oocysts as they multiplied and completed their developmental cycle in *Pseudomonas aeruginosa* biofilms (Koh et al. 2013).

Since large waterborne outbreaks are frequently observed for human cryptosporidiosis in developed countries, oocyst surveillance in water samples is of high importance, and improved control strategies are needed (Baldursson and Karanis 2011; Cacciò and Chalmers 2016). Detection of oocysts in water samples is based on screening of large volumes as well as a good recovery of oocysts. Typically, oocysts are enriched by filtration, eluted from the filter, and then separated by IMS.

There are very few reports that have studied and shown the occurrence of human cryptosporidiosis as a source of animals (Greene et al. 1990; Majewska et al. 1999; Chalmers et al. 2011a, b). In developed countries, immunosuppressed individuals living with dogs are particularly exposed; hence, dogs should be evaluated for infection with the zoonotic *C. parvum*. Farm animals as well as companion animals must be considered a potential source of human cryptosporidiosis. Besides calves, a zoonotic risk may be represented by goats since they seem to be predominantly infected with *C. parvum* (Santín and Trout 2008b). Furthermore, veterinarians, particularly when they have a suboptimal immune system, need to be aware of the zoonotic potential of *Cryptosporidium* spp.

2.2.2.1 Cattle

The first case of cryptosporidium infection in cattle was reported in a calf of 8 months of age with clinical signs of anorexia and cachexy (Panciera et al. 1971). Since then bovine cryptosporidiosis has been reported worldwide. At least four *Cryptosporidium* spp. infect bovines in an age-related manner. *C. parvum* infects predominantly pre-weaned calves causing diarrhea that may be fatal, *C. bovis* and *C. ryanae* post-weaned calves, and *C. andersoni* older and adult cattle (de Graaf et al. 1999; Fayer et al. 2006; Santín et al. 2008; Follet et al. 2011). The prevalence reported of cryptosporidiosis in dairy herds varies substantially among studies worldwide ranging from 13.6 to 78.0% (Trotz-Williams et al. 2005, 2007; Fayer et al. 2006, 2007; Silverlås et al. 2009a; Amer et al. 2013; Garro et al. 2016). Factors that affect prevalence are management practices, the geographic climate, and the season in which the sampling has been performed (Sturdee et al. 2003). The age of animals sampled is also relevant given that younger animals show commonly a higher prevalence (Santín et al. 2004; Maddox-Hyttel et al. 2006). Furthermore, the observed wide range of prevalence has been also attributed to the use of different diagnostic methods of a distinct degree of sensitivity as well as to different design and sampling schemes, such as sampling of herds with or without a history of diarrhea (see Sect. 2.2.1). Several risk factors have been reported to be associated with infection including the separation of calves from their dam, an animal age of under 20 days, not feeding colostrum to calves, dispensing of colostrum using a bucket, a large herd size, and multiple cow maternity facilities (Trotz-Williams et al. 2007; Duranti et al. 2009; Silverlås et al. 2009b; Delafosse et al. 2015; Garro et al. 2016).

2.2.2.2 Sheep and Goats

Ovine cryptosporidiosis has been first reported 1974 in Australia, whereas goat cryptosporidiosis has been described 7 years later (Barker and Carbonell 1974; Mason et al. 1981). Nowadays, the disease is considered an important cause of diarrhea, causing mortality in neonatal lamb and kid goats (Johnson et al. 1999; De Graaf et al. 1999). Especially in goat kids, morbidity and mortality can reach up to 100 and 50%, respectively (Santin 2013). *C. parvum* is the most frequent species reported to infected sheep and goat. Despite that more studies are needed to establish an association with human cryptosporidiosis, sheep and goats may be considered a risk for human infection since the most frequent species infecting these hosts is the most important zoonotic species. In addition, *C. ubiquitum* and *C. xiaoi* have also been frequently identified in sheep. In contrast, in goats, *C. ubiquitum* and *C. xiaoi* are rarely found, and only a single case of an infection with *C. hominis* has been reported (Fayer and Santín 2009; Giles et al. 2009; Fayer et al. 2010b; Imre et al. 2013). Other species and subtypes that have been reported only sporadically in sheep are *C. andersoni*, sheep genotype I, *C. hominis*, *C. fayeri*, *C. suis*, and pig genotype II (Ryan et al. 2005; Yang et al. 2009; Sweeny et al. 2011a).

Young animals seem to be more susceptible to the infection, and outbreaks are usually seen in animals under 2 weeks of age (De Graaf et al. 1999). Crowded flocks have been associated with a higher risk of occurrence of ovine cryptosporidiosis (Alonso-Fresán et al. 2005). In contrast, inadequate colostrum intake, contamination of kidding areas, and late kidding season have been associated with goat cryptosporidiosis (Sevinç et al. 2005; Giadinis et al. 2015).

2.2.2.3 Pig

The first description of cryptosporidiosis in pigs has been reported in 1977, and nowadays pigs are generally observed to be infected with *Cryptosporidium* spp. (Kennedy et al. 1977). The species and subtypes most commonly found are *C. parvum*, *C. suis*, and pig genotype II, while *C. muris* and mouse genotype are occasionally reported (Ryan et al. 2004; Xiao et al. 2006; Zintl et al. 2007). Recently, the species *C. scrofarum* has been reported in China and Poland (Rzezutka et al. 2014; Zou et al. 2017). An extensive study carried out on three Danish organic pig farms has revealed that the prevalence, intensity of infections, and the presence of the parasite varies significantly between age groups. The highest prevalence has been found in piglets and starter pigs, while infections in animals younger than 1 month, or in adults, are less frequent (Maddox-Hyttel et al. 2006; Petersen et al. 2015). Interestingly, a inverse age-related difference in the prevalence of *C. suis* and the pig genotype II has been also reported, whereas in piglets *C. suis* was more and the pig genotype II less frequently found; a vice versa prevalence of the two species was observed in weaners (Langkjaer et al. 2007).

Association with human cryptosporidiosis is rarely reported (Cama et al. 2003; Leoni et al. 2006); however, the zoonotic risk must be further investigated.

2.2.2.4 Horse

Equine cryptosporidiosis has been first reported in 1978 in immunodeficient foals and is nowadays worldwide observed in immunodeficient as well as in immunocompetent animals (Snyder et al. 1978; Santin 2013). *C. parvum* and *C. hominis* are the most frequent species reported in horses, suggesting that horses may represent a potential reservoir of human cryptosporidiosis (Grinberg et al. 2008; Deng et al. 2017). Although the *Cryptosporidium* horse genotype has been rarely reported in foals, it should be considered a zoonotic genotype as it has been found to be associated with human cryptosporidiosis (Burton et al. 2010). Risk factors reported to be associated with the infection was age—young animals are more susceptible than adult animals—and farm (Veronesi et al. 2010). Furthermore, *Cryptosporidium* oocysts appear to contaminate the walls and floors of equine perinatology units, posing a risk for the transmission of infection to newborn foals and a public health risk (Piva et al. 2016). Importantly, the association between the occurrence of cryptosporidiosis in foals and students attending these animals has been described (Galuppi et al. 2016).

2.2.2.5 Dog and Cat

Cryptosporidiosis in cats and dogs has been first described in 1979 and 1983, respectively (Iseki 1979; Wilson et al. 1983). Species most frequently reported in dogs are *C. parvum*, *C. canis*, and *C. meleagridis* while in cats *C. muris* and *C. felis* (Fayer et al. 2001; Irwin 2002; Hajdušek et al. 2004; Giangaspero et al. 2006; Pavlasek and Ryan 2007; Yoshiuchi et al. 2010). For the first time, the species *C. parvum*, *C. ryanae*, and *Cryptosporidium* rat genotype III have been reported in cats and *C. ubiquitum* in dog (Li et al. 2015; Yang et al. 2015).

The prevalence of cryptosporidiosis in dogs and cats has been determined in a few studies only. A study in dogs has reported a prevalence of 4.6 and 8.0% in Greece and South Africa, respectively (Saime et al. 2013; Kostopoulou et al. 2017). It has been reported that the prevalence in dogs and cats is significantly higher among younger animals (Yoshiuchi et al. 2010; Santin 2013). However, some studies could not support such an age correlation in dogs (Olabanji et al. 2016; Xu et al. 2016). Interestingly, a study from Nigeria has recently found that the confinement of dogs poses a risk factor for the infection with *Cryptosporidium* (Olabanji et al. 2016). *C. felis* has been involved as source of human infections, since it was found in immunocompromised as well as in immunocompetent individuals that had pet cats in many countries (Morgan et al. 2000; Pedraza-Díaz et al. 2001; Xiao et al. 2001; Caccio et al. 2002). However, a recent study from Spain does not support the zoonotic transmission between companion animals, dogs and cats, and humans (De Lucio et al. 2017).

2.2.2.6 Chicken

C. baileyi is the most commonly infecting species in avian animals, causing high morbidity and mortality in poultry (Lindsay and Blagburn 1990). *C. gallis* has been associated with chronic diarrhea and mortality of chicken coinfected with other enteropathogens, while *C. meleagridis* might be of minor veterinary importance in poultry farming (Santin 2013). Mortality of chicken due to *C. parvum* infection has been also reported.

2.2.2.7 Reptiles

In reptiles, cryptosporidiosis is of clinical importance since the disease is lethal and chronic. The infecting species *C. serpentis* and *C. varanii* are both not zoonotic (Table 2.1) (Santin 2013; Šlapeta 2013).

2.2.3 Economic Impact

Cryptosporidiosis of young farm ruminants like calves, goats, and lamb generates the neonatal diarrhea syndrome characterized by dehydration, impaired growth, and an increased mortality (Johnson et al. 1999; De Graaf et al. 1999). In bovine cryptosporidiosis the associated morbidity, weight loss and delayed growth, and sometimes mortality of young animals result in considerable economic losses. These losses comprise of direct costs that are brought about by a decreased performance and production and indirect costs due to the necessary veterinary assistance and extra care of ill animals, such as feeding of electrolyte solutions, fluid intravenous therapy, drug administration, or hygienic measures (De Graaf et al. 1999). In addition, calves that have recovered from disease are not able to compensate for weight loss and delayed growth.

Avian cryptosporidiosis, which leads to respiratory and digestive tract diseases, results in the decline of the performance of animal production—i.e., growth retardation and increased consumption index and a high mortality rate during severe infections—responsible for significant economic losses of the poultry industry (Lindsay and Blagburn 1990; De Graaf et al. 1999).

2.3 Clinical Effects and Pathology

Clinical disease in neonatal farm ruminants such as calves, goat kids, and lamb is commonly caused by the species *C. parvum*, but in sheep the species *C. xiaoi* and C. *ubiquitum* is occasionally also implied. In companion animals, clinical disease is observed in young horses and is associated with the infection of *C. parvum* and *C. hominis*. Cryptosporidiosis in other companion animals such as dog and cat is usually asymptomatic, and severe clinical disease has been only occasionally reported (Lucio-Forster et al. 2010).

The clinical signs of the disease are highly variable and depend on the infecting *Cryptosporidium* species and the animal host. Even within the same host, clinical signs range from asymptomatic to life-threatening according to the age and immune status of the individual animal (Johnson et al. 1999; de Graaf et al. 1999). The prepatent period, defined as the shortest time period between ingestion and shedding of the infective oocyst, and the patent period, defined as the duration of the infection and estimated by determining the duration of oocyst shedding, vary with the infective dose as well as with infecting *Cryptosporidium* sp. and host. Table 2.4 shows common pathogenic parasite species infecting farm and companion host species and the associated most typical physiopathological and clinical signs.

2.3.1 Farm Animals of Economic Importance

2.3.1.1 Cattle

At least four *Cryptosporidium* sp. infect bovines in an age-related manner. *C. parvum* infects predominantly pre-weaned calves, *C. bovis* and *C. ryanae* post-weaned calves, and *C. andersoni* older and adult cattle (Fayer et al. 2006; Santín et al. 2008; Follet et al. 2011). Neither *C. bovis* nor *C. ryanae* are associated with clinical signs or disease (Fayer et al. 2005, 2008; Kváč et al. 2008). Interestingly, no clinical signs are observed in *C. andersoni*-infected adult cows. However, pathophysiological

Table 2.4 Farm or companion animal host with the predominant often pathogenic *Cryptosporidium* species: physiopathology, clinical signs, prepatent and patent period, effective drugs, and association with human cryptosporidiosis

Host	Species	Prepatent period (days)	Patent period (days)	Physiopathology	Clinical signs	Drugs tested	Association with human cryptosporidiosis	Reference
Calves	*C. parvum*	2–7	1–12	Villous atrophy, reduction of surface area of the small intestinal mucosa. Secretion of PGI2 and PGE2, inhibition NaCl absorption. Fluid loss. Atrophy. Necrosis of enterocytes	Asymptomatic to self-limiting watery diarrhea, maldigestion, malabsorption, depression, weakness, and anorexia	Halofuginone NTZ, paromomycin, lasalocid, decoquinate, and β-cyclodextrin sulfadimethoxine	Commonly	Foster and Smith (2009) and Santin (2013)
Lambs	*C. parvum*	ND	ND	ND	Mild-to-severe diarrhea	Halofuginone, paromomycin	Possibly	Santin (2013), Viu et al. (2000) and Santín and Trout (2008b)
Goats	*C. parvum*	3[a]	10–11[b]	Atrophy, blunting, and fusion of villi	Asymptomatic to diarrhea, anorexia and prostration	Halofuginone NTZ, α-cyclodextrin, paromomycin, decoquinate	Possibly	Johnson et al. (2000), Mancassola et al. (1997), Santín and Trout (2008b) and Castro-Hermida et al. (2004)
Pigs	*C. suis*	2–9	9–15	Villous atrophy, villous fusion, cellular infiltration, necrosis of epithelial cells	Asymptomatic to mild diarrhea	No	Rarely	Argenzio et al. (1990), Vítovec et al. (2006), Enemark et al. (2003a) and Santín and Trout (2008b)

(continued)

Table 2.4 (continued)

Host	Species	Prepatent period (days)	Patent period (days)	Physiopathology	Clinical signs	Drugs tested	Association with human cryptosporidiosis	Reference
Dogs	*C. canis*	ND	ND	ND	Asymptomatic	No	Rarely	Greene et al. (1990), Lihua et al. (2007) and de Lucio et al. (2017)
Cats	*C. felis*	5–6	7–10	ND	Asymptomatic	Paromomycin (one case)	Occasionally	Morgan et al. (2000), Barr et al. (1994), Irwin (2002), Iseki (1979) and de Lucio et al. (2017)
Horses	*C. parvum*	2–5[a]	5–8[b]	Mild-to-moderate villous atrophy	Asymptomatic to diarrhea, dehydration and weakness	No	Occasionally	Xiao and Herd (1994), Cohen and Snowden (1996) and Bjorneby et al. (1991)
Chicken	*C. baileyi*	4–24	1–18	Hypertrophy and hyperplasia of epithelial cells. Mucus excess in the trachea, airsacculitis	Coughing, sneezing, dyspnea, depression	Enrofloxacin and paromomycin	No association	Sréter et al. (2002), Current et al. (1986) and Fernandez et al. (1990)

NTZ nitazoxanide; *ND* no data
[a]Start of oocyst shedding post-inoculation
[b]Time period of oocyst shedding

changes at the site of infection such as dilated gastric glands and changes in the gastric mucosa including pallor and thickening, hyperplasia, and inflammatory cell infiltration into the lamina propria of the infected area have been reported (Masuno et al. 2006).

C. parvum is the primary agent of neonatal diarrhea in calves and has been reported in all husbandries around the world. The duration and severity of diarrhea in dairy calves infected with *C. parvum* may lead to dehydration and death, but this is usually not seen in endemic herds (Santín and Trout 2008b). Clinical signs include profuse watery diarrhea that may be accompanied by depression, anorexia, and weakness, and a mortality of 5–10% has been reported (de Graaf et al. 1999). For *C. parvum* infection in calves, the prepatent period ranges from 1 to 12 days and the patent period from 2 to 7 days (Tzipori 1983) (Table 2.4).

2.3.1.2 Sheep

Diarrhea caused by cryptosporidiosis of lamb causes high morbidity and mortality worldwide (de Graaf et al. 1999; Quílez et al. 2008, Santin 2013). Of the most frequent sheep-infecting *Cryptosporidium* species —*C. parvum*, *C. ubiquitum*, and *C. xiaoi*—*C. parvum* has been suggested to be the main species associated with clinical illness, which varies between mild and severe diarrhea, though it has also been identified in healthy lamb (Table 2.4) (Majewska et al. 2000; Doris et al. 2008; Paoletti et al. 2009). Experimental infection of sheep with *C. ubiquitum* and *C. xiaoi* did not result in clinical disease (Fayer and Santín 2009; Fayer et al. 2010b). Further in-depth studies are needed to determine a significant association of these two parasites with regard to pathology and possible consequences for production (Sweeny et al. 2011b).

2.3.1.3 Goat

In goat kids, cryptosporidiosis is significantly associated with diarrhea followed by dehydration, anorexia, lethargy, prostration, and a very high morbidity ($\leq$100%) and mortality ($\leq$50%) (Santin 2013). The disease is distributed worldwide and considered to be of high economic importance. *C. parvum* has been identified in sick as well as in asymptomatic goats, while *C. xiaoi* was found only in diarrheic goats (Table 2.4) (Quílez et al. 2008; Muñoz et al. 1996; Díaz et al. 2010; Paraud et al. 2010; Santin 2013).

2.3.1.4 Chicken

In poultry, *C. baileyi* infections are considered the major cause of morbidity and mortality. In chicken, respiratory disease is a common form of clinical manifestation, while enteritis and renal disease are occasionally reported. Clinical signs of the respiratory manifestation are similar in all avian species and consist of rales, coughing, sneezing, and dyspnea (Lindsay and Blagburn 1990) (Table 2.4). It has been reported that younger broiler chicken are more susceptible than adults and may develop conjunctivitis, airsacculitis, sinusitis, rhinitis, and tracheitis. Renal cryptosporidiosis is an additional clinical form that occurs occasionally in *C. baileyi*-infected chicken. Concurrent occurrence of *Cryptosporidium* with other infective agents like Marek's disease virus, infectious bursal disease virus, or chicken anemia virus seems to promote synergistically this clinical manifestation including severe respiratory signs, growth retardation, and mortality. This has been experimentally confirmed by coinfection of chicken with *C. baileyi* and Marek's disease virus leading to enlarged and pale kidneys and urate crystals in surface tubules (Abbassi et al. 1999).

Intestine cryptosporidiosis of birds is characteristic for the infection of the small intestine and bursa with *C. meleagridis* (Ryan 2010). Infection can be accompanied with enteritis, diarrhea, and mortality (Pagès-Manté et al. 2007). However, in experimentally infected broiler chicken, neither final live weight nor mortality was influenced by the infection (Tůmová et al. 2002).

2.3.2 Other Hosts

2.3.2.1 Pig

Porcine cryptosporidiosis has been reported worldwide. Although it has been shown that *Cryptosporidium*-infected pigs show a higher prevalence of diarrhea than uninfected animals, the association between the infection and clinical signs are controversial (Quílez et al. 1996; Maddox-Hyttel et al. 2006; Vitovec et al. 2006; Hamnes et al. 2007). Piglets experimentally infected with *C. suis* developed mild or no clinical signs, whereas experimental infection with *C. parvum* resulted in diarrhea, oocyst shedding, body weight changes, and histological alterations among others (Enemark et al. 2003b) (Table 2.4). It has been observed that concomitant infection with rotavirus had an aggravating synergistic pathogenic effect on cryptosporidiosis in piglets resulting in prolonged diarrhea, an increased oocyst shedding, decreased weight gain, and an increased death rate (Enemark et al. 2003a).

2.3.2.2 Dog, Cat, and Horse

Cryptosporidiosis in companion animals such as dog, cat, and horse is most common in younger animals and has been reported worldwide with clinical signs ranging from asymptomatic to symptomatic (Santín and Trout 2008a). Usually, the course of disease of dogs infected with *C. canis* is asymptomatic, though severe diarrhea, malabsorption, and weight loss have occasionally been reported (Irwin 2002; Santin 2013). Also in cats, cryptosporidiosis is mostly asymptomatic and commonly associated with *C. felis* infection (Yoshiuchi et al. 2010). Clinical disease may develop with immunosuppressive conditions like feline leukemia virus infection or when coinfected with another enteric parasite (Monticello et al. 1987; Rambozzi et al. 2007; Santin 2013). Immunocompetent foals and to a lesser degree adult horses can develop cryptosporidiosis, but not always excretion of oocysts is accompanied with diarrhea (Xiao and Herd 1994; Majewska et al. 2004). However, a case where infection of foals resulted in severe morbidity, and three animals died, has been reported (Grinberg et al. 2003). In this case and in most other studies where the infecting species has been determined, the presence of *C. parvum* has been reported. Foals and horses that are immunosuppressed seem to attract the infection relatively easily and have been observed to develop severe diarrhea and shedding of oocysts, suggesting that the severity of the disease varies according to the immune status of the horses (Snyder et al. 1978; Bjorneby et al. 1991). After an experimental infection of immunodeficient foals with *C. parvum*, a severe, persistent, and watery diarrhea developed which was associated with oocyst excretion that lasted between 2 and 5 days (Table 2.4) (Bjorneby et al. 1991).

2.4 Prevention, Control, and Treatment

2.4.1 Chemotherapeutics

In animals, many chemicals have been screened and tested, but only halofuginone has been licensed for the treatment and control against cryptosporidiosis of calves. However, this drug is currently only available in Canada and indicated on emergency cases. So far, all tested drugs demonstrated only partial effectiveness on prophylaxis and therapy by reducing oocyst excretion as well as the severity of the disease. Most prominent drugs that have been shown to exhibit some efficiency in in vivo case control studies are (1) the coccidiostatic drugs, halofuginone and decoquinate, and the antiparasitic drug nitazoxanide (NTZ) and (2) the antibiotics, paromomycin, azithromycin, enrofloxacin, and lasalocid, as well as starch derivatives α- and β-cyclodextrin (Table 2.4).

2.4.1.1 Coccidiostatic and Antiparasitic Drugs

Although approved for veterinary use, the prophylactic and therapeutic effectiveness of halofuginone in calves is highly controversial, and a meta-analysis could neither determine clear benefits for prophylactic nor for therapeutic treatment (Silverlås et al. 2009b). Furthermore, halofuginone is toxic at twice the recommended dose and contraindicated for the use in diarrheic animals. Nonetheless, in one study, a preventive effect and a reduction of clinical signs and of environmental contamination could be observed when combined with strict hygienic measures—e.g., isolating calves in pens (De Waele et al. 2010).

In lamb, halofuginone has also been tested and was shown to prevent and treat cryptosporidiosis by reducing the shedding of oocysts and the incidence of diarrhea and mortality, though no effect on body weight gain was observed (Giadinis et al. 2008).

Recently, the efficacy of the drug in goat kids was demonstrated when given as a prophylactic treatment as it reduced oocyst excretion as well as diarrhea and mortality (Petermann et al. 2014).

Decoquinate did not decrease oocyst shedding in calves. However, shedding could be reduced in goat kids, yet animals did not recover weight (Moore et al. 2003; Mancassola et al. 1997; Lallemond et al. 2006).

In humans, nitazoxanide (NTZ) is the only drug approved by the FDA. While effective in the treatment of immunocompetent individuals, it shows only a moderate to no effect in children and immunoincompetent patients such as HIV-infected, lymphoma, and leukemia patients (Abubakar et al. 2007).

In cryptosporidiosis of calves, it has been found that NTZ reduced the duration of oocyst excretion and severity of disease in one study, yet no therapeutic effect could be demonstrated in another (Ollivett et al. 2009; Schnyder et al. 2009). Although in goat kids oocyst excretion could be somewhat reduced, the effective dose showed an acute toxicity (Viel et al. 2007).

2.4.1.2 Antibiotics

Paromomycin, an aminocyclitol antibiotic isolated from *Streptomyces*, showed a therapeutic effect in calves, goat kids, and neonatal lamb though in the latter an impaired growth was observed. In chickens, paromomycin has been found to be the most effective drug tested so far, and it has been suggested that in combination with appropriate sanitary procedures and disinfection, it may support the control of chicken cryptosporidiosis (Barr et al. 1994; Mancassola et al. 1995; Viu et al. 2000; Sréter et al. 2002). Also enrofloxacin—a fluoroquinolone antibiotic—has demonstrated a nearly similar efficacy as paromomycin, reducing oocyst excretion and showing a similar body weight gain as the control group (Sréter et al. 2002).

In dairy calves, azithromycin has been reported to reduce oocyst excretion and the severity of the disease, but its high cost prevents its implementation (Elitok et al. 2005). Lasalocid is banned in Europe owing to its use as a growth promoter; nonetheless it has been found to be effective in calf cryptosporidiosis in combination with supportive therapies such as adequate infusion (Sahal et al. 2005).

Castro-Hermida et al. (2002) have evaluated the therapeutic and prophylactic effect of the starch-derived excipient β-and α-cyclodextrin in lamb, calves, and goats. In lamb, β-cyclodextrin showed a highly effective preventive as well as therapeutic effect. A reduction of oocyst-infected animals and no mortality was seen after prophylactic treatment, whereas a reduction on oocyst shedding and severity of diarrhea was seen in animals that have received therapeutic treatment (Castro-Hermida et al. 2002). A lesser effect was seen when this drug was applied using prophylactic and therapeutic protocols in naturally infected calves, decreasing the severity of diarrhea and shortening the time of oocyst shedding (Castro-Hermida et al. 2001). The same authors also demonstrated a therapeutic efficiency of α-cyclodextrin in neonatal goat kids since a longer prepatent period was observed, whereas the patent period was found to be significantly shorter than in the control group (Castro-Hermida et al. 2004).

2.4.2 Vaccines

As to now, there are no vaccines available against cryptosporidiosis. Bovines possess a syndesmochorial placenta, and the fetus does not receive immunoglobulins from the dam via transplacental transmission. Passive transfer of immunoglobulins to calves is achieved by the feeding of the first milking colostrum. Accordingly, it has been demonstrated that high levels of antibodies in colostrum may protect calves fed with hyperimmune colostra (Fayer et al. 1990). Fayer et al. (1990) have demonstrated in an experimental neonatal mice model that passive immunotherapy is effective in reducing the number of intestinal infective oocysts. In this study, experimental neonatal mice were treated with the whole whey or purified specific immunoglobulin isotypes (IgG1 and IgA) from hyperimmune colostrum of a cow immunized with *C. parvum* oocysts. Oocyst shedding, and occurrence of diarrhea was significantly decreased, compared with mice that had received corresponding immunoglobulin isotypes of control colostrum (Fayer et al. 1990).

A number of immunodominant *Cryptosporidium* surface antigens have been identified and characterized as potential vaccine candidates (Boulter-Bitzer et al. 2007). Among others are the *C. parvum* antigens CP15/60 and P23 which are both expressed in the invasive zooite stages—i.e., sporozoite and merozoites (Jenkins et al. 1993). Against both these antigens, antibodies are generated in *C. parvum*-infected species and humans, and they have been therefore proposed as potential immunotherapeutic targets (Répérant et al. 1994). Importantly, nasal immunization of pregnant goat dams with a CP15-recombinant DNA vaccine showed a significant reduction of oocyst excretion in goat kid offspring. Furthermore, generation and application of a monoclonal antibody against CP15 have demonstrated protective immunity in experimentally infected mice (Jenkins et al. 1993; Sagodira et al. 1999).

Anti-recombinant P23-egg yolk antibodies (IgY) administered to *Cryptosporidium*-infected experimental mice could significantly reduce oocyst shedding (Shahbazi et al. 2009; Omidian et al. 2014). Also, P23-specific monoclonal antibodies passively applied to mice that had been experimentaly challenged with *C. parvum* oocyst reduced the infection (Perryman et al. 1999). In a further study, hyperimmune bovine colostrum was generated by immunizing pregnant cows with a recombinant form of P23. After challenge, neonatal calves showed substantial protection as exemplified by absence of diarrhea and nearly complete absence of oocyst shedding (Askari et al. 2016).

Another therapeutic approach of calve cryptosporidiosis might be vaccination with gamma irradiated oocysts. It has been shown that after inoculation of 1-day-old calves with 400 Gy irradiated oocysts, animals did not develop clinical signs and showed absence from oocyst excretion at 21 days of age. The applied irradiation doses ensured parasite infection allowing the development of a strong protective immune response but prevented the development of clinical diseases (Jenkins et al. 2004).

2.4.3 Control and Treatment

Since current prophylactic and therapeutic treatments using drugs are insufficient and vaccines are not available, the best approach to control cryptosporidiosis is by following strict hygienic measures in order to prevent dissemination of the infective oocysts and/or to eliminate them as much as possible from the environment. Especially, on large animal farms, this can be a challenging objective since animals are able to spread large number of oocyst—e.g., a single calve with diarrhea can shed up to 10^7 oocysts per gram of feces (Fayer et al. 1998). Continuously moving animals to clean areas when large numbers are concentrated is necessary to control the infection, but it is not always economically viable.

Other effective hygiene procedures include the disinfection of soil, calf housing, maternity pens, and tools like feeding equipment. Disinfection can be carried out by exposure of materials, utensils, or surfaces contaminated with oocyst to physical factors or by chemical treatment. Of the former, efficacy to reduce oocyst viability has been shown by heat, above 50 °C for 5 min; freezing, at −70 °C for 1 h; UV light; gamma radiation; solar irradiation; or pulsed light (Black et al. 1996; Jenkins et al. 2004; Quilez et al. 2005; Fayer 2008). Although oocysts are resistant to many commonly used disinfectants, some chemicals have shown to reduce oocyst viability such as 10% formol, aqueous or gaseous ammonia, or 3% hydrogen peroxide. Additionally, two commercial peroxygen based—Ox-Virin and Ox-Agua—and one based on 4-Chlor-m-kresol, Neopredisan, have been shown to be effective against oocysts (Quilez et al. 2005).

Furthermore, reducing risk factors of infection such as separation of cows and calves immediately after birth followed by managing calves within their respective age groups has also be recommended (Harp and Goff 1998). Other measures that have shown to reduce the *Cryptosporidium* burden and infection risk is the use of concrete flooring in calf housing areas, increase of the depth of the bedding to more than 11 cm, daily addition of bedding, feeding with fermented milk, and treatment with antibiotics (Mohammed et al. 1999; Brook et al. 2008; Trotz-Williams et al. 2008; Delafosse et al. 2015).

As there is no therapeutic drug, the preferred treatment in farm and domestic animals is supportive. Generally, this consists of replacement of fluid and electrolytes, nutritional support, and the application of antidiarrheal drugs.

References

Abbassi H, Coudert F, Chérel Y, et al. Renal Cryptosporidiosis (Cryptosporidium baileyi) in specific-pathogen-free chickens experimentally coinfected with Marek's disease virus. Avian Dis. 1999;43:738–44. https://doi.org/10.2307/1592742.

Abrahamsen M, Lancto C, Walcheck B, et al. Localization of a/b and g/d T Lymphocytes in *Cryptosporidium parvum*-infected tissues in Naive and Immune Calves. Infect Immun. 1997;65:2428–33.

Abubakar I, Aliyu SH, Arumugam C, et al. Treatment of cryptosporidiosis in immunocompromised individuals: systematic review and meta-analysis. Br J Clin Pharmacol. 2007;63:387–93. https://doi.org/10.1111/j.1365-2125.2007.02873.x.

Agnamey P, Sarfati C, Pinel C, et al. Evaluation of four commercial rapid immunochromatographic assays for detection of Cryptosporidium antigens in stool samples: a blind multicenter trial. J Clin Microbiol. 2011;49:1605–7. https://doi.org/10.1128/JCM.02074-10.

Aldeyarbi HM, Karanis P. The ultra-structural similarities between *Cryptosporidium parvum* and the Gregarines. J Eukaryot Microbiol. 2016a;63:79–85. https://doi.org/10.1111/jeu.12250.

Aldeyarbi HM, Karanis P. The fine structure of sexual stage development and sporogony of *Cryptosporidium parvum* in cell-free culture. Parasitology. 2016b;143(6):749–61. https://doi.org/10.1017/S0031182016000275.

Alonso-Fresán MU, García-Álvarez A, Salazar-García F, et al. Prevalence of Cryptosporidium spp. in asymptomatic sheep in family flocks from Mexico State. J Vet Med Ser B Infect Dis Vet Public Heal. 2005;52:482–3. https://doi.org/10.1111/j.1439-0450.2005.00889.x.

Alves M, Xiao L, Sulaiman I, et al. Subgenotype analysis of Cryptosporidium isolates from humans Cattle, and Zoo Ruminants in Portugal. J Clin Microbiol. 2003;41(6):2744–7. https://doi.org/10.1128/JCM.41.6.2744.

Alvarez-Pellitero P, Sitjà-Bobadilla A. Cryptosporidium molnari n. sp. (Apicomplexa: Cryptosporidiidae) infecting two marine fish species, Sparus aurata L. and Dicentrarchus labrax L. Int J Parasitol. 2002;32:1007–21.

Amer S, Zidan S, Adamu H, et al. Prevalence and characterization of Cryptosporidium spp. in dairy cattle in Nile river delta provinces, Egypt. Exp Parasitol. 2013;135:518–23. https://doi.org/10.1016/j.exppara.2013.09.002.

Argenzio R, Liacos J, Levy M, et al. Villous atrophy, crypt hyperplasia, cellular infiltration, and impaired glucose-Na absorption in enteric cryptosporidiosis of pigs. Gastroenterology. 1990;98:1129–40.

Askari N, Shayan P, Mokhber-Dezfouli MR, et al. Evaluation of recombinant P23 protein as a vaccine for passive immunization of newborn calves against *Cryptosporidium parvum*. Parasite Immunol. 2016;38:282–9. https://doi.org/10.1111/pim.12317.

Auray G, Lacroix-Lamande S, Mancassola R, et al. Involvement of intestinal epithelial cells in dendritic cell recruitment during *C. parvum* infection. Microbes Infect. 2007;9:574–82. https://doi.org/10.1016/j.micinf.2007.01.026.

Baldursson S, Karanis P. Waterborne transmission of protozoan parasites: review of worldwide outbreaks—an update 2004-2010. Water Res. 2011;45:6603–14.

Barker IK, Carbonell PL. Cryptosporidium agni sp.n. from lambs, and Cryptosporidium bovis sp.n. from a calf, with observations on the oocyst. Z Parasitenkd. 1974;44:289–98. https://doi.org/10.1007/BF00366112.

Barr SC, Jamrosz GF, Hornbuckle WE, et al. Use of paromomycin for treatment of cryptosporidiosis in a cat. J Am Vet Med Assoc. 1994;205:1742–3.

Barta JR, Thompson RCA. What is Cryptosporidium? Reappraising its biology and phylogenetic affinities. Trends Parasitol. 2006;22:463–8. https://doi.org/10.1016/j.pt.2006.08.001.

Bhat N, Joe A, PereiraRerrin M, Ward HD. Cryptosporidium p30, a galactose/N--acetylgalactosamine-specific lectin, mediates infection in vitro. J Biol Chem. 2007;282:34877–87. https://doi.org/10.1074/jbc.M706950200.

Bjorneby JM, Leach DR, Perryman LE. Persistent cryptosporidiosis in horses with severe combined immunodeficiency. Infect Immun. 1991;59:3823–6.

Black EK, Finch GR, Taghi-Kilani R, Belosevic M. Comparison of assays for *Cryptosporidium parvum* oocysts viability after chemical disinfection. FEMS Microbiol Lett. 1996;135:187–9. https://doi.org/10.1016/0378-1097(95)00447-5.

Borowski H, Clode PL, Thompson RCA. Active invasion and/or encapsulation? A reappraisal of host-cell parasitism by Cryptosporidium. Trends Parasitol. 2008;24:509–16. https://doi.org/10.1016/j.pt.2008.08.002.

Borowski H, Thompson RC, Armstrong T, Clode PL. Morphological characterization of *Cryptosporidium parvum* life-cycle stages in an in vitro model system. Parasitology. 2010;137:13–26. https://doi.org/10.1017/S0031182009990837.

Boulter-Bitzer JI, Lee H, Trevors JT. Molecular targets for detection and immunotherapy in *Cryptosporidium parvum*. Biotechnol Adv. 2007;25:13–44. https://doi.org/10.1016/j.biotechadv.2006.08.003.

Brook E, Hart CA, French N, Christley R. Prevalence and risk factors for Cryptosporidium spp. infection in young calves. Vet Parasitol. 2008;152:46–52. https://doi.org/10.1016/j.vetpar.2007.12.003.

Bruijnesteijn van Coppenraet LES, Wallinga JA, Ruijs GJHM, et al. Parasitological diagnosis combining an internally controlled real-time PCR assay for the detection of four protozoa in stool samples with a testing algorithm for microscopy. Clin Microbiol Infect. 2009;15:869–74. https://doi.org/10.1111/j.1469-0691.2009.02894.x.

Budu-Amoako E, Greenwood SJ, Dixon BR, et al. Giardia and Cryptosporidium on dairy farms and the role these farms may play in contaminating water sources in Prince Edward Island, Canada. J Vet Intern Med. 2012;26:668–73. https://doi.org/10.1111/j.1939-1676.2012.00930.x.

Bukhari Z, McCuin RM, Fricker CR, Clancy JL. Immunomagnetic separation of *Cryptosporidium parvum* from source water samples of various turbidities. Appl Environ Microbiol. 1998;64:4495–9.

Burton A, Nydam D, Dearen T, et al. The prevalence of Cryptosporidium and identification of the cryptosporidium horse genotype in foals in New York. Vet Parasitol. 2010;174(1-2):139–44. https://doi.org/10.1016/j.vetpar.2010.08.019.

Cacciò SM, Chalmers RM. Human cryptosporidiosis in Europe. Clin Microbiol Infect. 2016;22(6):471–80. https://doi.org/10.1016/j.cmi.2016.04.021.

Cacciò S, Pinter E, Fantini R, et al. Human infection with Cryptosporidium felis: case report and literature review. Emerg Infect Dis. 2002;8:85–6. https://doi.org/10.3201/eid0801.010269.

Cama V, Bern C, Sulaiman I, et al. Cryptosporidium species and genotypes in HIV-positive patients in Lima. J Eukaryot Microbiol. 2003;50:531–3. https://doi.org/10.1111/j.1550-7408.2003.tb00620.x.

Carraway M, Tzipori S, Widmer G. Identification of genetic heterogeneity in the *Cryptosporidium parvum* ribosomal repeat. Appl Environ Microbiol. 1996;62:712–6.

Castro-Hermida JA, González-Losada Y, Freire-Santos F, et al. Evaluation of β-cyclodextrin against natural infections of cryptosporidiosis in calves. Vet Parasitol. 2001;101:85–9. https://doi.org/10.1016/S0304-4017(01)00505-2.

Castro-Hermida JA, Gonzalez-Losada Y, Freire-Santos F, et al. Efficacy of beta-cyclodextrin against experimental cryptosporidiosis in neonatal lambs. J Parasitol. 2002;88:185–7. https://doi.org/10.1645/0022-3395(2002)088[0185:EOCAEC]2.0.CO;2.

Castro-Hermida JA, Pors I, Otero-Espinar F, et al. Efficacy of α-cyclodextrin against experimental cryptosporidiosis in neonatal goats. Vet Parasitol. 2004;120:35–41. https://doi.org/10.1016/j.vetpar.2003.12.012.

Cavalier-Smith T. Gregarine site-heterogeneous 18S rDNA trees, revision of gregarine higher classification, and the evolutionary diversification of Sporozoa. Eur J Protistol. 2014;50:472–95. https://doi.org/10.1016/j.ejop.2014.07.002.

Cevallos AM, Bhat N, Verdon R, et al. Mediation of *Cryptosporidium parvum* infection in vitro by mucin-like glycoproteins defined by a neutralizing monoclonal antibody. Infect Immun. 2000a;68:5167–75. https://doi.org/10.1128/IAI.68.9.5167-5175.2000.

Cevallos AM, Zhang X, Waldor MK, et al. Molecular cloning and expression of a gene encoding *Cryptosporidium parvum* glycoproteins gp40 and gp15. Infect Immun. 2000b;68:4108–16. https://doi.org/10.1128/IAI.68.7.4108-4116.2000.

Chalmers RM, Katzer F. Looking for Cryptosporidium: the application of advances in detection and diagnosis. Trends Parasitol. 2013;29:237–51. https://doi.org/10.1016/j.pt.2013.03.001.

Chalmers R, Smith R, Elwin K, et al. Epidemiology of anthroponotic and zoonotic human cryptosporidiosis in England and Wales, 2004-2006. Epidemiol Infect. 2011a;139:700–12. https://doi.org/10.1017/S0950268810001688.

Chalmers RM, Smith RP, Hadfield SJ, et al. Zoonotic linkage and variation in *Cryptosporidium parvum* from patients in the United Kingdom. Parasitol Res. 2011b;108(5):1321. https://doi.org/10.1007/s00436-010-2199-x.

Checkley W, White AC, Jaganath D, et al. A review of the global burden, novel diagnostics, therapeutics, and vaccine targets for cryptosporidium. Lancet Infect Dis. 2015;15:85–94. https://doi.org/10.1016/S1473-3099(14)70772-8.

Chen W, Harp J, Harmsen A. Requirements for CD4+ cells and gamma interferon in resolution of established *Cryptosporidium parvum* infection in mice. Infect Immun. 1993a;61:3928–32.

Chen WX, Harp JA, Harmsen AG, Havell EA. Gamma-interferon functions in resistance to *Cryptosporidium parvum* infection in severe combined immunodeficient mice. Infect Immun. 1993b;61:3548–51.

Chen XM, Levine SA, Tietz P, et al. *Cryptosporidium parvum* is cytopathic for cultured human biliary epithelia via an apoptotic mechanism. Hepatology. 1998;28:906–13. https://doi.org/10.1002/hep.510280402.

Chen XM, O'Hara SP, Huang BQ, et al. Apical organelle discharge by *Cryptosporidium parvum* is temperature, cytoskeleton, and intracellular calcium dependent and required for host cell invasion. Infect Immun. 2004;72:6806–16. https://doi.org/10.1128/IAI.72.12.6806-6816.2004.

Clode PL, Koh WH, Thompson RCA. Life without a host cell: what is Cryptosporidium? Trends Parasitol. 2015;31:614–24. https://doi.org/10.1016/j.pt.2015.08.005.

Cohen ND, Snowden K. Cryptosporidial diarrhea in foals. Compend Contin Educ Pract Vet. 1996;18:298–306.

Coupe S, Sarfati C, Hamane S, et al. Detection of Cryptosporidium and identification to the species level by nested PCR and restriction fragment length polymorphism. J Clin Microbiol. 2005;43:1017–23. https://doi.org/10.1128/JCM.43.3.1017.

Current WL, Upton SJ, Haynes TB. The life cycle of Cryptosporidium baileyi n. sp. (Apicomplexa, Cryptosporidiidae) infecting chickens. J Protozool. 1986;33:289–96. https://doi.org/10.1111/j.1550-7408.1986.tb05608.x.

De Graaf DC, Vanopdenbosch E, Ortega-Mora LM, et al. A review of the importance of cryptosporidiosis in farm animals. Int J Parasitol. 1999;29:1269–87. https://doi.org/10.1016/S0020-7519(99)00076-4.

De Lucio A, Bailo B, Aguilera M, et al. No molecular epidemiological evidence supporting household transmission of zoonotic Giardia duodenalis and Cryptosporidium spp. from pet dogs and cats in the province of Álava, Northern Spain. Acta Trop. 2017;170:48–56. https://doi.org/10.1016/j.actatropica.2017.02.024.

De Waele V, Speybroeck N, Berkvens D, et al. Control of cryptosporidiosis in neonatal calves: use of halofuginone lactate in two different calf rearing systems. Prev Vet Med. 2010;96:143–51. https://doi.org/10.1016/j.prevetmed.2010.06.017.

De Waele V, Berzano M, Berkvens D, et al. Age-stratified Bayesian analysis to estimate sensitivity and specificity of four diagnostic tests for detection of Cryptosporidium oocysts in neonatal calves. J Clin Microbiol. 2011;49:76–84. https://doi.org/10.1128/JCM.01424-10.

Delafosse A, Chartier C, Dupuy MC, et al. *Cryptosporidium parvum* infection and associated risk factors in dairy calves in western France. Prev Vet Med. 2015;118:406–12. https://doi.org/10.1016/j.prevetmed.2015.01.005.

Deng M, Templeton TJ, London NR, et al. *Cryptosporidium parvum* genes containing thrombospondin type 1 domains. Infect Immun. 2002;70:6987–95. https://doi.org/10.1128/IAI.70.12.6987-6995.2002.

Deng L, Wei L, Zhong Z, et al. Occurrence and genetic characteristics of Cryptosporidium hominis and Cryptosporidium andersoni in Horses from Southwestern China. J Eukaryot Microbiol. 2017;64(5):716–20. https://doi.org/10.1111/ijlh.12426.

Díaz P, Quílez J, Robinson G, et al. Identification of Cryptosporidium xiaoi in diarrhoeic goat kids (Capra hircus) in Spain. Vet Parasitol. 2010;172:132–4. https://doi.org/10.1016/j.vetpar.2010.04.029.

Díaz P, Hadfield SJ, Quílez J, et al. Assessment of three methods for multilocus fragment typing of *Cryptosporidium parvum* from domestic ruminants in north west Spain. Vet Parasitol. 2012;186:188–95. https://doi.org/10.1016/j.vetpar.2011.11.039.

Doris M-D, Michaela G, Kristin E, et al. Distribution of Cryptosporidium species in sheep in the UKNo Title. Vet Parasitol. 2008;154:214–9.

Drumo R, Widmer G, Morrison LJ, et al. Evidence of host-associated populations of *Cryptosporidium parvum* in Italy. Appl Environ Microbiol. 2012;78:3523–9. https://doi.org/10.1128/AEM.07686-11.

Duranti A, Cacciò SM, Pozio E, et al. Risk factors associated with *Cryptosporidium parvum* infection in cattle. Zoonoses Public Heal. 2009;56:176–82. https://doi.org/10.1111/j.1863-2378.2008.01173.x.

Elitok B, Elitok OM, Pulat H. Efficacy of azithromycin dihydrate in treatment of cryptosporidiosis in naturally infected dairy calves. J Vet Intern Med. 2005;19:590–3. https://doi.org/10.1892/0891-6640(2005)19[590:EOADIT]2.0.CO;2.

Ellis AE, Brown CA, Miller DL. Diagnostic exercise: chronic vomiting in a dog. Vet Pathol. 2010;47:991–3. https://doi.org/10.1177/0300985810375052.

Elwin K, Robinson G, Hadfield SJ, et al. A comparison of two approaches to extracting Cryptosporidium DNA from human stools as measured by a real-time PCR assay. J Microbiol Methods. 2012;89:38–40. https://doi.org/10.1016/j.mimet.2012.02.006.

Enemark HL, Ahrens P, Bille-Hansen V, et al. *Cryptosporidium parvum*: infectivity and pathogenicity of the "porcine" genotype. Parasitology. 2003a;126:407–16. https://doi.org/10.1017/S0031182003003032.

Enemark HL, Bille-Hansen V, Lind P, et al. Pathogenicity of *Cryptosporidium parvum*—evaluation of an animal infection model. Vet Parasitol. 2003b;113:35–57. https://doi.org/10.1016/S0304-4017(03)00034-7.

Fayer R. General biology. In: Fayer R, Xiao L, editors. Cryptosporidium & cryptosporidiosis. 2nd ed. Boca Raton: CRC; 2008. p. 1–42.

Fayer R. Taxonomy and species delimitation in Cryptosporidium. Exp Parasitol. 2010;124:90–7. https://doi.org/10.1016/j.exppara.2009.03.005.

Fayer R, Leek R. The effects of reducing conditions, medium, pH, temperature, and time on in vitro excystation of Cryptosporidium. J Protozool. 1984;31:567–9. https://doi.org/10.1111/j.1550-7408.1984.tb05504.x.

Fayer R, Santín M. Cryptosporidium xiaoi n. sp. (Apicomplexa: Cryptosporidiidae) in sheep (Ovis aries). Vet Parasitol. 2009;164:192–200. https://doi.org/10.1016/j.vetpar.2009.05.011.

Fayer R, Guidry A, Blagburn BL. Immunotherapeutic efficacy of bovine colostral immunoglobulins from a hyperimmunized cow against cryptosporidiosis in neonatal mice. Infect Immun. 1990;58:2962–5.

Fayer R, Gasbarre L, Pasquali P, et al. *Cryptosporidium parvum* infection in bovine neonates : dynamic clinical, parasitic and immunologic patterns. Int J Parasitol. 1998;28:49–56.

Fayer R, Morgan U, Upton SJ. Epidemiology of Cryptosporidium: transmission, detection and identification. Int J Parasitol. 2000;30:1305–22. https://doi.org/10.1016/S0020-7519(00)00135-1.

Fayer R, Trout JM, Xiao L, et al. Cryptosporidium canis n. sp. from domestic dogs. J Parasitol. 2001;87:1415–22. https://doi.org/10.1645/0022-3395(2001)087[1415:CCNSFD]2.0.CO;2.

Fayer R, Santín M, Xiao L. Cryptosporidium bovis n. sp. (Apicomplexa: Cryptosporidiidae) in cattle (Bos taurus). J Parasitol. 2005;91:624–9. https://doi.org/10.1645/GE-3435.

Fayer R, Santín M, Trout JM, Greiner E. Prevalence of species and genotypes of Cryptosporidium found in 1-2-year-old dairy cattle in the eastern United States. Vet Parasitol. 2006;135:105–12. https://doi.org/10.1016/j.vetpar.2005.08.003.

Fayer R, Santin M, Trout JM. Prevalence of Cryptosporidium species and genotypes in mature dairy cattle on farms in eastern United States compared with younger cattle from the same locations. Vet Parasitol. 2007;145:260–6. https://doi.org/10.1016/j.vetpar.2006.12.009.

Fayer R, Santín M, Trout JM. Cryptosporidium ryanae n. sp. (Apicomplexa: Cryptosporidiidae) in cattle (Bos taurus). Vet Parasitol. 2008;156:191–8. https://doi.org/10.1016/j.vetpar.2008.05.024.

Fayer R, Santín M, Dargatz D. Species of Cryptosporidium detected in weaned cattle on cow—calf operations in the United States. Vet Parasitol. 2010a;170:187–92. https://doi.org/10.1016/j.vetpar.2010.02.040.

Fayer R, Santín M, Macarisin D. Cryptosporidium ubiquitum n. sp. in animals and humans. Vet Parasitol. 2010b;172:23–32. https://doi.org/10.1016/j.vetpar.2010.04.028.

Fernandez A, Quezada M, Gomez MA, et al. Cryptosporidiosis in chickens from southern Spain. Avian Dis. 1990;34:224–7.

Follet J, Guyot K, Leruste H, et al. Cryptosporidium infection in a veal calf cohort in France: molecular characterization of species in a longitudinal study. Vet Res. 2011;42:116. https://doi.org/10.1186/1297-9716-42-116.

Foster DM, Smith GW. Pathophysiology of diarrhea in calves. Vet Clin North Am Food Anim Pract. 2009;25(13–36):xi. https://doi.org/10.1016/j.cvfa.2008.10.013.

Foster DM, Stauffer SH, Stone MR, Gookin JL. Proteasome inhibition of pathologic shedding of enterocytes to defend barrier function requires X-linked inhibitor of apoptosis protein and nuclear factor b. Gastroenterology. 2012;143(1):133–44.e4. https://doi.org/10.1053/j.gastro.2012.03.030.

Galuppi R, Piva S, Castagnetti C, et al. *Cryptosporidium parvum*: from foal to veterinary students. Vet Parasitol. 2016;219:53–6. https://doi.org/10.1016/j.vetpar.2016.02.001.

Garro CJ, Morici GE, Utges ME, et al. Prevalence and risk factors for shedding of Cryptosporidium spp. oocysts in dairy calves of Buenos Aires Province, Argentina. Parasite Epidemiol Control. 2016;1:36–41. https://doi.org/10.1016/j.parepi.2016.03.008.

Giadinis ND, Papadopoulos E, Lafi SQ, et al. Efficacy of halofuginone lactate for the treatment and prevention of cryptosporidiosis in goat kids: an extensive field trial. Small Rumin Res. 2008;76:195–200. https://doi.org/10.1016/j.smallrumres.2008.01.007.

Giadinis ND, Symeoudakis SPE, Lafi SQ, Karatzias H. Comparison of two techniques for diagnosis of cryptosporidiosis in diarrhoeic goat kids and lambs in Cyprus. Trop Anim Health Prod. 2012;44:1561–5. https://doi.org/10.1007/s11250-012-0106-4.

Giadinis N, Papadopoulos E, Lafi S, et al. Epidemiological observations on cryptosporidiosis in diarrheic goat kids in Greece. Vet Med Int. 2015;2015:764193. https://doi.org/10.1155/2015/764193.

Giangaspero A, Iorio R, Paoletti B, et al. Molecular evidence for Cryptosporidium infection in dogs in Central Italy. Parasitol Res. 2006;99:297–9. https://doi.org/10.1007/s00436-006-0169-0.

Giles M, Chalmers R, Pritchard G, et al. Cryptosporidium hominis in a goat and a sheep in the UK. Vet Rec. 2009;164(1):24–5.

Goñi P, Martín B, Villacampa M, et al. Evaluation of an immunochromatographic dip strip test for simultaneous detection of Cryptosporidium spp, Giardia duodenalis, and Entamoeba histolytica antigens in human faecal samples. Eur J Clin Microbiol Infect Dis. 2012;31(8):2077–82. https://doi.org/10.1007/s10096-012-1544-7.

Gookin JL, Duckett LL, Armstrong MU, et al. Nitric oxide synthase stimulates prostaglandin synthesis and barrier function in C. parvum-infected porcine ileum. Am J Physiol Gastrointest Liver Physiol. 2004;287:G571–81. https://doi.org/10.1152/ajpgi.00413.2003.

Greene CE, Jacobs GJ, Prickett D. Intestinal malabsorption and cryptosporidiosis in an adult dog. J Am Vet Med Assoc. 1990;197:365–7. https://doi.org/10.1016/S0020-7519(01)00361-7.

Grinberg A, Oliver L, Learmonth J, et al. Identification of Cryptosporidium parvum "'cattle'" genotype from a severe outbreak of neonatal foal diarrhoea. Vet Rec. 2003;153:628–31. https://doi.org/10.1136/vr.153.20.628.

Grinberg A, Learmonth J, Kwan E, et al. Genetic diversity and zoonotic potential of Cryptosporidium parvum causing foal diarrhea. J Clin Microbiol. 2008;46:2396–8. https://doi.org/10.1128/JCM.00936-08.

Hajdušek O, Ditrich O, Šlapeta J. Molecular identification of Cryptosporidium spp. in animal and human hosts from the Czech Republic. Vet Parasitol. 2004;122:183–92. https://doi.org/10.1016/j.vetpar.2004.04.005.

Hamnes IS, Gjerde BK, Forberg T, Robertson LJ. Occurrence of Cryptosporidium and Giardia in suckling piglets in Norway. Vet Parasitol. 2007;144:222–33. https://doi.org/10.1016/j.vetpar.2006.10.011.

Harp JA. Cryptosporidium and host resistance: historical perspective and some novel approaches. Anim Health Res Rev. 2003;4:53–62. https://doi.org/10.1079/AHRR200352.

Harp JA, Goff JP. Strategies for the control of *Cryptosporidium parvum* infection in calves. J Dairy Sci. 1998;81:289–94. https://doi.org/10.3168/jds.S0022-0302(98)75578-X

Heine J, Moon HW, Woodmansee DB. Persistent Cryptosporidium infection in congenitally athymic (nude) mice. Infect Immun. 1984;43:856–9.

Hijjawi NS, Meloni BP, Ng'anzo M, et al. Complete development of *Cryptosporidium parvum* in host cell-free culture. Int J Parasitol. 2004;34:769–77. https://doi.org/10.1016/j.ijpara.2004.04.001.

Huang BQ, Chen X-M, LaRusso NF. Cryptosporidium parvum attachment to and internalization by human biliary epithelia in vitro: a morphologic study. J Parasitol. 2004;90:212–21. https://doi.org/10.1645/GE-3204.

Imre K, Luca C, Costache M, et al. Zoonotic *Cryptosporidium parvum* in Romanian newborn lambs (Ovis aries). Vet Parasitol. 2013;191:119–22. https://doi.org/10.1016/j.vetpar.2012.08.020.

Irwin PJ. Companion animal parasitology: a clinical perspective. Int J Parasitol. 2002;32:581–93.

Iseki M. Cryptosporidium felis sp.n. (Protozoa: Eimeriorina) from the domestic cat. J Parasitol. 1979;28:285–307.

Jakobi V, Petry F. Differential expression of *Cryptosporidium parvum* genes encoding sporozoite surface antigens in infected HCT-8 host cells. Microbes Infect. 2006;8:2186–94. https://doi.org/10.1016/j.micinf.2006.04.012.

Jenkins MC, Fayer R, Tilley M, Upton SJ. Cloning and expression of a cDNA encoding epitopes shared by 15- and 60-kilodalton proteins of *Cryptosporidium parvum* sporozoites. Infect Immun. 1993;61:2377–82.

Jenkins M, Higgins J, Kniel K, et al. Protection of calves against cryptosporiosis by oral inoculation with gamma-irradiated *Cryptosporidium parvum* oocysts. J Parasitol. 2004;90:1178–80. https://doi.org/10.1645/GE-3333RN.

Jenkins MB, Eaglesham BS, Anthony LC, et al. Significance of wall structure, macromolecular composition, and surface polymers to the survival and transport of *Cryptosporidium parvum* oocysts. Appl Environ Microbiol. 2010;76:1926–34. https://doi.org/10.1128/AEM.02295-09.

Jex AR, Stanley KK, Lo W, et al. Detection of diarrhoeal pathogens in human faeces using an automated, robotic platform. Mol Cell Probes. 2012;26:11–5. https://doi.org/10.1016/j.mcp.2011.10.004.

Johnson EH, Muirhead DE, Windsor JJ, et al. Atypical outbreak of caprine cryptosporidiosis in the Sultanate of Oman. Vet Rec. 1999;145:521–4. https://doi.org/10.1136/vr.145.18.521.

Johnson EH, Windsor JJ, Muirhead DE, et al. Confirmation of the prophylactic value of paromomycin in a natural outbreak of caprine cryptosporidiosis. Vet Res Commun. 2000;24:63–7. https://doi.org/10.1023/A:1006381522986.

Karanis P, Plutzer J, Halim NA, et al. Molecular characterization of Cryptosporidium from animal sources in Qinghai province of China. Parasitol Res. 2007;101:1575–80. https://doi.org/10.1007/s00436-007-0681-x.

Karanis P, Kimura A, Nagasawa H, et al. Observations on Cryptosporidium life cycle stages during excystation. J Parasitol. 2008;94:298–300. https://doi.org/10.1645/GE-3348RN.

Kennedy GA, Kreitner GL, Strafuss AC. Cryptosporidiosis in three pigs. J Am Vet Med Assoc. 1977;170:348–50.

King BJ, Keegan AR, Phillips R, et al. Dissection of the hierarchy and synergism of the bile derived signal on *Cryptosporidium parvum* excystation and infectivity. Parasitology. 2012;139:1533–46. https://doi.org/10.1017/S0031182012000984.

Koh W, Clode PL, Monis P, Thompson RCA. Multiplication of the waterborne pathogen *Cryptosporidium parvum* in an aquatic biofilm system. Parasit Vectors. 2013;6:270. https://doi.org/10.1186/1756-3305-6-270.

Koh W, Thompson A, Edwards H, et al. Extracellular excystation and development of Cryptosporidium: tracing the fate of oocysts within Pseudomonas aquatic biofilm systems. BMC Microbiol. 2014;14:281. https://doi.org/10.1186/s12866-014-0281-8.

Kostopoulou D, Claerebout E, Arvanitis D, et al. Abundance, zoonotic potential and risk factors of intestinal parasitism amongst dog and cat populations: the scenario of Crete, Greece. Parasit Vectors. 2017;10:43. https://doi.org/10.1186/s13071-017-1989-8.

Kvác M, Hanzlíková D, Sak B, Kvetonová D. Prevalence and age-related infection of Cryptosporidium suis, C. muris and Cryptosporidium pig genotype II in pigs on a farm complex in the Czech Republic. Vet Parasitol. 2009;160:319–22.

Kváč M, Sak B, Květoňová D, et al. Infectivity, pathogenicity, and genetic characteristics of mammalian gastric Cryptosporidium spp. in domestic ruminants. Vet Parasitol. 2008;153:363–7. https://doi.org/10.1016/j.vetpar.2008.01.033.

Lallemond M, Villeneuve A, Belda J, Dubreuil P. Field study of the efficacy of halofuginone and decoquinate in the treatment of cryptosporidiosis in veal calves. Vet Rec. 2006;159:672–6.

Lalonde LF, Reyes J, Gajadhar AA. Application of a qPCR assay with melting curve analysis for detection and differentiation of protozoan oocysts in human fecal samples from Dominican Republic. Am J Trop Med Hyg. 2013;89:892–8. https://doi.org/10.4269/ajtmh.13-0106.

Langkjaer RB, Vigre H, Enemark HL, Maddox-Hyttel C. Molecular and phylogenetic characterization of Cryptosporidium and Giardia from pigs and cattle in Denmark. Parasitology. 2007;134:339–50. https://doi.org/10.1017/S0031182006001533.

Lemgruber L, Lupetti P. Crystalloid body, refractile body and virus-like particles in Apicomplexa: what is in there? Parasitology. 2012;139:285–93. https://doi.org/10.1017/S0031182011002034.

Lendner M, Daugschies A. Cryptosporidium infections: molecular advances. Parasitology. 2014;141:1511–32. https://doi.org/10.1017/S0031182014000237.

Leoni F, Amar C, Nichols G, et al. Genetic analysis of Cryptosporidium from 2414 humans with diarrhoea in England between 1985 and 2000. J Med Microbiol. 2006;55:703–7. https://doi.org/10.1099/jmm.0.46251-0.

Li W, Li Y, Song M, et al. Prevalence and genetic characteristics of Cryptosporidium, Enterocytozoon bieneusi and Giardia duodenalis in cats and dogs in Heilongjiang province, China. Vet Parasitol. 2015;208:125–34. https://doi.org/10.1016/j.vetpar.2015.01.014.

Lihua X, Cama VA, Lilia C, et al. Possible transmission of Cryptosporidium canis among children and a dog in a household. J Clin Microbiol. 2007;45:2014–6.

Lindsay BB. Cryptosporidiosis in birds. In: Cryptosporidiosis of man and animals. Boca Raton: CRC; 1990. p. 133–48.

Lucio-Forster A, Griffiths JK, Cama VA, Xiao L, Bowman DD. Minimal zoonotic risk of cryptosporidiosis from pet dogs and cats. Trends Parasitol. 2010;26:174–9.

Maddox-Hyttel C, Langkjær RB, Enemark HL, Vigre H. Cryptosporidium and Giardia in different age groups of Danish cattle and pigs-occurrence and management associated risk factors. Vet Parasitol. 2006;141:48–59. https://doi.org/10.1016/j.vetpar.2006.04.032.

Majewska AC, Werner A, Sulima P, Luty T. Survey on equine cryptosporidiosis in Poland and the possibility of zoonotic transmission. Ann Agric Environ Med. 1999;6:161–5.

Majewska AC, Werner A, Sulima P, Luty T. Prevalence of Cryptosporidium in sheep and goats bred on five farms in west-central region of Poland. Vet Parasitol. 2000;89:269–75. https://doi.org/10.1016/S0304-4017(00)00212-0.

Majewska AC, Solarczyk P, Tamang L, Graczyk TK. Equine *Cryptosporidium parvum* infections in western Poland. Parasitol Res. 2004;93:274–8. https://doi.org/10.1007/s00436-004-1111-y.

Mancassola R, Reperant JM, Naciri M, Chartier C. Chemoprophylaxis of *Cryptosporidium parvum* infection with paromomycin in kids and immunological study. Antimicrob Agents Chemother. 1995;39:75–8. https://doi.org/10.1128/AAC.39.1.75.

Mancassola R, Richard A, Naciri M. Evaluation of decoquinate to treat experimental cryptosporidiosis in kids. Vet Parasitol. 1997;69:31–7. https://doi.org/10.1016/S0304-4017(96)01094-1.

Mason RW, Hartley WJ, Tilt L. Intestinal cryptosporidiosis in a kid goat. Aust Vet J. 1981;57:386–8. https://doi.org/10.1111/j.1751-0813.1981.tb00529.x.

Masuno K, Yanai T, Hirata A, et al. Morphological and immunohistochemical features of Cryptosporidium andersoni in cattle. Vet Pathol. 2006;43:202–7. https://doi.org/10.1354/vp.43-2-202.

Mohammed HO, Wade SE, Schaaf S. Risk factors associated with *Cryptosporidium parvum* infection in dairy cattle in southeastern New York State. Vet Parasitol. 1999;83:1–13. https://doi.org/10.1016/S0304-4017(99)00032-1.

Monticello TM, Levy MG, Bunch SE, Fairley RA. Cryptosporidiosis in a feline leukemia virus-positive cat. J Am Vet Med Assoc. 1987;191:705–6.

Moore DA, Atwill ER, Kirk JH, et al. Prophylactic use of decoquinate for infections with *Cryptosporidium parvum* in experimentally challenged neonatal calves. J Am Vet Med Assoc. 2003;223:839–45. https://doi.org/10.2460/javma.2003.223.839.

Mor SM, Tzipori S. Cryptosporidiosis in children in Sub-Saharan Africa: a lingering challenge. Clin Infect Dis. 2009;47:915–21. https://doi.org/10.1086/591539.Cryptosporidiosis.

Morgan UM, Constantine CC, O'Donoghue P, et al. Molecular characterization of Cryptosporidium isolates from humans and other animals using random amplified polymorphic DNA analysis. Am J Trop Med Hyg. 1995;52:559–64.

Morgan U, Weber R, Xiao L, et al. Molecular characterization of Cryptosporidium isolates obtained from human immunodeficiency virus-infected individuals living in Switzerland, Kenya, and the United States. J Clin Microbiol. 2000;38:1180–3.

Muñoz M, Alvarez M, Lanza I, Cármenes P. Role of enteric pathogens in the aetiology of neonatal diarrhoea in lambs and goat kids in Spain. Epidemiol Infect. 1996;117:203–11.

Naitza S, Spano F, Robson KJH, Crisanti A. The thrombospondin-related protein family of apicomplexan parasites: the gears of the cell invasion machinery. Parasitol Today. 1998;14:479–84. https://doi.org/10.1016/S0169-4758(98)01346-5.

Noordeen F, Horadagoda NU, Faizal ACM, et al. Infectivity of *Cryptosporidium parvum* isolated from asymptomatic adult goats to mice and goat kids. Vet Parasitol. 2002;103:217–25. https://doi.org/10.1016/S0304-4017(01)00578-7.

O'Connor RM, Wanyiri JW, Cevallos AM, et al. *Cryptosporidium parvum* glycoprotein gp40 localizes to the sporozoite surface by association with gp15. Mol Biochem Parasitol. 2007;156:80–3. https://doi.org/10.1016/j.molbiopara.2007.07.010.

Olabanji G, Maikai M, Beatty V, Otolorin GR. Prevalence and risk factors associated with faecal shedding of Cryptosporidium oocysts in dogs in the Federal Capital Territory, Abuja, Nigeria. Vet Med Int. 2016;2016:4591238. https://doi.org/10.1155/2016/4591238.

Ollivett TL, Nydam DV, Bowman DD, et al. Effect of nitazoxanide on cryptosporidiosis in experimentally infected neonatal dairy calves. J Dairy Sci. 2009;92:1643–8. https://doi.org/10.3168/jds.2008-1474.

Omidian Z, Ebrahimzadeh E, Shahbazi P, et al. Application of recombinant *Cryptosporidium parvum* P23 for isolation and prevention. Parasitol Res. 2014;113:229–37. https://doi.org/10.1007/s00436-013-3648-0.

Pagès-Manté A, Pagès-Bosch M, Majó-Masferrer N, et al. An outbreak of disease associated with cryptosporidia on a red-legged partridge (Alectoris rufa) game farm. Avian Pathol. 2007;36:275–8. https://doi.org/10.1080/03079450701439389.

Panciera RJ, Thomassen RW, Garner FM. Cryptosporidial infection in a calf. Vet Path. 1971;8:479–84.

Paoletti B, Giangaspero A, Gatti A, et al. Immunoenzymatic analysis and genetic detection of *Cryptosporidium parvum* in lambs from Italy. Exp Parasitol. 2009;122:349–52. https://doi.org/10.1016/j.exppara.2009.05.006.

Paraud C, Pors I, Chartier C. Evaluation of oral tilmicosin efficacy against severe cryptosporidiosis in neonatal kids under field conditions. Vet Parasitol. 2010;170:149–52. https://doi.org/10.1016/j.vetpar.2010.01.024.

Park JH, Guk SM, Han ET, et al. Genotype analysis of Cryptosporidium spp. prevalent in a rural village in Hwasun-gun, Republic of Korea. Korean J Parasitol. 2006;44:27–33. https://doi.org/10.3347/kjp.2006.44.1.27.

Pavlasek I. Findings of cryptosporidia in the stomach of hens and of exotic and wild birds. Veterinarstvi. 2001;51:103–8.

Pavlasek I, Ryan U. The first finding of a natural infection of Cryptosporidium muris in a cat. Vet Parasitol. 2007;144:349–52. https://doi.org/10.1016/j.vetpar.2006.10.005.

Pavlasek I, Ryan U. Cryptosporidium varanii takes precedence over C. saurophilum. Exp Parasitol. 2008;118:434–7.

Pedraza-Díaz S, Amar C, Iversen A, et al. Unusual cryptosporidium species recovered from human faeces: first description of Cryptosporidium felis and Cryptosporidium "dog type" from patients in England. J Med Microbiol. 2001;50:293–6. https://doi.org/10.1099/0022-1317-50-3-293.

Peng MM, Xiao L, Freeman AR, et al. Genetic polymorphism among *Cryptosporidium parvum* isolates: evidence of two distinct human transmission cycles. Emerg Infect Dis. 1997;3:567–73. https://doi.org/10.3201/eid0304.970423.

Perryman LE, Kapil SJ, Jones ML, Hunt EL. Protection of calves against cryptosporidiosis with immune bovine colostrum induced by a *Cryptosporidium parvum* recombinant protein. Vaccine. 1999;17:2142–9. https://doi.org/10.1016/S0264-410X(98)00477-0.

Petermann J, Paraud C, Pors I, Chartier C. Efficacy of halofuginone lactate against experimental cryptosporidiosis in goat neonates. Vet Parasitol. 2014;202:326–9. https://doi.org/10.1016/j.vetpar.2014.02.027.

Petersen C, Gut J, Doyle PS, et al. Characterization of a >900,000-M(r) *Cryptosporidium parvum* sporozoite glycoprotein recognized by protective hyperimmune bovine colostral immunoglobulin. Infect Immun. 1992;60:5132–8.

Petersen H, Jianmin W, Katakam K, et al. Cryptosporidium and Giardia in Danish organic pig farms: seasonal and age-related variation in prevalence, infection intensity and species/genotypes. Vet Parasitol. 2015;214:29–39. https://doi.org/10.1016/j.vetpar.2015.09.020.

Petry F, Jakobi V, Tessema TS. Host immune response to *Cryptosporidium parvum* infection. Exp Parasitol. 2010;126:304–9. https://doi.org/10.1016/j.exppara.2010.05.022.

Piva S, Caffara M, Pasquali F, et al. Detection and quantification of Cryptosporidium oocysts in environmental surfaces of an equine perinatology unit. Prev Vet Med. 2016;131:67–74. https://doi.org/10.1016/j.prevetmed.2016.04.012.

Plutzer J, Karanis P. Genetic polymorphism in Cryptosporidium species: an update. Vet Parasitol. 2009;165:187–99. https://doi.org/10.1016/j.vetpar.2009.07.003.

Quílez J, Sánchez-Acedo C, Clavel A, et al. Prevalence of Cryptosporidium infections in pigs in Aragon (northeastern Spain). Vet Parasitol. 1996;67:83–8. https://doi.org/10.1016/S0304-4017(96)01026-6.

Quilez J, Sanchez-Acedo C, Cacho E, et al. Efficacy of two peroxygen-based disinfectants for inactivation of *Cryptosporidium parvum* oocysts. Appl Environ Microbiol. 2005;71:2479–83. https://doi.org/10.1128/AEM.71.5.2479.

Quílez J, Torres E, Chalmers RM, et al. Cryptosporidium genotypes and subtypes in lambs and goat kids in Spain. Appl Environ Microbiol. 2008;74:6026–31. https://doi.org/10.1128/AEM.00606-08.

Rambozzi L, Menzano A, Mannelli A, et al. Prevalence of cryptosporidian infection in cats in Turin and analysis of risk factors. J Feline Med Surg. 2007;9:392–6. https://doi.org/10.1016/j.jfms.2007.03.005.

Reduker DW, Speer CA, Blixt JA. Ultrastructure of *Cryptosporidium parvum* oocyst and excysting sporozoites as revealed by high resolution scanning electron microscopy. J Protozool. 1985;32:708–11. https://doi.org/10.1111/j.1550-7408.1985.tb03106.x.

Répérant JM, Naciri M, Iochmann S, et al. Major antigens of *Cryptosporidium parvum* recognised by serum antibodies from different infected animal species and man. Vet Parasitol. 1994;55:1–13. https://doi.org/10.1016/0304-4017(94)90051-5.

Robertson LJ, Campbell AT, Smith HV. Survival of *Cryptosporidium parvum* oocysts under various environmental pressures. Appl Environ Microbiol. 1992;58:3494–500.

Rosales MJ, Peréz Cordón G, Sánchez Moreno M, et al. Extracellular like-gregarine stages of *Cryptosporidium parvum*. Acta Trop. 2005;95:74–8. https://doi.org/10.1016/j.actatropica.2005.03.009.

Ryan U. Cryptosporidium in birds, fish and amphibians. Exp Parasitol. 2010;124:113–20. https://doi.org/10.1016/j.exppara.2009.02.002.

Ryan U, Hijjawi N. New developments in Cryptosporidium research. Int J Parasitol. 2015;45:367–73. https://doi.org/10.1016/j.ijpara.2015.01.009.

Ryan UM, Monis P, Enemark HL, et al. Cryptosporidium suis n. sp. (Apicomplexa: Cryptosporidiidae) in pigs (Sus scrofa). J Parasitol. 2004;90:769–73. https://doi.org/10.1645/GE-202R1.

Ryan UM, Bath C, Robertson I, et al. Sheep may not be an important zoonotic reservoir for Cryptosporidium and Giardia parasites. Appl Environ Microbiol. 2005;71:4992–7. https://doi.org/10.1128/AEM.71.9.4992-4997.2005.

Ryan U, Paparini A, Monis P, Hijjawi N. It's official—Cryptosporidium is a gregarine: what are the implications for the water industry? Water Res. 2016;105:305–13. https://doi.org/10.1016/j.watres.2016.09.013.

Rzezutka A, Kaupke A, Kozyra I, Pejsak Z. Molecular studies on pig cryptosporidiosis in Poland. Pol J Vet Sci. 2014;17:577–82. https://doi.org/10.2478/pjvs-2014-0086.

Sagodira S, Iochmann S, Mevelec MN, et al. Nasal immunization of mice with *Cryptosporidium parvum* DNA induces systemic and intestinal immune responses. Parasite Immunol. 1999;21:507–16. https://doi.org/10.1046/j.1365-3024.1999.00247.x.

Sahal M, Karaer Z, Yasa Duru S, et al. Cryptosporidiosis in newborn calves in Ankara region: clinical, haematological findings and treatment with Lasalocid-NA. DTW Dtsch Tierärztliche Wochenschrift. 2005;112:203–8, 210

Saime A, Tsipa MA, Bessong P. The epidemiology of cryptosporidium in cats and dogs in the Thohoyandou region, South Africa. Afr J Microbiol Res. 2013;7:2510–8. https://doi.org/10.5897/AJMR12.1391.

Santin M. Clinical and subclinical infections with Cryptosporidium in animals. N Z Vet J. 2013;61:1–10. https://doi.org/10.1080/00480169.2012.731681.

Santín M, Trout J. Companion animals. In: Ronald F, Lhiua X, editors. Cryptosporidium & cryptosporidiosis. 2nd ed. Boca Raton: CRC; 2008a. p. 437–50.

Santín M, Trout J. Livestock. In: Fayer R, Xiao L, editors. Cryptosporidium & cryptosporidiosis. 2nd ed. Boca Raton: CRC; 2008b. p. 455–83.

Santín M, Trout JM, Xiao L, et al. Prevalence and age-related variation of Cryptosporidium species and genotypes in dairy calves. Vet Parasitol. 2004;122:103–17. https://doi.org/10.1016/j.vetpar.2004.03.020.

Santín M, Trout JM, Vecino JAC, et al. Cryptosporidium, Giardia and Enterocytozoon bieneusi in cats from Bogota (Colombia) and genotyping of isolates. Vet Parasitol. 2006;141:334–9. https://doi.org/10.1016/j.vetpar.2006.06.004.

Santín M, Trout J, Fayer R. A longitudinal study of cryptosporidiosis in dairy cattle from birth to 2 years of age. Vet Parasitol. 2008;155:15–23. https://doi.org/10.1016/j.vetpar.2008.04.018.

Sasahara T, Maruyama H, Aoki M, et al. Apoptosis of intestinal crypt epithelium after *Cryptosporidium parvum* infection. J Infect Chemother. 2003;9:278–81. https://doi.org/10.1007/s10156-003-0259-1.

Savin C, Sarfati C, Menotti J, et al. Assessment of cryptodiag for diagnosis of cryptosporidiosis and genotyping Cryptosporidium species. J Clin Microbiol. 2008;46:2590–4. https://doi.org/10.1128/JCM.00226-08.

Schnyder M, Kohler L, Hemphill A, Deplazes P. Prophylactic and therapeutic efficacy of nitazoxanide against *Cryptosporidium parvum* in experimentally challenged neonatal calves. Vet Parasitol. 2009;160:149–54. https://doi.org/10.1016/j.vetpar.2008.10.094.

Sevinç F, Şimşek A, Uslu U. Massive *Cryptosporidium parvum* infection associated with an outbreak of diarrhoea in neonatal goat kids. Turkish. J Vet Anim Sci. 2005;29:1317–20.

Shahbazi P, Shayan P, Ebrahimzadeh E, Rahbari S. Specific egg yolk antibody against recombinant *Cryptosporidium parvum* P23 protein. Iran J Parasitol. 2009;4:15–24.

Sibley LD. Intracellular parasite invasion strategies. Science. 2004;304:248–53. https://doi.org/10.1126/science.1094717.

Silverlås C, Björkman C, Egenvall A. Systematic review and meta-analyses of the effects of halofuginone against calf cryptosporidiosis. Prev Vet Med. 2009a;91:73–84. https://doi.org/10.1016/j.prevetmed.2009.05.003.

Silverlås C, Bosaeus-Reineck H, Näslund K, Björkman C. Is there a need for improved Cryptosporidium diagnostics in Swedish calves? Int J Parasitol. 2013;43(2):155–61. https://doi.org/10.1016/j.ijpara.2012.10.009.

Silverlås C, Emanuelson U, de Verdier K, Björkman C. Prevalence and associated management factors of Cryptosporidium shedding in 50 Swedish dairy herds. Prev Vet Med. 2009b;90:242–53. https://doi.org/10.1016/j.prevetmed.2009.04.006.

Šlapeta J. Cryptosporidium species found in cattle: a proposal for a new species. Trends Parasitol. 2006;22:469–74.

Šlapeta J. Cryptosporidiosis and Cryptosporidium species in animals and humans: a thirty colour rainbow? Int J Parasitol. 2013;43:957–70. https://doi.org/10.1016/j.ijpara.2013.07.005.

Šlapeta J, Keithly J. *Cryptosporidium parvum* mitochondrial-type HSP70 targets homologous and heterologous mitochondria. Eukaryot Cell. 2004;3:483–94. https://doi.org/10.1128/EC.3.2.483.

Šlapeta J. The name *Cryptosporidium tyzzeri* Ren, Zhao, Zhang, Ning, Jian, Wang, Lv, Wang, Arrowood and Xiao, 2012 is permanently invalid. Exp Parasitol. 2012;130:306–7.

Šlapeta J, Müller N, Stack CM, Walker G, Lew-Tabor A, Tachezy J, Frey CF. Comparative analysis of Tritrichomonas foetus (Riedmuller, 1928) cat genotype, T. foetus (Riedmuller, 1928) cattle genotype and Tritrichomonas suis (Davaine, 1875) at 10 DNA loci. Int J Parasitol. 2012;42(13-14):1143–9. https://doi.org/10.1016/j.ijpara.2012.10.004.

Smith H. OIE manual of diagnostic tests and vaccines for terrestrial animals. 6th ed. Paris: OIE; 2008a. p. 1192–215.

Smith H. Diagnostics. In: Ronald F, Xiao L, editors. Cryptosporidium & cryptosporidiosis. 2nd ed. Boca Raton: CRC; 2008b. p. 173–208.

Smith HV, Nichols RAB, Grimason AM. Cryptosporidium excystation and invasion: getting to the guts of the matter. Trends Parasitol. 2005;21:133–42.

Snelling WJ, Lin Q, Moore JE, et al. Proteomics analysis and protein expression during sporozoite excystation of *Cryptosporidium parvum* (Coccidia, Apicomplexa). Mol Cell Proteomics. 2007;6:346–55. https://doi.org/10.1074/mcp.M600372-MCP200.

Snyder SP, England JJ, McChesney AE. Cryptosporidiosis in immunodeficient Arabian foals. Vet Pathol. 1978;15:12–7. https://doi.org/10.1177/030098587801500102.

Spano PC, Ranucci L, et al. Cloning of the entire COWP gene of *Cryptosporidium parvum* and ultrastructural localization of the protein during sexual parasite development. Parasitology. 1997;114:427–37.

Sréter T, Széll Z, Varga I. Anticryptosporidial prophylactic efficacy of enrofloxacin and paromomycin in chickens. J Parasitol. 2002;88:209–11. https://doi.org/10.1645/0022-3395(2002)088[0209:APEOEA]2.0.CO;2.

Sturdee AP, Bodley-Tickell AT, Archer A, Chalmers RM. Long-term study of Cryptosporidium prevalence on a lowland farm in the United Kingdom. Vet Parasitol. 2003;116:97–113. https://doi.org/10.1016/S0304-4017(03)00261-9.

Sweeny JPA, Ryan UM, Robertson ID, Jacobson C. Cryptosporidium and Giardia associated with reduced lamb carcase productivity. Vet Parasitol. 2011a;182:127–39. https://doi.org/10.1016/j.vetpar.2011.05.050.

Sweeny JPA, Ryan UM, Robertson ID, et al. Longitudinal investigation of protozoan parasites in meat lamb farms in southern Western Australia. Prev Vet Med. 2011b;101:192–203. https://doi.org/10.1016/j.prevetmed.2011.05.016.

Tarver AP, Clark DP, Diamond G, et al. Enteric beta-defensin: molecular cloning and characterization of a gene with inducible intestinal epithelial cell expression associated with *Cryptosporidium parvum* infection [published erratum appears in Infect Immun 1998 May;66(5):2399]. Infect Immun. 1998;66:1045–56.

Tessema T, Schwamb B, Lochner M, et al. Dynamics of gut mucosal and systemic Th1/Th2 cytokine responses in interferon-c and interleukin-12p40 knock out mice during primary and challenge *Cryptosporidium parvum* infection. Immunobiology. 2009;214:454–66.

Tetley L, Brown SMA, McDonald V, Coombs GH. Ultrastructural analysis of the sporozoite of *Cryptosporidium parvum*. Microbiology. 1998;144:3249–55. https://doi.org/10.1099/00221287-144-12-3249.

Thompson RCA, Olson ME, Zhu G, et al. Cryptosporidium and cryptosporidiosis. Adv Parasitol. 2005;59:77–158. https://doi.org/10.1016/S0065-308X(05)59002-X.

Thompson RCA, Koh WH, Clode PL. Cryptosporidium—what is it? Food Waterborne Parasitol. 2016;4:54–61. https://doi.org/10.1016/j.fawpar.2016.08.004.

Tomazic ML, Maidana J, Dominguez M, et al. Molecular characterization of Cryptosporidium isolates from calves in Argentina. Vet Parasitol. 2013;198:382–6. https://doi.org/10.1016/j.vetpar.2013.09.022.

Trotz-Williams LA, Jarvie BD, Martin SW, et al. Prevalence of *Cryptosporidium parvum* infection in southwestern Ontario and its association with diarrhea in neonatal dairy calves. Can Vet J. 2005;46:349–51.

Trotz-Williams LA, Wayne Martin S, Leslie KE, et al. Calf-level risk factors for neonatal diarrhea and shedding of *Cryptosporidium parvum* in Ontario dairy calves. Prev Vet Med. 2007;82:12–28. https://doi.org/10.1016/j.prevetmed.2007.05.003.

Trotz-Williams LA, Martin SW, Leslie KE, et al. Association between management practices and within-herd prevalence of *Cryptosporidium parvum* shedding on dairy farms in southern Ontario. Prev Vet Med. 2008;83:11–23. https://doi.org/10.1016/j.prevetmed.2007.03.001.

Tůmová E, Skřivan M, Marounek M, et al. Performance and oocyst shedding in broiler chickens orally infected with Cryptosporidium baileyi and Cryptosporidium meleagridis. Avian Dis. 2002;46:203–7. https://doi.org/10.1637/0005-2086(2002)046[0203:Paosib]2.0.Co;2.

Tyzzer EE. Cryptosporidium parvum (sp. nov.), a coccidium found in the small intestine of the common mouse. Arch Protisenkd. 1912;26:394–412.

Tzipori S. Cryptosporidiosis in animals and humans. Microbiol Rev. 1983;47:84–96.

Ungar BLP, Kao TC, Burris JA, Finkelman FD. Cryptosporidium infection in an adult-mouse model—independent roles for Ifn-gamma and Cd4+ lymphocytes-T in protective immunity. J Immunol. 1991;147:1014–22.

Valigurová A, Jirků M, Koudela B, et al. Cryptosporidia: epicellular parasites embraced by the host cell membrane. Int J Parasitol. 2008;38:913–22. https://doi.org/10.1016/j.ijpara.2007.11.003.

Veronesi F, Passamonti F, Cacciò S, et al. Epidemiological survey on equine Cryptosporidium and Giardia infections in Italy and molecular characterization of isolates. Zoonoses Public Health. 2010;57:510–7. https://doi.org/10.1111/j.1863-2378.2009.01261.x.

Viel H, Rocques H, Martin JCC. Efficacy of nitazoxanide against experimental cryptosporidiosis in goat neonates. Parasitol Res. 2007;102:163–6. https://doi.org/10.1007/s00436-007-0744-z.

Vítovec J, Hamadejová K, Landová L, et al. Prevalence and pathogenicity of Cryptosporidium suis in pre- and post-weaned pigs. J Vet Med B Infect Dis Vet Public Health. 2006;53:239–43. https://doi.org/10.1111/j.1439-0450.2006.00950.x.

Viu M, Quílez J, Sánchez-Acedo C, et al. Field trial on the therapeutic efficacy of paromomycin on natural *Cryptosporidium parvum* infections in lambs. Vet Parasitol. 2000;90:163–70. https://doi.org/10.1016/S0304-4017(00)00241-7.

Wagner S, Lynch NJ, Walter W, et al. Differential expression of the murine mannose-binding lectins A and C in lymphoid and nonlymphoid organs and tissues. J Immunol. 2003;170:1462–5.

Wanyiri JW, O'Connor R, Allison G, et al. Proteolytic processing of the Cryptosporidium glycoprotein gp40/15 by human furin and by a parasite-derived furin-like protease activity. Infect Immun. 2007;75:184–92. https://doi.org/10.1128/IAI.00944-06.

Weir SC, Pokorny NJ, Carreno RA, et al. Efficacy of common laboratory disinfectants on the infectivity of *Cryptosporidium parvum* oocysts in cell culture. Appl Environ Microbiol. 2002;68:2576–9. https://doi.org/10.1128/AEM.68.5.2576-2579.2002.

Wetzel DM, Schmidt J, Kuhlenschmidt MS, et al. Gliding motility leads to active cellular invasion by *Cryptosporidium parvum* sporozoites. Infect Immun. 2005;73:5379–87. https://doi.org/10.1128/IAI.73.9.5379-5387.2005.

Widmer G, Lee Y. Comparison of single- and multilocus genetic diversity in the protozoan parasites *Cryptosporidium parvum* and *C. hominis*. Appl Environ Microbiol. 2010;76:6639–44. https://doi.org/10.1128/AEM.01268-10.

Wilson R, Holscher M, Lyle S. Cryptosporidiosis in a pup. J Am Vet Med Assoc. 1983;183:1005–6, 965

Wyatt CR, Brackett EJ, Perryman LE, et al. Activation of intestinal intraepithelial T lymphocytes in calves infected with *Cryptosporidium parvum*. Infect Immun. 1997;65:185–90.

Wyatt C, Brackett E, Savidge J. Evidence for the emergence of a type-1-like immune response in intestinal mucosa of calves recovering from cryptosporidiosis. J Parasitol. 2001;87:90–5. https://doi.org/10.1645/0022-3395(2001)087[0090:EFTEOA]2.0.CO;2.

Wyatt CR, Lindahl S, Austin K, et al. Response of T lymphocytes from previously infected calves to recombinant *Cryptosporidium parvum* P23 vaccine antigen. J Parasitol. 2005;91:1239–42. https://doi.org/10.1645/GE-3446RN.1.

Wyatt CR, Riggs MW, Fayer R. Cryptosporidiosis in neonatal calves. Vet Clin North Am Food Anim Pract. 2010;26:89–103. https://doi.org/10.1016/j.cvfa.2009.10.001.

Xiao L. Molecular epidemiology of cryptosporidiosis: an update. Exp Parasitol. 2010;124:80–9. https://doi.org/10.1016/j.exppara.2009.03.018.

Xiao L, Herd RP. Epidemiology of equine Cryptosporidium and Giardia infections. Equine Vet J. 1994;26:14–7. https://doi.org/10.1111/j.2042-3306.1994.tb04323.x.

Xiao L, Escalante L, Yang C, et al. Phylogenetic analysis of Cryptosporidium parasites based on the small-subunit rRNA gene locas. Appl Environ Microbiol. 1999;65:1578–83.

Xiao L, Bern C, Limor J, et al. Identification of 5 types of Cryptosporidium parasites in children in Lima, Peru. J Infect Dis. 2001;183:492–7. https://doi.org/10.1086/318090.

Xiao L, Fayer R, Ryan U, Upton SJ. Cryptosporidium taxonomy: recent advances and implications for public health. Clin Microbiol Rev. 2004;17:72–97.

Xiao L, Moore JE, Ukoh U, et al. Prevalence and identity of Cryptosporidium spp. in pig slurry. Appl Environ Microbiol. 2006;72:4461–3. https://doi.org/10.1128/AEM.00370-06.

Xiao L, Ryan UM, Fayer R, Bowman DD, Zhang L. *Cryptosporidium tyzzeri* and *Cryptosporidium pestis*: Which name is valid? Exp Parasitol. 2012;130:308–9.

Xu H, Jin Y, Wu W, et al. Genotypes of Cryptosporidium spp., Enterocytozoon bieneusi and Giardia duodenalis in dogs and cats in Shanghai, China. Parasit Vectors. 2016;9:121. https://doi.org/10.1186/s13071-016-1409-5.

Yang R, Jacobson C, Gordon C, Ryan U. Prevalence and molecular characterisation of Cryptosporidium and Giardia species in pre-weaned sheep in Australia. Vet Parasitol. 2009;161:19–24. https://doi.org/10.1016/j.vetpar.2008.12.021.

Yang R, Ying J, Monis P, Ryan U. Molecular characterisation of Cryptosporidium and Giardia in cats (Felis catus) in Western Australia. Exp Parasitol. 2015;155:13–8. https://doi.org/10.1016/j.exppara.2015.05.001.

Yoshiuchi R, Matsubayashi M, Kimata I, et al. Survey and molecular characterization of Cryptosporidium and Giardia spp. in owned companion animal, dogs and cats, in Japan. Vet Parasitol. 2010;174:313–6. https://doi.org/10.1016/j.vetpar.2010.09.004.

Zaalouk TK, Bajaj-Elliott M, George JT, McDonald V. Differential regulation of beta-defensin gene expression during *Cryptosporidium parvum* infection. Infect Immun. 2004;72:2772–9. https://doi.org/10.1128/IAI.72.5.2772-2779.2004.

Zintl A, Neville D, Maguire D, et al. Prevalence of Cryptosporidium species in intensively farmed pigs in Ireland. Parasitology. 2007;134:1575–82. https://doi.org/10.1017/S0031182007002983.

Zou Y, Ma J, Yue D, et al. Prevalence and risk factors of Cryptosporidium infection in farmed pigs in Zhejiang, Guangdong, and Yunnan provinces, China. Trop Anim Health Prod. 2017;49:653–9. https://doi.org/10.1007/s11250-017-1230-y.

Eimeria

3

Berit Bangoura and Arwid Daugschies

Abstract

The genus *Eimeria* belongs to Coccidia, and its parasite species reside usually in the intestine of their respective host. They are strictly host-specific and characterized by a monoxenous life cycle. After endogenous development, oocysts are excreted into the environment where they mature into the infective stage. Due to their widespread distribution in populations of many different animal species and because one animal host can be infected by a variety of *Eimeria* spp., they are considered of great veterinary importance. Mainly poultry, cattle, and sheep industries incur significant economic losses due to eimeriosis. It is important to differentiate between *Eimeria* spp., since only some are highly pathogenic, while many are nonpathogenic. In farm animals and pets, mainly young animals are affected by a high parasite burden and the intracellular parasite replication results in pathology and the development of eimeriosis as clinical disease. *Eimeria* destroys host cells during asexual and sexual replication leading to marked tissue damage in susceptible hosts. The infection by intestinal *Eimeria* spp. is associated with diarrhea, exsiccosis, anorexia, and other symptoms entailing a reduced animal performance as well as cases of death. An *Eimeria* infection is detected by the demonstration of oocysts in fecal samples and subsequently differentiated based on oocyst morphology. In the case of chicken, the diagnosis is usually confirmed by diagnostic dissections. Currently, control of *Eimeria* spp. is mainly carried out by hygienic measures such as cleaning and disinfection, general herd health management, and application of pathogen-specific interventions like

B. Bangoura (✉)
Department of Veterinary Sciences, College of Agriculture and Natural Resources, University of Wyoming, Laramie, WY, USA
e-mail: bbangour@uwyo.edu

A. Daugschies
Faculty of Veterinary Medicine, Institute of Parasitology, Leipzig University, Leipzig, Germany

M. Florin-Christensen, L. Schnittger (eds.), *Parasitic Protozoa of Farm Animals and Pets*, https://doi.org/10.1007/978-3-319-70132-5_3

metaphylactic anticoccidial drug treatment. Currently, *Eimeria* species-specific vaccination is only available for poultry. Eimeriosis of farm animals is a herd-health issue requiring strategic control measures.

3.1 Morphology, Life Cycle, and Host-Pathogen Interactions

3.1.1 Morphology

3.1.1.1 The Oocyst

The major diagnostic stage of *Eimeria* is the oocyst. This stage is found in feces of hosts during patent infection and can be easily used for in vivo diagnosis. Oocysts are organized in a similar way in all *Eimeria* spp. They have a robust wall composed in some species of two and in others of three layers (Courtney et al. 1976; Fernandez et al. 1988). In three-layered oocyst walls, the outer layer is dense and thin; the median layer is the thickest component with a tubular network of cysteine- and tyrosine-rich proteins, while the inner layer is homogenous and osmophilic (Fernandez et al. 1988; Samuelson et al. 2013). The existence of an additional outer layer, the so-called ectocyst layer, has been discussed but could not be demonstrated (Stotish et al. 1978). In some species like *E. tenella* (host, chicken) and *E. wyomingensis* (host, cattle), a two-layered wall with a thin outer lipid layer—of approximately 10 nm and 100 to 275 μm in *E. tenella* and *E. nieschulzi*, respectively—and a thicker inner carbohydrate-based layer composed of disulfide-bondage linked 10 kDa glycoproteins—of approximately 90 nm and 180 nm in *E. bovis* and *E. nieschulzi*, respectively—have been described (Courtney et al. 1976; Stotish et al. 1978; Kefu et al. 2006).

Following excretion, the oocyst contains an undifferentiated sporont, which has the potential to further develop during exogenous sporulation or sporogony (Fig. 3.1). The sporont is composed of a nucleus with nucleolus, ribosomes, mitochondria, a smooth and a rough endoplasmic reticulum, as well as lipid and amylopectin bodies, yet does not contain wall-forming bodies and Golgi complexes (Sibert and Speer 1980). After the prepatent period—the time between oocyst uptake and their excretion—unsporulated oocysts are present in fresh feces. In feces samples stored for several days, also sporulated oocysts—which have undergone sporogony—may be observed.

Following sporogony, the oocyst is rendered infective containing four sporocysts each of which holds two sporozoites that are the host cell-invasive stages, i.e., each single oocyst is able to set free eight sporozoites each infecting a host cell (Fig. 3.2). Sporocysts are lined by a wall of approximately 20–25 nm thickness which holds a preformed opening at the more pointed end to enable sporozoite excystation after ingestion by the host. In addition, a residual body referred to as residuum is formed inside the sporocysts. The sporocyst residuum is assumed to contain nutrients for sporozoites resting in the environment. Sporozoites are connected to the residuum by a nanotubular network thought to transport nutrient-containing vesicles into the endoplasmic reticulum of the sporozoite. Particularly in moist environments, the

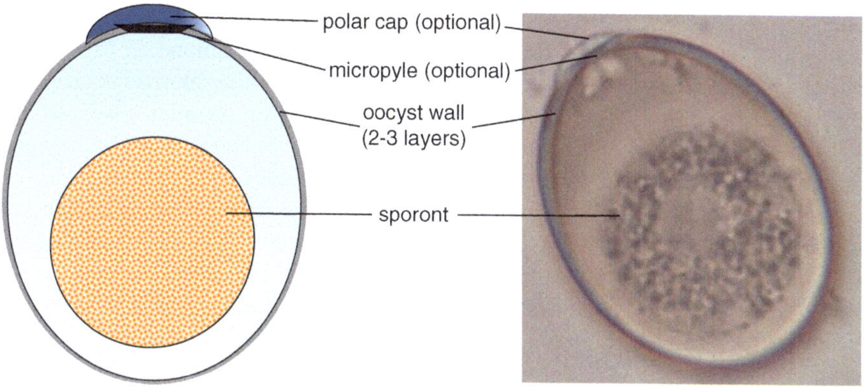

Fig. 3.1 Morphology of unsporulated oocysts (example on the right: *E. brasiliensis*; host, cattle; 400× magnification)

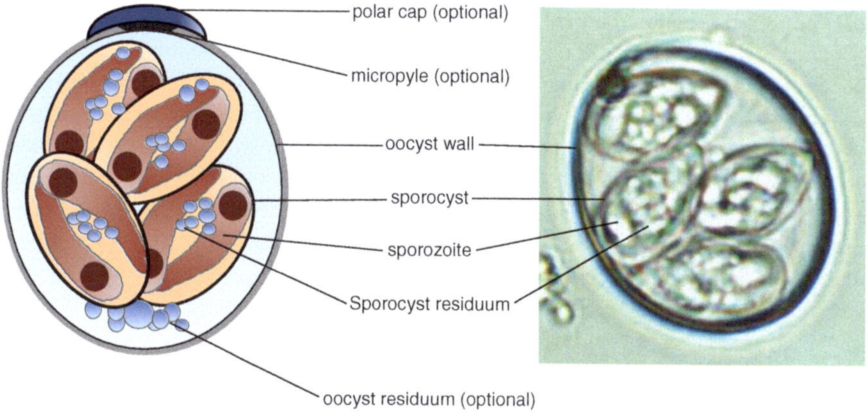

Fig. 3.2 Morphology of sporulated oocysts (example on the right: *E. maxima*; host, chicken; 400× magnification; note the absence of micropyle, polar cap, and oocyst residuum)

nanotube connection or sporocord allows survival and infectivity of the sporozoite for up to several years (Seemann et al. 2012).

3.1.1.2 Endogenous Stages

During endogenous development, first and subsequent generation meronts (asexual stages) and gamonts (sexual stages) are seen (Fig. 3.1). They are lying in a parasitophorous vacuole (PV) built during host cell invasion by the respective parasitic stage.

Sporozoites

Directly after uptake of the infective oocyst stage, invasive sporozoites are observed. These are slender, banana-shaped single motile cells that attach and invade suitable target host cells. The size of the sporozoite is species-specific—ranging between 10 and 13 μm—and their shape may vary between an elongated and a bluntly rounded appearance within the same *Eimeria* sp. (Roberts and Hammond 1970). Using phase-contrast microscopy, the pointed anterior part of the sporozoite appears darker than the posterior part, since in the former, the apical complex composed of several cellular organelles—polar rings, conoid, spirally arranged microtubules, and microneme and rhoptry vesicles—is located. Host cell adhesion and invasion depend on this apical complex. After invasion of host cells, sporozoites can be observed for several hours to several days, depending on the length of the prepatent period and the speed of multiplication. For example, in the slowly developing *E. bovis* (host, cattle), sporozoites are seen up to 6 days after host cell invasion (Dubremetz and Elsner 1979).

Meronts

Meronts represent an intracellular parasite stage that is wrapped into the PV—i.e., a host cell membrane hiding intracellular stages from host cell defense mechanisms. Depending on the *Eimeria* sp., meronts are ovoid to round in shape and contain between several hundred and more than 120,000 merozoites (Hamid et al. 2015). In general, first-generation meronts hold a higher number of merozoites than second-generation meronts. According to the number of merozoites contained, meront size differs and varies widely between *Eimeria* spp. and between the first- and second- or later generation meronts. Some species like *E. bovis* (host, cattle) form macromeronts during first asexual replication, which are macroscopically visible and may reach a size of up to several hundred micrometers. Merozoites vary in size according to the infecting *Eimeria* sp. and resemble in shape sporozoites. They also harbor the apical complex, a prerequisite for their ability to initiate, after their maturation and meront egress, host cell invasion before their next replication step.

Macro- and Microgamonts

After asexual proliferation, macro- and microgamonts are formed featuring sexual replication (see Sect. 3.1.2, life cycle). Macrogamonts contain a single macrogamete that represents a stationary parasite stage that remains within its host cell. In contrast, the microgamont releases a multitude of motile microgametes into the intestinal lumen invading macrogamete-infected host cells. Both macro- and microgamonts are ovoid in shape and can be distinguished from meronts by their morphology. A macrogamont contains a single nucleus and wall-forming bodies (WFBs); the latter represent cytoplasm-embedded vesicles that are involved in oocyst wall formation. There are two types of wall-forming bodies: WFBs of type I feature a maximum size of approximately 1.6 μm in diameter and are situated next to the parasite cell membrane. WFBs of type II are with up to 1.8 μm in diameter somewhat larger and are located deeper under the membrane surface, between WFBs of type I and the macrogamont nucleus (Scholtyseck and Hammond 1970). According to a recent study, WFBs of type II seem to play the major role in oocyst wall formation; however, this study is inconsistent with previous results on the impact of WFB of type I and II on wall formation (Kefu et al. 2006; Mouafo et al. 2002). In addition to WFBs, also lipid granula and a few mitochondria are found

(Scholtyseck and Hammond 1970). The macrogamont nucleus is irregular in shape and has been proposed to contain a nucleolus. During maturation and before fertilization of the macrogamont, the harboring host cell degenerates gradually by starting to vacuolize and disintegrate (Scholtyseck and Hammond 1970).

Zygote

Immediately after fertilization of macrogametes by microgametes and formation of the zygote, type I and II WFBs start to form a thick visible oocyst wall (Sibert and Speer 1980; Kefu et al. 2006). The immature oocyst has up to five membrane layers which later fuse into two or three final wall layers lining the unsporulated oocyst to be excreted into the environment (Sibert and Speer 1980; Kefu et al. 2006). The condensed sporont parasite stage can be observed in the zygote and the maturing oocyst as it leaves an optically empty space between its content and the developing oocyst wall (Sibert and Speer 1980).

3.1.2 Life Cycle

Eimeria sp. are known to be strongly host-specific, though a given host species can be infected by several different *Eimeria* sp. Most species parasitize the digestive tract; however, aberrantly, some species reside in the liver—e.g., *E. stiedae* in the rabbit—or in the kidneys, e.g., *E. truncata* in the goose, and in reptiles, also the gallbladder may be infested (Megía-Palma et al. 2015).

Eimeria are characterized by a monoxenic life history comprising of endogenous and exogenous developmental stages. Life cycle starts with the ingestion of the infective stage, the sporulated oocyst. The following endogenous phase includes intracellular asexual and sexual propagation at respective predilection sites (Shirley et al. 2005; Fig. 3.3). The duration of the endogenous replication or prepatent period may vary between different species. For example, for *E. acervulina* (host, chicken), it may take as short as 4 days, while for *E. bovis* (host, cattle), it may take up to 3 weeks.

Replication starts with excystation of sporozoites from oocysts and sporocysts. Some *Eimeria* spp. possess a micropyle, which is a predetermined breaking line at which oocyst rupture occurs. The four oocyst-contained sporocysts also feature and break at a preformed rupture site. Excystation depends on several factors, and there seem to be differences in the excystation process in different hosts. It has been proposed that the observed differences might be related to the oocyst wall structure; however, the exact reasons are not yet known.

In vivo, *Eimeria* excystation requires pancreatic fluid contained in the duodenal juice and bile salts (Doran and Farr 1962). In mammals, in vitro bile salts and trypsin have been reported to induce complete excystation of oocyst and sporocyst (Doran and Farr 1962). In rodents, excystation is comparatively easy to obtain by applying pepsin and hydrochloric acid to sporulated oocysts (Wiedmer et al. 2011). This is different from other mammalian species where pepsin is not assumed to play a major role in oocyst excystation. In contrast, for poultry *Eimeria* oocysts, none of the above factors is sufficient to induce excystation. Also, sodium taurocholate was

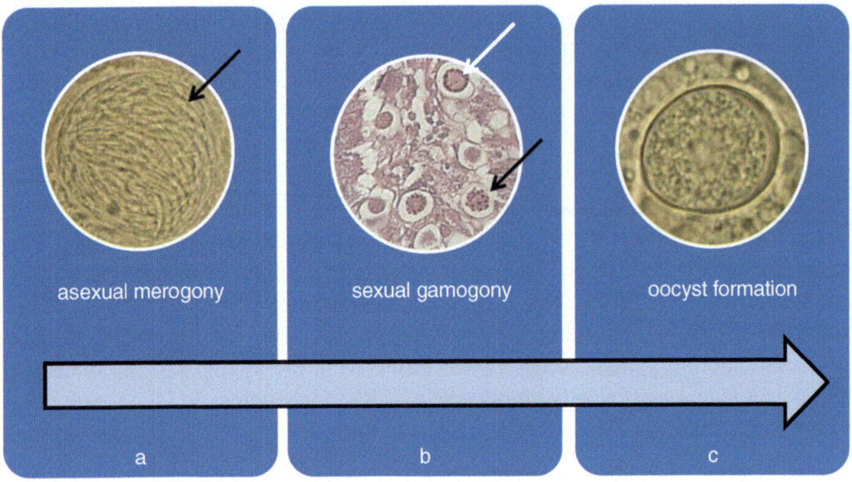

Fig. 3.3 Endogenous development of *Eimeria* comprises several stages: (**a**) asexual replication (merogony); the number of merogonies is species-dependent and varies between two and six merogonic cycles. Meront containing several hundred merozoites (*black arrow*); (**b**) sexual replication step (gamogony) with formation of macrogamonts (*white arrow*) and microgamonts; formation of zygote (*black arrow*); (**c**) oocyst released by host cell into gut lumen

not effective, and it is currently thought that a combination of yet unknown factors is required for host infection by sporozoites (Itagaki 1954). In gallinaceous birds, the gizzard is considered important for breakage of oocyst walls.

The oocyst-freed eight motile sporozoites are able to enter mucosal epithelial host cells. During invasion, the parasite envelops itself into the PV using a proportion of the host cell membrane in order to evade host cell defense mechanisms and its elimination. Inside the PV, the trophozoite, the first endogenous stage, is formed and grows subsequently into the schizont stage (Fig. 3.4).

Asexual replication starts with schizogony followed by merogony and the formation of multiple merozoites (Fig. 3.3a). Depending on the *Eimeria* sp., there may be one, two, or up to six subsequent merogony propagation cycles before sexual replication occurs (Shirley et al. 2005).

First, each schizont nucleus multiplies by a number of mitoses inside a growing intravacuolar PV. Newly formed nuclei are dispersed in the cytoplasm around the schizont membrane (Dubremetz and Elsner 1979). Subsequently, the multiplied nuclei, apical complexes—formed from intraparasitic vesicles filled with fibrillar and granular material—cytoplasm, and other organelles are split between budding merozoites a process referred to as merogony or merogenesis. In this process, each nucleus and a complete set of cellular organelles are surrounded with a newly formed membrane to generate a merozoite. In *Eimeria,* schizogony constitutes a specialized type of merogony that differs from other Apicomplexa rather using endodyogeny or endopolygeny for asexual replication. During merogony, the host cell grows enormously, and its metabolism alters to supply the parasite through micropores in the PV with nutrients. Especially in first-generation schizonts or meronts, a strong

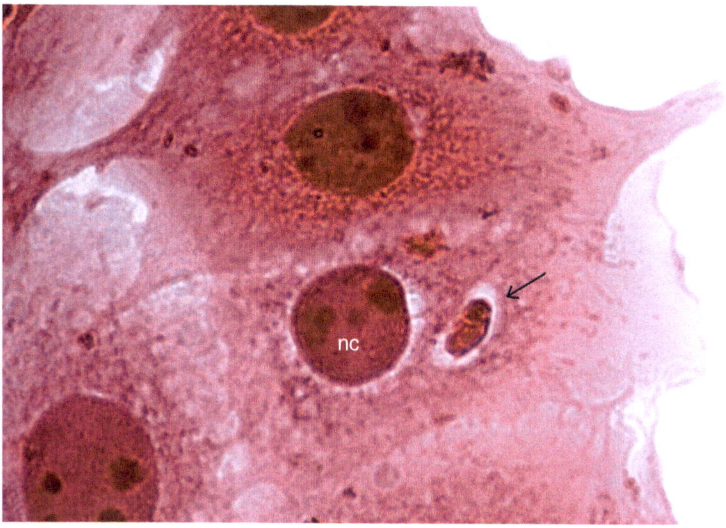

Fig. 3.4 *E. tenella* trophozoite formation in vitro. Parasitophorous vacuole with trophozoite developing into first-generation meront (*arrow*). *nc*, host cell nucleus

replication is observed, and single host cells may reach a size of several hundred micrometers. When merozoites have matured, they egress actively, invade secondary host cells, and continue their endogenous development. Following merogony, a host cell is typically invaded by a single merozoite. However, occasionally multi-infections of host cells are also observed (Kefu et al. 2006). The localization of the first and second merogony as well as that of the later gamogony differs between species.

During sexual replication, microgamonts containing a number of microgametes and macrogamonts—also called macrogametes—are formed. Mature microgametes are set free by destroying the host cell. They are motile stages able to move through the intestinal lumen toward macrogamonts. After penetration of the macrogamont-infected host cell, fusion of both haploid stages—i.e., oocyst fertilization—occurs. Already in the generated zygote, the oocyst wall develops into a solid shell protecting the oocyst in the environment (Fig. 3.3b). When the oocyst has formed, its host cell is destroyed, and the oocyst is shed with the feces marking the end of the prepatent period or prepatency (Fig. 3.3c). The following period of oocyst excretion is referred to as patent period or patency and can last for several days to weeks. Such a prolonged oocyst formation period is facilitated by the non-synchronous development of endogenous stages. In addition, repeated uptake of oocysts by the host from a contaminated environment may result in superinfection with the same species further prolonging the patent period.

Importantly, upon excretion, oocysts are not yet infective to a susceptible host. They need to mature in the environment, which includes a meiotic division and the formation of four sporocysts each containing two haploid sporozoites (see Sect. 3.1.1.1). This process of sporulation may take as short as 1 day under ideal—warm and moist—climatic conditions but is prolonged under suboptimal conditions (Allen and Fetterer 2002; Venkateswara et al. 2015).

Eimeria spp. infections are usually self-limiting. However, systemic extraintestinal stages have also been observed in chickens. For example, sporozoites were detected in the liver and spleen in the species *E. maxima*, and parasite stages have been also shown in the bursa of Fabricius in *E. tenella* (Anderson et al. 1976; Riley and Fernando 1988). These stages may be important in terms of *Eimeria* persistence as they may result in reactivation of the infection, e.g., during immune suppressive phases. For acute infections, no distinct pathology linked to extraintestinal stages of *Eimeria* spp. infecting the intestine has been shown.

3.1.3 Host-Pathogen Interaction

There is a tight host-pathogen interaction (HPI) in *Eimeria*. Parasite invasion and replication require resources of the host cell, and the parasite is able to modulate host cell metabolism and signaling pathways (Oakes et al. 2013). The first interaction occurs during host cell attachment. As has been described for Apicomplexa, motile parasitic stages secrete different signal proteins. Microneme (MIC) proteins are secreted during attachment to the host cell surface. Rhoptry neck (RON) proteins facilitate host cell membrane penetration and formation of a moving junction between the host cell membrane and the parasite. Rhoptry (ROP) proteins are secreted into the formed PV and the host cell cytoplasm. Especially ROP proteins are important virulence factors because they modulate host cell signal cascades and metabolism (Oakes et al. 2013).

During its replication, *Eimeria* exploits the cholesterol resources of the host cell by different pathways. In *E. bovis* (host, cattle), upregulation of the de novo synthesis of cholesterol and induction of a higher LDL (low-density lipoprotein)-mediated cholesterol uptake by the host cell have been shown (Hamid et al. 2015). This effect is most pronounced during first-generation meront formation (see Sect. 3.1.1.2, macromeronts in *E. bovis*). In addition, intracellular parasite stages exhibit evasion mechanism against the innate immune response of the host.

Proteolytic enzymes linked to invasion and remodeling of the host cell, its metabolic modulation, and egress of the mature parasitic stages are expressed in a stage-specific manner (Katrib et al. 2012). Many immunogenic proteins—e.g., MIC proteins, ROP and RON proteins, as well as surface antigens (SAG)—are secreted or expressed at the parasite surface and are accessible to cellular immune attack.

Due to their economic importance, most host immune defense reactions have been reported on *Eimeria* spp. infecting chickens, but these findings are supported by similar observations in other hosts. Immunocompetent hosts commonly develop immunity against *Eimeria* sp. infection (Witcombe and Smith 2014). Importantly, no cross-immunity is observed between different

Eimeria sp. in the same host. In addition, also an acquired immunity may be non-sterile in case of reinfection with the same species (Rose 1987; Daugschies and Najdrowski 2005). Different isolates of the same *Eimeria* species show antigenic diversity entailing incomplete immune protection between isolates as has been shown for *E. tenella* in chickens (Clark et al. 2017). Protective immunity is predominantly based on cellular reactions, but parameters of the humoral response—such as immunoglobulin concentrations in the blood—can be measured and are reliable indicators for presence of immunity. The contribution of humoral immunity is still not fully understood and under discussion (Wallach 2010).

In primary infections, innate immunity is considered most important due to the short prepatent period in many host species (Table 3.1) (Daszak 1999; Lillehoj et al. 2007). For instance, macrophages are able to detect sporozoites without prior stimulation and respond with the production of nitrogen monoxide (NO) (Lillehoj and Li 2004). However, it has been shown that the parasite is able to counteract macrophage activation by IL-10 induction (Collier et al. 2008). Epithelial host cells parasitized by first-generation meronts of *Eimeria* sp. pathogenic for ruminants react inter alia with an increased C-X-C motif chemokine 10 (CXCL-10) and granulocyte-macrophage colony-stimulating factor (GM-CSF) production activating the innate immune response by attracting macrophages, T cells, and NK cells (Hermosilla et al. 2015; Silva et al. 2015). Endothelial cells hosting early *Eimeria* stages are activated as displayed by intracellular adhesion molecule 1 (ICAM-1) upregulation. However, this host cell activation is quickly antagonized by the intracellular parasite leading to a decrease of ICAM-1 production and demonstrating that *Eimeria* modulates the interaction between host cells and leukocytes (Silva et al. 2015). In addition, mucosal mast cells are activated by the parasite as part of the innate immune response (Daszak 1999; Petrone et al. 2002).

In case of reinfections, the adaptive cellular immune response predominates and commonly results in protection (Lillehoj et al. 2007; Wallach 2010; Giambrone et al. 1981). Cellular Th1 immunity seems to be mainly directed toward asexual developmental stages such as meronts and merozoites (Rose and Hesketh 1976; Jenkins et al. 1991). However, sporozoite and micro- and macrogamete antigens also contribute to the development of immunity (Rose and Hesketh 1976; Jeurissen et al. 1996). Macrophages present sporozoite and merozoite antigens to CD4+ and CD8+ T cells but also to γδ-T-lymphocyte subpopulations. Zoite stages are also identified and attacked by natural killer cells (NK cells). In general, Th1-stimulating cytokines IFN-γ, IL-2, TGF-β, etc. are released (Yun et al. 2000; Rothwell et al. 2004). In parallel, B-cell epitopes present in surface antigen EtSAG1 or secretory EtMIC2 are recognized by parasite-neutralizing antibodies in the case of *E. tenella* (Trees et al. 1989; Jahn et al. 2009; Tomley et al. 1996).

Table 3.1 Oocyst morphology of pathogenic *Eimeria* spp. in domestic and farm animals of importance

Host	Pathogenic *Eimeria* sp.	Mean oocyst size and size range (µm)	Oocyst shape	Oocyst wall	Presence of micropyle	Sporulation time (h)	Special features	References
Cattle	*E. bovis*	28 × 20 (23–34 × 17–23)	Pear-shaped to ovoid	Smooth, colorless to yellow-brown	+	48–72	No oocyst residuum	Sommer (1998), Levine (1961), Schnieder (2006) and Florião et al. (2016)
	E. zuernii	18 × 16 (15–22 × 13–18)	Subspherical to spherical	Smooth, colorless	–	48–240	No oocyst residuum	Sommer (1998), Levine (1961), Schnieder (2006), Florião et al. (2016) and Bangoura and Daugschies (2007)
	E. alabamensis	19 × 13 (13–24 × 11–16)	Pear-shaped to ovoid	Smooth or granulated, colorless to pale yellow-brown	+	96–192	No oocyst residuum but polar granule	Sommer (1998), Levine (1961) and Schnieder (2006)
Sheep	*E. ovinoidalis*	19 × 13 (17–30 × 14–19)	Ellipsoid	Smooth, colorless to pale yellow	+	24–72	No oocyst residuum but several polar granules	Schnieder (2006) and Foreyt (1990)
	E. crandallis	22 × 19 (17–23 × 17–22)	Subspherical to ellipsoid	Smooth, colorless	+ (with polar cap)	24–72	No oocyst residuum but polar granule(s)	Schnieder (2006) and Foreyt (1990)
	E. ahsata	33 × 23 (29–37 × 17–28)	Ellipsoid to ovoid	Smooth, yellowish brown	+ (with polar cap)	48–72	No oocyst residuum but polar granule(s)	Schnieder (2006) and Levine and Ivens (1970)
	E. bakuensis	(23–36 × 15–24)	Ellipsoid	Smooth, pale yellowish brown	+ (with polar cap)	48–96	No oocyst residuum but polar granule(s)	Schnieder (2006) and Foreyt (1990)

Goat	*E. ninakohlyakimovae*	21 × 15 (20–22 × 14–16)	Ellipsoid	Thin-walled, colorless	–	48–72	No oocyst residuum	Schnieder (2006) and Levine and Ivens (1970)
	E. arloingi	27 × 18 (17–42 × 14–19)	Ellipsoid to ovoid	Smooth, thick, colorless	+ (with polar cap)	24–96	No oocyst residuum but polar granule(s)	Schnieder (2006) and Levine and Ivens (1970)
	E. christenseni	38 × 25 (27–44 × 17–31)	Ovoid to ellipsoid	Smooth, colorless to pale yellow	+ (with polar cap)	72–144	No oocyst residuum but polar granule(s)	Schnieder (2006), Foreyt (1990) and Levine and Ivens (1970)
	E. caprina	(27–40 × 19–26)	Ellipsoid to ovoid	Smooth, yellowish to dark brown	+	48–72	No oocyst residuum	Schnieder (2006) and Foreyt (1990)
Alpaca	*E. macusaniensis*	(81–107 × 61–80)	Ovoid to pear-shaped	Smooth, thick, dark brownish, slightly striated	+ (with polar cap)	13–21	No oocyst residuum	Schrey et al. (1991) and McKenna (2006)
Hog	*E. neodebliecki*[1]	21 × 16 (17–26 × 13–20)	Ellipsoid to ovoid	Smooth, colorless	–	13 days	No oocyst residuum but polar granule	Schnieder (2006) and Koudela and Vìtovec (1998)
	E. scabra[1]	32 × 23 (22–42 × 20–24)	Ovoid to ellipsoid	Thick, rough, striated yellow to brown	+	9–12 days	No oocyst residuum but polar granule(s)	Levine (1961), Schnieder (2006) and Hill et al. (1985)
	E. debliecki[1]	19 × 14 (15–23 × 11–18)	Ovoid to ellipsoid or subspherical	Smooth, colorless	–	6–9 days	No oocyst residuum but polar granule	Levine (1961) and Schnieder (2006)

(continued)

Table 3.1 (continued)

Host	Pathogenic *Eimeria* sp.	Mean oocyst size and size range (μm)	Oocyst shape	Oocyst wall	Presence of micropyle	Sporulation time (h)	Special features	References
Chicken	*E. acervulina*	18 × 14 (12–23 × 9–17)	Ovoid	Smooth, colorless	–	24	No oocyst residuum but polar granule	Levine (1961), Schnieder (2006) and Lee and Millard (1971)
	E. maxima	30 × 20 (21–42 × 16–30)	Ovoid	Smooth, yellowish	–	30–48	No oocyst residuum but polar granule	Levine (1961) and Schnieder (2006)
	E. tenella	22 × 19 (14–31 × 9–25)	Ovoid	Smooth, colorless	–	18–48	No oocyst residuum but polar granule	Levine (1961) and Schnieder (2006)
	E. mitis	16 × 15 (10–21 × 9–18)	Subspherical	Smooth, colorless	–	18–24	No oocyst residuum but polar granule	Levine (1961) and Schnieder (2006)
	E. praecox	21 × 17 (20–25 × 16–20)	Ovoid	Smooth, colorless	–	48	No oocyst residuum but polar granule	Levine (1961) and Schnieder (2006)
	E. brunette	26 × 22 (14–34 × 12–26)	Ovoid	Smooth, colorless	–	24–48	No oocyst residuum but polar granule	Levine (1961) and Schnieder (2006)
	E. necatrix	20 × 17 (12–29 × 11–24)	Ovoid	Smooth, colorless	–	18–24	No oocyst residuum but polar granule	Levine (1961) and Schnieder (2006)
Turkey	*E. adenoeides*	26 × 17 (19–31 × 13–21)	Ovoid to ellipsoidal	Smooth, colorless	+	24	No oocyst residuum but polar granule(s)	Levine (1961) and Schnieder (2006)
	E. meleagrimitis	19 × 16 (16–27 × 13–22)	Subspherical	Smooth, colorless	–		No oocyst residuum but polar granule(s)	Levine (1961) and Schnieder (2006)
	E. gallopavonis	26 × 21* (22–31 × 18–24)	Ovoid	Smooth, colorless	–	24	No oocyst residuum	Levine (1961) and Schnieder (2006)

Rabbit	*E. stiedae*	37 × 20 (30–41 × 15–24)	Ovoid to ellipsoid	Smooth, slightly pink	+	48–72	No oocyst residuum	Levine (1961) and Schnieder (2006)
	E. intestinalis	27 × 19 (21–30 × 16–21)	Pear-shaped	Smooth, yellowish to brownish	+	72–96	Oocyst residuum present	Levine (1961) and Schnieder (2006)
	E. flavescens	30 × 21 (25–35 × 18–24)	Ovoid	Smooth, yellowish	+	96	No oocyst residuum	Levine (1961) and Schnieder (2006)
	E. irresidua	39 × 23 (31–44 × 20–27)	Ellipsoid	Smooth, pale yellowish	+	96	No oocyst residuum but polar granule(s)	Levine (1961) and Schnieder (2006)
	E. magna	36 × 24 (31–42 × 20–28)	Ovoid	Smooth, dark yellow to brown	+ (with prominent surrounding)	48–72	Oocyst residuum present	Schnieder (2006)
	E. media	31 × 17 (25–35 × 15–20)	Ovoid to ellipsoid	Smooth, slightly pinkish	+	48	Oocyst residuum present	Levine (1961) and Schnieder (2006)
	E. piriformis	30 × 18 (25–33 × 16–21)	Pear-shaped, often slightly asymmetrical	Smooth, yellow-brown to dark brown	+	96	No oocyst residuum	Levine (1961) and Schnieder (2006)
Pigeon	*E. labbeana*	(13–24 × 12–23)	Subspherical to spherical	Smooth, colorless or light yellowish brown	–	Up to 4 days	No oocyst residuum but polar granule(s)	Levine (1961)
	E. columbae	16 × 14	Subspherical	Smooth, colorless or slightly colored	–	nd	Oocyst residuum present	Levine (1961)
Goose	*E. truncata*[*]	?(11–27 × 11–22)	Ovoid	Smooth, colorless	+ (with polar cap)	24–120	Oocyst residuum may be present	Levine (1961) and Schnieder (2006)
	E. anseris	21 × 17* (16–24 × 13–19)	Pear-shaped	Smooth, colorless	+	24–48	Oocyst residuum present	Levine (1961) and Schnieder (2006)
	E. nocens	29 × 20* (25–33 × 17–24)	Ovoid to ellipsoidal	Thick-walled, brown	+	60–72	No oocyst residuum	Levine (1961) and Schnieder (2006)

(continued)

Table 3.1 (continued)

Host	Pathogenic *Eimeria* sp.	Mean oocyst size and size range (μm)	Oocyst shape	Oocyst wall	Presence of micropyle	Sporulation time (h)	Special features	References
Duck	*E. mulardi*	21 × 17 (19–23 × 15–21)	Ovoid	Smooth, colorless	+	24–72	No oocyst residuum	Schnieder (2006)
Horse	*E. leuckarti*[1]	80 × 60 (70–90 × 49–69)	Ovoid to pear-shaped	Thick, dark brown shell	+	15–41	No oocyst residuum	Levine (1961), Schnieder (2006) and Barker and Remmler (1970)
Guinea pig	*E. caviae*	(13–26 × 12–23)	Ellipsoid to ovoid	Smooth, brownish	–	2–11	Oocyst residuum present	Schnieder (2006), Flausino et al. (2014) and Ellis and Wright (1961)

nd no data available to the authors
*Different values described in literature
[1]Nonpathogenic

3.2 Diagnosis, Epidemiology, and Economic Impact

3.2.1 Diagnosis

In general, fecal examination using the flotation method is applied to detect *Eimeria* oocysts in droppings of individual animals, while in herd animals, examination of a collective sample is preferred (Fig. 3.5). If oocysts are present, differentiation and identification of pathogenic and/or apathogenic species should be carried out to allow an informed decision on prophylaxis or treatment. In cattle, *Eimeria* species determination is feasible by microscopic examination of oocyst morphology (Fig. 3.6). In contrast, in chicken and some other hosts, it may be difficult to discriminate *Eimeria* spp., and species identification using polymerase chain reaction (PCR) is preferred (Fig. 3.7). Quantification of oocysts is mandatory if a herd health status is to be monitored over several successive production cycles. A quantitative approach allows for evaluation of management, hygienic measures, and drug efficacy to evaluate the development of drug resistance. For this reason, counting methods using FLOTAC chambers or McMaster slides are established in many diagnostic laboratories.

Especially in chickens, diagnostic histological sections and the scoring of intestinal lesions are commonly used to confirm tentative diagnoses of infections with pathogenic *Eimeria* species. The severity of eimeriosis may be estimated using a scoring system of intestinal lesions, which is available for the different pathogenic *Eimeria* sp. infecting chickens (Johnson and Reid 1970). Besides the scoring of gross intestinal lesions, also a microscopic scoring system has been established allowing evaluating the number of endogenous stages in intestinal mucosal scrapings.

Serologic detection has been established for a number of different pathogenic *Eimeria* spp. in several hosts (Faber et al. 2002; Constantinoiu et al. 2007). Currently,

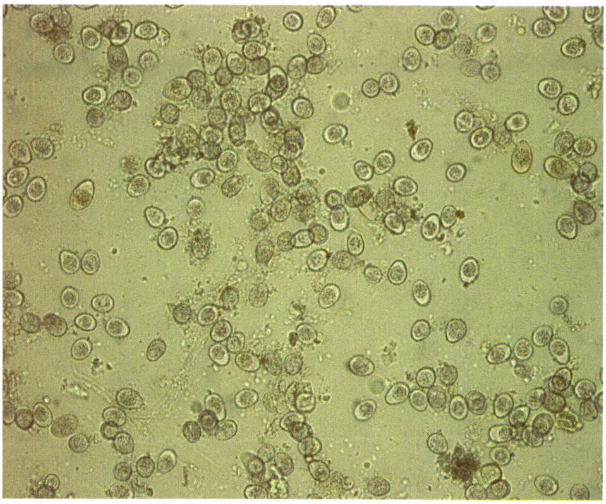

Fig. 3.5 Coproscopical finding in cattle with massive oocyst excretion using qualitative salt flotation method (unsporulated *E. bovis* and *E. auburnensis*, 100× magnification)

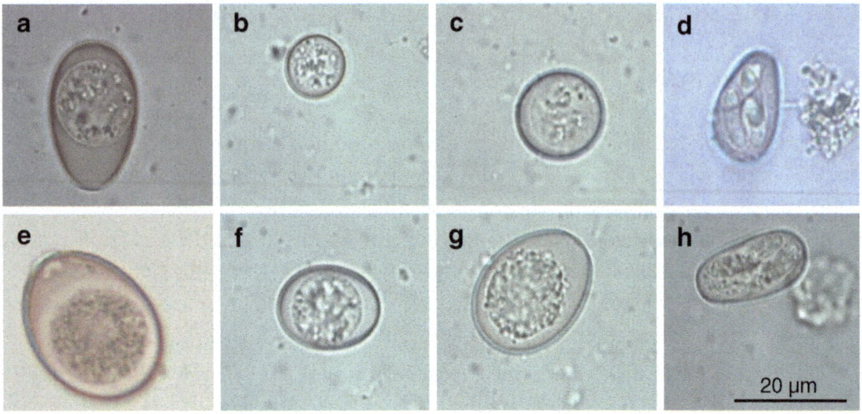

Fig. 3.6 Selection of commonly found *Eimeria* spp. oocysts from cattle. (**a**) *E. auburnensis*, (**b**) *E. subspherica*, (**c**) *E. zuernii*, (**d**) *E. alabamensis*, (**e**) *E. brasiliensis*, (**f**) *E. ellipsoidalis*, (**g**) *E. bovis*, (**h**) *E. cylindrica*. Note that only *E. bovis*, *E. zuernii*, and *E. alabamensis* are pathogenic. All other *Eimeria* spp. oocysts are incidental findings (100× magnification)

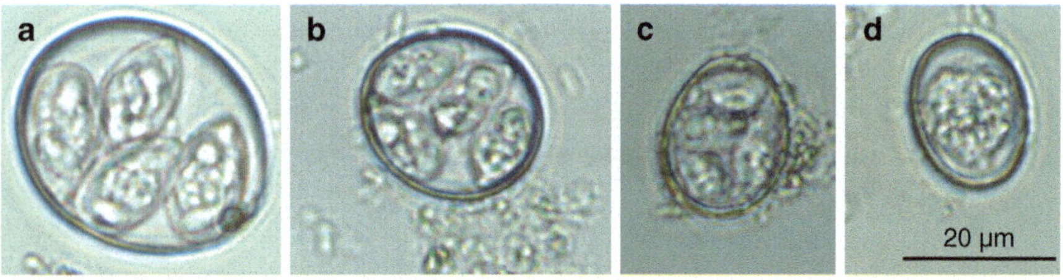

Fig. 3.7 Selection of commonly found *Eimeria* spp. oocysts from chicken. (**a**) *E. maxima*, (**b**) *E. tenella*, (**c**) *E. acervulina*, (**d**) *E. brunetti*. Note partially overlapping size and shape in different species impeding final microscopical diagnosis (400× magnification)

the most convenient assay is the enzyme-linked immunosorbent assay (ELISA) based on merozoite antigen. However, the detection of IgG or IgY in a mammalian or an avian host, respectively, indicates that the parasite has replicated and is already well established in the host. In mammalian hosts parasitized with *Eimeria* spp. with an endogenous development of several weeks—as, for example, *E. bovis* in cattle—a serologic response may be already detectable during the late prepatency. In contrast, serologic immune response in chicken is only measurable after the onset of patency or even later during the postpatent period. Due to the late and indirect detection of an infection, a serologic analysis for an individual pro- or a herd metaphylaxis is not suitable. However, ELISA is suitable for the detection of a secondary immune response after vaccination or to assess herd immunity in response to infection pressure and to determine a possible need of biosecurity measures (Constantinoiu et al. 2007).

3.2.2 Epidemiology

About 2000 *Eimeria* spp. have been described so far in mammalian, avian, reptilian, and other vertebrates as well as in invertebrate hosts (Megía-Palma et al. 2015). Following the distribution of their host, *Eimeria* sp. are ubiquitously distributed all over the world. Besides wildlife species, *Eimeria* spp. infect all major livestock species such as ruminants, pig, chicken, and other poultry. In Table 3.2 the economically important pathogenic *Eimeria* spp. in avian and mammalian hosts are shown. In general, *Eimeria* species are homoxenous—i.e., they infect one defined host species. One explanation for the narrow host range of *Eimeria* compared to other apicomplexans such as *Toxoplasma* or *Cryptosporidium* may be their comparatively small repertoire of microneme proteins. As these proteins are essential for host cell recognition, a small number of specific microneme proteins may translate into a limited recognition of host cell types (Cowper et al. 2012). However, it has also been proposed that host specificity of *Eimeria* is rather an adaptive phenomenon than a co-phylogenetic determination (Kvičerová and Hypša 2013). Interestingly, a limited oligoxenous host range has been observed in some avian *Eimeria* spp. a characteristic that may contribute to parasite distribution by wildlife and livestock cross-infection (Vrba and Pakandl 2015).

The usually observed homoxenous transmission limits the spread of the infection in an ecosystem or between different species within a mixed population. Nonetheless, since the infection dose is low—the ingestion of a single oocyst may establish an infection—the infection rate is commonly high in a given population of susceptible naïve hosts in a contaminated area or stable. However, the presence of pathogenic *Eimeria* spp. on a farm does not necessarily translate into clinical outbreaks in a herd or flock. Natural trickle infections—the repeated uptakes of low oocyst doses—are common in the field inducing protective immunity and endemic stability (Daugschies and Najdrowski 2005). The shift from a subclinical to a clinical infection is thought to be induced by additional factors such as malnutrition and/or concurrent diseases that undermine the animal health status. Except for porcine and some chicken-specific *Eimeria* spp., young animals are at highest risk to develop clinically relevant infections. This observation is rather due to the lack of immunity in younger and not to an age-specific resistance in older animals. Susceptibility to the infection is also genetically determined, as it has been shown to be dependent on the breed (Reeg et al. 2005; Passafaro et al. 2015). Accordingly, it has been reported that chicken breeds with an efficient innate immune response are less susceptible to *Eimeria* infections (Swaggerty et al. 2011).

Eimeriosis is a herd or flock disease building up over several infection cycles within a group of animals. Spreading within a herd or flock may occur within several days—e.g., chicken—or weeks, e.g., cattle. On endemic farms, on-site prevalences of up to 100% have been observed in susceptible populations (Matjila and Penzhorn 2002). In the environment, unsporulated *Eimeria* oocysts develop quickly into the infective sporulated oocyst at favorable moisture and temperatures of about

Table 3.2 Pathogenicity, localization, and infection course of *Eimeria* spp. of economic and clinical importance in farm and domestic host species

Host (no. of valid *Eimeria* spp.)	Pathogenic *Eimeria* spp.	Localization	Prepatent period (d)	Patent period (d)	Pathogenicity	References
Cattle (>20)	*E. bovis*	Intestine, mainly colon (including cecum)	21–23	6–10	+++	Daugschies et al. (1986)
	E. zuernii	Jejunum to colon (including cecum)	15–22	2–8	+++	Bangoura and Daugschies (2007)
	E. alabamensis	Jejunum to colon (including cecum)	6–11	1–13	++	Levine (1961) and Schnieder (2006)
Sheep (15)	*E. ovinoidalis*	Cecum, colon	12–15	9–10 (7–28)	+++	Levine (1961), Taylor and Catchpole (1994), Foreyt (1990) and Catchpole et al. (1993)
	E. crandallis	Small and large intestine (mainly cecum and colon)	13–20	Approx. 10	++	Schnieder (2006), Foreyt (1990) and Catchpole et al. (1993)
	E. ahsata	Small intestine	18–21	10–12	+++	Levine (1961), Schnieder (2006) and Foreyt (1990)
	E. bakuensis	Small intestine	19–29	Approx. 10	+++	Levine (1961) and Foreyt (1990)
Goat (9)	*E. ninakohlyakimovae*	Ileum, cecum, upper colon	10–14	10–14	+++	Levine (1961), Schnieder (2006), Foreyt (1990) and Yvoré et al. (1985)
	E. arloingi	Small intestine	14–20	14–15	+++	Levine (1961), Schnieder (2006) and Foreyt (1990)
	E. christenseni	Small intestine	14–23	3 to >30	++	Schnieder (2006) and Schreiner (2014)
	E. caprina	Small and large intestine	17–20	3–6	+++	Schnieder (2006)

Host (no. of valid *Eimeria* spp.)	Pathogenic *Eimeria* spp.	Localization	Prepatent period (d)	Patent period (d)	Pathogenicity	References
Alpaca (8)	*E. macusaniensis*	Small intestine	32–43	14–55 (few numbers)	++	Jarvinen (2008), McKenna (2006) and Gerken and Renieri (2006)
Hog (8)	*E. neodebliecki*	Jejunum	10	6–8	+ – ++	Koudela and Vitovec (1998)
	E. scabra	Small intestine	9–12	5		Levine (1961) and Hill et al. (1985)
	E. debliecki	Small intestine (partially large intestine)	7	10–15	+	Levine (1961)
Chicken (9)	*E. acervulina*	Duodenum (to proximal jejunum)	4	>10	++	Levine (1961) and Schnieder (2006)
	E. maxima	Middle jejunum	5–6	Few days	++	Levine (1961), Schnieder (2006) and Rose and Hesketh (1976)
	E. tenella	Large intestine (ceca)	6–7	>4	+++	Levine (1961) and Schnieder (2006)
	E. mitis	Distal jejunum, ileum	4–5	10	+	Levine (1961) and Schnieder (2006)
	E. necatrix	Small intestine (mainly jejunum)	5–7	12	+++	Levine (1961) and Schnieder (2006)
	E. praecox	Duodenum to proximal jejunum	4	4	+	Levine (1961) and Schnieder (2006)
	E. brunetti	Ileum and large intestine (ceca)	5	4–12	++	Levine (1961), Schnieder (2006) and Ryley et al. (1972)

(continued)

Table 3.2 (continued)

Host (no. of valid *Eimeria* spp.)	Pathogenic *Eimeria* spp.	Localization	Prepatent period (d)	Patent period (d)	Pathogenicity	References
Turkey (6)	*E. adenoeides*	Lower small intestine and ceca	4.5–5.5	7–20	++	Levine (1961) and Vrba and Pakandl (2014)
	E. meleagrimitis	Duodenum	5–6	>3	+++	Levine (1961) and Schnieder (2006)
	E. gallopavonis	Small and large intestine	6–7	nd	++	Levine (1961) and Vrba and Pakandl (2014)
Rabbit (10)	*E. stiedae*	Liver	14–16	Up to 36	+++	Schnieder (2006) and Hanada et al. (2003)
	E. intestinalis	Small intestine	9–10	>4	+++	Levine (1961), Schnieder (2006) and Shi et al. (2016)
	E. flavescens	Large intestine	8–11	nd	+++	Schnieder (2006)
	E. irresidua	Small intestine	8–10	nd	++	Levine (1961) and Schnieder (2006)
	E. magna	Small intestine	7–9	nd	++	Levine (1961) and Schnieder (2006)
	E. media	Small intestine	5–7	nd	++	Schnieder (2006)
	E. piriformis	Large intestine (colon)	9	nd	++	Levine (1961) and Schnieder (2006)
Pigeon (9)	*E. labbeana*	Small intestine	6–8	At least up to 27	+ – +++	Levine ((1961), Schnieder (2006), Balicka-Ramisz and Pilarczyk (2014) and Vercruysse (1990)

Host (no. of valid *Eimeria* spp.)	Pathogenic *Eimeria* spp.	Localization	Prepatent period (d)	Patent period (d)	Pathogenicity	References
Goose (5)	*E. truncata*	Kidneys	5–14	nd	+++	Levine (1961) and Schnieder (2006)
	E. nocens	Jejunum to rectum (including ceca)	4–9	nd	+++	Schnieder (2006) and Dai et al. (2005)
Duck (8)	*E. somateriae*	Kidneys	nd	nd	+ – +++	Persson et al. (1974) and Morner (1975)
	E. mulardi	Jejunum, ileum, ceca	5.5–7	nd	++	Schnieder (2006), Pakandl et al. (2002) and Chauve et al. (1994)
Horse (1)	*E. leuckarti*	Small intestine	31–34	5–12	(+)	Barker and Remmler (1970) and McQueary et al. (1977)
Guinea pig (1)	*E. caviae*	Small intestine to colon	7–11	4–5 (<9)	+	Kunstýr and Naumann (1981) and Daszak (1999)

nd no data available

−5–30 °C. Depending on the climate conditions, this may take 1 day to several weeks (Marquardt 1960). Especially temperatures below −10 °C and above 50 °C are detrimental for oocysts (Marquardt 1960; Matsui et al. 1989).

Eradication of oocysts from a contaminated site is difficult because of their high reproduction potential and their immense tenacity in the environment. Physical measures like ultraviolet light or heat application are most effective in oocyst inactivation, while chemical disinfection is restricted to the use of hazardous substances like chlorocresols (Daugschies et al. 2013). Hence, despite of implementation of hygienic measures, subsequent animal production cycles in the same environment will continue to suffer from eimeriosis since pathogen removal is incomplete and every infected animal contributes to further environmental contamination affecting the following animal group. The major reservoir for *Eimeria* is the young animal population where the pathogen is passed from one production cycle to the next. Due to high oocyst production, the young population is epidemiologically much more important than the parental population, which excretes low numbers of oocysts.

3.2.2.1 Chicken

Eimeria is often present in conventional and organic farming. However, there is a certain predisposition for severe infections in intensive farming due to the high stocking density of animals resulting in an increased environmental contamination of oocysts per area. Thus, clinical disease is more likely to affect intensively reared animals. In chickens, there are seven recognized *Eimeria* species, namely, *E. acervulina*, *E. necatrix*, *E. praecox*, *E. maxima*, *E. mitis*, *E. tenella*, and *E. brunetti*. Two additional species—*E. hagani* and *E. mivati*—have been described, but their validity is not yet accepted. The latter taxon may be a variant of *E. mitis* and has been considered to be a *nomen dubium* (Shirley et al. 1983; Vrba et al. 2011).

In general, mixed *Eimeria* spp. populations are present in most commercial farms though also mono-infections have been reported (Salisch et al. 1989; Graat et al. 1998; Schwarz et al. 2009; Sun et al. 2009; Shirzad et al. 2011; Györke et al. 2013). The prevalence of *Eimeria* spp. varies considerably between different geographical regions. On North American farms, *E. maxima* and *E. tenella* are highly prevalent followed by *E. mitis*, while *E. acervulina* is rarely observed (Schwarz et al. 2009). In contrast, in European broiler farms, the most widespread species is *E. acervulina* irrespective of the farm size. Especially in smaller farms, predominantly *E. tenella* is observed often with concomitant infections including *E. maxima* and/or *E. praecox*. In a multicentric Romanian study, 92% of all sampled broiler farms showed *Eimeria* spp. infections, of which 91% were found to be affected by *E. acervulina*. In other European countries, lower percentages of *Eimeria*-positive farms with a similar species spectrum are observed. No relation between farm size and *Eimeria* positivity has been found, yet mixed infections are more common on smaller broiler farms (Györke et al. 2013). On Asian broiler farms, the above four mentioned *Eimeria* species—*E. mitis*, *E. acervulina*, *E. maxima*, and *E. praecox*—and all other valid species are regularly found (Al-Natour et al. 2002; Sun et al. 2009; Shirzad et al. 2011; Chengat Prakashbabu et al. 2017). In layer husbandries,

E. tenella and *E. maxima* are most dominant, but eimeriosis due to other species like *E. acervulina* is also observed (Schneider and Haass 1967; Lundén et al. 2000).

3.2.2.2 Cattle

More than 20 valid *Eimeria* spp. are known to infect cattle (Daugschies and Najdrowski 2005). Of these, *E. bovis*, *E. zuernii*, and *E. alabamensis* are pathogenic; the former two species are found indoors as well as on pasture, while the latter is seen on pasture only. Although many species are named after geographical areas, they are distributed worldwide. Typically, susceptible young cattle are concomitantly infected by two to three different pathogenic *Eimeria* spp. (Ernst and Benz 1981). In severe infections with pathogenic species of high replication potential—but especially in the case of *E. bovis* due to its highly prolific macromeront stage—other *Eimeria* spp. may not be detectable. Bovine eimeriosis is frequently observed to impair herd health (Fitzgerald 1980; Cornelissen et al. 1995; Ovington et al. 1995). A few weeks after rehousing, calves typically shed pathogenic *Eimeria* oocysts with higher oocyst excretion numbers being observed in younger animals (Hiepe et al. 1978; Gräfner et al. 1985; Lassen et al. 2009). *E. zuernii* and *E. bovis*—both highly pathogenic species—are regarded as ubiquitous pathogens (Fitzgerald 1980; Daugschies and Najdrowski 2005). On-site prevalences of 100% are often reached on endemic farms (Matjila and Penzhorn 2002). Risk factor analyses have been carried out, and it was found that especially monthly water trough cleaning vs. no cleaning and a slatted floor vs. bedding significantly reduced the infection risk; also the shed climate was of significance to avoid oocyst excretion (Gräfner et al. 1978, 1985; Bangoura et al. 2012; Mitchell et al. 2012).

3.2.2.3 Sheep and Goat

There are 15 valid *Eimeria* species in sheep—of which *E. ovinoidalis*, *E. ahsata*, *E. bakuensis*, and *E. crandallis* are pathogenic—and 9 in goats—of which *E. ninakohlyakimovae*, *E. arloingi*, *E. caprina*, and *E. christenseni* are pathogenic (Table 3.2). For a long time, both these host species were thought to share *Eimeria* spp. due to the morphological similarities of excreted oocysts. However, it has been later found that no cross-transmission of *Eimeria* between both hosts occurs. *Eimeria* oocyst shedding is common in all age groups. Nonetheless, epidemiologically and economically important are infections of lambs and kids of approximately 4–5 weeks of age (Foreyt 1990). Often, pathogenic *Eimeria* species cycle within the offspring and oocyst excretion by older animals does only marginally contribute to environmental contamination.

3.2.2.4 Rabbit

Nine intestinal and the liver-residing *E. stiedae* species are known to infect rabbits. The infection with *E. stiedae*—causing liver coccidiosis—shows an age-specific resistance (Long 1973; Pellerdy 1974; Gomez-Bautista et al. 1987). Older animals are less prone to clinical eimeriosis or, in case of *E. stiedae*, to liver damage. Accordingly, parasite multiplication in older hosts contributes by far less to environmental contamination than in animals younger than approximately 3 months of age.

Wild rabbits harbor the same *Eimeria* spp. as domestic rabbits; nonetheless, their epidemiological importance as reservoir host is low since domestic rabbits do not get in contact with wild rabbits. However, transmission might be possible by contaminated fresh feed—grass, vegetables, etc.—from meadows, gardens, and farms. There are no shared *Eimeria* spp. in rabbits and hares.

3.2.2.5 Guinea Pig

E. caviae is frequently observed in pet guinea pigs as well as in those farmed for meat production or kept as laboratory animals (d'Ovidio et al. 2015; Kouam et al. 2015). The infection of young animals is clinically most important, but oocyst excretion may occur at all ages. There is no transmission of *Eimeria* from rabbit or chinchilla to guinea pig (de Vos 1970; Ming-Hsien and Hong-Kean 2010).

3.2.2.6 Horse

E. leuckarti is the only species known to infect the horse. Foals are more often affected by infections than older horses (Beelitz et al. 1996; Gülegen et al. 2016). Oocyst excretion is detected from about 1 month of age onward (Lyons and Tolliver 2004). In donkeys, also *E. leuckarti* is observed (Benbrook and Sloss 1962). *E. leuckarti* is not related to clinical disease and considered apathogenic.

3.2.2.7 Pigeon

There are nine *Eimeria* spp. in pigeons, of which *E. labbeana* and *E. columbarum* are considered of clinical importance. Most pigeons are subclinically infected excreting oocysts on a low level acting as reservoir host (Krautwald-Junghanns et al. 2009). Also wild pigeons pose a reservoir since contact with domestic animals, especially racing pigeons, cannot be prevented. Young pigeons and racing pigeons are most prone to clinical disease (Balicka-Ramisz and Pilarczyk 2014). In general, prevalence of *Eimeria* spp. in pigeon flocks is high and may reach up to 100%, and mixed infections are common.

3.2.2.8 Turkeys

Five *Eimeria* spp. have been described in wild and domestic turkeys—including the three pathogenic species *E. meleagrimitis*, *E. adenoeides*, and *E. gallopavonis*—and a sixth species named *E. innocua* has been lately described (Vrba and Pakandl 2014). There is little data on *Eimeria* prevalence in domestic turkeys. Studies from the USA indicate a prevalence of approximately 40–60% in commercial farms with a high proportion of pathogenic species, a situation which has also been observed in the UK (Clarkson and Gentles 1958; Chapman 2008). Cross-transmission of turkey *Eimeria* spp. to chicken has not been observed but has been demonstrated to other poultry—namely, the gray partridge (*Perdix perdix*) and the bobwhite quail (*Colinus virginianus*) (Vrba and Pakandl 2015). This finding may be of epidemiological impact for wild turkeys but is of limited importance for the domestic hosts.

3.2.2.9 Waterfowl

Eimeria infections in geese and ducks are widespread. Most species like *E. nocens*, *E. stigmosa*, and *E. anseris* in geese and *E. mulardi* in ducks reside in the intestine. An exception is *E. truncata* in geese and *E. somateriae* in ducks which develop in the kidneys. For both host species, prevalence studies are scarce. However, reports about high morbidity and mortality of geese and ducks due to eimeriosis outbreaks have been reported from many geographic regions (Nation and Wobeser 1977; Gajadhar et al. 1983; Skírnisson 1997; Dai et al. 2004).

3.2.2.10 Hog

Altogether eight valid *Eimeria* spp. have been reported in hogs: *E. debliecki*, *E. neodebliecki*, *E. perminuta*, *E. polita*, *E. porci*, *E. scabra*, *E. spinosa*, and *E. suis*. Interestingly, porcine eimeriosis widely remains subclinical and is mostly observed in gilts and sows. Under intensive rearing conditions, *Eimeria* spp. are sometimes observed in weaned piglets but particularly in older animals, e.g., boars and sows (Daugschies et al. 2004; Karamon et al. 2007). Porcine *Eimeria* spp. are seen more frequently on small than on large farms which is assumed to be associated with farm hygiene and management factors rather than the farm size itself (Daugschies et al. 2004; Karamon et al. 2007). Sows are often excreting oocysts, and prevalences may in some regions reach over 50%; on conventional piglet-producing farms, up to 82% of the sows were observed to be positive depending on their reproduction status (Daugschies et al. 2004). The most common species seems to be *E. neodebliecki* in the USA and *E. scabra*, *E. polita*, and *E. debliecki* in middle Europe though prevalence studies are scarce (Supperer 1961; Vetterling 1965; Pfister and Wolff 1975; Daugschies et al. 2004).

3.2.2.11 Reptiles

In reptiles, *Eimeria* has been reclassified into the genera *Acroeimeria* and *Choleoeimeria* leaving some *Eimeria* spp. *incertae sedis* (Paperna and Landsberg 1989; Megía-Palma et al. 2015). *Choleoeimeria* replicates in the gallbladder and biliary epithelium of the host. In contrast, *Acroeimeria* develops epicytoplasmatically in the microvillous zone of the reptilian intestine. Some species multiply in the cytoplasm of intestinal epithelial cells and are grouped into the genus *Eimeria*. Prevalence of *Eimeria* spp. strongly depends on animal management and acquisition as well as hygienic measures. Often, the prevalence is very low; in Poland, out of six reptilian hosts sampled under different private management conditions, only the bearded dragon (*Pogona viticeps*) showed occasional *Eimeria* oocyst shedding. For some *Eimeria*-like species, there seems to be cross-transmission between closely related reptilian genera; however, many *Eimeria*, *Choleoeimeria*, and *Acroeimeria* species need to be investigated much more thoroughly (McAllister et al. 2016).

3.2.3 Economic Impact

Owing to the fast spreading of the pathogen within a herd and the high perdurability of oocysts leading to long-term survival of infectious oocysts in the environment, eimeriosis will remain a continuing animal health threat and economic burden in an affected farm if no efficient control measures in form of treatment, hygiene, and herd management are established.

Clinical disease of eimeriosis leading to economic costs because of production losses and fatalities is held responsible for a relatively small proportion of lost revenues. Importantly, subclinical infections are considered to inflict a significantly larger economic damage. It has been reported that in the UK, poultry farming recorded losses of 38.6 million GBP in 1995, of which 80% resulted from subclinical infections—i.e., lower weight gain and poor feed conversion—followed by costs for prophylaxis and control (Williams 1999).

In cattle, losses of 8–9% were estimated for dairy farmers (Lassen and Ostergaard 2012). On one hand, this estimate includes direct costs due to an increased calf mortality and, on the other hand, indirect costs due to a delayed fertility caused by a retarded weight gain and lower final weight. It has been estimated that metaphylactic herd treatment shows great financial benefit for farmers irrespective of the treatment-related costs.

In addition, it has been demonstrated for chicken and is anticipated for alpacas that the higher susceptibility to coinfections with other pathogens has a high negative impact on the economic performance (Alnassan et al. 2014; Rojas et al. 2016).

3.3 Clinical Effects and Pathology

3.3.1 Overview

3.3.1.1 Clinical Effects

In a primary infection, especially young naïve animals are suffering from clinical eimeriosis. Mono- and mixed infections with pathogenic *Eimeria* species cause commonly diarrhea due to mucosal destruction. In cattle, *Eimeria* infections are thought to have prolonged adverse health effects resulting in a significant reduction of animal performance in terms of retarded growth and fertility (Lassen and Ostergaard 2012). Furthermore, *Eimeria* infections have been found to facilitate dysbacteriosis and secondary bacterial infections in calves (Stockdale 1977). In chickens, necrotic enteritis caused by pathogenic *C. perfringens* strains is promoted by *Eimeria* spp. infections (Van Immerseel et al. 2004). Also coinfections with other bacterial pathogens, such as *Listeria monocytogenes*, or enteral viruses and infectious bronchitis virus are promoted (Dhama et al. 2013; Koo et al. 2013). Thus, animal health is affected in a number of different ways during and following *Eimeria* infections. Due to its high contagiosity, eimeriosis has to be regarded as a herd disease rather than a disease of individual animals.

3.3.1.2 Pathology

The pathology of *Eimeria* spp. infections is species-specific and generally dose-dependent. It strongly depends on the replication rate of the species, induced tissue inflammation, predilection site, host immune response, and additional factors not yet completely understood. On one hand, typical pathological findings of intestinal eimeriosis are lesions partially produced by direct host cell damage of multiplying *Eimeria* stages visible in epithelial desquamation (Kouwenhoven and van der Horst 1973; Michael and Hodges 1971). On the other hand, the massive inflammatory reaction in the intestinal mucosa comprising of vasodilatation, edema, lymphocellular, and granulocytic infiltration is equally important for the observed pathology (Michael and Hodges 1971; Lillehoj and Trout 1996). Accordingly, pathological findings for liver coccidiosis by *E. stiedae* (host, rabbit) are characterized by hepatocyte necrosis and inflammatory cell invasion, while renal eimeriosis caused by *E. truncata* (host, goose) results in tubular cell destruction and interstitial inflammatory cell accumulation (Oksanen 1994; Jing et al. 2016)).

Furthermore, the predilection site of parasite development varies between *Eimeria* spp. resulting in species-specific manifestations of the infection. In poultry, asexual and sexual replication steps are principally limited to a restricted intestinal location. In contrast, the different developmental stages of many *Eimeria* spp. parasitizing ruminants tend to develop in different intestinal locations. Early asexual stages are mainly located in the small intestine and the upper colon, while later endogenous stages are detected in the colon, cecum, and rectum.

There is limited knowledge on *Eimeria* virulence factors. Important candidates are represented by ROP proteins—i.e., proteins that are involved in host cell colonization, remodeling, and metabolic modulation of the host cell. Other virulence factors may be represented by cathepsins—cysteine proteases related to papain-like enzymes—that have been reported as virulence factors in other Apicomplexa, and of which eimeripain is an important representative in *E. tenella* (Matsubayashi et al. 2014).

Pathology of *Eimeria* infection is strongly influenced by interactions with other pathogens. A general interaction between bacterial flora and *Eimeria*-induced pathology has been reported (Baba et al. 1990). In chicken eimeriosis, a mutual aggravation of pathological findings in coinfections with *Salmonella* spp. and *Clostridium perfringens*—both of which are common in commercial rearing systems—has been revealed (Chapman et al. 2002). Mixed infections of *Eimeria* with toxin-producing strains of *C. perfringens* are known to be responsible for necrotic enteritis in chickens and other avian and mammalian hosts. Also for mixed infections of *Eimeria* spp. with *Escherichia coli*, evidence for a synergistic pathological effect has been demonstrated with regard to the manifestation of intestinal lesions (Hegazy et al. 1999). Interestingly, a mutual pathological interaction is not limited to concurrent infections of *Eimeria* with other intestinal agents but has also been demonstrated for infectious bursal disease virus (IBDV) and is assumed for *Clostridium* spp. skin infections (Giambrone et al. 1977; Kabell et al. 2006; Li et al. 2010). Also coinfections of *Eimeria* with other endo- or ectoparasites are common—e.g., with roundworms, cestodes, and fleas—but no reliable data on the impact of these infections on the course of eimeriosis and vice versa are available (Thekisoe et al. 2003; Permin et al. 2002).

3.3.2 Farm Animals of Economic Importance

3.3.2.1 Chicken

E. necatrix and *E. tenella* represent the clinically and economically most important species both causing hemorrhagic enteritis. Low-level infections with other species like *E. acervulina*, *E. maxima*, and *E. brunetti* are highly common but remain usually subclinical (McDougald 1998).

Of the nine species that have been reported to infect chickens, each species prefers a certain intestinal section for endogenous development. In the upper intestine, mainly *E. acervulina*, *E. necatrix*, and *E. praecox* are present. In the midgut, *E. maxima* and *E. mitis* prevail. *E. tenella* and *E. brunetti* are typically found in the lower intestine—i.e., ileum, ceca, and rectum (Johnson and Reid 1970). Correspondingly, each *Eimeria* sp. provokes typical lesions in defined intestinal locations (Fig. 3.8):

E. acervulina Presence of a white, fibrous, striated mucosa in the small intestine.

E. tenella Shows hemorrhagic typhlitis.

E. brunetti Exhibition of necrotic lesions in the ileum, cecum, and rectum.

E. necatrix There are pinpoints to confluent hemorrhagic lesions in the jejunum and ileum—referred to as yolk stalk region—and the lumen is filled with a bright orange, mucoid material (Ryley et al. 1972).

E. praecox Display of catarrhal inflammation of the mucosa in the duodenum and upper jejunum, with reddening, and milky exudate (Allen and Jenkins 2010).

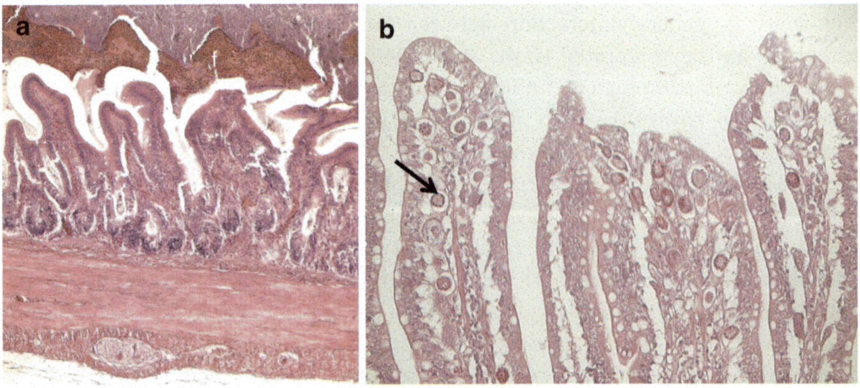

Fig. 3.8 Histopathological findings in *Eimeria* sp. infection in a chick. (**a**) Fibrinoid enteritis; f, fibrinous casts on epithelial surface; (**b**) unicellular macrogamont with wall-forming bodies lying in the host cell cytoplasm of mucosal cell (*arrow*)

E. maxima A ballooning is observed mainly for the midgut region. Petechiae are visible in thickened intestinal walls, and the intestinal content contains blood clots (Johnson and Reid 1970).

E. mitis No defined lesions are seen, but a general ballooning of the intestinal wall and an educed pigmentation and fluid accumulation are observed (Watkins et al. 1990).

In clinical eimeriosis diarrhea varies between pasty, watery, and hemorrhagic (McDougald 1998). A reduced general behavior, markedly reduced growth, a poor feed conversion rate, and cases of death are observed (Lillehoj and Trout 1996). For each *Eimeria* species, mortality rates vary greatly as they depend on the infecting field strain and are modulated by the infection dose, stress factors, feed quality, and concurrent disease (McDougald 1998; Hegazy et al. 1999; Lobago et al. 2005; Toulah 2007; Abu-Akkada and Awad 2012).

Different chicken breeds show different chromosome 1-encoded susceptibility to *Eimeria*-induced pathological changes and associated clinical eimeriosis (Pinard-Van Der Laan et al. 1998; Zhu et al. 2003; Kim et al. 2006). Especially widespread meat-type broiler chicken lines are prone to massive pathology (Pinard-Van der Laan et al. 1998). Currently it is assumed that the pro-inflammatory innate immune response confers protection against disease. Accordingly, for *E. tenella* a marked negative correlation between a strong innate immune response—characterized by a high basal IL-1β, IL-6, and IL-8 level—and the occurrence of intestinal lesions has been observed (Swaggerty et al. 2011). Mainly susceptible breeds exhibit a strong intestinal pathology that may result in substantial mortality.

3.3.2.2 Cattle

Exclusively three species are related to clinical disease—namely, *E. bovis*, *E. zuernii*, and *E. alabamensis*. *E. bovis* and *E. zuernii* are developing in the cytoplasm of the respective infected host cell. In contrast, *E. alabamensis* undergoes all endogenous stages until oocyst formation inside the host cell nuclei (Davis et al. 1957). Especially mixed infections with these species lead to pronounced pathology and clinical disease (Stromberg et al. 1986). However, often *E. bovis* and *E. zuernii* are also seen as single pathogens. Disease due to other *Eimeria* spp. has been sporadically reported, but such infections remain generally subclinical.

The clinical effect and pathology of *E. alabamensis* infection are clearly dose-dependent and are thus promoted by factors like spring turnout of calves to contaminated pasture (Hooshmand-Rad et al. 1994). Low-level infections may remain subclinical but long-term effects on performance are observed. Massive infections lead to severe watery diarrhea and reduced weight gain, and a general enteritis has been observed in the lower half of the small intestine. Furthermore, an extensive epithelial destruction accompanied by leukocytic infiltration and villi destruction occurs (Davis et al. 1957).

Infection with *E. bovis* and/or *E. zuernii* results in similar clinical symptoms. Severe diarrhea starts at the end of the prepatent period (Stockdale et al. 1982), its character varying from mucous to watery and hemorrhagic, often with lots of fibrin and tissue strands (Fig. 3.9, Daugschies et al. 1986). Calves show inappetence and

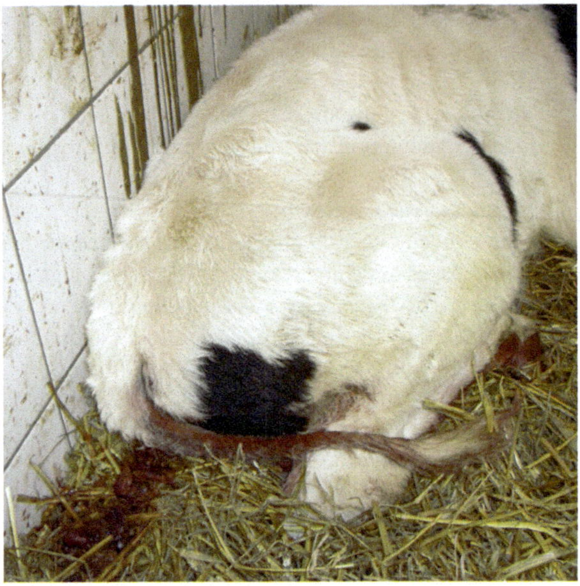

Fig. 3.9 Clinical disease in an *E. bovis*-infected calf. The animal displayed hemorrhagic diarrhea, anorexia, and reluctance to move

marked weight depression, which may result in long-term decrease of performance. Rarely observed is the phenomenon of *bovine nervous coccidiosis* caused by infection of either *E. bovis* or *E. zuernii* (Julian et al. 1976). The actual mechanism of this pathologic symptom remains unclear, though underlying toxin production and electrolyte imbalances are discussed (Radostits and Stockdale 1980). In Northern America there have been reports of temporal clustering in winter.

Pathology caused by *E. bovis* is characterized by early meront invasion of the posterior half of the small intestine (Hammond et al. 1946). However, the second meront and the gamont generations invade undifferentiated stem cells of large intestinal crypts in the cecum and colon (Hammond et al. 1963; Gregory 1982). This leads to epithelial destruction and a long regeneration period. First-generation meronts are macroscopically visible because of their large size of up to 435 μm. They are referred to as macromeronts and contain up to 120,000 merozoites per meront (Hammond et al. 1946; Hamid et al. 2015). Interestingly, these stages are not found in the epithelium but in central lacteals of the villi and are observed up to 20 days after infection. The consequence of this distinctive asexual replication is profound tissue destruction associated with subepithelial parasite development mainly in the ileum (Friend and Stockdale 1980). Subsequently, second-generation meront as well as gamont development in the colon and cecum confers hemorrhagic and fibrinoid typhlitis and colitis with severe thickening of the cecal and colon walls (Fig. 3.10). Histopathologically, the epithelium is lost completely, and the denuded lamina propria is infiltrated with leukocytes and covered by fibrin strands.

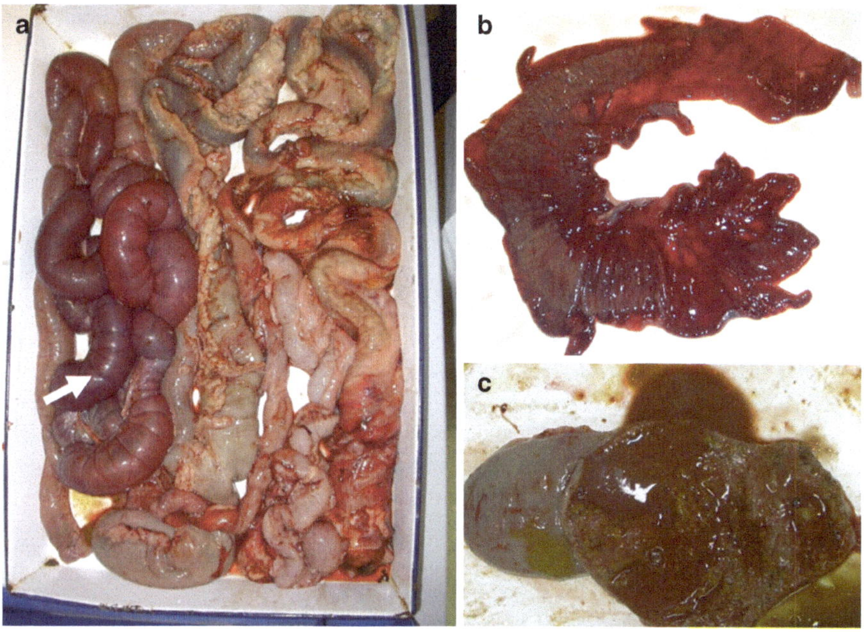

Fig. 3.10 Gross pathological findings in an *E. bovis*-infected calf. (**a**) Intestine with severe ballooning and reddening of the proximal colon (*arrow*), (**b**) hemorrhagic colitis, (**c**) hemorrhagic typhlitis with massive thickening and edema of the cecal wall

The species *E. zuernii* reveals most extensive changes 21 days after infection. Asexual replication causes only little pathological alterations, which are observed in the last 3 m of the small intestine. Also second-generation meronts cause little damage in the large intestine (Stockdale 1977). However, the sexual stages are responsible for severe damage in the large intestine, mainly proximal to the colon and cecum. Intestinal walls are thickened by edema of the submucosa and an extensive epithelial loss, and fibrin cast is present in high-dose infected animals. The intestinal content is bright or dark red because of severe hemorrhages. In less severe *E. zuernii* infections, moderate lesions are seen which are repaired within 1 or 2 weeks after the onset of oocyst excretion (Mundt et al. 2005).

3.3.3 Other Hosts

3.3.3.1 Sheep and Goat

In sheep eimeriosis, a dose-dependent subclinical to diarrheic disease—manifested by liquid feces and eventually fibrinous casts—is recognized though no linear relation between infection dose and induced clinical disease has been observed (Gregory and Catchipole 1990). In goat kids, infections with pathogenic species like *E. ninakohlyakimovae* provoke diarrhea of hemorrhagic character, recumbency, and anorexia at high infection doses resulting in a bad general health constitution (Ruiz et al. 2014).

Susceptibility to intestinal damage is predominantly observed at the age of 4–5 weeks (Foreyt 1990). In sheep, *E. ovinoidalis* infections cause severe enteritis that may lead to death. During early merogony, macromeronts are macroscopically visible in the lower small intestine, and in massive infections, their development leads to diphtheroid to hemorrhagic jejunitis with villus and crypt atrophy (Gregory and Catchpole 1987). Later parasitic stages may produce hemorrhagic typhlitis. After completion of parasite development, a prolonged tissue repair is seen. *E. crandallis* infection in lambs shows a similar pathology like *E. ovinoidalis* with the cecum and colon being most affected (Gregory and Catchipole 1990). Pathological findings in goat kids, infected by *E. ninakohlyakimovae* or *E. caprina*, are comparable to lesions that have been described in sheep (Taylor and Catchpole 1994). No lesions are observed in the liver, pancreas, or other organs (Hashemnia et al. 2012). In contrast to chickens, a relation between breed of sheep with the number of oocysts excreted and the severity of clinical disease cannot be found in young lambs (Reeg et al. 2005).

3.3.3.2 Rabbit

The infection with *E. stiedae*—causing hepatic eimeriosis—is highly pathogenic as manifested by diarrhea, weakness, severely reduced general behavior, and spontaneous cases of death (Jing et al. 2016). Intrahepatic *E. stiedae* replication in hepatobiliary duct epithelium leads to secondary damage of hepatocytes. Hepatic parenchymal cells show extensive necrosis; connective tissue hyperplasia and leukocyte invasion occur (Hanada et al. 2003; Jing et al. 2016). Macroscopically, the liver is greatly enlarged and shows a gray and white appearance. Liver consistency is hard, and many white nodules and nodes, which are foci of infection and variable in size, as well as diffuse lesions are present (Jing et al. 2016). Inflammatory reaction contributes to physical damage of hepatocytes finally resulting in globally increased liver enzyme levels of alanine aminotranferase (ALT), glutamine transferase (GGT), aspartate aminotransferase (AST), and lactate dehydrogenase (LDH). Although there is no intestinal replication, secondary lesions in the duodenum are found manifesting as catarrhal to hemorrhagic enteritis. Due to liver malfunction, systemic disease including enteritis is seen.

In contrast to hepatic eimeriosis, intestinal eimeriosis caused by *E. magna*, *E. irresidua*, *E. flavescens*, *E. piriformis*, and *E. intestinalis* often remains subclinical but can be pathogenic inflicting catarrhal enteritis as manifested by diarrhea and ballooning.

3.3.3.3 Turkeys

Of altogether five recognized species, *E. adenoeides*, *E. meleagrimitis*, and *E. gallopavonis* are considered to be highly pathogenic causing significant mortality in susceptible young poults and are responsible for reduced weight gain and bad performance (Hein 1969; Chapman 2008; Vrba and Pakandl 2014). In these pathogenic species, severe enteritis with formation of intestinal lesions is observed 5–7 days after infection. In *E. meleagrimitis* infection, catarrhal enteritis with watery intestinal contents develops, while in *E. gallopavonis* also hemorrhagic content is found in the ileum accompanied by high mortality (Vrba and Pakandl 2014). In the case of *E. adenoeides*, edema and petechial hemorrhages of the intestinal wall have been described (Chapman 2008).

Of the remaining two species, *E. meleagridis* has been described as nonpathogenic to mildly pathogenic, whereas *E. dispersa* is regarded as nonpathogenic causing transient enteritis for 1–2 days. All five *Eimeria* spp. display a site-specific development in the intestine inducing specific localized lesions. The three species *E. meleagrimitis*, E. *meleagridis*, and *E. dispersa* show asexual replication in the small intestine. Later stages—late asexual meronts and gamonts—tend to be found throughout the gut, with the main proportion developing in the large intestine. The remaining two species—*E. adenoeides* and *E. gallopavonis*—start their asexual replication in the ileum, rectum, and ceca, where later also their endogenous development takes place (Chapman 2008).

3.3.3.4 Waterfowl

Renal eimeriosis of geese and ducks may result in renal failure and eventual death (Oksanen 1994; Skírnisson 1997). In geese, especially *E. truncata* infections are known to entail cases of death due to renal eimeriosis (Oksanen 1994). *E. somateriae* infection of ducks causes great enlargement of kidneys, which show the presence of white or yellowish nodules (Skírnisson 1997). However, severe disease in ducks during renal eimeriosis is regarded to be of multifactorial genesis (Nation and Wobeser 1977; Skírnisson 1997).

Intestinal eimeriosis seems to be only important in geese, occasionally causing mortality in goslings (Oksanen 1994). It is caused by infection with *E. nocens* and is associated with acute enteritis located from duodenum to cloaca including ceca. Most affected by hemorrhagic lesions are the ileum and up- and downstream regions (Dai et al. 2005). Interestingly, the species *E. mulari* causes much more pronounced intestinal lesions in ducks of the genus *Anas* compared to those of the genus *Cairina* or mule ducks (Sercy et al. 1996).

3.3.3.5 Hog

Most infections are not related with any clinical symptoms (Daugschies et al. 2004). As pathogenicity of common species ranges between nonpathogenic and mildly pathogenic, infections remain widely unrecognized on endemic farms. However, if high infection doses are ingested or naïve animals are moved in from other stables resulting in coinfections with other intestinal pathogens, pathogenicity and cases of death may be observed (Gaudie et al. 2005). Acute disease is manifested by catarrhal pseudomembranous jejunitis and ileitis and can be induced by different *Eimeria* spp. (Vítovec et al. 1987; Gaudie et al. 2005). As to now, little is known about the relation between porcine *Eimeria* infections and long-term performance.

3.3.3.6 Reptiles

Little is known about pathogenicity of different *Choleoeimeria*, *Acroeimeria*, and *Eimeria* spp. in reptilian hosts and their clinical impact (see Sect. 3.2.2). A varying degree of pathogenicity may be assumed though data are scarce. *Choleoeimeria* infections are characterized by intraepithelial meronts and gamonts causing host cell remodeling—attachment to the basal membrane through a thin pedicle—hypertrophy, and detachment (Abdel-Haleem et al. 2016). Macroscopically, the gallbladder wall is thickened, and gallstones and debris are accumulating in the gallbladder (Szczepaniak et al. 2016). In contrast, in intestinal *Acroeimeria* or *Eimeria* spp., no gross alterations are seen providing little evidence for pathological changes (Lainson 2002).

3.4 Prevention, Control, and Treatment

On one hand, prevention is based on the reduction of the infection pressure by cleaning and disinfection measures, vaccination, and preventive drug treatment. On the other hand, it comprises measures to increase animal resistance by introducing measures to avoid stress and the use of resistant breeds.

3.4.1 Management

Animals susceptible to infection should be kept in clean houses and barns or on pastures without direct contact to animals that shed *Eimeria* oocysts. The quick sporulation of excreted oocysts should be taken into account by executing an all-in-all-out management for farm animals allowing for regular change of litter and hygienic measures. In addition, caretakers and veterinarians should change clothing to avoid mechanical transmission between herds or animal groups within a herd. Especially in broiler breeding, it is recommended to adhere strictly to biosecurity measures (Allen and Fetterer 2002). As eimeriosis is often aggravated by stress factors or secondary infections, animal husbandry should be optimized with regard to animal welfare including climate, nutrition, crowding or size of animal groups, and implementation of a general animal health prophylaxis—hygiene, vaccination against common pathogens, and screening of the herd health and herd immunity. Certain feed compositions are assumed to promote clinical eimeriosis—as, for example, high-protein diets in broilers—and should be avoided (Britton et al. 1964). Furthermore, different breeds of the same host species are known to exhibit different susceptibilities to clinical disease after an *Eimeria* infection by varying immune reactions (see Sects. 3.1.3 and 3.3.2.1). Thus, the selection and use of clinically resistant breeds can support prevention of eimeriosis on the farm.

Interestingly, in mammals, administration of colostrum is thought to confer a basal protection against early infections and coinfections with pathogens. However, in studies in sheep, such a protective effect of colostrum could not be demonstrated (Gregory and Catchipole 1990).

3.4.2 Cleaning and Disinfection

Removal and inactivation of sporulated oocysts from the environment are highly important to prevent eimeriosis. Correspondingly, it has been shown that regular cleaning of water troughs of livestock reduces the infection pressure and subclinical coccidiosis (Mitchell et al. 2012). However, limited other possibilities exist to eliminate and/or inactivate sporulated oocysts in free-ranging poultry and husbandry of farm animals on pasture.

Oocysts are highly resistant to chemical compounds commonly used for disinfection of animal housings. Although, chlorocresols are effective their use is prohibited in many countries (Shahiduzzaman et al. 2010). In addition, ammonium hydroxide and ammonia, carbon disulfide compounds, and hydrogen peroxide are active against oocysts if applied at high concentrations, and it has been also shown that ozone treatment inactivates oocysts (Hilbrich 1975, Williams 1997; Liou et al. 2002). In contrast, other common disinfectants like formaldehyde do not sufficiently inactivate sporulated *Eimeria* spp. oocysts (Blandino 1985).

Physical treatment by radioactive irradiation or with heat or freezing has been proposed to be highly effective in inactivating *Eimeria* oocysts, yet they are not of practical use in contaminated areas (Lee and Lee 2001). Importantly, UV light is highly efficient in oocyst inactivation and reduces the presence of infective stages on pasture particularly during the summer season (Farr and Wehr 1949; Marquardt 1960).

3.4.3 Vaccination

So far, vaccines are only available for chickens. Due to the lack of cross-immunity, vaccination has to be performed against every single *Eimeria* sp. suspected to contribute to eimeriosis on a given poultry farm. First- and second-generation vaccines as well as a third-generation vaccine are commercially available. First-generation vaccines contain a virulent *Eimeria* strain administered in a very low dose leading to a mild eimeriosis and protective immunity (Chapman et al. 2002; Blake and Tomley 2014). Second-generation attenuated vaccines are based on precocious *Eimeria* strains characterized by a reduced replication rate resulting in a lower number of merogonies and less intestinal damage as compared to the virulent wild-type strains (Chapman et al. 2002). In first- as well as second-generation vaccines, a protective immunity develops in response to antigens of several endogenous stages.

Currently, third-generation vaccines or subunit vaccines are under development. They either contain purified *Eimeria* or recombinant antigen, which may be combined with immunomodulators like IFN-γ (Lillehoj et al. 2000; Wallach et al. 2008). Promising candidate proteins are early stage antigens as represented by EtMIC2, merozoite, profilin, and rhomboid protein (Ding et al. 2005; Lillehoj et al. 2000; Li et al. 2012; Liu et al. 2013). In addition, gamogony proteins have been shown to be highly immunogenic in breeding layers (Wallach et al. 2008; Wallach 2010). It is assumed that in the field several antigens need to be combined for optimal protection and to avoid generation of resistances making a third-generation vaccine more elaborative and more expensive, limiting its use (Blake and Tomley 2014). A subunit vaccine against *E. maxima* based on a gametocyte antigen has been developed to a commercial product but appears not to be marketed anymore. Generally, genetic and antigenic diversity within a single *Eimeria* species are considered important in the future design of globally effective vaccines (Clark et al. 2017). The development of resistance against vaccination due to common polyclonal infections and cross-fertilization is discussed though it has not been observed so far (Blake and Tomley 2014; Clark et al. 2017).

3.4.4 Feed Additives

In order to prevent parasite multiplication, clinical disease, and subclinical losses, anticoccidial drugs are applied for several weeks as a long-term treatment via feed while animals are under infection risk. Thus, they are used prophylactically and metaphylactically—i.e., starting before infection and covering the potential infection period. Feed additives are used to prevent coccidiosis from whole herds irrespective of the presence of infection pressure in single animal groups.

Use of feed additives to prevent eimeriosis is a widespread tool in farm animal rearing—especially in poultry hatchery and feedlot cattle. There are several classes of chemical anticoccidial compounds available on the international market. They are grouped into synthetic, polyether antibiotics—ionophores—and mixed products (Peek and Landman 2011). The mode of action varies from coccidiostatic to coccidiocidal. Common targets of anticoccidials are cofactor synthesis, mitochondria, and cell membrane. Cofactor synthesis blockers are most effective during asexual replication. There are different substances acting by folate antagonism—e.g., sulfonamides or pyrimethamine—or as thiamine analogues, e.g., amprolium. Mitochondrial activity is impaired by quinolone drugs—e.g., decoquinate—pyridines (e.g., meticlorpindol), nicarbazin, or robenidine. Polyether ionophores impair cell membrane function and osmotic pressure predominantly damaging extracellular parasitic stages. A special mode of action is observed in the triazine drug toltrazuril, which develops coccidocidal activity by irreversibly damaging mitochondria and plastid-like organelles (Hackstein et al. 1995). In some efficacious drugs like halofuginone and diclazuril—targeting early meronts and a nucleoside analogue involved in failure of sexual replication—the exact mode of action is still unclear.

Resistance may develop against any of available drugs. In chicken under intensive rearing conditions, there are many *Eimeria* spp. resistances against all known long-term and widely used anticoccidials (Chapman and Jeffers 2014). In other hosts, little is known about resistance development because of the lack of investigation. In general, drug resistance is promoted by continuous and low-dose use of an anticoccidial compound.

To avoid the development of drug resistance in a flock, the applied anticoccidials should be changed regularly by introducing another anticoccidial with a different mode of action and thus shifting the selection pressure. There are two major forms of planned regular drug change—namely, rotation and shuttle programs. In the rotation system, drugs are changed between different successive animal groups. In a shuttle program, different drugs are used in the same animals during different rearing phases. These methods slow down development of parasite resistance but do not prevent it as attested by the presence of many multi-resistant *Eimeria* spp. strains that are seen in the field (Peek and Landman 2003). In poultry farms, drug resistance may be reversed by implementation of a mixed chemotherapy-vaccine program. Feed additives and live vaccine can be used in successive flocks in integrated control programs. The invention of drug-sensitive *Eimeria* strains antagonizing resistant field strains present in the flock may lead to a drug-susceptible *Eimeria* population after several cycles (Chapman and Jeffers 2014).

In addition to feed-additive drugs, also phytochemicals or the feeding of certain plant particles has been shown to have a beneficial effect on clinical eimeriosis and oocyst excretion. However, the anticoccidial potential of natural products is often much lower than that of commercial drugs. Consequently, studies show a limited efficacy of essential oils and extracts of a broad range of herbs and of traditional medicinal plants, respectively, against *Eimeria* spp. replication (Schreiner 2014; Alnassan et al. 2015; Pérez-Fonseca et al. 2016).

3.4.5 Treatment

Typically, in eimeriosis-diseased flocks, a targeted treatment of a defined age group is carried out to improve animal welfare and avoid economic losses. If implemented early enough—which would usually correspond to the prepatency period—this measure lowers infection pressure by suppressing oocyst excretion and environmental contamination. The choice of time point for drug application is crucial to render treatment effective. Medication should be best applied as early metaphylaxis, or latest, when first animals start oocyst excretion in the respective group under treatment (Epe et al. 2005; Ruiz et al. 2012; Enemark et al. 2015). The major challenge in scheduling metaphylactic treatment is the determination of the optimal time point because during prepatency, in vivo diagnosis of the current state of infection is of no use since at this early time point of infection, neither an *Eimeria* infection can be detected nor the endogenous parasitic phase determined. Hence, herd monitoring over several production cycles is indicated to schedule treatment accurately. As a rule of thumb, animals should be treated during the late asexual parasite replication. In the case of *Eimeria* infecting chickens, this corresponds to about 3 days and of *Eimeria* infecting cattle to approximately 14 days after exposure of animals to a contaminated environment—i.e., after stabling, beginning of the grazing period, etc.

Anticoccidials used for treatment of a defined number of animals at risk are represented by the triazines toltrazuril applied to cattle, chickens, sheep, etc. and diclazuril administered to cattle or sheep. Both drugs have to be administered only once, and the optimal time point of their administration is when the majority of infected animals of the treated group are in the prepatent period of infection.

In general, therapeutic treatment during the patency period is not recommended since endogenous *Eimeria* stages have already set lesions and the self-limiting character of the disease makes treatment at this time point obsolete. Nonetheless, sulfonamides are widely used for therapeutic treatment against eimeriosis of cattle, rabbit, etc.; however, the anticoccidial benefit may be limited since the long administration period potentially overlaps with the natural end of patency.

References

Abdel-Haleem HM, Al-Quraishy S, Abdel-Baki AA. Redescription and new locality record of *Choleoeimeria turcicus* from the gallbladder of *Hemidactylus turcicus* in Egypt, with notes on its endogenous development. Parasitol Res. 2016;115(5):2017–21. https://doi.org/10.1007/s00436-016-4945-1.

Abu-Akkada SS, Awad AM. Isolation, propagation, identification and comparative pathogenicity of five Egyptian field strains of *Eimeria tenella* from broiler chickens in five different provinces in Egypt. Res Vet Sci. 2012;92:92–5. https://doi.org/10.1016/j.rvsc.2010.10.023.

Allen PC, Fetterer RH. Recent advances in biology and immunobiology of *Eimeria* species and in diagnosis and control of infection with these coccidian parasites of poultry. Clin Microbiol Rev. 2002;15(1):58–65.

Allen PC, Jenkins MC. Observations on the gross pathology of *Eimeria praecox* infections in chickens. Avian Dis. 2010;54(2):834–40.

Alnassan AA, Kotsch M, Shehata AA, Krüger M, Daugschies A, Bangoura B. Necrotic enteritis in chickens: development of a straightforward disease model system. Vet Rec. 2014;174(22):555–61.

Alnassan AA, Thabet A, Daugschies A, Bangoura B. *In vitro* efficacy of allicin on chicken *Eimeria tenella* sporozoites. Parasitol Res. 2015;114(10):3913–5. https://doi.org/10.1007/s00436-015-4637-2.

Al-Natour MQ, Suleiman MM, Abo-Shehada MN. Flock-level prevalence of *Eimeria* species among broiler chicks in northern Jordan. Prev Vet Med. 2002;53(4):305–10.

Anderson WI, Giambrone JJ, Fletcher OJ Jr, Eidson CS, Reid WM. Demonstration of *Eimeria tenella* in bursa of Fabricius of chickens. Avian Dis. 1976;20(4):752–5.

Baba E, Kawamoto K, Fukata T, Arakawa A. Possibilities of bacteremia and toxemia in death of chickens infected with *Eimeria* tenella. Avian Dis. 1990;34:566–70.

Balicka-Ramisz A, Pilarczyk B. Occurrence of coccidia infection in pigeons in amateur husbandry. Diagnosis and prevention. Ann Parasitol. 2014;60(2):93–7.

Bangoura B, Daugschies A. Parasitological and clinical parameters of experimental *Eimeria zuernii* infection in calves and influence on weight gain and haemogram. Parasitol Res. 2007;100(6):1331–40.

Bangoura B, Mundt HC, Schmäschke R, Westphal B, Daugschies A. Prevalence of *Eimeria bovis* and *Eimeria zuernii* in German cattle herds and factors influencing oocyst excretion. Parasitol Res. 2012;110(2):875–81. https://doi.org/10.1007/s00436-011-2569-z.

Barker IK, Remmler O. Experimental *Eimeria leuckarti* infection in ponies. Vet Rec. 1970;86(15):448.

Beelitz P, Göbel E, Gothe R. Endoparasites of donkeys and horses kept in communal housing in upper Bavaria; species spectrum and incidence. Tierarztl Prax. 1996;24(5):471–5.

Benbrook EA, Sloss MW. Coccidial oocysts (*Eimeria leuckarti*) from north American donkeys (*Asinus asinus*). J Am Vet Med Assoc. 1962;140:817–8.

Blake DP, Tomley FM. Securing poultry production from the ever-present *Eimeria* challenge. Trends Parasitol. 2014;30:12–9. https://doi.org/10.1016/j.pt.2013.10.003.

Blandino T. Assessment of formaldehyde effectiveness on *Eimeria zurnii* oocysts under simulated field conditions. Rvsta Cub Cien Vet. 1985;16(1):67–71.

Britton WM, Hill CH, Barber CW. A mechanism of interaction between dietary protein levels and coccidiosis in chicks. J Nutr. 1964;82:306–10.
Catchpole J, Norton CC, Gregory MW. Immunisation of lambs against coccidiosis. Vet Rec. 1993;132(3):56–9.
Chapman HD. Coccidiosis in the turkey. Avian Pathol. 2008;37(3):205–23. https://doi.org/10.1080/03079450802050689.
Chapman HD, Jeffers TK. Vaccination of chickens against coccidiosis ameliorates drug resistance in commercial poultry production. Int J Parasitol Drugs Drug Resist. 2014;4(3):214–7. https://doi.org/10.1016/j.ijpddr.2014.10.002.
Chapman HD, Cherry TE, Danforth HD, Richards G, Shirley MW, Williams RB. Sustainable coccidiosis control in poultry production: the role of live vaccines. Int J Parasitol. 2002;32:617–29.
Chauve CM, Gastellu J, Reynaud MC. Pathogenicity of *Eimeria mulardi* Chauve et al., 1994 in experimentally infected mule-ducks. Precision on pathological stages. Avian Pathol. 1994;23(3):557–68.
Chengat Prakashbabu B, Thenmozhi V, Limon G, Kundu K, Kumar S, Garg R, Clark EL, Srinivasa Rao AS, Raj DG, Raman M, Banerjee PS, Tomley FM, Guitian J, Blake DP. *Eimeria* species occurrence varies between geographic regions and poultry production systems and may influence parasite genetic diversity. Vet Parasitol. 2017;233:62–72. https://doi.org/10.1016/j.vetpar.2016.12.003.
Clark EL, Tomley FM, Blake DP. Are *Eimeria* genetically diverse, and does it matter? Trends Parasitol. 2017;33(3):231–41. https://doi.org/10.1016/j.pt.2016.08.007.
Clarkson MJ, Gentles MA. Coccidiosis in turkeys. Vet Rec. 1958;70:211–4.
Collier CT, Hofacre CL, Payne AM, Anderson DB, Kaiser P, Mackie RI, Gaskins HR. Coccidia-induced mucogenesis promotes the onset of necrotic enteritis by supporting *Clostridium perfringens* growth. Vet Immunol Immunopathol. 2008;122:104–15.
Constantinoiu CC, Molloy JB, Jorgensen WK, Coleman GT. Development and validation of an ELISA for detecting antibodies to *Eimeria tenella* in chickens. Vet Parasitol. 2007;150(4):306–13.
Cornelissen AWCA, Verstegen R, van den Brand H, Perie NM, Eysker M, Lam TJGM, Pijpers A. An observational study of *Eimeria* species in housed cattle on Dutch dairy farms. Vet Parasitol. 1995;56:7–16.
Courtney CH, Ernst JV, Benz GW. Redescription of oocysts of the bovine coccidia *Eimeria bukidnonensis* Tubangui 1931 and *E. wyomingensis* Huizinga and winger 1942. J Parasitol. 1976;62(3):372–6.
Cowper B, Matthews S, Tomley F. The molecular basis for the distinct host and tissue tropisms of coccidian parasites. Mol Biochem Parasitol. 2012;186(1):1–10. https://doi.org/10.1016/j.molbiopara.2012.08.007.
Dai Y, Liu M, Tao JP, Li WL, Ding C, Liu XX. Survey on coccidia of domestic geese and their experimental infections. J Xiamen Univ (Nat Sci). 2004;42:228–33.
Dai Y, Liu X, Liu M, Tao J. The life cycle and pathogenicity of coccicium *Eimeria nocens* (Kotlán, 1933) in domestic goslings. J Parasitol. 2005;91(5):1122–6.
Daszak P. Zoite migration during *Eimeria tenella* infection: parasite adaptation to host defences. Parasitol Today. 1999;15:67–72.
Daugschies A, Najdrowski M. Eimeriosis in cattle: current understanding. J Vet Med B Infect Dis Vet Public Health. 2005;52(10):417–27.
Daugschies A, Akimaru M, Bürger HJ. Experimental *Eimeria bovis* infections in the calf: 1. Parasitologic and clinical findings. Dtsch Tierarztl Wochenschr. 1986;93(9):393–7.
Daugschies A, Imarom S, Ganter M, Bollwahn W. Prevalence of *Eimeria* spp. in sows at piglet-producing farms in Germany. J Vet Med B Infect Dis Vet Public Health. 2004;51(3):135–9.
Daugschies A, Bangoura B, Lendner M. Inactivation of exogenous endoparasite stages by chemical disinfectants: current state and perspectives. Parasitol Res. 2013;112(3):917–32. https://doi.org/10.1007/s00436-013-3324-4.
Davis LR, Bowman GW, Boughton DC. The endogenous development of *Eimeria alabamensis* Christensen, 1941, an intranuclear coccidium of cattle. J Protozool. 1957;4(4):219–25.

Dhama K, Verma AK, Rajagunalan S, Kumar A, Tiwari R, Chakraborty S, Kumar R. *Listeria monocytogenes* infection in poultry and its public health importance with special reference to food borne zoonoses. Pak J Biol Sci. 2013;16:301–8.

Ding X, Lillehoj HS, Dalloul RA, Min W, Sato T, Yasuda A, Lillehoj EP. *In ovo* vaccination with the *Eimeria tenella* EtMIC2 gene induces protective immunity against coccidiosis. Vaccine. 2005;23:3733–40.

Doran DJ, Farr MM. Excystation of the poultry coccidium, *Eimeria acervulina*. J Protozoo. 1962;l9:154–61.

Dubremetz JF, Elsner YY. Ultrastructural study of schizogony of *Eimeria bovis* in cell cultures. J Protozool. 1979;26(3):367–76.

Ellis PA, Wright AE. Coccidiosis in guinea-pigs. J Clin Pathol. 1961;14(4):394–6.

Enemark HL, Dahl J, Enemark JM. Significance of timing on effect of Metaphylactic Toltrazuril treatment against Eimeriosis in calves. Parasitol Res. 2015;114(Suppl 1):S201–12. https://doi.org/10.1007/s00436-015-4526-8.

Epe C, von Samson-Himmelstjerna G, Wirtherle N, von der Heyden V, Welz C, Beening J, Radeloff I, Hellmann K, Schnieder T, Krieger K. Efficacy of toltrazuril as a metaphylactic and therapeutic treatment of coccidiosis in first-year grazing calves. Parasitol Res. 2005;97(Suppl 1):S127–33.

Ernst JV, Benz GW. Coccidiosis. Curr Top Vet Med Anim Sci. 1981;6:377–92.

Faber JE, Kollmann D, Heise A, Bauer C, Failing K, Bürger HJ, Zahner H. *Eimeria* infections in cows in the periparturient phase and their calves: oocyst excretion and levels of specific serum and colostrum antibodies. Vet Parasitol. 2002;104(1):1–17.

Farr MM, Wehr EE. Survival of *Eimeria acervulina*, *E. tenella* and *E. maxima* oocysts on soil under various field conditions. Ann N Y Acad Sci. 1949;52:468–72.

Fernandez I, Benito J, Pardos F. Ultrastructure of the oocyst wall formation in *Eimeria beauchampi* leger and Duboscq, 1917, a coccidian parasite of *Glossobalanus minutus* (Kow.) (Enteropneusta, Hemichordata). Ann Parasitol Hum Comp. 1988;63(4):253–62.

Fitzgerald PR. The economic impact of coccidiosis in domestic animals. Adv Vet Sci Comp Med. 1980;24:121–43.

Flausino G, Berto BP, Mcintosh D, Furtado TT, Teixeira WL, Lopes CWG. Phenotypic and genotypic characterization of *Eimeria caviae* from Guinea pigs (*Cavia porcellus*). Acta Protozool. 2014;53(3):269–76.

Florião MM, Lopes Bdo B, Berto BP, Lopes CW. New approaches for morphological diagnosis of bovine *Eimeria* species: a study on a subtropical organic dairy farm in Brazil. Trop Anim Health Prod. 2016;48(3):577–84. https://doi.org/10.1007/s11250-016-0998-5.

Foreyt WJ. Coccidiosis and cryptosporidiosis in sheep and goats. Vet Clin North Am Food Anim Pract. 1990;6(3):655–70.

Friend SC, Stockdale PH. Experimental *Eimeria bovis* infection in calves: a histopathological study. Can J Comp Med. 1980;44(2):129–40.

Gajadhar AA, Wobeser G, Stockdale PHG. Coccidia of domestic and wild waterfowl (Anseriformes). Can J Zool. 1983;61:1–24.

Gaudie CM, Evans RJ, Done SH. Coccidiosis in replacement gilts. Vet Rec. 2005;157(17):517–8.

Gerken M, Renieri C. South American Camelids research: proceedings of the 4th European symposium on south American Camelids and DECAMA European seminar, Göttingen, 7–9 October 2004, vol. 1. Wageningen: Wageningen Academic; 2006.

Giambrone JJ, Anderson WI, Reid WM, Eidson CS. Effect of infectious bursal disease on the severity of *Eimeria tenella* infections in broiler chicks. Poult Sci. 1977;56:243–6.

Giambrone JJ, Klesius PH, Eckamn MK, Edgar SA. Influence of hormonal and chemical bursectomy on the development of acquired immunity to coccidia in broiler chickens. Poult Sci. 1981;60:2612–8.

Gomez-Bautista M, Rojo-Vazquez FA, Alunda JM. The effect of the host's age on the pathology of *Eimeria stiedai* infection in rabbits. Vet Parasitol. 1987;24(1-2):47–57.

Graat EA, van der Kooij E, Frankena K, Henken AM, Smeets JF, Hekerman MT. Quantifying risk factors of coccidiosis in broilers using on-farm data based on a veterinary practice. Prev Vet Med. 1998;33:297–308.

Gräfner G, Graubmann HD, Kron A. About the epizoology of cattle coccidiosis in raising and fattening farms. Mh Vet Med. 1978;33:910–2.
Gräfner G, Graubmann HD, Schwartz K, Hiepe T, Kron A. Further studies on the incidence, epidemiology and control of bovine *Eimeria* coccidiosis under intensive-rearing conditions. Mh Vet Med. 1985;40:41–4.
Gregory MW. Some factors in the pathogenesis of intestinal coccidiosis in mammals. Parasitology. 1982;85:R3.
Gregory MW, Catchipole J. Ovine coccidiosis: the pathology of *Eimeria crandallis* infection. Int J Parasitol. 1990;20(7):849–60.
Gregory MW, Catchpole J. Ovine coccidiosis: pathology of *Eimeria ovinoidalis* infection. Int J Parasitol. 1987;17(6):1099–111.
Gülegen E, Girisgin O, Girisgin AO, Cirak VY. *Eimeria leuckarti* infections in sport horses in Northwest Turkey. Berl Munch Tierärztl Wochenschr. 2016;129(1-2):52–5.
Györke A, Pop L, Cozma V. Prevalence and distribution of *Eimeria* species in broiler chicken farms of different capacities. Parasite. 2013;20:50. https://doi.org/10.1051/parasite/2013052.
Hackstein JHP, Mackenstedt U, Mehlhorn H, Meijerink JPP, Schubert H, Leunissen JAM. Parasitic apicomplexans harbor a chlorophyll a-D1 complex, the potential target for therapeutic triazines. Parasitol Res. 1995;81:207–16.
Hamid PH, Hirzmann J, Kerner K, Gimpl G, Lochnit G, Hermosilla CR, Taubert A. *Eimeria bovis* infection modulates endothelial host cell cholesterol metabolism for successful replication. Vet Res. 2015;46:100. https://doi.org/10.1186/s13567-015-0230-z.
Hammond DM, Bowman GW, Davis LR, Simms BT. The endogenous phase of the life cycle of *Eimeria bovis*. J Parasitol. 1946;32:409–27.
Hammond DM, Andersen FL, Miner ML. The occurrence of a second asexual generation in the life cycle of *Eimeria bovis* in calves. J Parasitol. 1963;49:428–34.
Hanada S, Omata Y, Umemoto Y, Kobayashi Y, Furuoka H, Matsui T, Maeda R, Saito A. Relationship between liver disorders and protection against *Eimeria stiedai* infection in rabbits immunized with soluble antigens from the bile of infected rabbits. Vet Parasitol. 2003;111(2-3):261–6.
Hashemnia M, Khodakaram-Tafti A, Razavi SM, Nazifi S. Experimental caprine coccidiosis caused by *Eimeria arloingi*: morphopathologic and electron microscopic studies. Vet Res Commun. 2012;36(1):47–55. https://doi.org/10.1007/s11259-011-9511-9.
Hegazy SH, Hassanein ZA, El-Sheshtawy EA, Awadalla SF. Effect of dual infections of Escherichia coli and pure caecal *Eimeria* sp in broiler chickens. J Egypt Soc Parasitol. 1999;29:859–72.
Hein H. *Eimeria adenoeides* and *E. meleagrimitis*: pathogenic effect in turkey poults. Exp Parasitol. 1969;24:163–70.
Hermosilla C, Stamm I, Menge C, Taubert A. Suitable in vitro culture of *Eimeria bovis* meront II stages in bovine colonic epithelial cells and parasite-induced upregulation of CXCL10 and GM-CSF gene transcription. Parasitol Res. 2015;114(8):3125–36. https://doi.org/10.1007/s00436-015-4531-y.
Hiepe T, Romeyke D, Jungmann R. Studies of coccidia infections in calves under intensive rearing conditions and recommendations for their control. Mh Vet Med. 1978;33:904–10.
Hilbrich P. Disinfection tests on *Eimeria tenella* oocytes. Berl Münch Tierärztl Wochenschr. 1975;88:144–8.
Hill JE, Lomax LG, Lindsay DS, Lynn BS. Coccidosis caused by *Eimeria scabra* in a finishing hog. J Am Vet Med Assoc. 1985;186(9):981–3.
Hooshmand-Rad P, Svensson C, Uggla A. Experimental *Eimeria alabamensis* infection in calves. Vet Parasitol. 1994;53(1-2):23–32.
Itagaki K. Further investigation on the mechanism of coccidial infection in fowl. J Fac Agric Tottori Univ. 1954;2:37–53.
Jahn D, Matros A, Bakulina AY, Tiedemann J, Schubert U, Giersberg M, Haehnel S, Zoufal K, Mock HP, Kipriyanov SM. Model structure of the immunodominant surface antigen of *Eimeria tenella* identified as a target for sporozoite-neutralizing monoclonal antibody. Parasitol Res. 2009;105(3):655–68. https://doi.org/10.1007/s00436-009-1437-6.
Jarvinen JA. Infection of llamas with stored *Eimeria macusaniensis* oocysts obtained from guanaco and alpaca feces. J Parasitol. 2008;94(4):969–72.

Jenkins MC, Castle MD, Danforth HD. Protective immunization against the intestinal parasite *Eimeria acervulina* with recombinant coccidial antigen. Poult Sci. 1991;70(3):539–47.

Jeurissen SH, Janse EM, Vermeulen AN, Vervelde L. *Eimeria tenella* infections in chickens: aspects of host-parasite interaction. Vet Immunol Immunopathol. 1996;54(1-4):231–8.

Jing J, Liu C, Zhu SX, Jiang YM, LC W, Song HY, Shao YX. Pathological and ultrastructural observations and liver function analysis of *Eimeria stiedai*-infected rabbits. Vet Parasitol. 2016;223:165–72. https://doi.org/10.1016/j.vetpar.2016.04.020.

Johnson J, Reid WM. Anticoccidial drugs: lesion scoring techniques in battery and floor-pen experiments with chickens. Exp Parasitol. 1970;28(1):30–6.

Julian RJ, Harrison KB, Richardson JA. Nervous signs in bovine coccidiosis. Mod Vet Pract. 1976;57(9):711–2; 716, 718

Kabell S, Handberg KJ, Bisgaard M. Impact of coccidial infection on vaccine- and vvIBDV in lymphoid tissues of SPF chickens as detected by RT-PCR. Acta Vet Scand. 2006;48:17.

Karamon J, Ziomko I, Cencek T. Prevalence of *Isospora suis* and *Eimeria* spp. in suckling piglets and sows in Poland. Vet Parasitol. 2007;147(1-2):171–5.

Katrib M, Ikin RJ, Brossier F, Robinson M, Slapetova I, Sharman PA, Walker RA, Belli SI, Tomley FM, Smith NC. Stage-specific expression of protease genes in the apicomplexan parasite, *Eimeria tenella*. BMC Genomics. 2012;13:685. https://doi.org/10.1186/1471-2164-13-685.

Kefu Z, Yingying W, Mei C, Lihong W, Shuichun H, Jun Z, Renhai L, Hong X. *Eimeria tenella*: further studies on the development of the oocyst. Exp Parasitol. 2006;113(3):174–8.

Kim ES, Hong YH, Min W, Lillehoj HS. Fine-mapping of coccidia-resistant quantitative trait loci in chickens. Poultry Sci. 2006;85(11):2028–30.

Koo BS, Lee HR, Jeon EO, Han MS, Min KC, Lee SB, Mo IP. Molecular survey of enteric viruses in commercial chicken farms in Korea with a history of enteritis. Poultry Sci. 2013;92:2876–85. https://doi.org/10.3382/ps.2013-03280.

Kouam MK, Meutchieye F, Nguafack TT, Miegoué E, Tchoumboué J, Theodoropoulos G. Parasitic fauna of domestic cavies in the western highlands of Cameroon (Central Africa). BMC Vet Res. 2015;2015(11):288. https://doi.org/10.1186/s12917-015-0605-4.

Koudela B, Vitovec J. Biology and pathogenicity of *Eimeria neodebliecki* Vetterling, 1965 in experimentally infected pigs. Parasitol Int. 1998;47(4):249–56.

Kouwenhoven B, van der Horst CJ. Histological observations with respect to the immune mechanism in *Eimeria acervulina* infection in the domestic fowl. Z Parasitenkd. 1973;42(1):11–21.

Krautwald-Junghanns ME, Zebisch R, Schmidt V. Relevance and treatment of coccidiosis in domestic pigeons (*Columba livia* forma *domestica*) with particular emphasis on toltrazuril. J Avian Med Surg. 2009;23(1):1–5.

Kunstýr I, Naumann S. Laboratory animal diseases in pictures. Coccidiosis in guinea pigs; with emphasis on diagnosis. Z Versuchstierkd. 1981;23(4):255–7.

Kvičerová J, Hypša V. Host-parasite incongruences in rodent *Eimeria* suggest significant role of adaptation rather than cophylogeny in maintenance of host specificity. PLoS One. 2013;8:e63601.

Lainson R. Intestinal coccidia (Apicomplexa: Eimeriidae) of Brazilian lizards. *Eimeria carmelinoi* n.sp., from *Kentropyx calcarata* and *Acroeimeria paraensis* n.sp. from *Cnemidophorus lemniscatus lemniscatus* (Lacertilia: Teiidae). Mem Inst Oswaldo Cruz. 2002;97(2):227–37.

Lassen B, Ostergaard S. Estimation of the economical effects of *Eimeria* infections in Estonian dairy herds using a stochastic model. Prev Vet Med. 2012;106(3-4):258–65. https://doi.org/10.1016/j.prevetmed.2012.04.005. Epub 2012 May 17.

Lassen B, Viltrop A, Raaperi K, Järvis T. *Eimeria* and *Cryptosporidium* in Estonian dairy farms in regard to age, species, and diarrhoea. Vet Parasitol. 2009;166(3–4):212–9.

Lee MB, Lee EH. Coccidial contamination of raspberries: mock contamination with *Eimeria acervulina* as a model for decontamination treatment studies. J Food Prot. 2001;64(11):1854–7.

Lee DL, Millard BJ. The structure and development of the macrogamete and oocyst of *Eimeria acervulina*. Parasitology. 1971;62(1):31–4.

Levine ND. Protozoan parasites of domestic animals and of man. 1st ed. Minneapolis, MN: Burgess; 1961.

Levine ND, Ivens V. The coccidian parasites (protozoa, sporozoa) of ruminants. Urbana, IL: University of Illinois Press; 1970.

Li G, Lillehoj HS, Lee KW, Jang SI, Marc P, Gay CG, Ritter GD, Bautista DA, Phillips K, Neumann AP, Rehberger TG, Siragusa GR. An outbreak of gangrenous dermatitis in commercial broiler chickens. Avian Pathol. 2010;39:247–53. https://doi.org/10.1080/03079457.2010.487517.

Li J, Zheng J, Gong P, Zhang X. Efficacy of *Eimeria tenella* rhomboid-like protein as a subunit vaccine in protective immunity against homologous challenge. Parasitol Res. 2012;110:1139–45. https://doi.org/10.1007/s00436-011-2603-1.

Lillehoj HS, Li G. Nitric oxide production by macrophages stimulated with coccidia sporozoites, lipopolysaccharide, or interferon-gamma, and its dynamic changes in SC and TK strains of chickens infected with *Eimeria tenella*. Avian Dis. 2004;48:244–53.

Lillehoj HS, Trout JM. Avian gut-associated lymphoid tissues and intestinal immune responses to *Eimeria* parasites. Clin Microbiol Rev. 1996;9:349–60.

Lillehoj HS, Choi KD, Jenkins MC, Vakharia VN, Song KD, Han JY, Lillehoj EP. A recombinant *Eimeria* protein inducing interferon-gamma production: comparison of different gene expression systems and immunization strategies for vaccination against coccidiosis. Avian Dis. 2000;44:379–89.

Lillehoj HS, Kim CH, Keeler CL Jr, Zhang S. Immunogenomic approaches to study host immunity to enteric pathogens. Poultry Sci. 2007;86:1491–500.

Liou CT, Wang JS, Ooi HK. Effect of ozone treatment on *Eimeria colchici* oocysts. J Parasitol. 2002;88:159–62.

Liu Y, Zheng J, Li J, Gong P, Zhang X. Protective immunity induced by a DNA vaccine encoding *Eimeria tenella* rhomboid against homologous challenge. Parasitol Res. 2013;112:251–7. https://doi.org/10.1007/s00436-012-3132-2.

Lobago F, Worku N, Wossene A. Study on coccidiosis in Kombolcha poultry farm, Ethiopia. Trop Anim Health Prod. 2005;37:245–51.

Long PL. Pathology and pathogenicity of coccidial infection. In: Hammond DM, Long PL, editors. *Eimeria, Isospora, Toxoplasma* and related genera. Baltimore: University Park Press; 1973. p. 253–94.

Lundén A, Thebo P, Gunnarsson S, Hooshmian-Rad P, Tauson R, Uggla A. *Eimeria* infections in litter-based, high stocking density systems for loose-housed laying hens in Sweden. Br Poult Sci. 2000;41:440–7.

Lyons ET, Tolliver SC. Prevalence of parasite eggs (*Strongyloides westeri, Parascaris equorum*, and strongyles) and oocysts (*Emeria leuckarti*) in the feces of thoroughbred foals on 14 farms in central Kentucky in 2003. Parasitol Res. 2004;92(5):400–4.

Marquardt WC. Effect of high temperature on sporulation of *Eimeria zurnii*. Exp Parasitol. 1960;10:58–65.

Matjila PT, Penzhorn BL. Occurrence and diversity of bovine coccidia at three localities in South Africa. Vet Parasitol. 2002;104:93–102.

Matsubayashi M, Hatta T, Miyoshi T, Anisuzzaman SK, Yamaji K, Shimura K, Isobe T, Kita K, Tsuji N. Localization of eimeripain, an *Eimeria tenella* cathepsin B-like cysteine protease, during asexual and sexual intracellular development in chicken ceca. J Vet Med Sci. 2014;76(4): 531–7.

Matsui T, Morii T, Iijima T, Kobayashi F, Fujino T. Transformation of oocysts from several coccidian species by heat treatment. Parasitol Res. 1989;75(4):264–7.

McAllister CT, Scott Seville R, Hartdegen R. Two new species of coccidia (Apicomplexa: Eimeriidae) from leaf-tailed geckos, *Uroplatus* spp. (Sauria: Gekkonidae) from Madagascar, including a new host of *Eimeria brygooi* Upton & Barnard, 1987. Syst Parasitol. 2016;93(8):815–23. https://doi.org/10.1007/s11230-016-9660-x.

McDougald LR. Intestinal protozoa important to poultry. Poultry Sci. 1998;77:1156–8.

McKenna PB. *Eimeria macusaniensis* in camelids—a brief review. Surveillance. 2006;33(4): 8–10.

McQueary CA, Worley DE, Catlin JE. Observations on the life cycle and prevalence of *Eimeria leuckarti* in horses in Montana. Am J Vet Res. 1977;38(10):1673–4.

Megía-Palma R, Martínez J, Acevedo I, Martín J, García-Roa R, Ortega J, Peso-Fernández M, Albaladejo G, Cooper RD, Paranjpe DA, Sinervo BR, Merino S. Phylogeny of the reptilian *Eimeria*: are *Choleoeimeria* and *Acroeimeria* valid generic names? Zool Scr. 2015;44(6):684–92. https://doi.org/10.1111/zsc.12126.

Michael E, Hodges RD. The pathogenic effects of *Eimeria acervulina*: a comparison of single and repeated infections. Vet Rec. 1971;89(12):329–33.

Ming-Hsien L, Hai-I H, Hong-Kean O. Prevalence, infectivity and oocyst sporulation time of rabbit-coccidia in Taiwan. Trop Biomed. 2010;27(3):424–9.

Mitchell ES, Smith RP, Ellis-Iversen J. Husbandry risk factors associated with subclinical coccidiosis in young cattle. Vet J. 2012;193(1):119–23. https://doi.org/10.1016/j.tvjl.2011.09.017.

Morner R. Endoparasites of eider-ducklings (Somateria mollissima) on the Swedish east coast. In: Anderesson A, Fredga S, editors. Proceedings of the symposium on sea ducks, Stockholm, Sweden. Gloucester: International Waterfowl Research Bureau; 1975.

Mouafo AN, Weck-Heimann A, Dubremetz JF, Entzeroth R. Monoclonal antibodies speciWc for the two types of wall-forming bodies of *Eimeria tenella* macrogametes (Coccidia, Apicomplexa). Parasitol Res. 2002;88:217–24.

Mundt HC, Bangoura B, Rinke M, Rosenbruch M, Daugschies A. Pathology and treatment of *Eimeria zuernii* coccidiosis in calves: investigations in an infection model. Parasitol Int. 2005;54(4):223–30.

Nation PN, Wobeser G. Renal coccidiosis in wild ducks in Saskatchewan. J Wildl Dis. 1977;13(4):370–5. https://doi.org/10.7589/0090-3558-13.4.370.

Oakes RD, Kurian D, Bromley E, Ward C, Lal K, Blake DP, Reid AJ, Pain A, Sinden RE, Wastling JM, Tomley FM. The rhoptry proteome of *Eimeria tenella* sporozoites. Int J Parasitol. 2013;43(2):181–8. https://doi.org/10.1016/j.ijpara.2012.10.024.

Oksanen A. Mortality associated with renal coccidiosis in juvenile wild greylag geese (*Anser anser anser*). J Wildl Dis. 1994;30(4):554–6.

d'Ovidio D, Noviello E, Ianniello D, Cringoli G, Rinaldi L. Survey of endoparasites in pet guinea pigs in Italy. Parasitol Res. 2015;114(3):1213–6. https://doi.org/10.1007/s00436-014-4289-7.

Ovington KS, Alleva LM, Kerr EA. Cytokines and immunological control of *Eimeria* spp. Int J Parasitol. 1995;25:1331–51.

Pakandl M, Reynaud MC, Chauve CM. Electron microscopic study on the endogenous development of *Eimeria mulardi*, Chauve, Reynaud and Gounel, 1994: a coccidium from the mule duck. Parasitol Res. 2002;88(2):160–4.

Paperna I, Landsberg JH. Description and taxonomic discussion of eimerian coccidia from African and Levantine geckoes. S Afr J Zool. 1989;24:345–55.

Passafaro TL, Carrera JP, dos Santos LL, Raidan FS, dos Santos DC, Cardoso EP, Leite RC, Toral FL. Genetic analysis of resistance to ticks, gastrointestinal nematodes and *Eimeria* spp. in Nellore cattle. Vet Parasitol. 2015;210(3-4):224–34. https://doi.org/10.1016/j.vetpar.2015.03.017.

Peek HW, Landman WJM. Resistance to anticoccidial drugs of Dutch avian *Eimeria* spp. field isolates originating from 1996, 1999 and 2001. Avian Pathol. 2003;32:391–401.

Peek HW, Landman WJ. Coccidiosis in poultry: anticoccidial products, vaccines and other prevention strategies. Vet Quart. 2011;31(3):143–61. https://doi.org/10.1080/01652176.2011.605247.

Pellerdy LP. Coccidia and Coccidiosis. 2nd ed. Berlin and Hamburg: Verlag Paul Parey; 1974. p. 434–45.

Pérez-Fonseca A, Alcala-Canto Y, Salem AZ, Alberti-Navarro AB. Anticoccidial efficacy of naringenin and a grapefruit peel extract in growing lambs naturally-infected with *Eimeria* spp. Vet Parasitol. 2016;232:58–65. https://doi.org/10.1016/j.vetpar.2016.11.009.

Permin A, Esmann JB, Hoj CH, Hove T, Mukaratirwa S. Ecto-, endo- and haemoparasites in free-range chickens in the Goromonzi District in Zimbabwe. Prev Vet Med. 2002;54:213–24.

Persson L, Borg K, Falt H. On the occurrence of Endoparasites in eider ducks in Sweden. Viltrevy. 1974;9:1–24.

Petrone VM, Constantino CF, Pradal-Roa P. Identification and quantification of granulocytes in caecal mucosa and submucosa of chickens experimentally infected with *Eimeria tenella* and *Salmonella enteritidis*. Br Poult Sci. 2002;43(5 Suppl):653–61.

Pfister F, Wolff K. Incidence of endoparasites in swine bred under "specific-pathogen-free" ("SPF") and conventional conditions. Schweiz Arch Tierheilkd. 1975;117(10):585–95.

Pinard-Van Der Laan MH, Monvoisin JL, Pery P, Hamet N, Thomas M. Comparison of outbred lines of chickens for resistance to experimental infection with coccidiosis (*Eimeria tenella*). Poultry Sci. 1998;77(2):185–91.

Radostits OM, Stockdale PHG. A brief review of bovine coccidiosis in western Canada. Can Vet J. 1980;21(8):227–30.

Reeg KJ, Gauly M, Bauer C, Mertens C, Erhardt G, Zahner H. Coccidial infections in housed lambs: oocyst excretion, antibody levels and genetic influences on the infection. Vet Parasitol. 2005;127(3-4):209–19.

Riley D, Fernando MA. Eimeria maxima (Apicomplexa): A comparison of sporozoite transport in naive and immune chickens. J Parasitol. 1988;74(1):103.

Roberts WL, Hammond DM. Ultrastructural and cytologic studies of the sporozoites of four *Eimeria* species. J Protozool. 1970;17(1):76–86.

Rojas M, Manchego A, Rocha CB, Fornells LA, Silva RC, Mendes GS, Dias HG, Sandoval N, Pezo D, Santos N. Outbreak of diarrhea among preweaning alpacas (*Vicugna pacos*) in the southern Peruvian highland. J Infect Dev Ctries. 2016;10(3):269–74. https://doi.org/10.3855/jidc.7398.

Rose ME. Immunity to *Eimeria* infections. Vet Immunol Immunopathol. 1987;17(1-4):333–43.

Rose ME, Hesketh P. Immunity to coccidiosis: stages of the life-cycle of *Eimeria maxima* which induce, and are affected by, the response of the host. Parasitology. 1976;73(1):25–57.

Rothwell L, Young JR, Zoorob R, Whittaker CA, Hesketh P, Archer A, Smith AL, Kaiser P. Cloning and characterization of chicken IL-10 and its role in the immune response to *Eimeria maxima*. J Immunol. 2004;173(4):2675–82.

Ruiz A, Guedes AC, Muñoz MC, Molina JM, Hermosilla C, Martín S, Hernández YI, Hernández A, Pérez D, Matos L, López AM, Taubert A. Control strategies using diclazuril against coccidiosis in goat kids. Parasitol Res. 2012;110(6):2131–6. https://doi.org/10.1007/s00436-011-2746-0.

Ruiz A, Muñoz MC, Molina JM, Hermosilla C, Andrada M, Lara P, Bordón E, Pérez D, López AM, Matos L, Guedes AC, Falcón S, Falcón Y, Martín S, Taubert A. Immunization with *Eimeria ninakohlyakimovae*-live attenuated oocysts protect goat kids from clinical coccidiosis. Vet Parasitol. 2014;199(1-2):8–17. https://doi.org/10.1016/j.vetpar.2013.09.032.

Ryley JF, Millard BJ, Long PL. Further studies on the life cycle of *Eimeria brunetti* Levine 1942. Z Parasitenkd. 1972;40(1):35–48.

Salisch H, Erhorn I, Behr KP, Siegmann O. Subclinical coccidiosis in broilers and pullets. Dtsch Tierärztl Wochenschr. 1989;96:493–6.

Samuelson J, Bushkin GG, Chatterjee A, Robbins PW. Strategies to discover the structural components of cyst and oocyst walls. Eukaryot Cell. 2013;12(12):1578–87. https://doi.org/10.1128/EC.00213-13.

Schneider J, Haass K. Occurrence of cecum (*Eimeria tenella*) coccidiosis in flocks of laying hens. Berl Münch Tierärztl Wochenschr. 1967;80(6):111–4.

Schnieder T. In: Schnieder T, editor. Veterinaermedizinische Parasitologie. 6th ed. Berlin: Paul Parey Verlag; 2006.

Scholtyseck E, Hammond DM. Electron microscope studies of macrogametes and fertilization in *Eimeria bovis*. Z Parasitenkd. 1970;34(4):310–8.

Schreiner LS. Field testing of the efficacy of additives based on essential oils to the milk replacer and litter treatment to control goat eimeriosis. Thesis. 2014. http://www.qucosa.de/recherche/frontdoor/?tx_slubopus4frontend%5bid%5d=urn:nbn:de:bsz:15-qucosa-144699.

Schrey CF, Abbott TA, Stewart VA, Marquardt WC. Coccidia of the llama, *Lama glama*, in Colorado and Wyoming. Vet Parasitol. 1991;40(1-2):21–8.

Schwarz RS, Jenkins MC, Klopp S, Miska KB. Genomic analysis of *Eimeria* spp. populations in relation to performance levels of broiler chicken farms in Arkansas and North Carolina. J Parasitol. 2009;95(4):871–80. https://doi.org/10.1645/GE-1898.1.

Seemann E, Kurth T, Entzeroth R. Insight into the ultrastructural organisation of sporulated oocysts of *Eimeria nieschulzi* (Coccidia, Apicomplexa). Parasitol Res. 2012;111(5):2143–7. https://doi.org/10.1007/s00436-012-3066-8.

Sercy O, Nie K, Pascalon A, Fort G, Yvoré P. Receptivity and susceptibility of the domestic duck (*Anas platyrhynchos*), the Muscovy duck (*Cairina moschata*), and their hybrid, the mule duck, to an experimental infection by *Eimeria mulardi*. Avian Dis. 1996;40(1):23–7.

Shahiduzzaman M, Dyachenko V, Keidel J, Schmäschke R, Daugschies A. Combination of cell culture and quantitative PCR (cc-qPCR) to assess disinfectants efficacy on *Cryptosporidium* oocysts under standardized conditions. Vet Parasitol. 2010;167(1):43–9. https://doi.org/10.1016/j.vetpar.2009.09.042.

Shi T, Tao G, Bao G, Suo J, Hao L, Fu Y, Suo X. Stable transfection of *Eimeria intestinalis* and investigation of its life cycle, reproduction and immunogenicity. Front Microbiol. 2016;7:807. https://doi.org/10.3389/fmicb.2016.00807.

Shirley MW, Jeffers TK, Long PL. Studies to determine the taxonomic status of *Eimeria mitis*, Tyzzer 1929 and *E. mivati*, Edgar and Seibold 1964. Parasitology. 1983;87:185–98.

Shirley MW, Smith AL, Tomley FM. The biology of avian *Eimeria* with an emphasis on their control by vaccination. Adv Parasitol. 2005;60:285–330.

Shirzad MR, Seifi S, Gheisari HR, Hachesoo BA, Habibi H, Bujmehrani H. Prevalence and risk factors for subclinical coccidiosis in broiler chicken farms in Mazandaran province, Iran. Trop Anim Health Prod. 2011;43(8):1601–4. https://doi.org/10.1007/s11250-011-9876-3.

Sibert GJ, Speer CA. Fine structure of zygotes and oocysts of *Eimeria nieschulzi*. J Protozool. 1980;27(4):374–9.

Silva LM, Vila-Viçosa MJ, Cortes HC, Taubert A, Hermosilla C. Suitable *in vitro Eimeria arloingi* macromeront formation in host endothelial cells and modulation of adhesion molecule, cytokine and chemokine gene transcription. Parasitol Res. 2015;114(1):113–24. https://doi.org/10.1007/s00436-014-4166-4.

Skírnisson K. Mortality associated with renal and intestinal coccidiosis in juvenile eiders in Iceland. Parassitologia. 1997;39(4):325–30.

Sommer C. Quantitative characterization, classification and reconstruction of oocyst shapes of Eimeria species from cattle. Parasitology. 1998;116(Pt 1):21–8.

Stockdale PH. The pathogenesis of the lesions produced by *Eimeria zuernii* in calves. Can J Comp Med. 1977;41(3):338–44.

Stockdale PH, Sheard A, Tiffin GB. Resistance to *Eimeria bovis* produced after chemotherapy of experimental infections in calves. Vet Parasitol. 1982;9(3-4):171–7.

Stotish RL, Wang CC, Meyenhofer M. Structure and composition of the oocyst wall of *Eimeria tenella*. J Parasitol. 1978;64(6):1074–81.

Stromberg BE, Schlotthauer JC, Hamann KJ, Saatara Oz H, Bemrick WJ. Experimental bovine coccidiosis: control with monensin. Vet Parasitol. 1986;22(1-2):135–40.

Sun XM, Pang W, Jia T, Yan WC, He G, Hao LL, Bentué M, Suo X. Prevalence of *Eimeria* species in broilers with subclinical signs from fifty farms. Avian Dis. 2009;53(2):301–5.

Supperer R. Die Bedeutung der Sauen und Eber für die Verparasitierung der Schweinebestände. Wien Tierärztl Monatsschr. 1961;48:201–10.

Swaggerty CL, Genovese KJ, He H, Duke SE, Pevzner IY, Kogut MH. Broiler breeders with an efficient innate immune response are more resistant to *Eimeria tenella*. Poultry Sci. 2011;90(5):1014–9. https://doi.org/10.3382/ps.2010-01246.

Szczepaniak KO, Tomczuk K, Lojszczyk-Szczepaniak A, Lopuszynski W. Reclassification of *Eimeriapogonae* Walden (2009) as *Choleoeimeria pogon*ae comb. nov. (Apicomplexa: Eimeriidae). Parasitol Res. 2016;115(2):681–5.

Taylor MA, Catchpole J. Review article: coccidiosis of domestic ruminants. Appl Parasitol. 1994;35(2):73–86.

Thekisoe MM, Mbati PA, Bisschop SP. Diseases of free-ranging chickens in the Qwa-Qwa District of the northeastern free state province of South Africa. J S Afr Vet Assoc. 2003;74:14–6.

Tomley FM, Bumstead JM, Billington KJ, Dunn PP. Molecular cloning and characterization of a novel acidic microneme protein (Etmic-2) from the apicomplexan protozoan parasite, *Eimeria tenella*. Mol Biochem Parasitol. 1996;79(2):195–206.

Toulah FH. Effect of aflatoxin on the coccidial infection in broilers. J Egypt Soc Parasitol. 2007;37:785–92.

Trees AJ, Karim MJ, McKellar SB, Carter SD. *Eimeria tenella*: local antibodies and interactions with the sporozoite surface. J Protozool. 1989;36(4):326–33.

Van Immerseel F, De Buck J, Pasmans F, Huyghebaert G, Haesebrouck F, Ducatelle R. *Clostridium perfringens* in poultry: an emerging threat for animal and public health. Avian Pathol. 2004;33(6):537–49.

Venkateswara Rao P, Raman M, Gomathinayagam S. Sporulation dynamics of poultry *Eimeria* oocysts in Chennai. J Parasit Dis. 2015;39(4):689–92. https://doi.org/10.1007/s12639-013-0403-5.

Vercruysse J. Efficacy of toltrazuril and clazuril against experimental infections with *Eimeria labbeana* and *E. columbarum* in racing pigeons. Avian Dis. 1990;34(1):73–9.

Vetterling JM. Coccidia (protozoa: Eimeriidae) of swine. J Parasitol. 1965;51(6):897–912.

Vítovec J, Koudela B, Stĕrba J. Pathology and pathogenicity of *Eimeria scabra* (Henry, 1931) in experimentally infected pigs. Folia Parasitol (Praha). 1987;34(4):299–304.

de Vos AJ. Studies on the host range of *Eimeria chinchillae* de Vos & van der Westhuizen, 1968. Onderstepoort J Vet Res. 1970;37(1):29–36.

Vrba V, Pakandl M. Coccidia of turkey: from isolation, characterisation and comparison to molecular phylogeny and molecular diagnostics. Int J Parasitol. 2014;44(13):985–1000. https://doi.org/10.1016/j.ijpara.2014.06.004.

Vrba V, Pakandl M. Host specificity of turkey and chicken *Eimeria*: controlled cross-transmission studies and a phylogenetic view. Vet Parasitol. 2015;208(3-4):118–24. https://doi.org/10.1016/j.vetpar.2015.01.017.

Vrba V, Poplstein M, Pakandl M. The discovery of the two types of small subunit ribosomal RNA gene in *Eimeria mitis* contests the existence of *E. mivati* as an independent species. Vet Parasitol. 2011;183(1-2):47–53. https://doi.org/10.1016/j.vetpar.2011.06.020.

Wallach M. Role of antibody in immunity and control of chicken coccidiosis. Trends Parasitol. 2010;26(8):382–7. https://doi.org/10.1016/j.pt.2010.04.004.

Wallach MG, Ashash U, Michael A, Smith NC. Field application of a subunit vaccine against an enteric protozoan disease. PLoS One. 2008;3(12):e3948. https://doi.org/10.1371/journal.pone.0003948.

Watkins KL, Bafundo KW, Donovan DJ. Anticoccidial effect of monensin against *Eimeria mitis* and *Eimeria dispersa*. Poul Sci. 1990;69(6):1009–11.

Wiedmer S, Stange J, Kurth T, Bleiss W, Entzeroth R, Kurth M. New insights into the excystation process and oocyst morphology of rodent *Eimeria* species. Protist. 2011;162(4):668–78. https://doi.org/10.1016/j.protis.2011.03.001.

Williams RB. Laboratory tests of phenolic disinfectants as oocysticides against the chicken coccidium *Eimeria tenella*. Vet Rec. 1997;141(17):447–8.

Williams RB. A compartmentalised model for the estimation of the cost of coccidiosis to the world's chicken production industry. Int J Parasitol. 1999;29:1209–29.

Witcombe DM, Smith NC. Strategies for anti-coccidial prophylaxis. Parasitology. 2014;141(11):1379–89. https://doi.org/10.1017/S0031182014000195.

Yun CH, Lillehoj HS, Lillehoj EP. Intestinal immune responses to coccidiosis. Dev Comp Immunol. 2000;24(2-3):303–24.

Yvoré P, Esnault A, Naciri M. La coccidiose caprine: Effect de contaminations mono ou multispécifiques. Rec Med Vet. 1985;161(4):347–51.

Zhu JJ, Lillehoj HS, Allen PC, Van Tassell CP, Sonstegard TS, Cheng HH, Pollock D, Sadjadi M, Min W, Emara MG. Mapping quantitative trait loci associated with resistance to coccidiosis and growth. Poul Sci. 2003;82(1):9–16.

Sarcocystis

4

Cecilia Decker Franco, Leonhard Schnittger, and Monica Florin-Christensen

Abstract

Sarcocystis (derived from the Greek words *sarx*, which means flesh, and *kystis*, which means bladder) are apicomplexan protozoans that cause sarcocystosis or sarcocystiosis. Infections are characterized by the formation of numerous sarcocysts, which are essentially parasite-full sacs ranging in size from micrometers to several centimeters, in the muscles or nervous tissue of a great variety of animals. The genus is composed of more than 100 species that differ in pathogenicity, host specificity, and sarcocyst structure and location. *Sarcocystis* are obligatory intracellular, with a typical coccidian life cycle, consisting of merogony, gametogony, and sporogony. The life cycle involves an intermediate and a definitive host, usually an herbivore and a carnivore, respectively. At first, a series of asexual reproduction steps culminate with sarcocyst formation. Ingestion of cyst-infected tissues by the definitive host leads to sexual reproduction of the parasite in the digestive tract, followed by excretion of infective forms in the feces. The cycle is closed when an intermediate host becomes infected by the fecal-oral route. Most *Sarcocystis* are species-specific for intermediate and family-specific for definitive hosts. Infection of farm animals is sometimes associated with the reduction in quality and quantity of meat, wool, and fiber, resulting in important economic losses. Additionally, some *Sarcocystis* species are zoonotic. Thus, the study of sarcocystosis constitutes an active field of research.

C. Decker Franco (✉) • L. Schnittger • M. Florin-Christensen
Institute of Pathobiology, Center for Research on Veterinary and Agronomic Sciences (CICVyA), National Institute of Agricultural Technology (INTA-Argentina), Hurlingham, Province of Buenos Aires, Argentina

National Council of Scientific and Technological Research (CONICET), Buenos Aires, Argentina
e-mail: decker.cecilia@inta.gob.ar

M. Florin-Christensen, L. Schnittger (eds.), *Parasitic Protozoa of Farm Animals and Pets*, https://doi.org/10.1007/978-3-319-70132-5_4

4.1 Morphology, Life Cycle, and Host-Pathogen Interactions

Sarcocystis was first reported in Switzerland in 1843 by Miescher, who found white threads in the skeletal muscle of a house mouse—*Mus musculus*—which came to be known as tubules of Miescher (Levine 1986). Until the 1970s, the taxonomic position of the group was not clear, and the major criteria for naming new species were cyst structure and host species. However, studies based on intermediate host specificity indicated that some *Sarcocystis* parasites with structurally similar cysts are actually different species, for example, *S. tenella* and *S. capracanis*. Additionally, some species of *Sarcocystis*, such as *S. falcatula* and *S. neurona*, can infect numerous host species (Dubey 2015). This information, as well as sequencing data, is currently helping to change the status of some species and rename others.

Some of the most important species of *Sarcocystis* that affect farm animals and pets are shown in Table 4.1. Even though sarcocystosis is mainly a veterinary problem, some species are pathogenic to man, such as *S. hominis* and *S. suihominis*. Humans are the definitive hosts of these parasites and become infected by the ingestion of raw or undercooked meat of cattle and pig, respectively, which act as intermediate hosts. Humans can also serve as accidental intermediate or aberrant hosts for several species of *Sarcocystis*, through the ingestion of oocysts (Fayer 2004).

Table 4.1 *Sarcocystis* species in farm animals and pets

Intermediate host	*Sarcocystis* species	Definitive host	References
Pig (*Sus scrofa*)	*S. miescheriana*	Dog (*Canis familiaris*), raccoon (*Procyon lotor*), wolf (*Canis lupus*), red fox (*Vulpes vulpes*), and jackal (*Canis aureus*)	Golubkov et al. (1974), Dubey (2015) and Meshkov (1980)
	S. suihominis	Human (*Homo sapiens*) and primates (*Macaca mulatta*, *Macaca irus*, *Pan troglodytes*, and *Papio cynocephalus*)	
	S. porcifelis	Cat (*Felis catus*)	
Cattle (*Bos taurus*)	*S. cruzi*	Dog (*Canis familiaris*), coyote (*Canis latrans*), red fox (*Vulpes vulpes*), crab-eating fox (*Cerdocyon thous*), raccoon dog (*Procyon lotor*, *Nyctereutes procyonoides*), and wolf (*Canis lupus*)	Bucca et al. (2011), Dubey et al. (1989), Dubey (2015), Gjerde (2016), Nourani et al. (2010), Nourollahi Fard et al. (2009), Rodrigues et al. (2008), Saito et al. (1994) and Soulsby (1987)
	S. hirsuta	Cat (*Felis catus*)	
	S. hominis	Human (*Homo sapiens*), rhesus monkey (*Macaca mulatta*), cynomolgus monkey (*Macaca fascicularis*), baboon (*Papio cynocephalus*), and possibly chimpanzee (*Pan troglodytes*)	
	S. rommeli	Unknown	

Table 4.1 (continued)

Intermediate host	*Sarcocystis* species	Definitive host	References
Sheep (*Ovis aries*)	*S. tenella*	Dog (*Canis familiaris*), coyote (*Canis latrans*), and red fox (*Vulpes vulpes*)	Dubey et al. (1989), Dubey (2015), Levine (1986) and Munday and Obendorf (1984)
	S. arieticanis	Dog (*Canis familiaris*)	
	S. mihoensis		
	S. microps		
	S. gigantea	Cat (*Felis catus*)	
	S. medusiformis		
Goat (*Capra hircus*)	*S. capracanis*	Dog (*Canis familiaris*), coyote (*Canis latrans*), red fox (*Vulpes vulpes*), and crab-eating fox (*Cerdocyon thous*)	Dubey (2015) and Levine (1986)
	S. hircicanis	Dog (*Canis familiaris*)	
	S. moulei	Cat (*Felis catus*)	
Equine (*Equus spp.*)	*S. betrami*	Dog (*Canis familiaris*)	Heydorn et al. (1975), Levine (1986), Soulsby (1987) and Dubey (2015)
	S. fayeri		
	S. equicanis		
	S. neurona	Opossum (*Didelphis virginiana*, *Didelphis albiventris*)	
South American camelids	*S. aucheniae*	Dog (*Canis familiaris*)	Carletti et al. (2013), Martín et al. (2016) and Moré et al. (2016)
	S. masoni	Unknown	
Dog (*Canis familiaris*)	*S. caninum*	Unknown	Dubey (2015)
	S. svanai		
Cat (*Felis catus*)	*S. felis*	Unknown	
Chicken (*Gallus gallus*)	*S. wenzeli*	Dog (*Canis familiaris*) and cat (*Felis catus*)	Dubey (2015) and Mansfield et al. (2001)
	S. horvathi	Unknown	

4.1.1 Morphology

Sarcocystis parasites undergo numerous morphological changes that allow distinct functions necessary to carry out host cell invasion, asexual multiplication, or sexual reproduction along their life cycle. Stages that undergo asexual multiplication are *schizonts* and *metrocytes*; host cell invasion is carried out by *merozoites, bradyzoites*, and *sporozoites*, while sexual reproduction involves *micro-* and *macrogametes* produced by *micro-* and *macrogamonts* and leads to the formation of *oocysts*. Importantly, sporozoites and bradyzoites are produced in wall-enclosed structures containing multiple parasites—the *sporocyst* and the *sarcocyst*—which guarantee an efficient transmission to the intermediate and definitive host, respectively.

The main features of *Sarcocystis* stages and structures are described below. As a rule, the phylum-characteristic apical complex is clearly observed in infective stages that interact with and internalize in host cells, while it is not apparent in stages engaged in asexual or sexual multiplication.

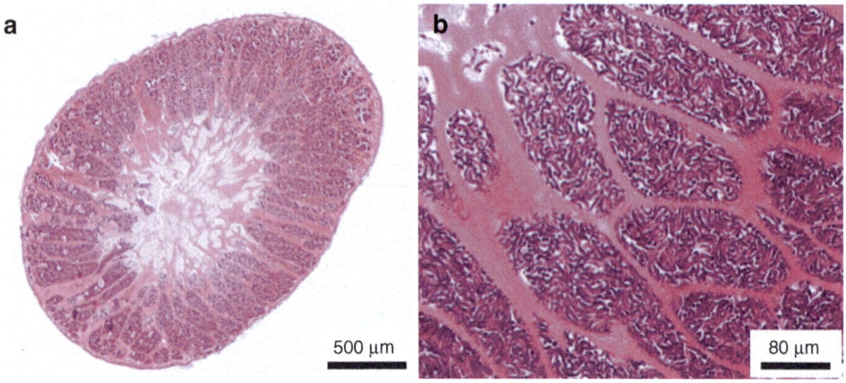

Fig. 4.1 Cross-section of a *Sarcocystis aucheniae* sarcocyst dyed with hematoxylin-eosin. Cytophaneres directed to the center (**a**), 4×; sectioned part of a cyst (**b**), 40×; thousands of bradyzoites arranged in compartments enclosed by thick septa

The schizont is the first parasite stage found in the intermediary host. It develops after a sporozoite has invaded an intermediary host cell, which is generally, but not exclusively, an endothelial cell of a mesenteric lymph node. Early schizonts are ovoid and contain a large nucleus and a single nucleolus. In a process known as endopolygeny, the nucleus gets lobulated and shows several nucleoli. A spindle apparatus with microtubules and two centromeres is associated to each lobe. It guides the genetic material to the lobe end, where a merozoite is formed, and eventually buds, giving the schizont the appearance of a rosette of merozoites. Schizonts develop free in the host cell cytoplasm and are not contained in a parasitophorous vacuole (PV).

Merozoites disseminate the infection in the intermediary host. They are motile, crescent-shaped organisms, with a rhoptry-less apical complex. After budding from a schizont, they can be found free in the blood or located within mononuclear cells. In the latter case, they can divide to form two merozoites by endodyogeny. Upon invasion of a suitable host cell, they start a new schizogony cycle (Fayer 2004).

Sarcocysts are the most characteristic structures produced by *Sarcocystis* parasites. They constitute the last stage of the asexual phase in the intermediary host and are generated after a merozoite has invaded a myocyte or a nervous cell. The membrane of the PV that encloses the parasite and the material underlying it form a wall, providing a safe microenvironment for multiplication. According to the species, sarcocysts can be found in skeletal, cardiac, or smooth myocytes or in neural cells. They can present different shapes—globular, filamentous, and fusiform—and sizes, from a few microns to several centimeters. These and other physical features, such as the presence or absence of internal partitions and the ultrastructure of their walls, aid in species identification (Fayer 2004; Dubey 2015). A certain degree of size variation according to the age of the cyst and the type of parasitized host cell is sometimes observed. For example, within the same species, cysts in cardiac muscles are always smaller than those in skeletal muscles. The wall can invaginate forming villar protrusions, or cytophaneres, of different shapes and sizes. There are over 80 distinct types of cyst wall structures (Dubey 2015). Immediately underneath, there is a granular layer from which septa generally arise, separating the sarcocyst into compartments (Fig. 4.1a). Some sarcocysts, however, have no septa. Table 4.2 and

Table 4.2 Examples of sarcocysts and their special features

Species	Intermediate host	Sarcocyst Shape	Size (μm)	Type of wall	Distribution	References
Sarcocystis tenella	Sheep	Round, elliptical, or elongated	≤700	Palisade-like, 1–3 μm thick	Worldwide	Dubey (2015) and Saito and Itagaki (1994)
S. arieticanis	Sheep	Oval	≤900	Hairlike of variable shape, thin (<1.0)	Worldwide	Dubey (2015)
S. neurona	Cat, skunk, raccoon, armadillo, sea otter, horse (aberrant)	Round	≤700	Striated	Worldwide	Dubey (2015) and Dubey et al. (2015)
S. aucheniae	South American camelids	Rice grain-like	≤5000	Cauliflower-like, 50 μm thick	South America	Carletti et al. (2013)
S. masoni	South American camelids	Oval	≤800	Cauliflower-like, 2.5–3.5 μm thick	South America	Moré et al. (2016)
S. cruzi	Cattle	Oval	≤420	Ribbonlike, thin (<1.0)	Worldwide	Dubey (2015) and Moré et al. (2010)
S. hominis	Cattle	Globular or oval	≤2,600	Fingerlike, thick (>6)	Worldwide	Dubey (2015)
S. gigantea	Sheep	Globular	≤10,000	Cauliflower-like, <2 μm thick		Saito and Itagaki (1994)
S. capracanis	Goat	Oval	≤1000	Honeycomb-like, 3 μm thick	Worldwide	Dubey (2015) and Saito and Itagaki (1994)

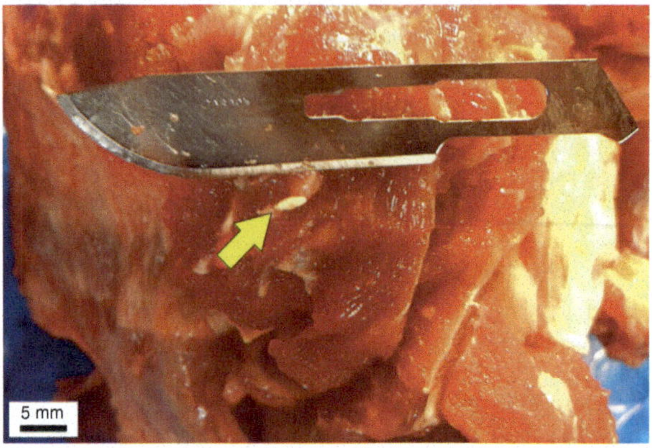

Fig. 4.2 Cyst of *Sarcocystis aucheniae* (*arrow*) found in muscle tissues of a llama (*Lama glama*)

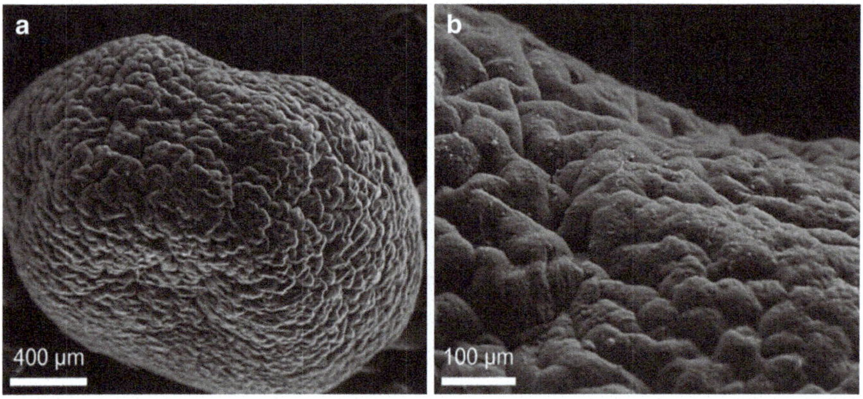

Fig. 4.3 Scanning electron microscopy of a *Sarcocystis aucheniae* cyst. View of the whole cyst showing a brain-like wall (**a**); detailed image of the cyst wall (**b**)

Figs. 4.2 and 4.3 show examples of sarcocysts and their special features. Parasites are located in the fluid contained between partitions (Fig. 4.1b) or, when no partitions are present, free within the cyst. The number of parasites contained in a sarcocyst varies with the species and the stage of maturation: young cysts as small as 5 µm in diameter might contain only two parasites, while a mature cyst can contain over 10^7 parasites, as is the case of *S. aucheniae* macrocysts (Carletti et al. 2013).

Metrocytes and bradyzoites are the parasite stages found in sarcocysts. Metrocytes—mother cells—are rapidly multiplying forms dominant in immature cysts. They are round to oval and have a variable size according to the stage of division. During the transformation of a merozoite into a metrocyte, many of the organelles of the apical complex, such as micronemes, conoid, and polar and apical rings, disappear, while ribosomes, endoplasmic reticulum, and mitochondria become more abundant, and the nucleus becomes larger. Bradyzoites, slow cells or also known as cystozoites, are the dominant forms in mature cysts. They are approximately 17 by 4 µm in size and display gliding motility and a characteristic apical complex (Figs. 4.4 and 4.5). In mature cysts, metrocytes localize in the cortex and

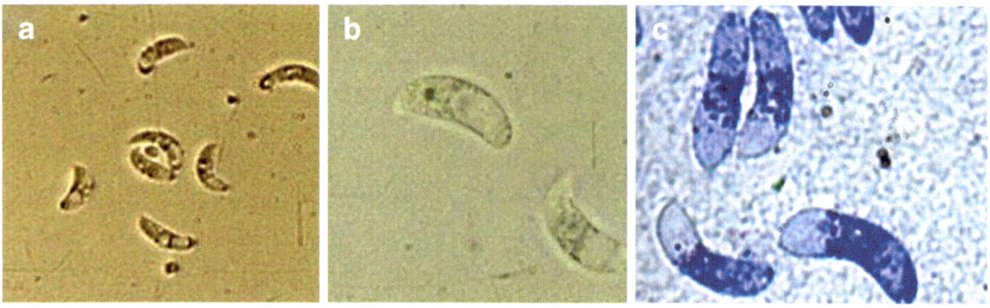

Fig. 4.4 Micrographs of *Sarcocystis aucheniae* bradyzoites. Phase-contrast microscopy (**a**), 400× and (**b**) 1000×, fixed drop stained with hematoxylin-eosin (**c**), 1000×; note the polarity of granulous material

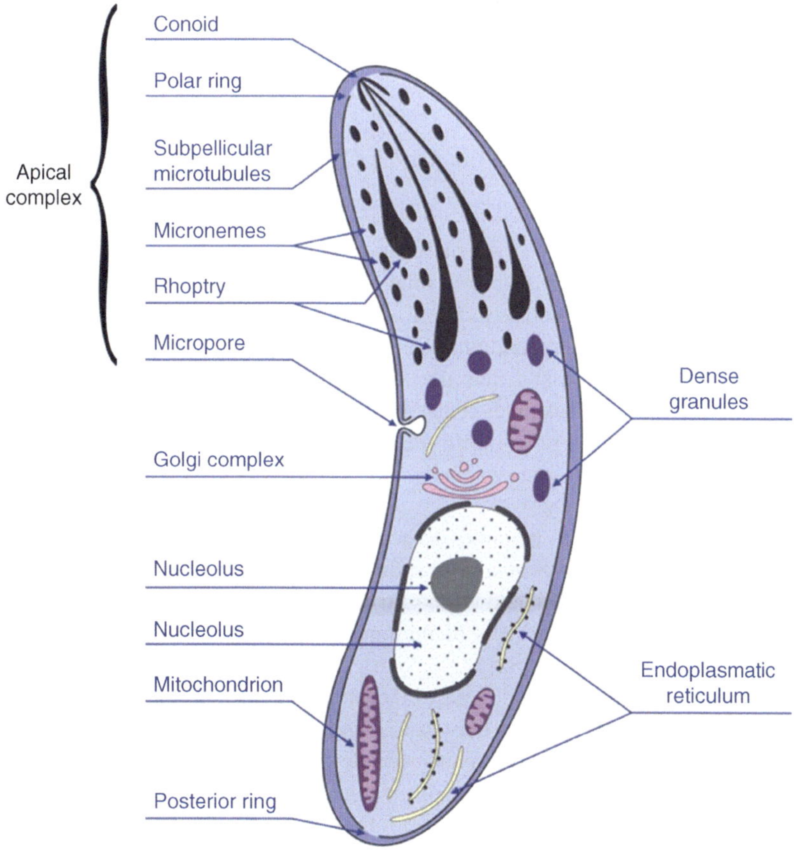

Fig. 4.5 Scheme of a *Sarcocystis* bradyzoite

stain lightly with hematoxylin and eosin, while bradyzoites are found in the medulla and get heavily stained (Dubey 2015).

Micro- and macrogamonts are formed upon bradyzoite infection of goblet cells of the small intestine of the definitive host. Macrogamonts are round or ovoid, measure up to 20 μm in diameter, and contain a single large nucleus. Initially intracellular, they are usually freed into the lamina propria after lysis of the host cell. Microgamonts are elongated, slightly smaller than macrogamonts, and contain several nuclei that move to the periphery. Slender microgametes with two flagella, measuring up to 10 μm in diameter, are formed around each nucleus (Dubey 2015).

Oocysts result from the fusion of micro- and macrogametes. They have an ellipsoid shape, measure around 20 μm long, and are surrounded by a thin wall with a dense external layer and an internal layer of one to four membranes. When eliminated in the feces, they are sporulated and contain two sporocysts. Sporocysts measure 10 by 15 μm and are indistinguishable between species. Each sporocyst has four sporozoites arranged lengthwise (Fig. 4.6) (Dubey 2015; Fayer et al. 2015).

Sporozoites are banana-shaped cells measuring 11–19 μm long by 7–10 μm wide, with all the structural features of bradyzoites. In addition, they possess one or more virus-like crystalloid bodies that consist of electron-dense and electron-lucent granules. These structures likely represent a source of energy or amino acids, as was postulated for *Eimeria* sp. (Dubremetz and Torpier 1978).

4.1.2 Life Cycle

The biological cycle of *Sarcocystis* parasites remained unknown until 1972 when it was recognized that the predator-prey relationship corresponded to a definitive and an intermediate host, respectively. The life cycle consists largely of schizogony—also known as merogony—gametogony, and sporogony. The latter two comprise the sexual phase of the cycle and take place in the intestine of the definitive host—predator. Schizogony corresponds to the asexual phase and occurs in various tissues of the intermediate host—prey—until the formation of bradyzoite-containing cysts that are mainly located in muscle fibers, as mentioned before (Dubey et al. 1989) (Fig. 4.6).

A compatible herbivore becomes infected by ingesting pastures or water contaminated with sporulated *Sarcocystis* sporocysts. Exposure to trypsin and bile in the small intestine causes the liberation of four motile sporozoites from each sporocyst that invades endothelial cells of mesenteric lymph node arteries. Here, first-generation schizogony takes place, giving rise to numerous motile merozoites that bud from a schizont and are released to the bloodstream. Peripheral blood smears show the presence of merozoites between 24 and 46 days postinfection. Merozoites invade endothelial cells of downstream arterioles, capillaries, and veins, distributing throughout the body and producing additional generations by schizogony. Schizonts of some *Sarcocystis* species can also be found in connective tissue cells, macrophages, neural cells, and cells of many different organs. The last schizogony cycle takes place when a merozoite invades a muscle cell—skeletal, smooth, or cardiac—or, exceptionally, a nervous cell and forms a sarcocyst within its surrounding PV. Intracellularly, the invading merozoite first differentiates into a metrocyte, which reproduces by endodyogeny inside the PV. In this type of asexual reproduction, two daughter cells arise from an existing one, which is consumed during the

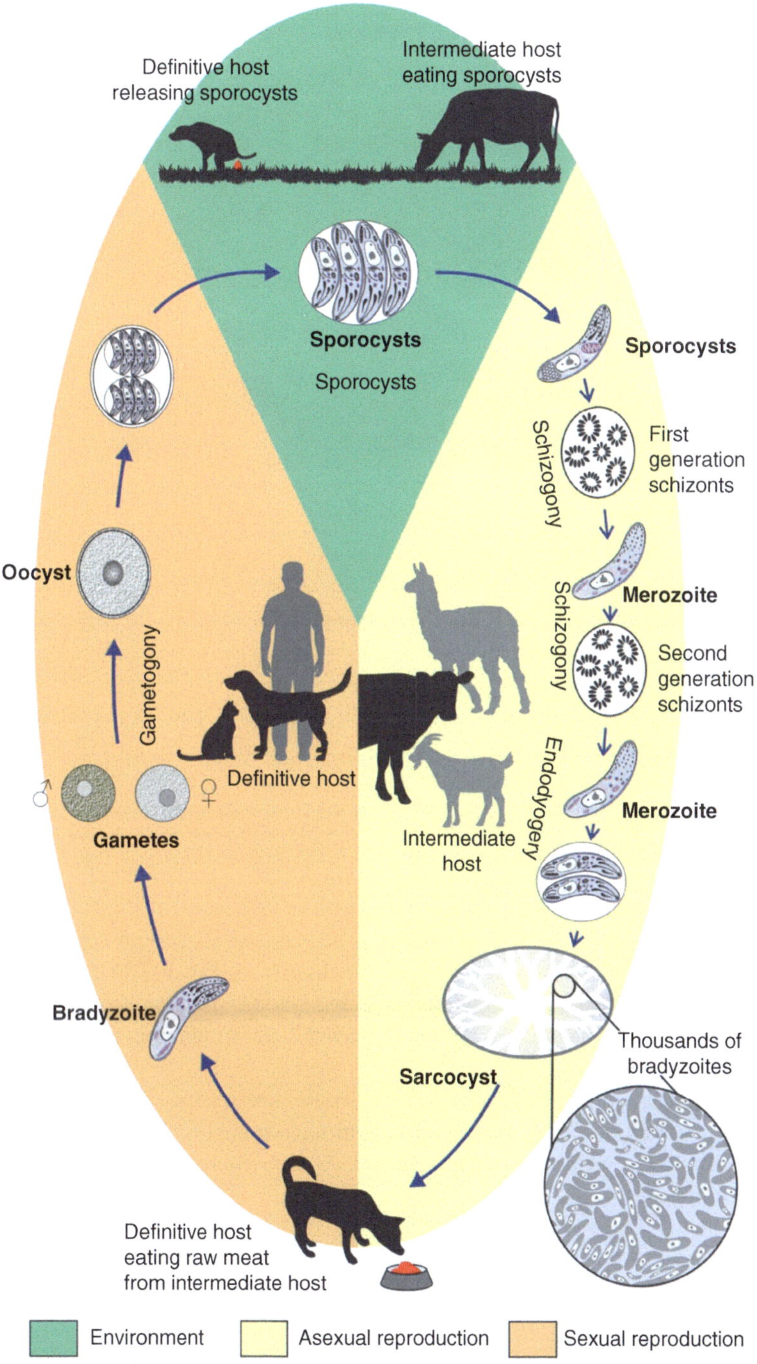

Fig. 4.6 Life cycle of *Sarcocystis*

Fig. 4.7 Dog feeding on raw meat during the slaughter of a llama (Catamarca, Argentina)

process. Concomitantly, a wall develops, isolating the nascent sarcocyst from the surrounding tissues. Eventually, metrocytes stop division and differentiate into infective bradyzoites, which display an apical complex that will aid in the invasion of definitive host cells. A sarcocyst full of bradyzoites is considered mature. The time elapsed between infection and formation of a mature sarcocyst varies between species but lasts in general around 2 months. Cysts can then persist in the tissues for months or even years. Bradyzoites are infectious for the definitive host, while schizonts and immature sarcocysts are not. The number and distribution of cysts in a particular host depend on different factors, including the amount of sporozoites ingested, the *Sarcocystis* and host species involved, the stage of infection, and the immune status of the animal. Gametogony is possible when a compatible carnivore feeds on cyst-containing tissues of an intermediate host (Fig. 4.7). Upon digestion of the cyst wall in the stomach and intestine, bradyzoites are liberated from the sarcocyst and penetrate host cells, generally goblet cells, or enterocytes of the small intestine. Each bradyzoite differentiates intracellularly into either a macro- or a macrogamont. Each macrogamont yields a single macrogamete, while microgamonts become multinucleated and yield several microgametes. The latter are motile and migrate to the surface of a macrogamete. After membrane fusion, the microgamete nucleus enters and fertilizes the macrogamete, yielding a zygote. A thin wall (<1 μm) then develops around the zygote giving place to the formation of the oocyst. The whole process of gametogony and fertilization can be completed in 1 day. The infected cells move to the lamina propria, where sporulation happens, giving rise to two elongated sporocysts containing four sporozoites each. The timing of sporocyst excretion by the definitive host after ingestion of sarcocysts is highly variable within the same species but in general starts after 7 and 14 days (Dubey et al. 1989; Dubey 2015; Fayer et al. 2015).

4.1.3 Host-Pathogen Interactions

Even though *Sarcocystis* is a very broad genus that includes some of the most prevalent parasites of vertebrate animals, they are understudied compared to other members of the Apicomplexa. For this reason, information on host-pathogen interactions is rather scarce. Since *Sarcocystis* sp. are obligate intracellular parasites, surface molecules that participate in recognition and invasion of host cells are likely essential for their survival. A prominent group of coccidian surface proteins are the SAGs, a family of glycosylphosphatidylinositol (GPI)-anchored surface antigens, initially characterized in *Toxoplasma gondii* and *Neospora* spp. (Howe et al. 2005). *T. gondii* SAGs have been implicated in receptor-ligand interactions with the host cell surface and in the stimulation of immune responses during infection, suggesting they are attractive targets for anti-coccidian drugs or immunotherapy approaches (Jacquet et al. 2001; Rachimel et al. 2004). Homology searches in an expressed sequence tag (EST) database of *S. neurona* allowed the identification of four SAG family members in this parasite, and other two members were later discovered (Howe et al. 2005; Crowdus et al. 2008). These proteins were demonstrated to be expressed on the surface of merozoites and to be highly immunogenic. The presence of SAG family proteins in different coccidian genera suggests a conserved essential function (Howe et al. 2005). Interestingly, studies performed with *S. neurona* merozoites, bradyzoites, and sporozoites showed that expression of individual SAGs is stage-specific. This is consistent with findings in *T. gondii* and suggests that surface antigen switching could be essential for the completion of the parasite life cycle (Gautam et al. 2011). Another GPI-anchored protein, surface protein 1 or SnSPR1, has been also identified in the *S. neurona* EST database. Contrary to SAGs, SnSPR1 shows no orthologs in other coccidian genera. It is expressed at the surface of merozoites in all stages of schizont development, is immunogenic, and might also participate in host-pathogen interactions (Zhang and Howe 2008).

Microneme proteins are also key elements in the invasion process of apicomplexan parasites, likely involved in the attachment and entry into the host cell (Dubremetz et al. 1998). The ortholog of a *T. gondii* microneme protein was identified in the *S. neurona* EST database and named SnMIC10. This protein has been shown to be differentially expressed in the apical end of merozoites during endopolygeny, supporting the view that micronemes are only needed during cell invasion (Hoane et al. 2003). A large array of conserved and species-specific proteins of *Sarcocystis* parasites await characterization and could serve as targets for control strategies.

Effector cells of the host immune system are mobilized during a *Sarcocystis* infection. The predominant cells infiltrating visceral and muscular tissues are lymphocytes and macrophages (Dubey et al. 1982). The cell infiltration of mononuclear cells starts during the third week of infection and can last for several months, even after the parasite is no longer detectable in visceral tissues (Gasbarre et al. 1984). Whether these cellular events participate in the recovery of the host from sarcocystosis has not been established, and passive transfer of resistance via cells or antibodies has not been reported. The intense cellular response seen in immune animals that survive lethal challenges indicates a cell-mediated immunity against the parasite. Protective immunity has been shown to be induced only by homologous *Sarcocystis* species. As an example, in *S. hirsuta* experimentally infected cattle, no protection against challenge with *S. cruzi* was shown (Dubey 2015; Fayer and Dubey 1984; Ford 1985).

S. neurona-experimentally infected horses develop clinical disease and generate antibodies in serum and cerebrospinal fluid. Alterations in the immune cell subset expression that changed during disease progression were observed. Infected horses showed decreased antigen-specific proliferation responses compared to nonexperimentally infected horses, suggesting that the process between antigen-presenting cells—monocyte/dendritic cells—and/or T-cell antigen recognition may be damaged in *S. neurona*-infected horses (Lewis et al. 2014).

4.2 Diagnosis and Epidemiology

4.2.1 Diagnosis

Diagnosis of acute sarcocystosis is difficult since symptoms are not very specific and, therefore, easily confused with other pathological processes. Regularly, a diagnosis of sarcocystosis is based on the elimination of other causative agents, a good epidemiologic evaluation of the intermediate host, and its relationship to definitive hosts, as well as clinical findings (Cordero del Campillo et al. 1999). However, there are some techniques that have been employed or developed over the years and are described below.

4.2.1.1 Definitive Host

The diagnosis of sarcocystosis in the definitive host is based mostly on the identification of *Sarcocystis* sporocysts in the feces. This can be achieved by using a coproparasitological assay, consisting of flotation in zinc sulfate or other high-density solutions, followed by microscopic observation (Dubey et al. 2015). However, species cannot be discriminated by this method because of morphological similarities of sporocysts, thus molecular tests need to be employed.

4.2.1.2 Intermediate Host

Diagnosis can sometimes be inferred from epidemiological data of the region of interest, as well as information obtained by coproparasitological analysis of definitive hosts. Usually, confirmatory diagnosis is achieved by postmortem examination of the skeletal muscle. Cysts of some species are visible to the naked eye, as in the case of *S. aucheniae* and *S. gigantea*, that infect South American camelids and sheep, respectively (Fig. 4.2). In a great number of species, however, cysts are microscopic, so other examination methods are applied.

Microscopy and electron microscopy allow to diagnose *Sarcocystis* sp. and to differentiate morphological features between species. They are specific but time-consuming, limiting their application on large numbers of samples (Moré et al. 2010).

Artificial digestion is a sensitive method that allows analyzing large amounts of tissue. It detects bradyzoites released from cysts, but it does not allow to differentiate between species of *Sarcocystis* (Savini et al. 1996). In the digestion procedure, tissues are incubated with proteases, such as trypsin or pepsin. Released bradyzoites can be used as antigen, and the species can be identified by molecular methods.

Histopathology and examination of fresh tissue allows the differentiation between thick- and thin-walled cysts, but not between species. The sensitivity of histopathological examinations is lower due to the smaller volume of sample that can be processed (Dubey et al. 1989).

Polymerase chain reaction (PCR) is an important tool for epidemiological studies. It allows to detect DNA of the parasite in small volumes of sample and also to differentiate between *Sarcocystis* and related organisms, such as *Toxoplasma* and *Neospora,* or discriminate between *Sarcocystis* species (Ortega-Mora et al. 2007). Different PCR protocols have been developed over the years. For example, recent studies showed that it is possible to detect DNA of *S. aucheniae* in the blood of South American camelids employing a semi-nested PCR (Martín et al. 2016). Other studies allowed the differentiation of *Sarcocystis* species affecting cattle using multiplex real-time PCR (Moré et al. 2013) or applied PCR followed by RFLP—*restriction fragment length polymorphism*—to determine the *Sarcocystis* species affecting sheep (Hamidinejat et al. 2014). In addition to PCR, another highly sensitive DNA amplification technique that takes place at constant temperature, and thus does not require the use of a thermocycler, is being increasingly applied to the diagnosis of different pathogens (Notomi et al. 2015). This technique—known as *loop-mediated isothermal amplification* or LAMP—has been successfully applied to the molecular detection of *S. fayeri* in horsemeat (Furukawa et al. 2016).

In immunohistochemistry, anatomical, immunological, and biochemical techniques are combined to identify discrete tissue components. The method is based on the interaction of target antigens with specific antibodies tagged with a label that allows visualizing the distribution and localization of specific components within cells and in the proper tissue context. A number of immunohistochemical methods have been developed for the improvement of the sensitivity and specificity of the histological detection of life cycle stages of *Sarcocystis* species. However, these methods are complicated due to the high cross-reactivity among *Sarcocystis* species when polyclonal antisera are used. In addition, cross-reactivity of anti-*Sarcocystis* sp. antibodies is sometimes extended to other cyst-forming coccidia such as *T. gondii* (Uggla and Buxton 1990). Monoclonal antibodies have been produced for *S. cruzi, S. tenella,* and *S. arieticanis*; however also in this case, most are cross-reactive with antigens of heterologous *Sarcocystis* species (Ortega-Mora et al. 2007).

Enzyme-linked immunosorbent assay (ELISA) detects and measures antibodies and is the most commonly used serological test for the diagnosis of *Sarcocystis*. However, cross-reactivity with heterologous *Sarcocystis* species is also a problem here (Tenter 1995). Different types of antigens have been employed in ELISA tests. As an example, an indirect ELISA (iELISA) based on *S. cruzi* bradyzoite antigens was developed in Sri Lanka for the detection of anti-*Sarcocystis* spp. antibodies in cattle (Kalubowila et al. 2004). In addition, in Argentina, an iELISA based on an immunogenic protein fraction extracted from *S. aucheniae* bradyzoites was applied to the detection of anti-*Sarcocystis* antibodies in South American camelids (Romero et al. 2014). More recently, in the USA, recombinant forms of *S. neurona* SnSAG surface antigens were used in an iELISA format to measure antibodies in serum and cerebrospinal fluid. The latter revealed active infection in the central nervous system (Yeargan et al. 2015).

In a Western blot, a mixture of proteins is separated by gel electrophoresis and then transferred to a membrane. The membrane is incubated with specific antibodies against the protein or proteins of interest. Detection is achieved by reaction with an enzyme-conjugated anti-species antibody, followed by incubation with a colorimetric or chemiluminescent substrate (Mahmood and Yang 2012). Detection of antibodies against two *S. neurona*-specific antigens, of 29 and 17 kDa, is currently the standard serological diagnostic method for infections with this parasite in horses (Rossano et al. 2000; Hamir and Dubey 2001).

4.2.2 Epidemiology

Sarcocystis infections of farm animals are worldwide distributed and often present high prevalences, both in developing and industrialized countries (Dubey et al. 1989). The percentages of infected animals depend on various aspects, including host, viability of the sporocysts in the environment, number of sporocysts released by the definitive host, immune status of the intermediate host, hygiene, and proximity between definitive host and intermediate hosts, among others (McKenna and Charleston 1994; Savini et al. 1996). According to the species, infections can cause mortality, morbidity, abortions, lower meat yield, and economic losses due to confiscation of meat when macroscopic cysts are found (Poulsen and Stensvold 2014). In cattle, *Sarcocystis* infections are regularly asymptomatic, with prevalences over 90%, being *S. cruzi* the most commonly found species (Moré et al. 2010). In Argentina, recent studies showed a direct connection between the type of breeding of llamas and prevalence of anti-*Sarcocystis* sp. antibodies (Romero et al. 2017). In Bolivia, a study made in abattoirs showed that 23–50% of llama carcasses contained macroscopic cysts, with infection rates higher in females and in older animals (Rooney et al. 2013).

Seroprevalences reported for *S. neurona* in US horses vary from 15 to 89%, depending on the geographic region. They are also lower during the winter season compared to the rest of the year (Pusterla et al. 2014; Reed et al. 2016). Seroprevalences of around 35% were reported in Brazil and Argentina, indicating that the parasite is also present in South America (Dubey et al. 1999a, b). Most infections have been observed in young animals of 1–5 years or older than 13 years of age. Interestingly, the likelihood of infection was significantly reduced in farms where wildlife had no access to feed and when a creek or river was present as a water source. On the other hand, stress related to heavy exercise, transport, injury, surgery, or parturition was found to increase the risk of disease caused by *S.*

neurona. In addition, racehorses and show horses had higher infection risks than breeding and pleasure horses (Reed et al. 2016).

Sarcocystosis is usually acquired horizontally through ingestion of contaminated food or water. In addition, anti-*Sarcocystis* sp. antibodies were detected in horse fetuses and newborn foals, indicating the occurrence of transplacental transmission. However, this event appears to be rare (Duarte et al. 2004).

A number of molecular typing techniques, including PCR-RFLP, microsatellite, and whole-genome fingerprinting, have been developed to differentiate between *S. neurona* isolates. Application of these techniques can aid in understanding parasite transmission and epidemiology and reveal its population structure (Elsheikha and Mansfield 2007).

4.3 Clinical Effects, Prevention, and Treatment

4.3.1 Clinical Effects

Sarcocystis natural infections of intermediate hosts are in most cases asymptomatic. However, *S. neurona* is a special case, due to the parasite tropism for horse nervous tissues, infecting both gray and white matter, which provokes focal or multifocal signs of neurological disease. This syndrome was initially known as segmental myelitis and, later, focal encephalitis-myelitis, until the presence of protozoa in characteristic lesions led to the current name of equine protozoal myeloencephalitis (EPM). The disease can actually also be caused by *Neospora hughesi*, although the majority of cases are due to *S. neurona*. Clinical EPM signs include dysphagia, upper airway dysfunction, muscle weakness and atrophy, ataxia, weakness of limbs, and even seizures. Severely affected horses show difficulty in standing, walking, or swallowing and present head tilt and facial nerve paralysis. In some cases, signs stabilize but relapse in a few days or weeks (Dubey et al. 2015; Reed et al. 2016).

Experimental infections of different farm animals and pets, on the other hand, have shown a number of severe signs (Table 4.3). As a rule in *Sarcocystis* infections, necrosis of cells and tissues produced by multiplication of schizonts is very common, but it does not appear to be extensive enough to cause severe illness or death in large animals—cattle, sheep, goats, and pigs. However, an intense inflammatory reaction is usually associated to second-generation schizont maturation. Eosinophilic myositis (EM) is a specific inflammatory condition of striated muscles. It happens during the penetration of myocytes by merozoites and might be related to products liberated from merozoites or myocytes. It has been found mostly in cattle, occasionally in sheep, and rarely in pigs and horses. With the progression of EM, eosinophils and myocytes degenerate, resulting in granulomas with a central area of necrosis. Later, the tissue becomes surrounded by zones of giant cells, epithelial cells, lymphocytes, and fibrocytes (Dubey 2015).

Abortion can result when animals become infected with pathogenic species of *Sarcocystis* during pregnancy, and there are many unanswered questions and observations that appear contradictory about the effect of sarcocystosis on fetal health.

Table 4.3 Clinical effects of *Sarcocystis* spp. infections of farm animals and pets

Intermediate host	*Sarcocystis* species	Clinical effects (depending on the number of sporocysts ingested)
Pig (*Sus scrofa*)	*S. miescheriana*	Weight loss, purpura of the skin, dyspnea, muscle tremors, diarrhea, myositis, abortion, and death
	S. suihominis	
	S. porcifelis	
Cattle (*Bos taurus*)	*S. cruzi*	Fever, weight loss, anemia, hair loss, weakness, prostration, abortion, reduced milk yield, hypersalivation, neurologic signs, and death
	S. hirsuta	Fever, diarrhea, and anemia
	S. hominis	Anemia
Sheep (*Ovis aries*)	*S. tenella*	Anorexia, weight loss, fever, anemia, loss of wool, abortion, premature birth, neural signs, myositis, and death
	S. arieticanis	Fever, less wool growth, and death
	S. gigantea	
Goat (*Capra hircus*)	*S. capracanis*	Fever, weakness, anorexia, weight loss, tremors, irritability, abortion, and death
Equine (*Equus* spp.)	*S. fayeri*	Mild anemia, hair loss, and fever
	S. equicanis	Fever, apathy, and anorexia
	S. neurona	Ataxia, general weakness with muscle spasticity, loss of appetite, decreased tongue tone, and facial paresis
South American camelids (*Lama* spp.)	*S. aucheniae*	Necrosis, hemorrhage, eosinophilic myositis, abortion, and death
Dog (*Canis familiaris*)	*S. caninum*	Fever, apathy, anorexia, muscle weakness, ataxia, and elevated liver and muscle enzymes
Chicken (*Gallus gallus*)	*S. wenzeli*	Myositis, muscular weakness, and neurologic signs
	S. horvathi	

With the exception of *S. neurona*, where signs correspond to natural or experimental infections, all other effects were registered exclusively in experimental infection studies (Dubey 2015; Fayer et al. 2015)

When infection is induced experimentally, most animals can develop clinical sarcocystosis and abortion, showing parasites and lesions in maternal placentomes but rarely infecting the fetus or fetal membranes. Unlike experimentally infected animals, those with natural infections show parasites, lesions, or both in the fetuses (Jerrett et al. 1984).

Definitive hosts usually do not present clinical signs. Dogs, cats, coyotes, foxes, and raccoons fed with tissues infected with different *Sarcocystis* species-excreted sporocysts but were otherwise asymptomatic. However, a few dogs and coyotes vomited or were anorexic for 1–2 days following ingestion of meat. In addition, in trials made in human volunteers who ingested beef and pork infected with *S. hominis* or *S. suihominis*, respectively, clinical symptoms were observed, including vomiting, diarrhea, and respiratory distress (Dubey 2015). Indeed, soluble extracts prepared from the tissue cysts of various *Sarcocystis* species have been shown to contain powerful toxins—sarcotoxins—which have even proven lethal when administered to laboratory animals (Hiepe et al. 1981; Harada et al. 2013, Kamata et al. 2014). Accordingly, experimental inoculation of rabbits with an extract derived from *S. fusiformis* or *S. cruzi* cysts developed a shock-like state and/or death, likely caused by the sudden exposure to high doses of sarcotoxins (Saleque et al. 1991; Nakamura et al. 1999).

4.3.2 Prevention

Currently, there is no vaccine to protect animals against sarcocystosis, but experimental studies indicate that cattle, sheep, goats, and pigs develop a humoral response when inoculated with small numbers of live sporocysts. For this reason, there is hope of developing a vaccine for sarcocystosis in the future (Dubey 2015). In the USA, a killed whole *S. neurona* merozoite vaccine was marketed by Fort Dodge, but the product has been retired from the market, because no differences between vaccinated and control horses were found during the trials (Dubey et al. 2015).

For the time being, interrupting the cycle of the parasite is the only practical method of control. This can be achieved by preventing definitive hosts to consume raw or insufficiently cooked meat infected with *Sarcocystis* or to cohabitate with intermediate hosts.

It is important to take into account that during veterinary inspections in slaughterhouses, meat containing macroscopic cysts can be confiscated, but microscopic cysts pass unnoticed. For this reason, it can be assumed that most meat that is consumed is infected with *Sarcocystis* (Godoy et al. 2007). It has been demonstrated that freezing meat is effective to prevent the occurrence of food poisoning when consuming raw meat containing sarcocysts. Indeed, freezing *S. fayeri* sarcocyst-infected horsemeat for 48 h at −20 °C resulted in the disappearance of bradyzoites, as well as of a 15 kDa sarcotoxin found to be responsible for the clinical signs associated with food poisoning (Kamata et al. 2014).

Confiscated carcasses should be buried or incinerated to prevent definitive hosts from eating infected meat, and the prophylactic use of anticoccidials in definitive and intermediate hosts could help to control sarcocystosis in farm animals and pets (Dubey 2015).

4.3.3 Treatment

Currently there is no specific prophylactic or therapeutic treatment for sarcocystosis. Infections usually go undetected, but in case of the appearance of clinical signs, some drugs have proved effective to partially reduce illness and parasitic load (Table 4.4). In horses, treatment with an anticoccidial drug has been shown to increase ten times the likelihood of recovery from clinical EPM signs and survival. Treatment, however, was more effective when milder, rather than severe, clinical signs were present (Saville et al. 2000; Pusterla et al. 2014; Reed et al. 2016).

Table 4.4 Drugs used for the treatment of sarcocystosis (Cordero del Campillo et al. 1999; Dubey 2015; Fayer 2004; Lindsay and Dubey 1999; Reed et al. 2016)

Drug	Target host	Action
Amprolium	Cattle, sheep, goat, pig, chicken, and dog	Coccidiostat Blocks the thiamine transporter
Decoquinate	Poultry, cattle, sheep, goat, and equine	Coccidiostat
Diclazuril	Poultry, cattle, sheep, goat, and equine	Coccidiostat
Dihydrofolate reductase inhibitors	Equine	Inhibits folate's actions
Salinomycin	Animals	Coccidiostat
Halofuginone	Animals	Coccidiostat. Inhibits T-helper 17 cells
Lasalocid	Cattle, sheep, and poultry	Coccidiostat
Ponazuril	Equine	Coccidiostat. Inhibits mitochondrial function
Primaquina	Definitive hosts	Gametocide. Inhibits the synthesis of polyamines
Robenidine	Sheep and goat	Coccidiostat
Sulfadoxine and pyrimethamine	Definitive hosts, equine	Inhibits dihydropteroate synthetase activity
Sulfadoxine and trimethoprim	Cattle, sheep, goat, and pig	Inhibits dihydropteroate synthase

References

Bucca M, Brianti E, Giuffrida A, Ziino G, Cicciari S, Panebianco A. Prevalence and distribution of *Sarcocystis* spp. cysts in several muscles of cattle slaughtered in Sicily, Southern Italy. Food Control. 2011;22:105–8. https://doi.org/10.1016/j.foodcont.2010.05.015.

Carletti T, Martin M, Romero S, Morrison DA, Marcoppido G, Florin-Christensen M, Schnittger L. Molecular identification of *Sarcocystis aucheniae* as the macrocyst-forming parasite of llamas. Vet Parasitol. 2013;198:396–400. https://doi.org/10.1016/j.vetpar.2013.09.007.

Cordero del Campillo M, Rojas F, Fernández M, Sánchez M, Rodríguez S, López I. Parasitología veterinaria. Madrid: McGraw-Hill; 1999. p. 968.

Crowdus CA, Marsh AE, Saville WJ, Lindsay DS, Dubey JP, Granstrom DE, Howe DK. SnSAG5 is an alternative surface antigen of *Sarcocystis neurona* strains that is mutually exclusive to SnSAG1. Vet Parasitol. 2008;158(1–2):36–43. https://doi.org/10.1016/j.vetpar.2008.08.012.

Duarte PC, Conrad PA, Barr BC, Wilson WD, Ferraro GL, Packham AE, Carpenter TE, Gardner IA. Risk of transplacental transmission of *Sarcocystis neurona* and *Neospora hughesi* in California horses. J Parasitol. 2004;90(6):1345–51. https://doi.org/10.1645/GE-3372.

Dubey JP. Sarcocystosis of animals and humans. 2nd ed. Boca Raton: CRC Press; 2015.

Dubey JP, Speer CA, Epling GP. Sarcocystosis in newborn calves fed *Sarcocystis cruzi* sporocysts from coyotes. Am J Vet Res. 1982;43:2147–64. PMID: 6819793.

Dubey JP, Speer CA, Fayer R. Structure and life cycle. Sarcocystosis of animals and man. Boca Raton: CRC; 1989.

Dubey JP, Kerber CE, Granstrom DE. Serologic prevalence of Sarcocystis neurona, Toxoplasma gondii, and Neospora caninum in horses in Brazil. J Am Vet Med Assoc. 1999b;215:970–2. PMID: 10511862.

Dubey JP, Venturini MC, Venturini L, McKinney J, Pecoraro M. Prevalence of antibodies to *Sarcocystis neurona*, *Toxoplasma gondii* and *Neospora caninum* in horses from Argentina. Vet Parasitol. 1999a;86:59–62. https://doi.org/10.1016/S0304-4017(99)00127-2.

Dubey JP, Howe DK, Furrc M, Savilled WJ, Marshd AE, Reede SM, Griggf ME. An update on *Sarcocystis neurona* infections in animals and Equine Protozoal Myeloencephalitis (EPM). Vet Parasitol. 2015;209(0):1–42. https://doi.org/10.1016/j.vetpar.2015.01.026.

Dubremetz JF, Torpier G. Freeze fracture study of the pellicle of an eimerian sporozoite (protozoa, coccidia). J Ultrastruct Res. 1978;62:94–109. PMID: 418187.

Dubremetz JF, Garcia-Reguet N, Conseil V, Fourmaux MN. Apical organelles and host-cell invasion by Apicomplexa. Int J Parasitol. 1998;28:1007–13. PMID: 9724870.

Elsheikha HM, Mansfield LS. Molecular typing of *Sarcocystis neurona*: current status and future trends. Vet Parasitol. 2007;149:43–55. https://doi.org/10.1016/j.vetpar.2007.06.039.

Fayer R. *Sarcocystis* spp. in human infections. Clin Microbiol Rev. 2004;17(4):894–902. https://doi.org/10.1128/CMR.17.4.894-902.2004.

Fayer R, Douglas H, Dubey JP. Human infections with *Sarcocystis* species. Clin Microbiol Rev. 2015;28(2):295–311. https://doi.org/10.1128/CMR.00113-14.

Fayer R, Dubey JP. Protective immunity against clinical sarcocystosis in cattle. Vet Parasitol. 1984;15:187–201. PMID: 6437053.

Ford GE. Immunity of sheep to homologous challenge with dog borne *Sarcocystis* species following varying levels of prior exposure. Int J Parasitol. 1985;15:629–34. PMID: 3937817.

Furukawa M, Minegishi Y, Izumiyama S, Yagita K, Mori H, Uemura K, Etoh Y, Maeda E, Sasaki M, Ichinose K, Harada S, Kamata Y, Otagiri M, Sugita-Konishi Y, Ohnishi T. The development of a novel, validated, rapid and simple method for the detection of *Sarcocystis fayeri* in horse meat in the sanitary control setting. Biocontrol Sci. 2016;21(2):131–4. https://doi.org/10.4265/bio.21.131.

Gautam A, Dubey JP, Saville WJ, Howe DK. The SnSAG merozoite surface antigens of *Sarcocystis neurona* are expressed differentially during the bradyzoite and sporozoite life cycle stages. Vet Parasitol. 2011;183(1–2):37–42. https://doi.org/10.1016/j.vetpar.2011.06.024.

Gasbarre LC, Suter P, Fayer R. Humoral and cellular immune responses in cattle and sheep inoculated with *Sarcocystis*. Am J Vet Res. 1984;45:1592–6. PMID: 6433756.

Gjerde B. Molecular characterization of *Sarcocystis bovifelis*, *Sarcocystis bovini* n. sp, *Sarcocystis hirsuta* and *Sarcocystis cruzi* from cattle (*Bos taurus*) and *Sarcocystis sinensis* from water buffaloes (*Bubalus bubalis*). Parasitol Res. 2016;115:1473. https://doi.org/10.1007/s00436-015-4881-5.

Godoy R, Vilca M, Gonzáles A, Leyva V, Sam R. Saneamiento y detoxificación de carne de llama (*Lama glama*) infectada con *Sarcocystis aucheniae* mediante cocción, horneado, fritura y congelado. Rev Investig Vet del Perú. 2007;18(1):51–6. 10.15381/rivep.v18i1.1275.

Golubkov VI, Rybaltovskii DV, Kislyakova ZI. The source of infection for swine *Sarcocystis*. Veterinarya. 1974;11:85–7. (In Russian).

Hamidinejat H, Moetamedi H, Alborzi A, Hatami A. Molecular detection of *Sarcocystis* species in slaughtered sheep by PCR–RFLP from south-western of Iran. J Parasitol Dis. 2014;38(2):233–7. https://doi.org/10.1007/s12639-012-0231-z.

Hamir AN, Dubey JP. Myocarditis and encephalitis associated with *Sarcocystis neurona* infection in raccoons (*Procyonlotor*). Vet Parasitol. 2001;95:335–40. PMID: 11223214.

Harada S, Furukawa M, Tokuoka E, Matsumoto K, Yahiro S, Miyasaka J, Saito M, Kamata Y, Watanabe M, Irikura D, Matsumoto H, Sugita-Konishi Y. Control of toxicity of *Sarcocystis fayeri* in horsemeat by freezing treatment and prevention of food poisoning caused by raw consumption of horsemeat. Shokuhin Eiseigaku Zasshi. 2013;54(3):198–203. https://doi.org/10.3358/shokueishi.54.198.

Heydorn AO, Gestrich R, Mehlhorn H, Rommel M. Proposal for a new nomenclature of the Sarcosporidia. Z Parasitenkd. 1975;48:73–82. https://doi.org/10.1007/BF00389639.

Hiepe F, Litzke LF, Scheibner G, Jungmann R, Hiepe T, Montag T. Ultersuchungen zur toxischen Wirkung von Extrakten aus *Sarcocystis ovifelis* Makrozysten auf Kaninchen. Mh Vet Med. 1981;36:908–10.

Hoane JS, Carruthers VB, Striepen B, Morrison DP, Entzeroth R, Howe DK. Analysis of the *Sarcocystis neurona* microneme protein SnMIC10: protein characteristics and expression during intracellular development. Int J Parasitol. 2003;33:671–9. PMID: 12814647.

Howe DK, Gaji RY, Mroz-Barrett M, Gubbels M, Striepen B, Stamper S. *Sarcocystis neurona* merozoites express a family of immunogenic surface antigens that are orthologues of the *Toxoplasma gondii* surface antigens (SAGs) and SAG-related sequences. Infect Immun. 2005;73(2):1023–33. https://doi.org/10.1128/IAI.73.2.1023–1033.2005.

Jacquet A, Coulon L, De Neve J, Daminet V, Haumont M, Garcia L, Bollen A, Jurado M, Biemans R. The surface antigen SAG3 mediates the attachment of *Toxoplasma gondii* to cell-surface proteoglycans. Mol Biochem Parasitol. 2001;116:35–44. PMID: 11463464.

Jerrett IV, McOrist S, Waddington J, Browning JM, Malecki JC, McCausland IP. Diagnostic studies of the fetus, placenta and maternal blood from 265 bovine abortions. Cornell Vet. 1984;74:8–20.

Kalubowila DGW, Udagama-Randeniya PV, Perera NAN, Rajapakse RPV. Seroprevalence of *Sarcosystis* spp. in cattle and buffaloes from the wet and dry zones of Sri Lanka: a preliminary study. J Vet Med. 2004;51:89–93. https://doi.org/10.1111/j.1439-0450.2004.00726.x.

Kamata Y, Saito M, Irikura D, Yahata Y, Ohnishi T, Bessho T, Watanabe M, Sugita-Konishi Y. A toxin isolated from *Sarcocystis fayeri* in raw horsemeat may be responsible for food poisoning. J Food Prot. 2014;77(5):814–9. https://doi.org/10.4315/0362-028X.JFP-13-351.

Lewis SR, Ellison SP, Dascanio JJ, Lindsay DS, Gogal RM Jr, Werre SR, Surendran N, Breen ME, Heid BM, Andrews FM, Buechner-Maxwell VA, Witonsky SG. Effects of experimental *Sarcocystis neurona*-induced infection on immunity in an equine model. J Vet Med. 2014;2014:239495. https://doi.org/10.1155/2014/239495.

Levine N. The taxonomy of *Sarcocystis* (Protozoa: Apicomplexa) species. Parasitol Today. 1986;7:54–6. PMID: 3091802.

Lindsay DS, Dubey JP. Determination of the activity of pyrimethamine, trimethoprim, and sulfonamides and combinations of pyrimethamine and sulfonamides against *Sarcocystis neurona* in cell cultures. Vet Parasitol. 1999;82:205–21. PMID: 10348099.

Mahmood T, Yang PC. Western blot: technique, theory, and trouble shooting. N Am J Med Sci. 2012;4(9):429–34. https://doi.org/10.4103/1947-2714.100998.

Mansfield LS, Schott HC, Murphy AJ, Rossano MG, Tanhauser SM, Patterson JS, Nelson K, Ewart SL, Marteniuk JV, Bowman DD, Kaneene JB. Comparison of *Sarcocystis neurona* isolates derived from horse neural tissue. Vet Parasitol. 2001;95(2–4):167–78. https://doi.org/10.1016/S0304-4017(00)00388-5.

Martín M, Decker C, Romero S, Carletti T, Schnittger L, Florin-Christensen M. Molecular detection of *Sarcocystis aucheniae* in the blood of llamas from Argentina. Rev Argent Microbiol. 2016;48(3):200–5. https://doi.org/10.1016/j.ram.2016.03.009.

McKenna PB, Charleston WAG. The outdoor survival of *Sarcocystis gigantea* sporocysts. Vet Parasitol. 1994;55:21–7. PMID: 7886918.

Meshkov S. The jackal (*Canis aureus*) as a new host of *Sarcocystis* infecting swine. Vet Sbirka. 1980;78:20–1. (In Russian).

Moré G, Abrahamovich P, Jurado S, Bacigalupe D, Marin JC, Rambeaud M, Venturini L, Venturini MC. Prevalence of *Sarcocystis* spp. in Argentinean cattle. Vet Parasitol. 2010;177(1–2):162–5. https://doi.org/10.1016/j.vetpar.2010.11.036.

Moré G, Regensburger C, Gos ML, Pardini L, Verma SK, Ctibor J, Serrano-Martínez ME, Dubey JP, Venturini MC. *Sarcocystis masoni*, n. sp. (Apicomplexa: Sarcocystidae), and redescription of *Sarcocystis aucheniae* from llama (*Lama glama*), guanaco (*Lama guanicoe*) and alpaca (*Vicugna pacos*). Parasitology. 2016;143(5):617–26. https://doi.org/10.1017/S003118201600007X.

Moré G, Schares S, Maksimov A, Conraths FJ, Venturini MC, Schares G. Development of a multiplex real time PCR to differentiate *Sarcocystis* spp. affecting cattle. Vet Parasitol. 2013;197(1–2):85–94. https://doi.org/10.1016/j.vetpar.2013.04.024.

Munday BL, Obendorf DL. Development and growth of *Sarcocystis gigantea* in experimentally infected sheep. Vet Parasitol. 1984;15:203–11. PMID: 6437054.

Nakamura Y, Saito M, Shibata Y, Itagaki H. Induction of tumor necrosis factor α and nitric oxide in rabbits inoculated with a cyst extract of *Sarcocystis cruzi*. Vet Parasitol. 1999;85:235–43. https://doi.org/10.1016/S0304-4017(99)00128-4.

Notomi T, Mori Y, Tomita N, Kanda H. Loop-mediated isothermal amplification (LAMP): principle, features, and future prospects. J Microbiol. 2015;53:1–5. https://doi.org/10.1007/s12275-015-4656-9.

Nourani H, Matin S, Nouri A, Azizi H. Prevalence of thin-walled *Sarcocystis cruzi* and thick-walled *Sarcocystis hirsuta* or *Sarcocystis hominis* from cattle in Iran. Trop Anim Health Prod. 2010;42:1225–7. https://doi.org/10.1007/s11250-010-9552-z.

Nourollahi Fard SR, Asghari M, Nouri F. Survey of *Sarcocystis* infection in slaughtered cattle in Kerman, Iran. Trop Anim Health Prod. 2009;41:1633–6. https://doi.org/10.1007/s11250-009-9358-z.

Ortega-Mora LM, Gottstein B, Conraths FJ, Buxton D. Protozoal abortions in farm ruminants: guidelines for diagnosis and control. Wallingford: CAB International; 2007. https://doi.org/10.1079/9781845932114.0000.

Poulsen CS, Stensvold CR. Current status of epidemiology and diagnosis of human sarcocystosis. J Clin Microbiol. 2014;52(10):3524–30. https://doi.org/10.1128/JCM.00955-14.

Pusterla N, Mackie S, Packham A, Conrad PA. Serological investigation of transplacental infection with *Neospora hughesi* and *Sarcocystis neurona* in broodmares. Vet J. 2014;202(3):649–50. https://doi.org/10.1016/j.tvjl.2014.09.015.

Rachimel N, Buzoni-Gatel D, Dutta C, Mennechet FJ, Luangsay S, Minns LA, Grigg ME, Tomavo S, Boothroyd JC, Kasper LH. The induction of acute ileitis by a single microbial antigen of *Toxoplasma gondii*. J Immunol. 2004;173:2725–35. https://doi.org/10.4049/jimmunol.173.4.2725.

Reed SM, Furr MD, Howe K, Johnson AL, MacKay RJ, Morrow JK, Pusterla N, Witonsky S. Equine Protozoal Myeloencephalitis: an updated consensus statement with a focus on parasite biology, diagnosis, treatment, and prevention. J Vet Intern Med. 2016;30:491–502. https://doi.org/10.1111/jvim.13834.

Rodrigues JS, Meireles GS, Carvalho Filho PR, Ribeiro CT, Flausino W, Lopes CWG. *Sarcocystis cruzi* (Apicomplexa: Sarcocystidae) no cachorro do mato (*Cerdocyon thous*). Pesqui Vet Bras. 2008;28:561–4. https://doi.org/10.1590/S0100-736X2008001100004.

Romero SA, Carletti T, Decker Franco C, Moré G, Schnittger L, Florin-Christensen M. Seropositivity to *Sarcocystis* infection of llamas correlates with breeding practices. Vet Parasitol Reg Studies and Reports. 2017;10:68–70.

Rooney AL, Limon G, Vides H, Cortez A, Guitian J. *Sarcocystis* spp. in llamas (*Lama glama*) in Southern Bolivia: a cross sectional study of the prevalence, risk factors and loss in income caused by carcass downgrades. Prev Vet Med. 2013;116(3):296–304. https://doi.org/10.1016/j.prevetmed.2013.11.014.

Rossano MG, Mansfield LS, Kaneene JB, Murphy AJ, Brown CM, Schott HC, Fox JC. Improvement of western blot test specificity for detecting equine serum antibodies to *Sarcocystis neurona*. J Vet Diagn Investig. 2000;12:28–32. https://doi.org/10.1177/104063870001200105.

Saito M, Itagaki H. Experimental infection of raccoon dogs with *Sarcocystis cruzi* and *S. miescheriana*. J Vet Anim Sci. 1994;56:671–4. PMID: 7999889.

Saleque A, Bhatia BB, Juyal PD, Rahman H. Toxicity of cyst extract of *Sarcocystis fusiformis* from buffalo in rabbits and mice. Vet Parasitol. 1991;38:61–5. https://doi.org/10.1016/0304-4017(91)90009-K.

Saville WJ, Morley PS, Reed SM, Granstrom DE, Kohn CW, Hinchcliff KW, Wittum TE. Evaluation of risk factors associated with clinical improvement and survival of horses with equine protozoal myeloencephalitis. J Am Vet Med Assoc. 2000;217(8):1181–5. PMID: 11043689.

Savini G, Robertson ID, Dunsmore JD. Viability of the sporocysts of *Sarcocystis cruzi* after exposure to different temperatures and relative humidities. Vet Parasitol. 1996;67:153–60. https://doi.org/10.1016/S0304-4017(96)01046-1.

Soulsby EJ. Parasitología y enfermedades parasitarias en los animales domésticos. 7th ed. México: Interamericana; 1987. p. 823.

Tenter AM. Current research on *Sarcocystis* species of domestic animals. Int J Parasitol. 1995;25:1311–30. https://doi.org/10.1016/0020-7519(95)00068-D.

Uggla A, Buxton D. Immune responses against *Toxoplasma* and *Sarcocystis* infection in ruminants: diagnosis and prospect for vaccination. Rev Sci Tech. 1990;2:441–619. PMID: 2132690.

Yeargan M, Rocha I, Morrow J, Graves A, Reed SM, Howe DK. A new trivalent SnSAG surface antigen chimera for efficient detection of antibodies against *Sarcocystis neurona* and diagnosis of equine protozoal myeloencephalitis. J Vet Diagn Investig. 2015;27(3):377–81. https://doi.org/10.1177/1040638715584995.

Zhang D, Howe DK. Investigation of SnSPR1, a novel and abundant surface protein of *Sarcocystis neurona* merozoites. Vet Parasitol. 2008;152(3–4):210–9. https://doi.org/10.1016/j.vetpar.2007.12.036.

5 Neospora

Dadin Prando Moore and Maria Cecilia Venturini

Abstract

This chapter emphasizes the research performed on canine and bovine neosporosis during almost three decades after its description. This parasitic disease caused by the protozoan *Neospora caninum* is associated with neuromuscular disorders in dogs and abortions in cattle. Neosporosis was first recognized in dogs in Norway, but its description of *N. caninum* as a new genus and species was proposed in the USA by Dubey and collaborators in 1988. After the protozoan was isolated from dogs suffering the disease, neosporosis was recognized as the first cause of bovine abortion in dairy cattle from California, USA. Noteworthy, *N. caninum* is one of the most efficiently congenitally-transmitted organisms. *N. caninum*-infected cows deliver up to 90% of congenitally infected calves. After the indirect fluorescent antibody test was described for serological diagnosis of neosporosis, many serological tests were developed for detection of antibodies in cattle and other domestic and wild species. The definitive host was unknown until 1998, when *N. caninum* oocysts were identified and characterized in dogs experimentally fed with infected mouse tissues. Experimental oral infection of cattle with *N. caninum* oocysts was achieved later on. Nowadays, additional definitive hosts have been described: the coyote (*Canis latrans*), the Australian dingo (*Canis lupus dingo*), and the wolf (*Canis lupus*). Treatments of neuromuscular neosporosis in dogs have been successfully described, but no vaccine or drugs are available for controlling neosporosis in cattle even though economic losses have been estimated in over 1 billion dollars in both beef and dairy industry worldwide.

D.P. Moore (✉)
National Council of Scientific and Technological Research (CONICET), Buenos Aires, Argentina
e-mail: moore.dadin@inta.gob.ar

M.C. Venturini
Laboratory of Immunoparasitology (LAINPA), School of Veterinary Sciences, National University of La Plata, La Plata, Argentina

M. Florin-Christensen, L. Schnittger (eds.), *Parasitic Protozoa of Farm Animals and Pets*, https://doi.org/10.1007/978-3-319-70132-5_5

5.1 Morphology, Life Cycle, and Host-Pathogen Interactions

Neosporosis caused by the protozoan *Neospora caninum* is a parasitic disease associated with neuromuscular disorders in dogs and abortions in cattle. Probably misdiagnosed as toxoplasmosis until the late 1980s, neosporosis was first described as a neuromuscular syndrome in dogs (Bjerkås et al. 1984; Dubey et al. 1988a, b). Because most research groups working on toxoplasmosis worldwide extrapolated their knowledge and technology in knowing this parasitic disease, many advances were possible.

Neosporosis was first recognized in dogs in Norway (Bjerkås et al. 1984). The description of *N. caninum* as a new genus and species was proposed in the USA (Dubey et al. 1988a), and the parasite was isolated from dogs suffering the disease (Dubey et al. 1988b; Lindsay and Dubey 1989). Then, neosporosis was identified as cause of abortion in dairy cattle in New Mexico, USA (Thilsted and Dubey 1989). *N. caninum* transplacental transmission was experimentally induced in cats, dogs, sheep, and cattle (Dubey and Lindsay 1989a, b, 1990; Dubey et al. 1992). Recognized as the first cause of bovine abortion in dairy cattle from California, USA (Barr et al. 1991), the disease was experimentally induced in cattle using parasites isolated from bovine aborted fetuses (Barr et al. 1994). Noteworthy *N. caninum* is one of the most efficiently congenitally-transmitted organisms. *N. caninum*-infected cows deliver up to 90% infected calves (Anderson et al. 1997). After the indirect fluorescent antibody test was described for serological diagnosis of neosporosis, many ELISAs were developed for detection of antibodies in cattle and other domestic and wild species (Björkman et al. 1994; Paré et al. 1995; Baszler et al. 1996; Dubey et al. 1996). The definitive host was unknown until 1998, when *N. caninum* oocysts were identified and characterized in dogs experimentally fed with infected mouse tissues (McAllister et al. 1998; Lindsay et al. 1999a). Indeed experimental oral infection of cattle with *N. caninum* oocysts was achieved later on (de Marez et al. 1999). Three years later, after the dog was recognized experimentally as definitive host, *N. caninum* oocysts were isolated and identified in the feces from a naturally infected dog for the first time (Basso et al. 2001a, b). Additional studies described the coyote, *Canis latrans*; the Australian dingo, *Canis lupus dingo*; and the wolf, *Canis lupus* as definitive hosts (Gondim et al. 2004; King et al. 2010; Dubey et al. 2011). Nowadays, the entire genome sequence information on *N. caninum* Nc-Liverpool isolate is available (http://www.genedb.org/Homepage/Ncaninum).

N. caninum belongs to the **phylum** Apicomplexa, **class** Sporozoasida, **subclass** Coccidiasina, **order** Eucoccidiorida, **suborder** Eimeriorina, **family** Sarcocystidae, **genus** *Neospora* (Dubey et al. 1988a, 2002; Goodswen et al. 2013). Another species of *Neospora*, *N. hughesi*, was described from a horse in California, USA, in 1998 (Marsh et al. 1998).

5.1.1 Morphology

There are three infectious stages of *N. caninum*: tachyzoites, bradyzoites, and sporozoites. Tachyzoites are crescent-shaped, approximately 2 × 7.5 μm with a pointed anterior—conoidal—end and a rounded posterior end and represent the rapid multiplying stage. In the host cell cytoplasm, tachyzoites are located within a parasitophorous vacuole (PV), surrounded by a parasitophorous vacuolar membrane (PVM). The tachyzoite has an outer plasma membrane; two inner membranes; cytoskeletal elements such as microtubules, apical and polar rings, and conoid; and secretory organelles, such as rhoptries, micronemes, dense granules, a long single mitochondrion, lipid body, a Golgi complex, ribosomes, rough and smooth endoplasmic reticula, micropore, nucleus, amylopectin granules, and an apicoplast, a non-photosynthetic plastid organelle (Dubey and Lindsay 1996).

Almost all cell types in the body, including neural cells, endothelial cells, dermal cells, retinal cells, macrophages, hepatocytes, and fibroblasts can be hosts for tachyzoites (reviewed in Dubey and Lindsay 1996).

It is generally known that tachyzoites (Fig. 5.1) move by gliding, flexing, undulating, and rotating. The glideosome is an actin-myosin motor complex which provides their motility. In a few seconds, initial host cell contact is mediated by parasite surface antigens (SAGs). Then, the discharge of secretory organelles (MICs, ROPs,

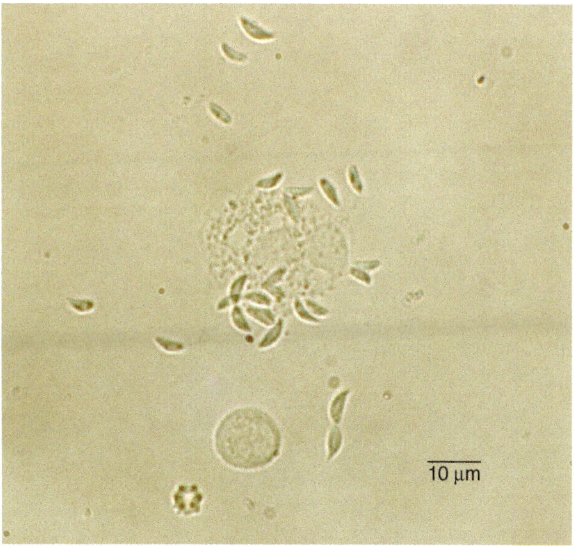

Fig. 5.1 *N. caninum* tachyzoites strain NC Argentina LP-1, bovine isolate from a congenitally-infected asymptomatic dairy male calf. Microphotograph gently provided by Dr. LM Campero, LAINPA, FCV, UNLP. 400×

RONs) governs host cell invasion. Once the invasion process is achieved, a set of secretory proteins—dense granule (GRA) proteins—are released into the PV (Carruthers and Blackman 2005).

Studied in *T. gondii*, the host cell plasmalemma forms the PVM. Tachyzoites secrete several molecules which ultimately give origin to a tubular membranous network (TMN) (Sibley et al. 1995). Tachyzoites multiply asexually within the PV by repeated endodyogeny. Both invasion and egression steps are regulated by Ca^{2+} (Behrendt et al. 2008).

Bradyzoites represent the slow multiplying stage, and they are enclosed in host tissue cysts. They also remain intracellular and divide by endodyogeny. In early stages, cysts may be as small as 5 μm in diameter, but as bradyzoites divide, they may contain more than 200 organisms and grow over 100 μm in diameter (reviewed in Dubey and Lindsay 1996). The cyst wall is 0.5–4.0 μm thick in histological sections. In contrast, intramuscular cysts found in striated muscle are elongated and have a thin cyst wall (Dubey et al. 2004; Peters et al. 2001). The tissue cyst wall is elastic, argyrophilic, and PAS negative. Bradyzoites are loosely arranged inside the cyst (Fig. 5.2, gently provided by Dr. Walter Basso). Both bradyzoites and tissue cysts autofluorescence under UV light (Lei et al. 2005). The size of bradyzoites is variable, averaging 6.5 × 1.5 μm (Dubey et al. 2004). Bradyzoites contain several amylopectin granules and fewer rhoptries as compared to tachyzoites, while they contain more micronemes, also posterior to the nucleus. Tissue cysts and

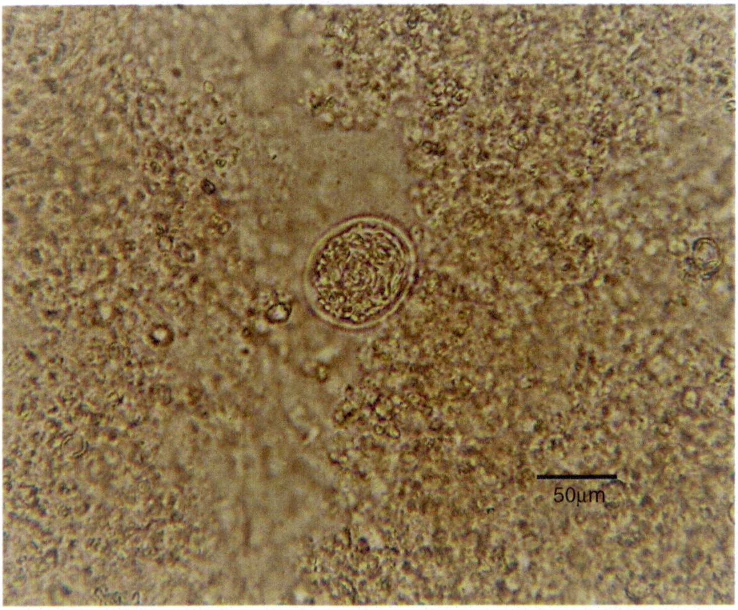

Fig. 5.2 Thick-walled *N. caninum* cyst observed in fresh brain samples from axis deer (*Axis axis*). Microphotograph gently provided by Drs Venturini and Basso, LAINPA, FCV, UNLP. 400×

bradyzoites can survive up to 14 days at refrigeration temperature—4 °C—but they can be killed by freezing (reviewed in Dubey and Lindsay 1996; Lindsay et al. 1992).

The stages of schizogony and gametogony remain unknown, and oocysts represent the only known sexual stage of *N. caninum*. Unsporulated oocysts, 10–11 μm in diameter, are shed in canid feces. Their wall measures 0.6–0.8 μm thick and is colorless. Sporulated oocysts contain two sporocysts and a residual body and measure 8.4 × 6.1 μm (Lindsay et al. 1999b). Each sporocyst contains four sporozoites. It has been shown that oocysts can remain infective almost after 4-year storage at 4 °C (Uzeda et al. 2007).

5.1.2 Life Cycle

The life cycle of the parasite is partially known and a scheme is shown in Fig. 5.3. Transmission can occur through carnivorism, transplacental infection, and the fecal-oral route (reviewed in Dubey et al. 2007).

The definitive hosts for *N. caninum* are the dog (*Canis familiaris*) (McAllister et al. 1998), the coyote (*Canis latrans*) (Gondim et al. 2004), the dingo (*Canis lupus dingo*) (King et al. 2010), and the gray wolf (*Canis lupus*) (Dubey et al. 2011).

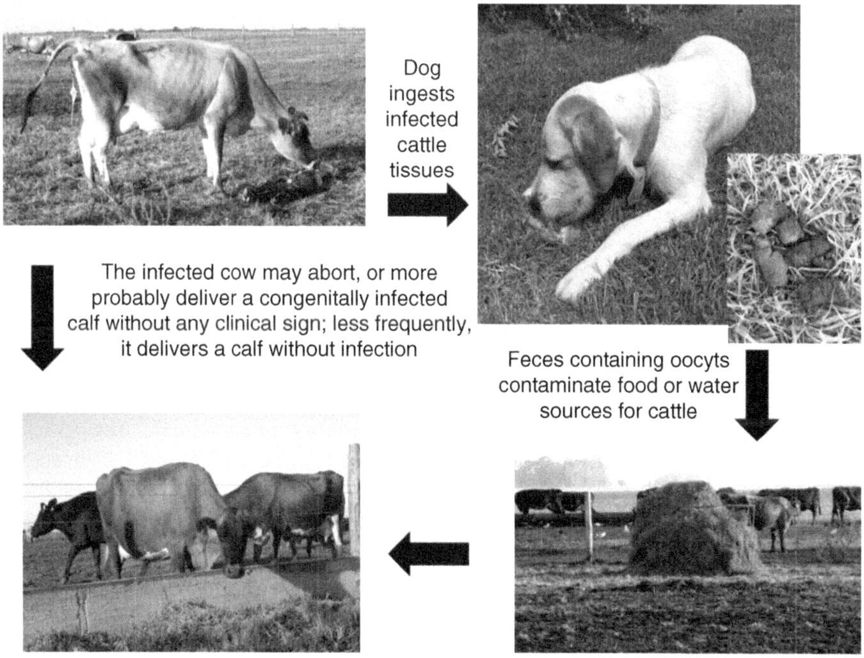

Fig. 5.3 Scheme of the parasite cycle of *N. caninum*

Recognized intermediate hosts based on isolation of viable parasites are axis deer, white-tailed deer, cow, dog, European bison, gray wolf, sheep, and water buffalo. Additionally, *Neospora* DNA has been identified in avian species, badger, brown bear, capybaras, ferret, hoary fox, mink, mouse, otter, marten, rabbit, rat, red fox, shrew, squirrel, stoat, and voles (Dubey and Schares 2011).

N. caninum is transmitted postnatally—horizontally—to carnivorous hosts by ingestion of tissues infected with tachyzoites or tissue cysts and to herbivorous hosts by ingestion of food or drinking water contaminated by sporulated oocysts. The parasite can be transmitted transplacentally—vertically—from an infected dam to her progeny during pregnancy. To describe precisely the origin of the transplacental infection, the terms "exogenous transplacental transmission" and "endogenous transplacental transmission" have been proposed (Trees and Williams 2005). The first one occurs after a primary, oocyst-derived infection of a pregnant dam, while the second takes place in a persistently infected dam after recrudescence of the infection during pregnancy.

How postnatal transmission occurs in nature remains unclear. Indeed, it is unknown how oocysts can be easily spread and disseminated in pasture. Contamination of water sources is a more plausible explanation, but evidence is still lacking. Not only *N. caninum* oocysts have been identified in only a few dogs worldwide, but also the numbers of oocysts shed by dogs is usually low (Dubey and Schares 2011).

Natural infections can be acquired by ingestion of tissue cysts in infected meat or oocysts from food and water contaminated with dog feces. Then, sporozoites excyst in the gut and spread locally to mesenteric lymph nodes and via lymphatic circulation reach the blood and, eventually, distant organs (Dubey et al. 2006). Although *N. caninum* has been detected in semen (Ortega-Mora et al. 2003), venereal transmission is unlikely. Transplacental and lactogenic transmission has been only demonstrated in mice (Cole et al. 1995). The sylvatic cycle for *N. caninum* has been partially described (Almería 2013). A complete picture of all possible routes requires further studies.

5.1.3 Host-Pathogen Interactions

Once infection occurs, parasite-host cell cross talk takes place. Infection by apicomplexan parasites modulates the host cell cytoskeleton dynamics (Cardoso et al. 2016). Microtubules start to surround the parasite upon the first minutes of invasion by forming a cone-shaped microtubule network, and a microtubule ring on the host cell is observed around the parasite entrance site. *N. caninum* and *T. gondii* also affect the positioning of the centrosome, which is closely associated with the nucleus in uninfected cells, but during *N. caninum* and *T. gondii* infection, the host cell centrosome is recruited away from the nuclear membrane and positioned near the PVM. *N. caninum* infection also modulates the spatial organization of host cell actin microfilaments and intermediate filaments, as visualized in astrocytes of organotypic rat brain slice cultures. F-actin bundles and glial fibrillary acid protein filaments have been found in close juxtaposition to the cytoplasmic side of the PVM (Vonlaufen et al. 2002).

Dendritic cells (DC) exposed to *N. caninum* tachyzoites have an increased expression of IL-12p40, IL-10, and TNF-α (Strohbusch et al. 2009). In agreement, *N. caninum* lysate antigen preparations were less effective in inducing IL-12, IFN-γ, and TNF-α than whole tachyzoites (Feng et al. 2010). Macrophage-depleted mice are more susceptible to *N. caninum* infection. Increased mortality and neurological impairment were observed in the *N. caninum*-infected mice lacking the CCR5 chemokine receptor. Poor migration of DC and natural killer T (NKT) cells to the site of infection was observed in $CCR5^{-/-}$ mice, and higher levels of IFN-γ and CCL5 expression, which are associated with brain tissue damage, were observed in the brain tissue of $CCR5^{-/-}$ mice during the acute phase of infection (Abe et al. 2015).

N. caninum triggers a type I IFN α/β response in infected host cells (Beiting et al. 2014). In both mice and cattle, host responses to *N. caninum* are dependent on the Toll-like receptor TLR3 (Beiting et al. 2014; Marín et al. 2017). C57BL/10ScCr mice that lack Toll-like receptor 4 and a functional IL-12 receptor are highly susceptible to neosporosis (Botelho et al. 2007). Nucleotide-binding oligomerization domain (NOD)-like receptors NOD1 and NOD2 seem to be another important component governing the innate immune response. NOD2 is involved in macrophage responsiveness against *N. caninum* infection in vitro and in vivo, but NOD2-triggered responses are required but not essential for *N. caninum* growth restriction during sublethal infection (Davoli-Ferreira et al. 2016).

Humoral and cellular adaptive immune responses are important to control the infection (Hemphill and Gottstein 2006). The principal protection mechanisms involve IFN-γ and IL-12 (Khan et al. 1997; Kasper and Khan 1998; Baszler et al. 1999a; Nishikawa et al. 2001a, b; Ritter et al. 2002). Noteworthy, CD4+ T-cells are crucial for protection against *N. caninum* infection, while CD8+ T-cells are not (Tanaka et al. 2000). Since μMT-antibody KO mice are more susceptible to *N. caninum* infection, humoral immune responses also appear important (Eperon et al. 1999). It has been proposed that an appropriate Th population balance is required (Bartley et al. 2009; Nishikawa et al. 2003). In pregnant mice, similar to the situation in cattle, immunomodulation toward a Th population response associated with high IL-4 production is usually observed during pregnancy, thus favoring parasite proliferation and vertical transmission (Innes et al. 2002).

Earlier studies focused on the relationship between host cell apoptosis and *N. caninum* infection. It was demonstrated that in this case, apoptosis is associated to an increased caspase-3 and caspase-8 activity (Nishikawa et al. 2001a, b). Interestingly, *N. caninum* bradyzoites, as well as their host cells, are predestined for long-term survival. *N. caninum* inhibits host cell apoptosis in the absence of discernible Nf-kappaB activation (Herman et al. 2007).

5.2 Diagnosis and Epidemiology

5.2.1 Diagnosis

Many diagnostic serological tests have been used in individual animals and/or in sero-epidemiologic observations (Dubey and Lindsay 1996). However, since most studies have used tachyzoite-derived antigens, infections might have been underdiagnosed in cases when antibodies against antigens derived from other parasites stages—bradyzoites and sporozoites—were not detected. Ancillary tests, like cytology, immunohistochemistry (IHC), electron microscopical examinations, and detection of parasitic DNA by PCR, have also been extensively used. Although a definitive diagnosis can be achieved only by complementary techniques, a relevant role is played by pathologists when observing histological sections, because characteristic lesions crucially show the unbalance between host and parasite (Anderson et al. 2000).

Serological tests are applied antemortem and can provide information on the stage of infection. Indirect fluorescent antibody test (IFAT), *Neospora* agglutination test (NAT), latex agglutination test (LAT), and various enzyme-linked immunosorbent assays (ELISAs) have been validated. Also, an applied printing immunoassay with recombinant Nc-SAG1 for detection of antibodies to *N. caninum* in cattle has been recently developed (Wilkowsky et al. 2011).

IFAT, considered a gold standard test, is based on the use of intact tachyzoites which can be fixed to glass slides (Dubey et al. 1988b). Positive sera should result in continuous peripheral fluorescence of the tachyzoites, while a partial, only polar fluorescence, must be interpreted as negative. IFAT has disadvantages because the reading is subjective and may vary when using different microscopes, protocols, and reagents. In addition, there is no established diagnostic cut-off titer. A serum dilution of 1:200 has been proposed as the best option to detect infected cattle, having the highest sensitivity and specificity (Reichel and Drake 1996).

In NAT, whole formalin-killed tachyzoites are used as antigen. No special equipment or conjugates are needed in this test (Romand et al. 1998). The NAT was validated using a larger number of sera from several species. Also examination of a large variety of animal species is possible by using LATs based on recombinant antigens (NCSAG1) (Moraveji et al. 2012). Specific antibodies from several wildlife species have been reported using cELISA because no specific conjugates are available. Others have replaced species-specific conjugates by protein G and A conjugates.

A large number of sera can be examined rapidly by ELISA, and numerous variations of this type of test have been published, including assays based on ISCOM-incorporated antigens or recombinant antigens and competitive and blocking ELISAs. Some are commercially available (Björkman et al. 1994; Baszler et al. 1996; Aguado-Martínez et al. 2008; Hosseininejad et al. 2010; Alvarez-García et al. 2013). Interestingly, by using recombinant antigen-based ELISAs, it is possible to differentiate the stage of infection by examining the antibody responses against different antigens to which a host is differentially exposed—acute or chronic stage (Aguado-Martínez et al. 2008).

During the course of infection, the avidity of specific antibodies to parasite antigens changes. Avidity values are low in the acute stage of infection, and, during infection progression, avidity values increase. The difference in titers of sera assessed with or without urea treatment can be used to determine whether an infection is new or chronic (Björkman et al. 1999; McAllister et al. 2000).

Gel electrophoresis (SDS-PAGE) has been largely used to generate well-characterized reference samples for the validation of new tests. In addition, immunoblots are important to confirm inconclusive findings and to detect antibody reactions to particular antigens (Álvarez-García et al. 2002; Campero et al. 2015a, b). Conformational epitopes are predominantly involved in the *N. caninum*-specific antibody response since stronger reactions have been observed against nonreduced antigens of around 17–19, 29–30, and 35–37 kDa (Campero et al. 2015a, b). Also, patterns of recognized antigens change during the course of infection.

Dogs may not seroconvert for tachyzoite antigens after excreting oocysts. Nevertheless, serological assays are important tools in epidemiological studies to examine the exposure of dogs to *N. caninum*. In dogs suffering clinical signs, the demonstration of serum antibodies to *N. caninum* by any of the different tests can help confirm the diagnosis. Although IFAT titers were generally high in cases with severe neosporosis, there is no correlation found between the magnitude of titers and clinical signs. Based on serological findings, clinical signs, and diagnostic imaging procedures, a proper treatment can be defined (Dubey 2013).

Because in ruminants there is no passive transfer of antibodies across the placenta, serology provides relevant information in prenatally infected animals. Sampling of blood before colostrum intake is used for diagnosis of congenital infection when testing newborn calves. Also, other body fluids (pleural or peritoneal fluids or exudates) can be used for detection of specific antibodies (Anderson et al. 2000; Caspe et al. 2012).

Gross lesions are rare and the fetuses are autolyzed and mummified (Fig. 5.4). Hydrocephalus may occur and pale white foci may be present in skeletal muscles and the heart (Dubey et al. 1988a). Focal discolorated areas have been reported in the placenta (Fioretti et al. 2003).

Rapid diagnosis can be achieved by direct observation and/or conventional Giemsa stain after cytospin or impression smears, mostly from mice. Conventional histopathology is very useful in recognition of lesions and arriving at provisional diagnosis. Tissue cysts of *N. caninum* have a thicker wall—up to 4 μm—and are more easily found in dogs than in cattle. TEM has been used to differentiate *N. caninum* rhoptries from those in other protozoans. *N. caninum* tachyzoite rhoptries are electrodense with an amorphous matrix, and few extend posterior to the nucleus. IHC staining is necessary to confirm the presence of the parasite in tissues showing characteristic lesions such as multifocal necrotizing nonsuppurative meningoencephalitis (Fig. 5.5). Formalin-fixed, paraffin-embedded tissues have been commonly used for IHC staining for specimens submitted to diagnostic laboratories. Although *N. caninum* antigens can be detected even in tissues preserved in formalin for years, fixation for short periods (24 h) is recommended. Both the avidin-biotin complex (ABC) indirect immunoperoxidase method and the peroxidase-antiperoxidase (PAP) technique are good,

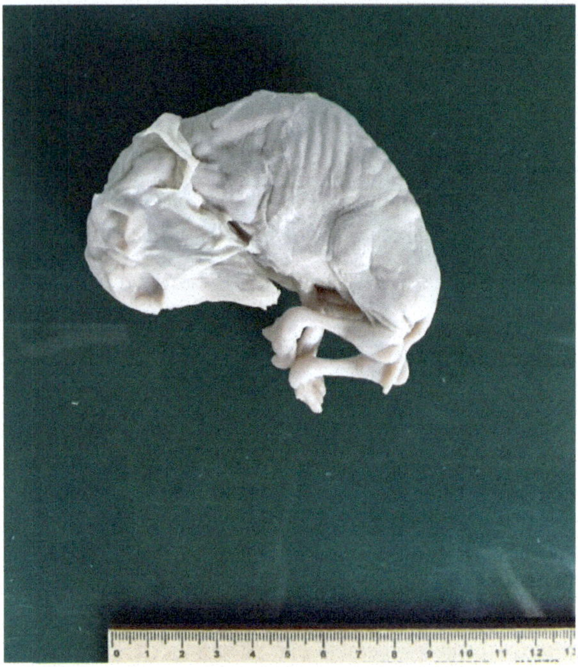

Fig. 5.4 Bovine fetal mummification due to *N. caninum* infection

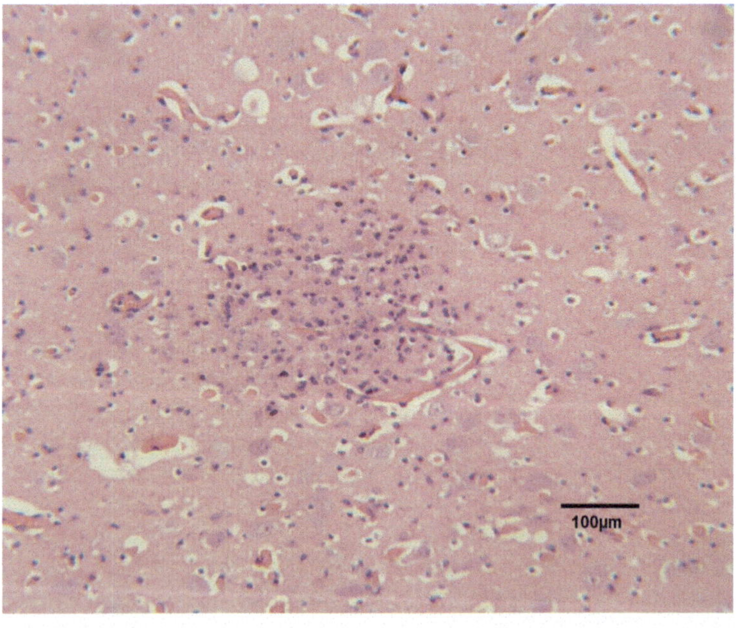

Fig. 5.5 Focal nonsuppurative encephalitis due to *N. caninum* infection in a bovine fetus 100×

although proper studies to compare their concordance are still lacking. For diagnostic purposes, polyclonal antibodies are equally good (Campero et al. 1998). Isolation of viable *N. caninum* by bioassay has little practical importance for routine diagnosis but is essential to study the population structure of *N. caninum* (Regidor-Cerrillo et al. 2013). Indeed, the number of viable parasites in naturally infected tissues is usually low, and isolates are required in order to study differences in expression of various proteins between strains of *N. caninum* to explain differences in virulence and immunogenicity (Regidor-Cerrillo et al. 2013).

Several gene targets from *N. caninum* DNA can be detected by several methods. Moreover, PCR protocols were developed that not only detect but also quantify *N. caninum* DNA. Quantitative PCR has become one of the key methodologies to examine the course of infection and to assess the activity of vaccines or prophylactic drugs (Pereira et al. 2014; Pinitkiatisakul et al. 2008). Different target DNAs were chosen to establish *N. caninum*-specific primer pairs. Genes coding for rRNA and the pNc5 gene are important targets for diagnostic and quantitative PCRs because of their repetitive character. On the other hand, many PCR protocols using the ITS1 region as target have been published. One study compared the performance of an ITS1-based (Holmdahl and Mattsson 1996) and a pNc5-based PCR (Np4/Np7) (Basso et al. 2001a, b) and observed a higher sensitivity in the former (Sánchez et al. 2009). Although two-step nested PCRs are superior in sensitivity, they have the disadvantage of having a higher risk of contamination. As alternative, a single-tube nested PCR combines the higher sensitivity of a nested PCR with the lower risk of contamination. An ITS1-based one-tube nested PCR for *N. caninum* has been developed (Ellis et al. 1999), and an analytical sensitivity of 1–10 fg genomic DNA of *N. caninum* tachyzoites was reported, which is considered equivalent to the genomic DNA of 0.1–0.01 tachyzoites (Ellis et al. 1999). Modified primers, Np6plus and Np21plus (Müller et al. 1996), were included into a multiplex PCR to identify a number of infectious agents from aborted bovine clinical samples (Tramuta et al. 2011).

All quantitative PCRs developed so far are based on the pNc5 gene. PCR approaches based on SYBR Green incorporation are less laborious (Collantes-Fernández et al. 2002; Ghalmi et al. 2008; Okeoma et al. 2005; Pinitkiatisakul et al. 2008). By using a SYBR Green I-based real-time PCR, the number of tachyzoites in brain samples of aborted bovine fetuses was estimated 2.9–26.6 per mg of brain. The disadvantage of SYBR Green assays is that also unspecific amplification products are labeled, and amplicons need to be carefully checked for specificity. On the other hand, real-time PCRs employing TaqMan probes, which increase the specificity of the reaction (Reitt et al. 2007; Pereira et al. 2014), were employed for diagnostic purposes in cattle (Pessoa et al. 2016) and in dogs (Legnani et al. 2015).

Microsatellite markers are used to characterize new *N. caninum* isolates (Campero et al. 2015a, b) and confirm the identity of strains applied in experimental animal infections (Bacigalupe et al. 2013). In addition, microsatellite typing was used to clarify epidemiological situations. In clinical material, only identical or almost identical microsatellite patterns were observed in DNA collected from several fetuses aborted in individual herds during bovine abortion storms (Basso et al. 2010).

5.2.2 Epidemiology

Dogs have a key role in the epidemiology because they play both the role of definitive and intermediate hosts in the life cycle of *N. caninum* (Dubey 2013). Many studies showed that the serological prevalence is high. Risk factors for *N. caninum* infection are discussed as follows:

1. Age: most dogs become infected postnatally, probably by the ingestion of tissue cysts containing material from intermediate hosts. It should be taken into account that only a small portion of dogs seroconvert and develop antibodies against *N. caninum* tachyzoites (Bandini et al. 2011).
2. Gender: reactivation of *N. caninum* during gravidity may explain why non-spayed female dogs are significantly more often seropositive than males (Goždzik et al. 2011).
3. Breed: with still some controversy, two studies showed higher seroprevalences in Siberian huskies in Japan (Kubota et al. 2008) or Boxer breed in Italy (Cringoli et al. 2002). There is a need to further establish the susceptibility of particular breeds.
4. Presence of intermediate hosts of *N. caninum:* cattle farm dogs have a significantly higher risk of being positive compared to urban dogs (Basso et al. 2001b).
5. Habitat and type of dogs: street dogs are more prone to infection, because they have access to outdoors food, including waste or prey animals (Gennari et al. 2002). Also dogs used for hunting showed an increased risk to be *N. caninum*-seropositive (Nogueira et al. 2013; Sicupira et al. 2012).

In cattle, seroprevalence studies do not necessarily reflect the true prevalence, because in most of them, the sensitivity and specificity of the applied tests was not established. However, there are general conclusions that can be drawn: (a) The seroprevalence is higher in cattle with history of abortions; (b) seroprevalence is higher in dairy cattle than in beef cattle; however, whether breed or management—or an interaction of both—is responsible needs to be clarified; and (c) seroprevalence increases with age.

The epidemiological risk factors in cattle have been reviewed and grouped in those associated to infection and to abortion (Dubey et al. 2007). Most of them are discussed as follows:

1. History of reproduction problems: cattle having abortion/stillbirths were reported as risk factors for seropositivity (Moore et al. 2009).
2. Cattle biotype: In some studies, dairy cattle showed a higher risk to be seropositive (Moore et al. 2002; Quintanilla-Gozalo et al. 1999); nevertheless, there is no general rule on which type of cattle has a higher risk of infection.
3. Cattle stocking density and size of farm land: high stocking density was identified as a potential risk factor for seropositivity (Barling et al. 2000). Seroprevalence decreased in herds with increasing sizes of farm lands used for forage production in dairy cattle in southern Brazil and Canada (Corbellini et al. 2006; VanLeeuwen et al. 2010).

4. Breed: although different humoral mechanisms against *N. caninum* infection and abortion were observed in pure-breed pregnancies compared with cross-breed pregnancies in beef/dairy cattle (Santolaria et al. 2011), studies regarding an increased susceptibility of different cattle breeds are still lacking.
5. Age: the risk of being seropositive may increase with the age and/or parity number in beef and dairy cattle suggesting postnatal infection (Moore et al. 2009, 2014). Interestingly, the higher the parity number of seropositive dams, the lower the chance of progeny to be prenatally infected was (Dijkstra et al. 2003).
6. Source of replacement heifers: rearing replacement heifers in farms with endemic infections may be associated with persistent *N. caninum* infections in the herd. In contrast, purchasing replacement heifers in herds with low rates of infection represents a risk (Woodbine et al. 2008; Asmare et al. 2013).
7. Presence of definitive hosts: the number of farm dogs was a risk factor for seropositivity in cattle in most epidemiological studies in dairy herds (Hobson et al. 2005). Inversely, in the case of beef cattle, the presence of dogs resulted in increased protection, probably due to the chasing off wild canids from farm lands (Barling et al. 2001). In this regard, farmers should be aware that their dogs must be fed with commercial or cooked food.
8. Other intermediate hosts: naturally infected mice and rats may be important sources of infection for definitive hosts. The presence of domestic poultry was found to be a risk factor for seropositivity (Sierra et al. 2011). Moreover, experimentally infected embryonated chicken eggs induced oocyst excretion in dogs (Furuta et al. 2007).
9. Grazing, fodder, and drinking water: oocyst-contaminated pastures, fodder, and drinking water are regarded as potential sources for postnatal infection of cattle (Barling et al. 2001; Dubey et al. 2007).
10. Feeding colostrum or milk: cross-suckling of calves born to seronegative mothers on seropositive cows did not lead to an infection (Davison et al. 2001). But *N. caninum* DNA was demonstrated in bovine milk (Moskwa et al. 2007). Just one study in dairy cattle suggested that feeding of pooled colostrum is a putative risk factor for seropositivity (Corbellini et al. 2006).
11. Factors related to concurrent infections: a statistically significant association between antibodies against *N. caninum* and bovine viral diarrhea virus (BVDV) has been reported (Björkman et al. 2000). Another study showed positive association between antibodies against bovine herpesvirus 1 (BHV-1) and antibodies against *N. caninum* (Rinaldi et al. 2007). The rationale is that both viral infections induced immunosuppression, favoring the recrudescence of *N. caninum* infections in latently infected dams.
12. Climatic season: a higher temperature—up to not yet defined limits—may favor a faster sporulation of oocysts in the environment surrounding cattle. Mild temperatures and humidity not only favor sporulation and survival of coccidian oocysts but also support the growth of fungi. Fungal toxins are suspected to cause immune suppression in cattle (Bartels et al. 1999; Thurmond et al. 1995; Wouda et al. 1999).

13. Stress and body condition: heat stress may reflect the general susceptibility of pregnant cows to any type of stress during the second trimester of gestation (Yániz et al. 2010). A body condition score ≤4 increased, together with the level of *N. caninum*-specific antibodies, the chance of beef cows to experience a reproductive failure (Waldner 2005).

5.3 Clinical Signs, Prevention, and Treatment

5.3.1 Clinical Signs

Since *N. caninum* is one of the best host-adapted protozoa, most infections do not cause clinical disease (Moré et al. 2009). Eventually, acute or chronic infection can lead to disease and death. However, the host frequently recovers with the acquisition of immunity, coincident with the appearance of humoral antibodies. Chronic latent neosporosis may be reactivated by rupture of tissue cysts—bradyzoites convert to tachyzoites—but factors related to this change and rupture are unknown. Nevertheless, chronic neosporosis has been reactivated in experimentally infected dogs by administration of high doses of corticosteroids (Dubey and Lindsay 1990). Also, the physiopathology and the immune mechanisms involved in the reactivation—depression of the cellular-mediated immunity—during the second trimester of gestation in pregnant cattle have been postulated and demonstrated (Williams et al. 2000; Innes et al. 2002; Cantón et al. 2014). Tissue cyst rupture usually results in granulomatous inflammation. Even wild animals may suffer fatal disease, usually associated to stress and/or immune suppression (Donahoe et al. 2015).

5.3.1.1 Canine Neosporosis

In dogs, *N. caninum* infections are common but clinical disease is relatively rare. Although over 25 viable isolates have been reported worldwide, most of them are derived from dogs with clinical illness. Clinical neosporosis is common in young, very old, and/or immunosuppressed dogs (Dubey 2013).

Several clinical forms of canine neosporosis can be differentiated, although they may overlap:

1. Transplacentally infected puppies develop the most severe signs: ascending rear limb paresis and paralysis and atrophy and rigid contracture of the muscles. As rapid diagnosis is unusual, these dogs may progress to polyradiculoneuritis, polymyositis, and, eventually, meningoencephalomyelitis (Dubey 2013).
2. Adult dogs may suffer a wider range of manifestations like encephalitis and myositis, but paraparesis is the most common neurological sign (Dubey 2013).
3. Dermatitis due to *N. caninum* affects mainly adult dogs (Perl et al. 1998; La Perle et al. 2001).

4. Stillbirths and neonatal deaths have been reported (Barber and Trees 1996). Subclinically infected bitches can transmit the parasite to their fetuses, and successive litters from the same bitch may be born infected, but alternating generations of infected and noninfected litters have also been reported (Barber and Trees 1998; Dubey et al. 1990; Heckeroth and Tenter 2007). The stage of gestation at which *N. caninum* is transmitted from the bitch to the fetus remains unknown.
5. Severe hepatitis has been also reported in dogs (Hoon-Hanks et al. 2013; Cochrane and Dubey 1993; Jardine and Dubey 1992).
6. Dogs can die suddenly of severe myocarditis and heart failure without obvious involvement of other tissues (Odin and Dubey 1993; Meseck et al. 2005).
7. An unusual case of diffuse peritonitis with peritoneal effusion has been reported in an 11-year-old Rhodesian Ridgeback (Holmberg et al. 2006).
8. Pneumonitis due to *N. caninum* was diagnosed in an 11-year-old dog with persistent cough and tachypnea (Greig et al. 1995).

5.3.1.2 Cattle Neosporosis

In both beef and dairy cattle, *N. caninum* is a primary pathogen. The main clinical sign is abortion, but ataxia and weakness of limbs and permanent hyperextension could be underdiagnosed in congenitally infected calves (Micheloud et al. 2015). Fetuses may die in utero or be resorbed, mummified, autolyzed, stillborn, or born alive, clinically normal but being persistently infected. Mummification is common in neosporosis (Fig. 5.4) (Moore et al. 2002; Ghanem et al. 2009). Cows of any age may abort, but persistently infected heifers are particularly vulnerable during their first gestation (Thurmond and Hietala 1997). Cows may abort repeatedly in consecutive pregnancies, but repetition of the abortion has been estimated in 5% (Anderson et al. 1995). Abortions in cattle can have an epidemic or endemic—sporadic—character (Calandra et al. 2014). Abortions are considered epidemic if more than 10% of cows at risk abort within 6–8 weeks (Anderson et al. 2000). Within few weeks, abortion rates can reach >18% of the pregnant herd (McAllister et al. 1996).

5.3.2 Prevention

Nowadays, the options for controlling bovine neosporosis are based in the interruption of the *N. caninum* parasitic cycle (McAllister 2016): (1) limiting access of definitive hosts to water and food supplies; (2) removing tissues produced by abortions; (3) selective culling through the use of diagnostic tools; (4) although difficult, controlling any species involved in the sylvatic life cycle of this protozoan—including rodents and many other mammals—and (5) transferring embryos, a procedure that can be performed without any risk for the recipient dam; moreover, embryos can be obtained from a given seropositive cow and transferred to a seronegative recipient dam, and the progeny will be born free of *N. caninum* infection (Campero et al. 2003).

A commercial, inactivated vaccine was introduced to the market more than a decade ago, but due to its low efficacy, it was subsequently withdrawn from sales. Interestingly, live vaccines are used as effective tools for the prevention of other apicomplexan infections, such as toxoplasmosis, theileriosis, babesiosis, and coccidiosis (McAllister 2016). In cattle, *N. caninum* live tachyzoites inoculated previous to mating have higher efficacy compared to inactivated vaccines (Hecker et al. 2013). Potential of pathogenicity reversion, costly production and distribution, and latency in the intermediate host are considered the main disadvantages of currently available live vaccines. A desired goal is to achieve a vaccine similar to Toxovax®, in which *T. gondii* tachyzoites appear to have lost the ability to encyst in the vaccinated host. Another disadvantage of live vaccines is the impossibility to distinguish between vaccinated and naturally infected cattle, at least with currently available serological tests. Additionally, no drugs have been developed neither to prevent transplacental transmission nor to avoid abortion (Reichel et al. 2015).

5.3.3 Treatment

Many drugs have been tested to treat neosporosis (reviewed by Hemphill and Gottstein 2006). Nowadays sulfonamides and pyrimethamine have been widely used for therapy of neosporosis in dogs. Acting synergistically, these drugs block the metabolic pathway involving p-aminobenzoic acid and the folic-folinic acid cycle, respectively. Because the metabolites from these drugs are excreted shortly after administration, daily doses are required for several weeks or even months. Trimethoprim, another folic acid antagonist, was used in combination with sulfonamides in in vitro tests (Lindsay et al. 1996). Also, the antibiotic clindamycin, available in many formulations and presentations, is used to treat neosporosis in dogs. Toltrazuril and ponazuril, anticoccidial drugs that damage the parasite apicoplast and mitochondrion, are effective against neosporosis in mice (Gottstein et al. 2001). Although the effect of toltrazuril was investigated intensively in mice with promising results, the efficacy results in cattle were inconclusive (Haerdi et al. 2006). Recently, artemisinin, another drug used against *Plasmodium*, had lethal effects on in vitro cultured *N. caninum*; however, such effect was not observed in the mouse model (Müller et al. 2016).

To reduce the chance of illness in dogs, precise diagnosis should be performed as soon as possible. Because of their rapid penetration to the central nervous system, trimethoprim-sulfonamide or pyrimethamine and sulfonamide are the better treatment options when neurological signs are observed. Although clindamycin is effective in suppressing the replication and dissemination of tachyzoites, it is not effective against bradyzoites contained in tissular cysts. When planning the breeding of bitches, it must be considered that the parasite can be transmitted repeatedly through successive litters of progeny (Dubey 2013).

References

Abe C, Tanaka S, Nishimura M, Ihara F, Xuan X, Nishikawa Y. Role of the chemokine receptor CCR5-dependent host defense system in *Neospora caninum* infections. Parasit Vectors. 2015;8:e5.

Aguado-Martínez A, Álvarez-García G, Fernández-García A, Risco-Castillo V, Arnaiz-Seco I, Rebordosa-Trigueros X, Navarro-Lozano V, Ortega-Mora LM. Usefulness of rNcGRA7- and rNcSAG4-based ELISA tests for distinguishing primo-infection, recrudescence, and chronic bovine neosporosis. Vet Parasitol. 2008;157(3-4):182–95.

Almería S. *Neospora caninum* and wildlife. ISRN Parasitol. 2013;947347. doi: https://doi.org/10.5402/2013/947347.

Álvarez-García G, Pereira-Bueno J, Gómez-Bautista M, Ortega-Mora LM. Pattern of recognition of Neospora caninum tachyzoite antigens by naturally infected pregnant cattle and aborted foetuses. Vet Parasitol. 2002;107:15–27.

Alvarez-García G, García-Culebras A, Gutiérrez-Expósito D, Navarro-Lozano V, Pastor-Fernández I, Ortega-Mora LM. Serological diagnosis of bovine neosporosis: a comparative study of commercially available ELISA tests. Vet Parasitol. 2013;198:85–95.

Anderson ML, Andrianarivo AG, Conrad PA. Neosporosis in cattle. Anim Reprod Sci. 2000;60-61:417–31.

Anderson ML, Palmer CW, Thurmond MC, Picanso JP, Blanchard PC, Breitmeyer RE, Layton AW, McAllister M, Daft B, Kinde H, Read DH, Dubey JP, Conrad PA, Barr BC. Evaluation of abortions in cattle attributable to neosporosis in selected dairy herds in California. J Am Vet Med Assoc. 1995;207:1206–10.

Anderson ML, Reynolds JP, Rowe JD, Sverlow KW, Packham AE, Barr BC, Conrad PA. Evidence of vertical transmission of *Neospora* sp. infection in dairy cattle. J Am Vet Med Assoc. 1997;210:1169–72.

Asmare K, Regassa F, Robertson LJ, Skijerve E. Seroprevalence of *Neospora caninum* and associated risk factors in intensive or semi-intensively managed dairy and breeding cattle of Ethiopia. Vet Parasitol. 2013;193:85–94.

Bacigalupe D, Basso W, Caspe SG, Moré G, Lischinsky L, Gos ML, Leunda M, Campero L, Moore DP, Schares G, Campero CM, Venturini MC. *Neospora caninum* NC-6 Argentina induces fetopathy in both serologically positive and negative experimentally inoculated pregnant dams. Parasitol Res. 2013;112:2585–92.

Bandini LA, Neto AFA, Pena HFJ, Cavalcante GT, Schares G, Nishi SM, Gennari SM. Experimental infection of dogs (*Canis familiaris*) with sporulated oocysts of *Neospora caninum*. Vet Parasitol. 2011;176:151–6.

Barber JS, Trees AJ. Clinical aspects of 27 cases of neosporosis in dogs. Vet Rec. 1996;139:439–43.

Barber JS, Trees AJ. Naturally occurring vertical transmission of *Neospora caninum* in dogs. Int J Parasitol. 1998;28:57–64.

Barling KS, Sherman M, Peterson MJ, Thompson JA, McNeill JW, Craig TM, Adams LG. Spatial associations among density of cattle, abundance of wild canids, and seroprevalence to *Neospora caninum* in a population of beef calves. J Am Vet Med Assoc. 2000;217:1361–5.

Barling KS, McNeill JW, Paschal JC, McCollum FT III, Craig TM, Adams LG, Thompson JA. Ranch-management factors associated with antibody seropositivity for *Neospora caninum* in consignments of beef calves in Texas, USA. *Prev Vet Med.* 2001;52:53–61.

Barr BC, Anderson ML, Dubey JP, Conrad PA. *Neospora*-like protozoal infections associated with bovine abortions. Vet Pathol. 1991;28:110–6.

Barr BC, Rowe JD, Sverlow KW, BonDurant RH, Ardans AA, Oliver MN, Conrad PA. Experimental reproduction of bovine fetal *Neospora* infection and death with a bovine *Neospora* isolate. J Vet Diagn Investig. 1994;6:207–15.

Bartels CJM, Wouda W, Schukken YH. Risk factors for *Neospora caninum*-associated abortion storms in dairy herds in the Netherlands (1995 to 1997). Theriogenology. 1999;52:247–57.

Bartley PM, Wright SE, Maley SW, Buxton D, Nath M, Innes EA. The development of immune responses in Balb/c mice following inoculation with attenuated or virulent *Neospora caninum* tachyzoites. Parasite Immunol. 2009;31:392–401.

Basso W, Venturini L, Venturini MC, Hill DE, Kwok OCH, Shen SK, Dubey JP. First isolation of *Neospora caninum* from the feces of a naturally infected dog. J Parasitol. 2001a;87:612–8.

Basso W, Venturini L, Venturini MC, Moore P, Rambeau M, Unzaga JM, Campero C, Bacigalupe D, Dubey JP. Prevalence of *Neospora caninum* infection in dogs from beef-cattle farms, dairy farms, and from urban areas of Argentina. J Parasitol. 2001b;87:906–7.

Basso W, Schares S, Minke L, Bärwald A, Maksimov A, Peters M, Schulze C, Müller M, Conraths FJ, Schares G. Microsatellite typing and avidity analysis suggest a common source of infection in herds with epidemic *Neospora caninum*-associated bovine abortion. Vet Parasitol. 2010;173:24–31.

Baszler TV, Knowles DP, Dubey JP, Gay JM, Mathison BA, McElwain TF. Serological diagnosis of bovine neosporosis by *Neospora caninum* monoclonal antibody-based competitive inhibition enzyme-linked immunosorbent assay. J Clin Microbiol. 1996;34:1423–8.

Baszler TV, Gay LJC, Long MT, Mathison BA. Detection by PCR of *Neospora caninum* in fetal tissues from spontaneous bovine abortions. J Clin Microbiol. 1999a;37:4059–64.

Behrendt JH, Taubert A, Zahner H, Hermosilla C. Studies on synchronous egress of coccidian parasites (*Neospora caninum, Toxoplasma gondii, Eimeria bovis*) from bovine endothelial host cells mediated by calcium ionophore A23187. Vet Res Commun. 2008;32:325–32.

Beiting DP, Peixoto L, Akopyants NS, Beverley SM, Wherry EJ, Christian DA, Hunter CA, Brodsky IE, Roos DS. Differential induction of TLR3-dependent innate immune signaling by closely related parasite species. PLoS One. 2014;9:e88398.

Bjerkås I, Mohn SF, Presthus J. Unidentified cyst-forming sporozoon causing encephalomyelitis and myositis in dogs. Z Parasitenkd. 1984;70:271–4.

Björkman C, Lundén A, Holmdahl J, Barber J, Trees AJ, Uggla A. *Neospora caninum* in dogs: detection of antibodies by ELISA using an iscom antigen. Parasite Immunol. 1994;16: 643–8.

Björkman C, Näslund K, Stenlund S, Maley SW, Buxton D, Uggla A. An IgG avidity ELISA to discriminate between recent and chronic *Neospora caninum* infection. J Vet Diagn Investig. 1999;11:41–4.

Björkman C, Alenius S, Emanuelsson U, Uggla A. *Neospora caninum* and bovine virus diarrhoea virus infections in Swedish dairy cows in relation to abortion. Vet J. 2000;159:201–6.

Botelho AS, Teixeira L, Correia da Costa JM, Faustino AMR, Castro AG, Vilanova M. Neospora caninum: high susceptibility to the parasite in C57BL/10ScCr mice. Exp Parasitol. 2007;115:68–75.

Calandra PM, Di Matía JM, Cano DB, Odriozola EA, García JA, Späth EAJ, Odeón AC, Paolicchi FA, Morrell EL, Campero CM, Moore DP. Neosporosis epidémica y endémica: descripción de dos eventos en bovinos para cría. Rev Argent Microbiol. 2014;46:315–9.

Campero CM, Anderson ML, Conosciuto G, Odriozola H, Bretschneider G, Poso MA. *Neospora caninum*-associated abortion in a dairy herd in Argentina. Vet Rec. 1998;143:228–9.

Campero CM, Moore DP, Lagomarsino H, Odeón AC, Castro M, Visca H. Serological status and abortion rate in progeny obtained by natural service or embryo transfer from *Neospora caninum*-seropositive cows. J Veterinary Med Ser B. 2003;50:458–60.

Campero LM, Minke L, Moré G, Rambeaud M, Bacigalupe D, Moore DP, Hecker Y, Campero CM, Schares G, Venturini MC. Evaluation and comparison of serological methods for the detection of bovine neosporosis in Argentina. Rev Argent Microbiol. 2015a;47:295–301.

Campero LM, Venturini MC, Moore DP, Massola L, Lagomarsino H, García B, Bacigalupe D, Rambeaud M, Pardini L, Leunda MR, Schares G, Campero CM. Isolation and molecular characterization of a new *Neospora caninum* isolate from cattle in Argentina. Exp Parasitol. 2015b;155:8–12.

Cantón GJ, Katzer F, Maley SW, Bartley PM, Benavides-Silván J, Palarea-Albaladejo J, Pang Y, Smith SH, Rocchi MS, Buxton D, Innes EA, Chianini F. Inflammatory infiltration into placentas of *Neospora caninum* challenged cattle correlates with clinical outcome of pregnancy. Vet Res. 2014;45:e11.

Cardoso R, Soares H, Hemphill A, Leitão A. Apicomplexans pulling the strings: manipulation of the host cell cytoskeleton dynamics. Parasitology. 2016;143(8):957–70. https://doi.org/10.1017/S0031182016000524.

Carruthers VB, Blackman MJ. A new release on life: emerging concepts in proteolysis and parasite invasion. Mol Microbiol. 2005;55:1617–30.

Caspe SG, Moore DP, Leunda MR, Cano DB, Lischinsky L, Regidor-Cerrillo J, Álvarez García G, Echaide IG, Bacigalupe D, Ortega Mora LM, Odeón AC, Campero CM. The *Neospora caninum*-Spain 7 isolate induces placental damage, fetal death and abortion in cattle when inoculated in early gestation. Vet Parasitol. 2012;189:171–81.

Cochrane SM, Dubey JP. Neosporosis in a golden retriever dog from Ontario. Can Vet J. 1993;34:232–3.

Cole RA, Lindsay DS, Blagburn BL, Dubey JP. Vertical transmission of *Neospora caninum* in mice. J Parasitol. 1995;81:730–2.

Collantes-Fernández E, Zaballos Á, Álvarez-García G, Ortega-Mora LM. Quantitative detection of *Neospora caninum* in bovine aborted fetuses and experimentally infected mice by real-time PCR. J Clin Microbiol. 2002;40:1194–8.

Corbellini LG, Pescador CA, Frantz F, Wunder E, Steffen D, Smith DR, Driemeier D. Diagnostic survey of bovine abortion with special reference to *Neospora caninum* infection: importance, repeated abortion and concurrent infection in aborted fetuses in Southern Brazil. Vet J. 2006;172:114–20.

Cringoli G, Rinaldi L, Capuano F, Baldi L, Veneziano V, Capelli G. Serological survey of Neospora caninum and Leishmania infantum co infectionin dogs. Vet Parasitol. 2002;106(4):307–13.

Davison HC, Guy CS, McGarry JW, Guy F, Williams DJL, Kelly DF, Trees AJ. Experimental studies on the transmission of *Neospora caninum* between cattle. Res Vet Sci. 2001;70: 163–8.

Davoli-Ferreira M, Fonseca DM, Mota CM, Dias MS, Lima-Junior DS, da Silva MV, Quirino GFS, Zamboni DS, Silva JS, Mineo TWP. Nucleotide-binding oligomerization domain-containing protein 2 prompts potent inflammatory stimuli during *Neospora caninum* infection. Sci Rep. 2016;6:29289.

de Marez T, Liddell S, Dubey JP, Jenkins MC, Gasbarre L. Oral infection of calves with *Neospora caninum* oocysts from dogs: humoral and cellular immune responses. Int J Parasitol. 1999;29:1647–57.

Dijkstra T, Barkema HW, Eysker M, Beiboer ML, Wouda W. Evaluation of a single serological screening of dairy herds for *Neospora caninum* antibodies. Vet Parasitol. 2003;110:161–9.

Donahoe SL, Lindsay SA, Krockenberger M, Phalen D, Slapeta J. A review of neosporosis and pathologic findings of *Neospora caninum* infection in wildlife. Int J Parasitol Parasites Wildl. 2015;4:216–38.

Dubey JP. Neosporosis in dogs. Commonwealth Agric Bureau Rev. 2013;8:055.

Dubey JP, Lindsay DS. Transplacental *Neospora caninum* infection in cats. J Parasitol. 1989a;75:765–71.

Dubey JP, Lindsay DS. Transplacental *Neospora caninum* infection in dogs. Am J Vet Res. 1989b;50:1578–9.

Dubey JP, Lindsay DS. *Neospora caninum* induced abortion in sheep. J Vet Diagn Investig. 1990;2:230–3.

Dubey JP, Lindsay DS. A review of *Neospora caninum* and neosporosis. Vet Parasitol. 1996;67:1–59.

Dubey JP, Schares G. Neosporosis in animals—the last five years. Vet Parasitol. 2011;180:90–108.

Dubey JP, Carpenter JL, Speer CA, Topper MJ, Uggla A. Newly recognized fatal protozoan disease of dogs. J Am Vet Med Assoc. 1988a;192:1269–85.

Dubey JP, Hattel AL, Lindsay DS, Topper MJ. Neonatal *Neospora caninum* infection in dogs: isolation of the causative agent and experimental transmission. J Am Vet Med Assoc. 1988b;193:1259–63.

Dubey JP, Koestner A, Piper RC. Repeated transplacental transmission of *Neospora caninum* in dogs. J Am Vet Med Assoc. 1990;197:857–60.

Dubey JP, Lindsay DS, Anderson ML, Davis SW, Shen SK. Induced transplacental transmission of *Neospora caninum* in cattle. J Am Vet Med Assoc. 1992;201:709–13.

Dubey JP, Lindsay DS, Adams DS, Gay JM, Baszler TV, Blagburn BL, Thulliez P. Serologic responses of cattle and other animals infected with *Neospora caninum*. Am J Vet Res. 1996;57:329–36.

Dubey JP, Barr BC, Barta JR, Bjerkås I, Björkman C, Blagburn BL, Bowman DD, Buxton D, Ellis JT, Gottstein B, Hemphill A, Hill DE, Howe DK, Jenkins MC, Kobayashi Y, Koudela B, Marsh AE, Mattsson JG, McAllister MM, Modrý D, Omata Y, Sibley LD, Speer CA, Trees AJ, Uggla A, Upton SJ, Williams DJL, Lindsay DS. Redescription of *Neospora caninum* and its differentiation from related coccidia. Int J Parasitol. 2002;32:929–46.

Dubey JP, Sreekumar C, Knickman E, Miska KB, Vianna MCB, Kwok OCH, Hill DE, Jenkins MC, Lindsay DS, Greene CE. Biologic, morphologic, and molecular characterisation of *Neospora caninum* isolates from littermate dogs. Int J Parasitol. 2004;34:1157–67.

Dubey JP, Buxton D, Wouda W. Pathogenesis of bovine neosporosis. J Comp Pathol. 2006;134:267–89.

Dubey JP, Schares G, Ortega-Mora LM. Epidemiology and control of neosporosis and *Neospora caninum*. Clin Microbiol Rev. 2007;20:323–67.

Dubey JP, Jenkins MC, Rajendran C, Miska K, Ferreira LR, Martins J, Kwok OCH, Choudhary S. Gray wolf (*Canis lupus*) is a natural definitive host for *Neospora caninum*. Vet Parasitol. 2011;181:382–7.

Ellis JT, McMillan D, Ryce C, Payne S, Atkinson R, Harper PAW. Development of a single tube nested polymerase chain reaction assay for the detection of *Neospora caninum* DNA. Int J Parasitol. 1999;29:1589–96.

Eperon B, Hemphill G. Susceptibility of B-cell deficient C57BL/6 (muMT) mice to Neospora caninum infection. Parasite Immunol. 1999;21(5):225–36.

Feng X, Zhang N, Tuo W. *Neospora caninum* tachyzoite –and antigen-stimulatedcytokine production by bone marrow-derived dendritic cells and spleen cells of naïve BALB/c mice. J Parasitol. 2010;96:717–23.

Fioretti DP, Pasquai P, Diaferia M, Mangili V, Rosignoli L. Neospora caninum infection and congenital transmission: serological and parasitological study of cows up to the fourth gestation. J Vet Med B. 2003;50:399–404.

Furuta PI, Mineo TWP, Carrasco AOT, Godoy GS, Pinto AA, Machado RZ. *Neospora caninum* infection in birds: experimental infections in chicken and embryonated eggs. Parasitology. 2007;134:1931–9.

Gennari SM, Yai LEO, D'Áuria SNR, Cardoso SMS, Kwok OCH, Jenkins MC, Dubey JP. Occurrence of *Neospora caninum* antibodies in sera from dogs of the city of São Paulo, Brazil. Vet Parasitol. 2002;106:177–9.

Ghalmi F, China B, Kaidi R, Daube G, Losson B. Detection of *Neospora caninum* in dog organs using real time PCR systems. Vet Parasitol. 2008;155:161–7.

Ghanem ME, Suzuki T, Akita M, Nishibori M. *Neospora caninum* and complex vertebral malformation as possible causes of bovine fetal mummification. Can Vet J. 2009;50:389–92.

Gondim LFP, McAllister MM, Pitt WC, Zemlicka DE. Coyotes (*Canis latrans*) are definitive hosts of *Neospora caninum*. Int J Parasitol. 2004;34:159–61.

Goodswen SJ, Kennedy PJ, Ellis JT. A review of the infection, genetics, and evolution of *Neospora caninum*: from the past to the present. Infect Genet Evol. 2013;13:133–50.

Gottstein B, Eperon S, Dai WJ, Cannas A, Hemphill A, Greif G. Efficacy of toltrazuril and ponazuril against experimental *Neospora caninum* infection in mice. Parasitol Res. 2001;87:43–8.

Goždzik K, Wrzesień R, Wielgosz-Ostolska A, Bien J, Kozak-Ljunggren M, Cabaj W. Prevalence of antibodies against *Neospora caninum* in dogs from urban areas in Central Poland. Parasitol Res. 2011;108:991–6.

Greig B, Rossow KD, Collins JE, Dubey JP. *Neospora caninum* pneumonia in an adult dog. J Am Vet Med Assoc. 1995;206:1000–1.

Haerdi C, Haessig M, Sager H, Greif G, Staubli D, Gottstein B. Humoral immune reaction of newborn calves congenitally infected with *Neospora caninum* and experimentally treated with toltrazuril. Parasitol Res. 2006;99:534–40.

Hecker YP, Moore DP, Quattrocchi V, Regidor-Cerrillo J, Verna A, Leunda MR, Morrell E, Ortega-Mora LM, Zamorano P, Venturini MC, Campero CM. Immune response and protection provided by live tachyzoites and native antigens from the NC-6 Argentina strain of *Neospora caninum* in pregnant heifers. Vet Parasitol. 2013;197:436–46.

Heckeroth AR, Tenter AM. Immunoanalysis of three litters born to a Doberman bitch infected with *Neospora caninum*. Parasitol Res. 2007;100:837–46.

Hemphill A, Gottstein B. *Neospora caninum* and neosporosis—recent achievements in host and parasite cell biology and treatment. Acta Parasitol. 2006;51:15–25.

Herman RK, Molestina RE, Sinai AP, Howe DK. The Apicomplexan pathogen *Neospora caninum* inhibits host cell apoptosis in the absence of discernible NF-κB activation. Infect Immun. 2007;75:4255–62.

Hobson JC, Duffield TF, Kelton D, Lissemore K, Hietala SK, Leslie KE, McEwen B, Peregrine AS. Risk factors associated with *Neospora caninum* abortion in Ontario Holstein dairy herds. Vet Parasitol. 2005;127:177–88.

Holmberg TA, Vernau W, Melli AC, Conrad PA. *Neospora caninum* associated with septic peritonitis in an adult dog. Vet Clin Pathol. 2006;35:235–8.

Holmdahl OJM, Mattsson JG. Rapid and sensitive identification of *Neospora caninum* by *in vitro* amplification of the internal transcribed spacer 1. Parasitology. 1996;112:177–82.

Hoon-Hanks LL, Regan D, Dubey JP, Porter MC, Duncan CG. Hepatic neosporosis in a dog treated for pemphigus foliaceus. J Vet Diagn Investig. 2013;25:807–10.

Hosseininejad M, Hosseini F, Mosharraf M, Shahbaz S, Mahzounieh M, Schares G. Development of an indirect ELISA test using an affinity purified surface antigen (P38) for sero-diagnosis of canine *Neospora caninum* infection. Vet Parasitol. 2010;171:337–42.

Innes EA, Andrianarivo AG, Björkman C, Williams DJL, Conrad PA. Immune responses to *Neospora caninum* and prospects for vaccination. Trends Parasitol. 2002;18:497–504.

Jardine JE, Dubey JP. Canine neosporosis in South Africa. Vet Parasitol. 1992;44:291–4.

Kasper LH, Khan IA. Antigen-specific CD8+ T cells protect against lethal toxoplasmosis in mice infected with *Neospora caninum*. Infect Immun. 1998;66:1554–60.

Khan IA, Schwartzman JD, Fonseka S, Kasper LH. *Neospora caninum*: role for immune cytokines in host immunity. Exp Parasitol. 1997;85:24–34.

King JS, Šlapeta J, Jenkins DJ, Al-Qassab SE, Ellis JT, Windsor PA. Australian dingoes are definitive host of *Neospora caninum*. Int J Parasitol. 2010;40:945–50.

Kubota N, Sakata Y, Miyazaki N, Itamoto K, Bannai H, Nishikawa Y, Xuan X, Inokuma H. Serological survey of *Neospora caninum* infection among dogs in Japan through species-specific ELISA. J Vet Med Sci. 2008;70:869–72.

La Perle KMD, Del Piero F, Carr RF, Harris C, Stromberg PC. Cutaneous neosporosis in two adult dogs on chronic immunosuppressive therapy. J Vet Diagn Investig. 2001;13:252–5.

Legnani S, Pantchev N, Forlani A, Zini E, Schares G, Balzer J, Roccabianca P, Ferri F, Zanna G. Emergence of cutaneous neosporosis in a dog receiving immunosuppressive therapy: molecular identification and management. Vet Dermatol. 2015;27:49–e14.

Lei Y, Davey M, Ellis J. Autofluorescence of *Toxoplasma gondii* and *Neospora caninum* cysts in vitro. J Parasitol. 2005;91:202–8.

Lindsay DS, Dubey JP. In vitro development of *Neospora caninum* (Protozoa: Apicomplexa) from dogs. J Parasitol. 1989;75:163–5.

Lindsay DS, Blagburn BL, Dubey JP. Factors affecting the survival of *Neospora caninum* bradyzoites in murine tissues. J Parasitol. 1992;78:70–2.

Lindsay DS, Butler JM, Rippey NS, Blagburn BL. Demonstration of synergistic effects of sulfonamides and dihydrofolate reductase/thymidylate synthase inhibitors against *Neospora caninum* tachyzoites in cultured cells, and characterization of mutants resistant to pyrimethamine. Am J Vet Res. 1996;57:68–72.

Lindsay DS, Dubey JP, Duncan RB. Confirmation that the dog is a definitive host for *Neospora caninum*. Vet Parasitol. 1999a;82:327–33.

Lindsay DS, Upton SJ, Dubey JP. A structural study of the *Neospora caninum* oocyst. Int J Parasitol. 1999b;29:1521–3.

Marín MS, Hecker YP, Quintana S, Pérez SE, Leunda MR, Cantón GJ, Cobo ER, Moore DP, Odeón AC. Toll-like receptors 3, 7 and 8 are upregulated in the placental caruncle and fetal spleen of *Neospora caninum* experimentally infected cattle. Vet Parasitol. 2017;236:58–61.

Marsh AE, Barr BC, Packham AE, Conrad PA. Description of a new *Neospora* species (Protozoa: Apicomplexa: Sarcocystidae). J Parasitol. 1998;84:983–91.

McAllister MM, Björkman C, Anderson-Sprecher R, Rogers DG. Evidence of point-source exposure to and protective immunity in a herd of beef cows. J Am Vet Med Assoc. 2000;217(6):881–7.

McAllister MM. Diagnosis and control of bovine Neosporosis. Vet Clin North Am Food Anim Pract. 2016;32:443–63.

McAllister M, Huffman EM, Hietala SK, Conrad PA, Anderson ML, Salman MD. Evidence suggesting a point source exposure in an outbreak of bovine abortion due to neosporosis. J Vet Diagn Investig. 1996;8:355–7.

McAllister MM, Dubey JP, Lindsay DS, Jolley WR, Wills RA, McGuire AM. Dogs are definitive hosts of *Neospora caninum*. Int J Parasitol. 1998;28:1473–8.

McAllister MM, Björkman C, Anderson-Sprecher R, Rogers DG. Evidence of point-source exposure to and protective immunity in a herd of beef cows. J Am Vet Med Assoc. 2000;217(6):881–7.

Meseck EK, Njaa BL, Haley NJ, Park EH, Barr SC. Use of a multiplex polymerase chain reaction to rapidly differentiate *Neospora caninum* from *Toxoplasma gondii* in an adult dog with necrotizing myocarditis and myocardial infarct. J Vet Diagn Investig. 2005;17:565–8.

Micheloud JF, Moore DP, Canal AM, Lischinsky L, Hecker YP, Canton GJ, Odriozola E, Odeon AC, Campero CM. First report of congenital *Neospora caninum* encephalomyelitis in two newborn calves in the Argentinean Pampas. J Vet Sci Technol. 2015;6:251.

Moore DP, Campero CM, Odeón AC, Posso MA, Cano D, Leunda MR, Basso W, Venturini MC, Späth E. Seroepidemiology of beef and dairy herds and fetal study of *Neospora caninum* in Argentina. Vet Parasitol. 2002;107:303–16.

Moore DP, Perez A, Agliano S, Brace M, Canton G, Cano D, Leunda MR, Odeón AC, Odriozola E, Campero CM. Risk factors associated with *Neospora caninum* infections in cattle in Argentina. Vet Parasitol. 2009;161:122–5.

Moraveji M, Hosseini A, Moghaddar N, Namavari MM, Eskandari MH. Development of latex agglutination test with recombinant NcSAG1 for the rapid detection of antibodies to *Neospora caninum* in cattle. Vet Parasitol. 2012;189:211–7.

Moré G, Bacigalupe D, Basso W, Rambeaud M, Beltrame F, Ramirez B, Venturini MC, Venturini L. Frequency of horizontal and vertical transmission for *Sarcocystis cruzi* and *Neospora caninum* in dairy cattle. Vet Parasitol. 2009;160:51–4.

Moskwa B, Pastusiak K, Bien J, Cabaj W. The first detection of *Neospora caninum* DNA in the colostrum of infected cows. Parasitol Res. 2007;100:633–6.

Müller N, Zimmermann V, Hentrich B, Gottstein B. Diagnosis of *Neospora caninum* and *Toxoplasma gondii* infection by PCR and DNA hybridization immunoassay. J Clin Microbiol. 1996;34:2850–2.

Müller J, Aguado-Martínez A, Manser V, Wong HN, Haynes RK, Hemphill A. Repurposing of antiparasitic drugs: the hydroxy-naphthoquinone buparvaquone inhibits vertical transmission in the pregnant neosporosis mouse model. Vet Res. 2016;47:32.

Nishikawa Y, Mishima M, Nagasawa H, Igarashi I, Fujisaki K, Otsuka H, Mikami T. Interferon-gamma-induced apoptosis in host cells infected with *Neospora caninum*. Parasitology. 2001a;123:25–31.

Nishikawa Y, Tragoolpua K, Inoue N, Makala L, Nagasawa H, Otsuka H, Mikami T. In the absence of endogenous gamma interferon, mice acutely infected with *Neospora caninum* succumb to a lethal immune response characterized by inactivation of peritoneal macrophages. Clin Diagn Lab Immunol. 2001b;8:811–7.

Nishikawa Y, Inoue N, Makala L, Nagasawa H. A role for balance of interferon-gamma and interleukin-4 production in protective immunity against *Neospora caninum* infection. Vet Parasitol. 2003;116:175–84.

Nogueira CI, Mesquita LP, Abreu CC, Nakagaki KYR, Seixas JN, Bezerra PS, Rocha CMBM, Guimaraes AM, Peconick AP, Varaschin MS. Risk factors associated with seroprevalence of

Neospora caninum in dogs from urban and rural areas of milk and coffee production in Minas Gerais state, Brazil. Epidemiol Infect. 2013;141:2286–93.

Odin M, Dubey JP. Sudden death associated with *Neospora caninum* myocarditis in a dog. J Am Vet Med Assoc. 1993;203:831–3.

Okeoma CM, Stowell KM, Williamson NB, Pomroy WE. *Neospora caninum*: quantification of DNA in the blood of naturally infected aborted and pregnant cows using real-time PCR. Exp Parasitol. 2005;110:48–55.

Ortega-Mora LM, Ferre I, del Pozo I, Caetano da Silva A, Collantes-Fernández E, Regidor-Cerrillo J, Ugarte-Garagalza C, Aduriz G. Detection of *Neospora caninum* in semen of bulls. Vet Parasitol. 2003;117:301–8.

Paré J, Hietala SK, Thurmond MC. An enzyme-linked immunosorbent assay (ELISA) for serological diagnosis of *Neospora* sp. infection in cattle. J Vet Diagn Investig. 1995;7:352–9.

Pereira GR, Vogel FSF, Bohrer RC, da Nóbrega JE, Ilha GF, da Rosa PRA, Glanzner WG, Camillo G, Braunig P, de Oliveira JFC, Gonçalves PBD. *Neospora caninum* DNA detection by TaqMan real-time PCR assay in experimentally infected pregnant heifers. Vet Parasitol. 2014;199:129–35.

Pessoa GA, Martini AP, Trentin JM, Dalcin VC, Leonardi CEP, Vogel FSF, de Sá Filho MF, Rubin MIB, Silva CAM. Impact of spontaneous *Neospora caninum* infection on pregnancy loss and subsequent pregnancy in grazing lactating dairy cows. Theriogenology. 2016;85:519–27.

Perl S, Harrus S, Satuchne C, Yakobson B, Haines D. Cutaneous neosporosis in a dog in Israel. Vet Parasitol. 1998;79(3):257–61.

Peters M, Lütkefels E, Heckeroth AR, Schares G. Immunohistochemical and ultrastructural evidence for *Neospora caninum* tissue cysts in skeletal muscles of naturally infected dogs and cattle. Int J Parasitol. 2001;31:1144–8.

Pinitkiatisakul S, Mattsson JG, Lundén A. Quantitative analysis of parasite DNA in the blood of immunized and naive mice after infection with *Neospora caninum*. Parasitology. 2008;135:175–82.

Quintanilla-Gozalo A, Pereira-Bueno J, Tabarés E, Innes EA, González-Paniello R, Ortega-Mora LM. Seroprevalence of *Neospora caninum* infection in dairy and beef cattle in Spain. Int J Parasitol. 1999;29:1201–8.

Regidor-Cerrillo J, Díez-Fuertes F, García-Culebras A, Moore DP, González-Warleta M, Cuevas C, Schares G, Katzer F, Pedraza-Díaz S, Mezo M, Ortega-Mora LM. Genetic diversity and geographic population structure of bovine *Neospora canium* determined by microsatellite genotyping analysis. PLoS One. 2013;8:e72678.

Reichel MP, Drake JM. The diagnosis of *Neospora* abortions in cattle. N Z Vet J. 1996;44:151–4.

Reichel MP, Moore DP, Hemphill A, Ortega-Mora LM, Dubey JP, Ellis JT. A live vaccine against *Neospora caninum* abortions in cattle. Vaccine. 2015;33:1299–301.

Reitt K, Hilbe M, Voegtlin A, Corboz L, Haessig M, Pospischil A. Aetiology of bovine abortion in Switzerland from 1986 to 1995—A retrospective study with emphasis on detection of *Neospora caninum* and *Toxoplasma gondii* by PCR. J Vet Med A. 2007;54:15–22.

Rinaldi L, Pacelli F, Iovane G, Pagnini U, Veneziano V, Fusco G, Cringoli G. Survey of *Neospora caninum* and bovine herpes virus 1 coinfection in cattle. Parasitol Res. 2007;100:359–64.

Ritter DM, Kerlin R, Sibert G, Brake D. Immune factors influencing the course of infection with Neospora caninum in the murine host. J Parasitol. 2002;88(2):271.

Romand S, Thulliez P, Dubey JP. Direct agglutination test for serologic diagnosis of *Neospora caninum* infection. Parasitol Res. 1998;84:50–3.

Sánchez GF, Banda RVM, Sahagun RA, Ledesma MN, Morales SE. Comparison between immunohistochemistry and two PCR methods for detection of *Neospora caninum* in formalin-fixed and paraffin-embedded brain tissue of bovine fetuses. Vet Parasitol. 2009;164:328–32.

Santolaria P, Almería S, Martínez-Bello D, Nogareda C, Mezo M, Gonzalez-Warleta M, Castro-Hermida JA, Pabón M, Yániz JL, López-Gatius F. Different humoral mechanisms against *Neospora caninum* infection in purebreed and crossbreed beef/dairy cattle pregnancies. Vet Parasitol. 2011;178:70–6.

Sibley LD, Niesman IR, Parmley SF, Cesbron-Delauw MF. Regulated secretion of multi-lamellar vesicles leads to formation of a tubulovesicular network in host-cell vacuoles occupied by *Toxoplasma gondii*. J Cell Sci. 1995;108:1669–77.

Sicupira PML, de Magalhães VCS, Galvão GS, Pereira MJS, Gondim LFP, Munhoz AD. Factors associated with infection by *Neospora caninum* in dogs in Brazil. Vet Parasitol. 2012;185: 305–8.

Sierra RC, Medina-Esparza L, Parra MR, García-Vázquez Z, Cruz-Vázquez C. Risk factors associated with *Neospora caninum* antibody seroprevalence in dairy cattle in Aguascalientes, Mexico. Rev Mex Cienc Pecu. 2011;2:15–24.

Strohbusch M, Müller N, Hemphill A, Margos M, Grandgirard D, Leib S, Greif G, Gottstein B. *Neospora caninum* and bone marrow-derived dendritic cells: parasite survival, proliferation, and induction of cytokine expression. Parasite Immunol. 2009;31:366–72.

Tanaka T, Hamada T, Inoue N, Nagasawa H, Fujisaki K, Suzuki N, Mikami T. The role of CD4+ or CD8+ T cells in the protective immune response of BALB/c mice to *Neospora caninum* infection. Vet Parasitol. 2000;90:183–91.

Thilsted JP, Dubey JP. Neosporosis-like abortions in a herd of dairy cattle. J Vet Diagn Investig. 1989;1:205–9.

Thurmond M, Hietala S. Strategies to control Neospora infection in cattle. Bovine Pract. 1995;29:60–3.

Thurmond MC, Hietala SK. Effect of congenitally acquired *Neospora caninum* infection on risk of abortion and subsequent abortions in dairy cattle. Am J Vet Res. 1997;58:1381–5.

Tramuta C, Lacerenza D, Zoppi S, Goria M, Dondo A, Ferroglio E, Nebbia P, Rosati S. Development of a set of multiplex standard polymerase chain reaction assays for the identification of infectious agents from aborted bovine clinical samples. J Vet Diagn Investig. 2011;23:657–64.

Trees AJ, Williams DJL. Endogenous and exogenous transplacental infection in *Neospora caninum* and *Toxoplasma gondii*. Trends Parasitol. 2005;21:558–61.

Uzeda RS, Costa KS, Santos SL, Pinheiro AM, de Almeida MAO, McAllister MM, Gondim LFP. Loss of infectivity of *Neospora caninum* oocyts maintained for a prolonged time. Korean J Parasitol. 2007;45:295–9.

VanLeeuwen JA, Haddad JP, Dohoo IR, Keefe GP, Tiwari A, Scott HM. Risk factors associated with *Neospora caninum* seropositivity in randomly sampled Canadian dairy cows and herds. Prev Vet Med. 2010;93:129–38.

Vonlaufen N, Gianinazzi C, Müller N, Simon F, Björkman C, Jungi TW, Leib SL, Hemphill A. Infection of organotypic slice cultures from rat central nervous tissue with *Neospora caninum*: an alternative approach to study host-parasite interactions. Int J Parasitol. 2002;32:533–42.

Waldner CL. Serological status for *Neospora caninum*, bovine viral darrhea virus, and infectious bovine rhinotracheitis virus at pregnancy testing and reproductive performance in beef herds. Anim Reprod Sci. 2005;90:219–42.

Wilkowsky SE, Bareiro GG, Mon ML, Moore DP, Caspe G, Campero C, Fort M, Romano MI. An applied printing immunoassay with recombinant Nc-SAG1 for detection of antibodies to *Neospora caninum* in cattle. J Vet Diagn Investig. 2011;23:971–6.

Williams DJL, Guy CS, McGarry JW, Guy F, Tasker L, Smith RF, MacEachern K, Cripps PJ, Kelly DF, Trees AJ. *Neospora caninum*-associated abortion in cattle: the time of experimentally-induced parasitaemia during gestation determines foetal survival. Parasitology. 2000;121:347–58.

Woodbine KA, Medley GF, Moore SJ, Ramirez-Villaescusa A, Mason S, Green LE. A four year longitudinal sero-epidemiology study of *Neospora caninum* in adult cattle from 114 cattle herds in south west England: associations with age, herd and dam-offspring pairs. BMC Vet Res. 2008;4:35.

Wouda W, Bartels CJM, Moen AR. Characteristics of *Neospora caninum*-associated abortion storms in dairy herds in The Netherlands (1995 to 1997). Theriogenology. 1999;52:233–45.

Yániz JL, López-Gatius F, García-Ispierto I, Bech-Sàbat G, Serrano B, Nogareda C, Sanchez-Nadal JA, Almería S, Santolaria P. Some factors affecting the abortion rate in dairy herds with high incidence of *Neospora*-associated abortions are different in cows and heifers. Reprod Domest Anim. 2010;45:699–705.

Toxoplasma

6

Gastón Moré, Maria Cecilia Venturini, Lais Pardini,
and Juan Manuel Unzaga

Abstract

Toxoplasmosis is a zoonotic infection caused by the protozoon *Toxoplasma gondii* affecting animals worldwide. *T. gondii* has a facultative heteroxenous life cycle with felids as definitive hosts and a wide range of mammalian and avian species as intermediate hosts. Infectious stages are oocysts present in felid feces, bradyzoites forming tissue cysts, and tachyzoites. While two major clonal lineages (type II and III) dominate clinical and natural isolates in Europe and North America, other parts of the world, like Brazil and Argentina, are dominated by non-clonal or other clonal *T. gondii* lineages, representing a greater genetic diversity. Different animal species show a variable degree of susceptibility to *T. gondii* infection and to the development of clinical signs. Humans are considered susceptible, and about one third of the human population in the world is estimated to be infected. Toxoplasmosis is an important abortive disease in small ruminants. It is mainly asymptomatic in cattle and chicken, but the latter is an epidemiological sentinel, and isolation of viable parasites from this source is frequent. Infected pigs can show reproductive failure, and, in addition, their tissues are considered, together with infected small ruminant tissues, a relevant source of human infection. Cats and dogs can show neuromuscular disease mainly associated with other immunosuppressive conditions, such as viral infec-

G. Moré (✉) • L. Pardini
Laboratory of Immunoparasitology (LAINPA), School of Veterinary Sciences, National University of La Plata, La Plata, Argentina

National Council of Scientific and Technological Research (CONICET), CABA, Buenos Aires, Argentina
e-mail: gastonmore@fcv.unlp.edu.ar

M.C. Venturini • J.M. Unzaga
Laboratory of Immunoparasitology (LAINPA), School of Veterinary Sciences, National University of La Plata, La Plata, Argentina

M. Florin-Christensen, L. Schnittger (eds.), *Parasitic Protozoa of Farm Animals and Pets*, https://doi.org/10.1007/978-3-319-70132-5_6

tions. Some species like New World monkeys, lemurs, Pallas' cats, slender-tailed meerkats, and some Australian marsupials are highly susceptible to fatal generalized toxoplasmosis. General control measures are presented, focusing on the prevention of human toxoplasmosis.

6.1 Morphology, Life Cycle, and Host-Pathogen Interactions

6.1.1 Morphology and Life Cycle

The protozoan *Toxoplasma gondii* is an intracellular parasite which can infect mammals and birds. As other apicomplexan parasites, *T. gondii* shows a banana-shaped eukaryotic cell with a complete apical complex. It was detected initially in a hamster-like rodent (*Ctenodactylus gundi*) and due to its shape was designated as *Toxoplasma* (*toxo*, arc; *plasma*, life), while *gondii* is a deformation of the host species name. Each protozoon measures about 3–6 μm by 0.5–2 μm. They can parasitize virtually any nucleated cell. *T. gondii* is a cyst-forming coccidian from the order Eimeriorina and the unique species of the genus *Toxoplasma*.

T. gondii has a facultative heteroxenous life cycle with felids as definitive hosts and a wide range of mammalian and avian species as intermediate hosts (Dubey et al. 1998).The term facultative refers to the facts that transmission can occur among intermediate hosts, without a definitive host, and that felids can also act as intermediate hosts. In domestic cats and wild felids as definitive hosts, the parasites reproduce by asexual and sexual multiplication in intestinal cells. This is known as intestinal or entero-epithelial life cycle. Macro- and microgametes are formed—gametogony—by differentiation, and, as a result of gamete fusion, immature oocysts are produced and excreted with feces. The prepatent period can vary from 3 to 18 dpi in accordance with the infective stage ingested—tissue cysts and oocysts, respectively. Oocysts are spherical and measure 10–12 μm in diameter (Fig. 6.1). In the environment, under suitable conditions of temperature and humidity, oocysts undergo a division—sporogony or post-zygotic division—producing sporulated or mature oocysts which contain two sporocysts with four sporozoites. The regular zoites of *T. gondii* are haploids, and the formed zygote undergoes a meiotic division. At this stage, recombination of genetic material among sporozoites in formation can take place. Oocysts can remain viable for periods up to 18 months, especially with regular to high humidity (Dubey 1998b).

In intermediate hosts, including cats and humans, the parasites reproduce asexually, and an extraintestinal cycle takes place. If an intermediate host becomes infected by oocyst ingestion, sporozoites are released following oocyst wall disruption by enzymatic digestion. The zoites enter host cells inside a parasitophorous vacuole, which derives from the host cell membrane and components secreted by the protozoans. Then, *T. gondii* multiplies asexually in the host cell by endodyogeny—merogony—producing merozoites. Merogony takes place within any nucleated cell, including macrophages, and, initially, consists of a rapid division producing

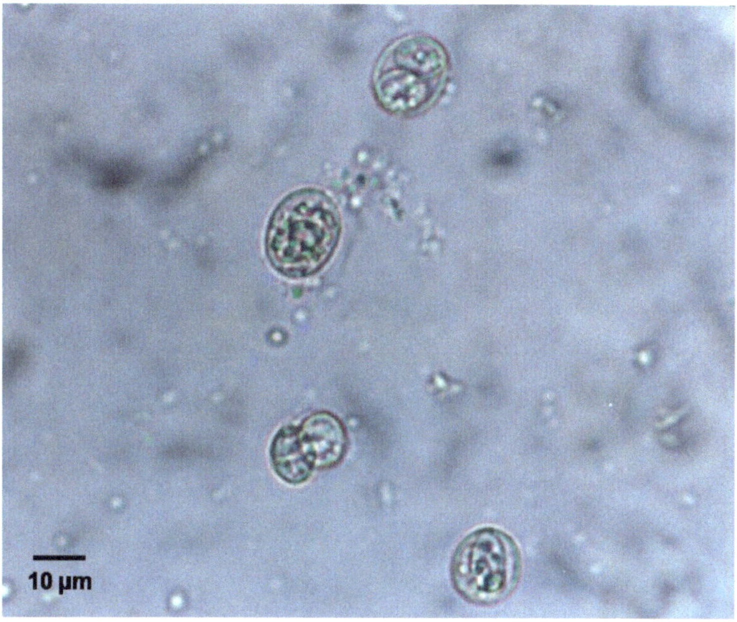

Fig. 6.1 *Toxoplasma gondii* oocysts in a cat stool sample. Note three sporulated oocysts with two sporocysts each and one immature/unsporulated oocyst. 400×

host cell disruption. Merozoites that result from this accelerated division are named tachyzoites, from the Greek *tachos*: speed. After dividing rapidly, tachyzoites enter cells, mainly of the neuromuscular system, and multiply slowly forming bradyzoites—*brady*: slow, in Greek—within tissue cysts. The brain, liver, lungs, skeletal muscle, and eyes are the main sites of appearance of tissue cysts. Cysts are spherical or elongated, limited by a cyst wall, and measure about 70 or 100 µm in diameter in brain and skeletal muscles, respectively (Figs. 6.2 and 6.3).

6.1.2 Infection Routes and Host-Pathogen Interactions

One of the principal sources of infection for both definitive and intermediate hosts is the ingestion of tissue cysts. It can take place through the consumption of raw or not well-cooked tissues, cannibalism, or scavenger behavior and can affect carnivores, omnivores, or scavenger animals (Tenter et al. 2000).

Another important route of infection is the ingestion of mature oocysts present in food, soil, water, and pastures, contaminated with infected cat feces. All animal species can be infected through this route, but it is considered particularly relevant for herbivores (Tenter et al. 2000). Oral infection with mature oocysts or with tissue cysts containing bradyzoites is known as horizontal infection.

A special infection route of *T. gondii* is the transplacental or vertical transmission through the passage of tachyzoites from mother to offspring. This process can

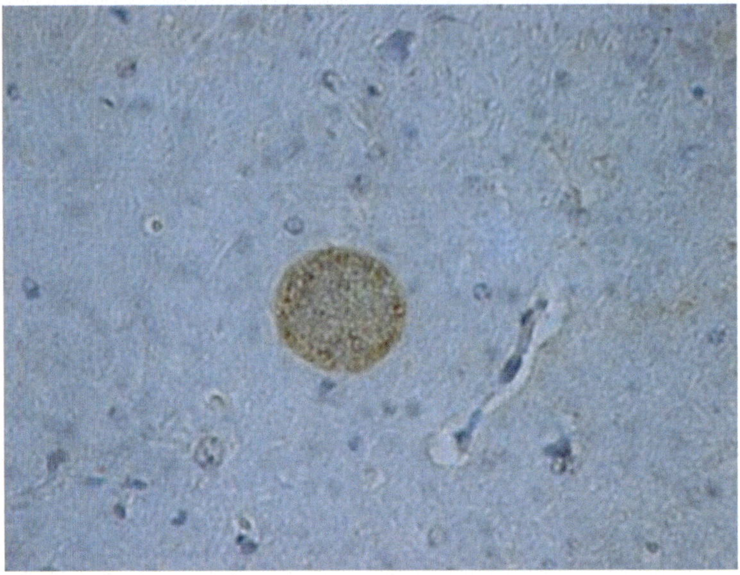

Fig. 6.2 Tissue cyst of *Toxoplasma gondii* in the brain. Positive specific immunohistochemistry stain. 400×

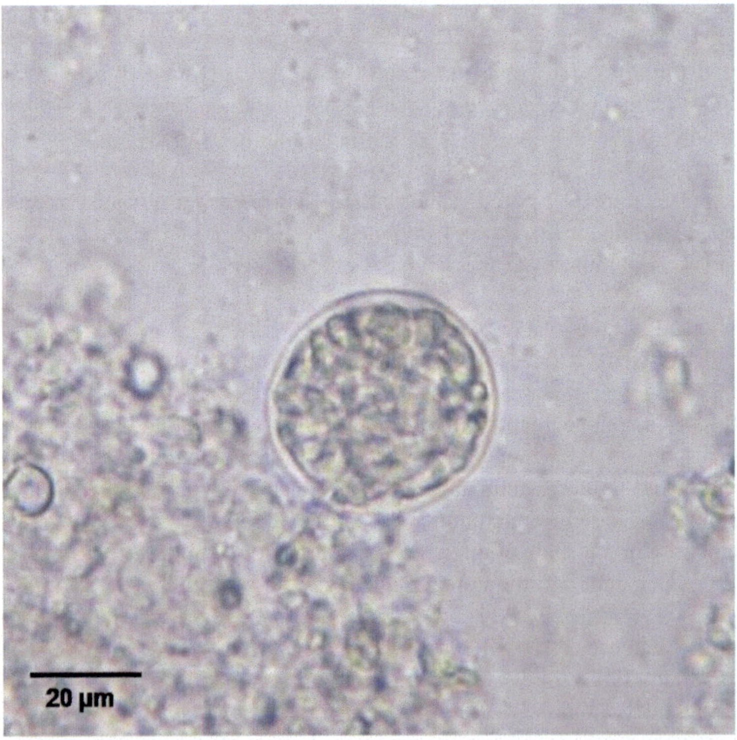

Fig. 6.3 *Toxoplasma gondii* tissue cyst from a mouse brain homogenate. Unstained. 400×

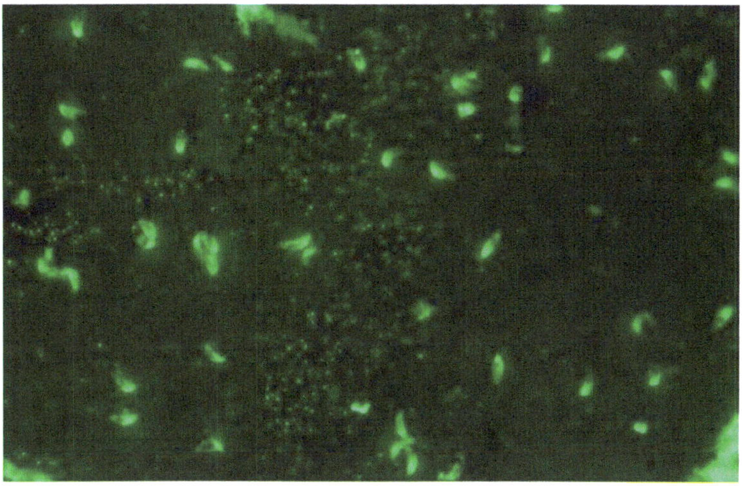

Fig. 6.4 Positive IFAT reaction using *T. gondii* RH strain as antigen. 400×

The ELISA test allows detection of IgM, IgG, and IgA antibodies and is as sensitive as the IFAT. There are different commercially available tests produced using whole antigens, native proteins, or recombinant proteins; thus a solid validation is required to use them (Dubey 2010b; Pardini et al. 2012).

The modified agglutination test (MAT) consists in a direct agglutination which uses 2-mercaptoethanol to treat the sera and to destroy nonspecific immunoglobulins and IgM. It is useful for serological diagnosis in different animal species, particularly in wild animals for which specific conjugates are not available. The antigen is prepared with whole *T. gondii* tachyzoites treated with formalin that are recognized by the test sera. The reaction is performed in microplates using different buffers and Evans blue staining. Diluted sera are incubated with the antigen preparation overnight at room temperature. In addition, the indirect agglutination test, in which antigenic fractions are adsorbed to particles, such as red blood cells or latex particles, can be used for prevalence studies. However, it is not advisable for serological diagnosis, due to its low sensitivity (Dubey 2010b). Immunoblotting is not commonly employed as routine diagnosis but can be useful to confirm results of other serological tests. Since there is no serological method considered as a gold standard, measuring antibodies by two tests or using a widely validated test is advisable for prevalence studies and for a proper diagnosis in different animal species.

6.2.2 Epidemiology

Toxoplasmosis is a worldwide distributed disease affecting a wide range of hosts. Different animal species show different degrees of susceptibility to the parasite and a variety of clinical signs. Humans are considered susceptible to *T. gondii* infection, and about one third of the human population in the world is estimated to be asymptomatic carriers of the parasite (Dubey 2010b). In addition to congenital transmission, humans can acquire the parasite by the ingestion of undercooked or raw tissue cyst-containing meat mainly from pigs, small ruminants, and birds, as well as food or vegetables contaminated with oocyst-infected felid feces.

Serological studies of *T. gondii* in small ruminants are useful to evaluate the infection status of herds and individual animals and to relate it with the occurrence of abortions. They are also important to identify possible sources of infection for other hosts. Antibodies to *T. gondii* have been detected in goats worldwide with prevalence rates between 17 and 71% (Deng et al. 2016; Faria et al. 2007; Garcia-Bocanegra et al. 2013; Iovu et al. 2012; Lopes et al. 2013; Mancianti et al. 2013; Stormoen et al. 2012; Tzanidakis et al. 2012). The relative importance of the vertical transmission in the epidemiology of goat toxoplasmosis is under study. Also in sheep, antibodies to *T. gondii* have been recorded worldwide, being higher in adult ewes, suggesting a postnatal/horizontal infection as the main route of infection.

T. gondii is known to infect a large spectrum of avian species, ranging from passerine birds, like sparrows, to domesticated birds such as chickens (Dubey 2002). Adult backyard chickens can show infection rates up to 100%, while chickens raised indoors show very low prevalence rates of 0–5% (Dubey 2010a). Since parasite isolation from chicken tissues is relatively easier than from other animals, most parasite isolates worldwide have been obtained from this species. Moreover, due to the high susceptibility of infection observed in free-range chickens, they are an excellent environmental indicator of *T. gondii* presence in a given region (Moré et al. 2012; Pardini et al. 2016). In addition, migratory birds have been postulated to contribute to the dissemination of different *T. gondii* genotypes in their migratory routes.

The prevalence rates of *T. gondii* in pigs are remarkably variable according to raising conditions (indoor versus outdoor farms) and age (market age of fattening pigs versus breeder pigs) (Dubey 1986). Feral pigs can also be infected; however, the few studies conducted revealed lower prevalence than that observed in domestic pigs. Reported prevalence rates of *T. gondii* in pigs in the 1980s ranged between 22–78%, 19–58%, and 12–31% in Latin American countries, Argentina and Brazil; in European countries, Belgium and Italy; and in Southeast Asia, Indonesia and Japan, respectively (Dubey 2010b). More recent studies detected seroprevalence values of 59%, 37–58%, 51–72%, 4–39%, and 15.8% in Argentina, Central European countries, Latin American countries, Southeast Asia, and the USA, respectively (Jones et al. 2001a; Tenter et al. 2000).

Seroepidemiological studies in various geographical regions have detected positive bovines. However, the specificity of some serological tests for *T. gondii* is lowered by cross-reactivity with *Neospora caninum* and other related protozoa (Dubey 2010b). Many studies in the past probably misdiagnosed bovine abortions as due to *T. gondii,* while the protozoon involved might have been *N. caninum*. Seroprevalence rates ranging from 0 to 60% have been reported in different countries; however, care should be taken concerning the interpretation of these results (Dubey 2010b). In a national survey in the USA, antibodies to *T. gondii* or viable parasites were not detected in 2049 beef samples (Dubey et al. 2005). Parasite DNA could be detected in cattle tissues; however, the importance of bovine tissues as source of infection for other hosts is uncertain (Moré et al. 2008).

Dogs are commonly infected by the ingestion of *T. gondii* oocysts from soil, vegetables, and water contaminated with infected feline feces, by the ingestion of raw or not well-cooked cyst-containing tissues from infected animals, and by transplacental infection to puppies (Dubey 2010b). Since dogs are usually living closely with humans and have a similar opportunity for *T. gondii* infection, prevalence of *T. gondii* in dogs might reflect human infection. Tenter et al. (2000) summarized prevalence rates to *T. gondii* in dogs of various countries and found strong regional differences. Recorded rates were of 60% by IFAT in Argentina, 46–84% by IFAT in Brazil, 36% by IFAT in Israel, 47% by IFAT in Spain, 5% by ELISA in China, and 8–25% by ELISA and LAT in Taiwan. In Trinidad and Tobago, rates measured by LAT were 60.5% in stray dogs, 30.5% in hunting dogs, and 25.5% in pet dogs (Ali et al. 2003).

Felids are important in the epidemiology of toxoplasmosis since they act as both definitive and intermediate hosts. Since *T. gondii* has been detected in most studied regions, it is assumed that felids (wild and domestic) could be infected all around the world. When cats are infected with tissue cysts, oocysts, or tachyzoites, prepatent periods of 3–10, 19–41, or 9–11 days, respectively, are followed by oocyst excretion in the feces (Dubey 2010b). As peak levels of specific IgG antibodies take place at 14 days post-infection, IgG antibodies might not be detected during oocyst excretion, especially after ingestion of tissue cysts. It is possible that a chronically infected animal, which has already shed oocysts, re-sheds after an immunosuppressive condition, which is commonly induced by feline immunodeficiency virus—FIV—infection, other chronic concomitant diseases, or age (Dubey and Carpenter 1993). Since sexual reproduction of the parasite is not associated with clinical signs, detection of oocysts in stool samples from symptomatic animals submitted to laboratory analysis is rare. Studies conducted using high number of stool samples submitted to copro-parasitological analysis revealed a prevalence of *T. gondii* oocysts in around 0.1 and 1% in domestic cats (Epe et al. 1993; Schares et al. 2008).

As mentioned previously, *T. gondii* can infect mammals and birds; therefore, any wild animal could be potentially infected. Additionally, it is assumed that the parasites have a *domestic* cycle and a *wild or sylvatic* cycle. Apparently, coevolution of parasites and host tends to the equilibrium, and the adaptation is expressed as an asymptomatic infection. However, when a host adapted to one type of cycle is infected with a parasite adapted to other hosts, the infection tends to be severe or symptomatic (Carme et al. 2009).

6.2.2.1 Population Structure

The global population structure of *T. gondii* is very diverse and varies between continents. However, its differential appearance in distinct parts of the world is only partially understood. Most *T. gondii* isolates from humans and animals from North America and Europe have been classified into one of three genetic lineages—identified as type I, II, or III—based on PCR-RFLP analysis (Howe et al. 1997; Howe and Sibley 1995). These lineages have different virulence phenotypes in mouse infection models: type I strain is highly lethal in outbred mice, while type II and III strains are significantly less virulent (Sibley and Boothroyd 1992). However, it is not known if this virulence phenotype described in mice may also be observed in other animal species. In the last 10 years, *T. gondii* isolates from South America were characterized as atypical, considering the abovementioned classification, as well as sequencing and microsatellite typing (Ajzenberg et al. 2004; Beck et al. 2009; Pena et al. 2008). While two major clonal lineages—type II and III—dominate clinical and natural isolates in Europe and North America, other parts of the world, like Brazil and Argentina, are dominated by non-clonal or other clonal *T. gondii* lineages representing a greater genetic diversity of *T. gondii* (Ajzenberg 2015; Ajzenberg et al. 2004, Shwab et al. 2014). Until now, 15 haplogroups have been reported worldwide (Su et al. 2012), but the existence of as yet unknown ones in more remote areas of the globe is expected (Ajzenberg 2015). Molecular and genomic *T. gondii* information from different strains reported worldwide can be retrieved from the *Toxoplasma* genomic resource (http://toxodb.org/toxo/). It has been hypothesized that genetic diversity in *T. gondii* is driven by selective pressures due to its adaptation to different intermediate or definitive hosts, included in domestic and wildlife cycles in different parts of the world (VanWormer et al. 2014).

6.3 Clinical Effects, Prevention, and Treatment

6.3.1 Clinical Effects

One of the principal clinical signs is the loss of vision due to a toxoplasmic retinochoroiditis known as ocular toxoplasmosis (Labalette et al. 2002). Congenital toxoplasmosis—the primo-infection during pregnancy—can produce neurological and ocular disorders in fetuses (Jones et al. 2001b). Another clinical presentation is encephalitis in immunocompromised patients caused by reactivation of *T. gondii*.

Toxoplasmosis is an important cause of small ruminant abortions, and the ingestion of undercooked meat and unpasteurized milk from infected animals can be a source of infection for humans (Dubey 2010b). *T. gondii*-caused abortion in sheep and goats is frequently associated with necrosis and calcification of the fetal cotyledons and normal inter-cotyledonary areas (Dubey 2010b). However, since similar lesions can be also observed in *N. caninum*-associated abortion, discrimination between these infections out of histopathological lesions is difficult (McAllister et al. 1996; Ortega-Mora et al. 2007). Repeated abortions in *T. gondii*-infected goats are frequently observed, suggesting a potential reactivation of the disease during successive gestations (Unzaga et al. 2014). Infection during early pregnancy may lead to fetal death and reabsorption; therefore, the ewe appears barren. An infection that takes place between approximately 50 and 120 days of gestation leads to abortion, stillborns, mummified fetuses, or the birth of weak lambs. The infected ewes acquire a protective immunity which prevents future abortions. This fact has been considered for the development and application of a vaccine to prevent sheep abortion due to *T. gondii* which is available in a few countries like New Zealand, Great Britain, Ireland, Portugal, and Spain.

Clinical effects in avian species vary considerably. Severe toxoplasmosis has been reported in canaries, which can develop ophthalmitis (including cataracts and blindness). Chickens are considered highly susceptible to become infected; however, they are resistant to the development of clinical signs. Nevertheless, a recent case report showed symptomatic chicken, and the highest rates of affected animals were adults with central nervous signs. Further histological examination revealed encephalitis, neuritis, myocarditis, and retina lesions (Dubey et al. 2007). Recently, a study that involved the experimental infection with *T. gondii* of 1-day old chicken was conducted. A significantly higher mortality was observed in white layer lines, as compared to brown layer lines, suggesting that differences in the genetic background may influence the susceptibility of chicken to toxoplasmosis (Schares et al. 2016). Many other avian species have been reported as carrying *T. gondii* infections, although only few cases were reported as suffering a severe disease, most of which were birds raised in captivity.

Pigs infected with *T. gondii* are the most important source of human infection in several countries, especially in the USA (Dubey 2010b). Humans are commonly infected by the ingestion of tissue cysts through the consumption of raw or not well-cooked pork. Clinical toxoplasmosis in pigs is considered rare; however, a few reports from Taiwan, Japan, Switzerland, and the USA have indicated increases in abortions and neonatal mortality and/or febrile and neurological symptoms (Basso et al. 2015; Dubey 2010b). In addition, experimental infection with *T. gondii* in pigs demonstrated that some neonatal infected pigs died after birth, although abortion and congenital infections were rarely observed (Dubey 2010b; Wingstrand et al. 1997).

Cattle are considered one of the least susceptible animal species to develop clinical toxoplasmosis. Moreover, although animals can be infected experimentally, the parasite is eliminated or reduced to undetectable levels in a few weeks post-infection (Dubey 1983). On the other hand, some researchers have detected parasite DNA in bovine fetuses (Gottstein et al. 1998), and the parasite was successfully isolated from two aborted bovine fetuses (Canada et al. 2002). From the above mentioned, it is possible to assume that *T. gondii* can be transmitted vertically in cattle; however, this appears to be a rare occurrence.

Domestic cats and other felids acting as intermediate hosts can develop neurological or ocular signs due to the extraintestinal multiplication of the protozoon. Susceptibility for developing clinical toxoplasmosis increases with feline immunodeficiency virus (FIV) infection, indicating cooperative interaction between *T. gondii* and FIV, as well as in older cats, suggesting a potential reduction of the immunological control of the infection (Dubey and Carpenter 1993). Pallas' cats (*Felis manul manul*) and sand cats (*Felis margarita*) are highly susceptible to develop generalized toxoplasmosis (Basso et al. 2005; Dubey et al. 2010). Intestinal cycle regularly occurs without clinical signs, where an infected feline could be eliminating millions of oocysts.

In dogs, *T. gondii* infections lead to respiratory or neuromuscular clinical signs. However, the latter signs are difficult to differentiate from those produced by *N. caninum* (Dubey et al. 1988a). Therefore, it is important to remark the need of differential diagnosis between these two parasitic infections. Concurrent infections with canine distemper virus could potentiate or reactivate a *T. gondii* infection, and animals can suffer a severe associated disease (Dubey 2010b).

Some species like New World monkeys (Dietz et al. 1997), lemurs (Spencer et al. 2004), Pallas' cats (Basso et al. 2005), slender-tailed meerkats, and some Australian marsupials (Basso et al. 2007; Dubey et al. 1988b) are considered highly susceptible to clinical toxoplasmosis. Most of these animal species suffer a multi-organic or generalized fatal toxoplasmosis. Apparently, animals with short evolutionary contact with *T. gondii* (desert- or tree-living animals) can suffer a severe disease, even if a low-virulence parasite strain is involved (Basso et al. 2009). As example, a colony of slender-tailed meerkats in a zoo was reported to suffer severe and disseminated toxoplasmosis due to a genotype III strain of *T. gondii*, which shows low virulence in mice (Basso et al. 2009). Wallabies, especially females, are considered more susceptible to acute toxoplasmosis than kangaroos (Basso et al. 2007; Dubey et al. 1988b). However, sudden death and similar lesions to those reported previously in the wallaby *Macropus rufogriseus* have been reported in male kangaroos (Adkesson et al. 2007; Basso et al. 2007; Bermudez et al. 2009; Dubey and Crutchley 2008). In addition, a large amount of *T. gondii* cysts was detected in muscles of a *Macropus rufus* kangaroo, which could be related to species or individual susceptibility and/or related to poor immune control of parasite multiplication (Moré et al. 2010). Additionally, the *T. gondii* strain affecting this animal was identified as genotype III, and the case occurred in the same zoo where the meerkats abovementioned died, suggesting that this genotype may be frequent in the zoo environment. Fatal toxoplasmosis has been described in most genera of New World primates in several parts of the world (Gyimesi et al. 2006; Pardini et al. 2015). The infection generally takes the form of acute disease, and animals die suddenly, sometimes without preceding signs. If signs are present, they are nonspecific and include lethargy, malaise, depression, anorexia, diarrhea, hypothermia, serosanguineous nasal discharge, and respiratory distress (Gyimesi et al. 2006; Pardini et al. 2015).

It is possible to assume that stressed animals in some environments suffer more severe toxoplasmosis due to a poor immunity control of parasite multiplication.

6.3.2 Prevention

Since *T. gondii* can persist in different animal species and uses different transmission routes, a complete control and eradication appears utopic. Most recommendations to prevent infection with this protozoon are related to the predominant route of infection in each animal species. Bradyzoites from tissues are inactivated by incubation at 65 °C, freezing at −20 °C during a week, or gamma irradiation at 0.5 kGy (Dubey 1996). Therefore, cooking or freezing potentially infected animal tissues before ingestion by other species is important to reduce parasite viability and minimize infection. Consumption of raw meat or undercooked products is a frequent cultural behavior; therefore, slaughtered animals destined to human consumption should be analyzed for the presence of *T. gondii* and/or proceed as previously mentioned to minimize the risk of infection.

Once excreted by felids, oocysts are immature. Therefore, daily disposing of stools from domestic cats is useful to reduce potential oocyst dispersion in the environment. Mature sporulated oocysts are long-term resistant, but they are destroyed by boiling water. This can be used for disinfection of surfaces in contact with cat feces. To minimize the chances of a cat being infected and lately contaminate the environment with oocysts, the use of commercial dry or canned feed is recommended. Neutering of cats is recommended to reduce their behavior of hunting birds and rodents. Washing vegetables with safe water is a main recommendation to eliminate or decrease the amounts of potentially present oocysts, as well as preventing felids to take contact with farms or markets that produce or sell vegetables. Humans in contact with soil (potentially contaminated with oocysts) should wash their hands as frequently as possible and/or protect them with garden gloves.

Gamble et al. (1999) reported regional differences in prevalence of *T. gondii* in pigs and suggested that education on farm management practices should be targeted to reduce toxoplasmosis.

Keeping animals in zoo environments is a controversial issue. Several wild felid species can act as definitive hosts, contaminating the environment with oocysts. Unfortunately, the presence of stray domestic cats within zoos around the world is frequently observed, a fact which can have a negative impact on environment contamination. Moreover, several animal species are extremely susceptible to developing fatal toxoplasmosis. In spite of this, a main recommendation to zoo authorities is to build blackout-type enclosures to maintain susceptible species protected from accidental ingestion of oocysts, as well as to minimize stress to avoid potential reactivation of chronic infections.

A live vaccine (Toxovax), using a nonpersistent modified strain of *T. gondii* (S48), available in New Zealand, the UK, and Europe, prevents *T. gondii* abortion in sheep (Buxton et al. 1991). Also, a live vaccine using a mutant strain of *T. gondii* (T-263) is being developed in the USA to reduce oocyst shedding by cats (Verma and Khanna 2013).

6.3.3 Treatment

As yet, there are no drugs able to kill *T. gondii* tissue cysts in humans or animals. Treatment is recommended to reduce tachyzoite multiplication. Drugs such as clindamycin, sulfadiazine, pyrimethamine, toltrazuril, or ponazuril are effective both in vitro and in vivo on *T. gondii* multiplication. These drugs need to be applied for periods of about 30 days to be effective in clinical cases. Corticosteroids are indicated to reduce the extension of inflammatory ocular lesions. A prophylactic treatment of ewes with monensin has been proposed; however, its application is controversial considering current international protocols to avoid antimicrobial resistance (Dubey 2010b).

It is now apparent that many atypical genotypes exist besides the typical three genotypes first described. These genotypes can differ in virulence and transmissibility from the typical genotypes that have been used in the majority of scientific research over the past 70 years. Recommendations to prevent congenital toxoplasmosis (CT) have been determined based on the information produced by researching with these typical strains (Lindsay and Dubey 2011). It is important to identify the biological behavior of atypical strains and to develop new recommendations for the prevention and the monitoring of women at risk for developing CT, especially in South American countries where a higher diversity of *T. gondii* genotypes has been reported.

References

Adkesson MJ, Gorman ME, Hsiao V, Whittington JK, Langan JN. *Toxoplasma gondii* inclusions in peripheral blood leukocytes of a red-necked wallaby (*Macropus rufogriseus*). Vet Clin Pathol. 2007;36:97–100.

Ajzenberg D. 1995-2015: it is time to celebrate 20 years of (intensive) genotyping of *Toxoplasma gondii* strains. Future Microbiol. 2015;10:689–91. https://doi.org/10.2217/fmb.15.23.

Ajzenberg D, Banuls AL, Su C, Dumetre A, Demar M, Carme B, Darde ML. Genetic diversity, clonality and sexuality in *Toxoplasma gondii*. Int J Parasitol. 2004;34:1185–96. https://doi.org/10.1016/j.ijpara.2004.06.007.

Ali CN, Harris JA, Watkins JD, Adesiyun AA. Seroepidemiology of *Toxoplasma gondii* in dogs in Trinidad and Tobago. Vet Parasitol. 2003;113:179–87.

Basso W, Edelhofer R, Zenker W, Mostl K, Kubber-Heiss A, Prosl H. Toxoplasmosis in Pallas' cats (*Otocolobus manul*) raised in captivity. Parasitology. 2005;130:293–9.

Basso W, Venturini MC, More G, Quiroga A, Bacigalupe D, Unzaga JM, Larsen A, Laplace R, Venturini L. Toxoplasmosis in captive Bennett's wallabies (*Macropus rufogriseus*) in Argentina. Vet Parasitol. 2007;144:157–61. https://doi.org/10.1016/j.vetpar.2006.09.030.

Basso W, et al. Isolation and molecular characterization of *Toxoplasma gondii* from captive slender-tailed meerkats (*Suricata suricatta*) with fatal toxoplasmosis in Argentina. Vet Parasitol. 2009;161:201–6. https://doi.org/10.1016/j.vetpar.2009.01.006.

Basso W, Handke M, Sydler T, Borel N, Grimm F, Sidler X, Deplazes P. Involvement of *Toxoplasma gondii* in reproductive disorders in Swiss pig farms. Parasitol Int. 2015;64:157–60. https://doi.org/10.1016/j.parint.2014.11.017.

Beck HP, et al. Molecular approaches to diversity of populations of apicomplexan parasites. Int J Parasitol. 2009;39:175–89. https://doi.org/10.1016/j.ijpara.2008.10.001.

Bermudez R, Failde LD, Losada AP, Nieto JM, Quiroga MI. Toxoplasmosis in Bennett's wallabies (*Macropus rufogriseus*) in Spain. Vet Parasitol. 2009;160:155–8. https://doi.org/10.1016/j.vetpar.2008.10.082.

Blader IJ, Saeij JP. Communication between *Toxoplasma gondii* and its host: impact on parasite growth, development, immune evasion, and virulence. APMIS. 2009;117:458–76. https://doi.org/10.1111/j.1600-0463.2009.02453.x.

Buxton D, Thomson K, Maley S, Wright S, Bos HJ. Vaccination of sheep with a live incomplete strain (S48) of *Toxoplasma gondii* and their immunity to challenge when pregnant. Vet Rec. 1991;129:89–93.

Canada N, Meireles CS, Rocha A, da Costa JM, Erickson MW, Dubey JP. Isolation of viable *Toxoplasma gondii* from naturally infected aborted bovine fetuses. J Parasitol. 2002;88:1247–8. https://doi.org/10.1645/0022-3395(2002)088[1247:IOVTGF]2.0.CO;2.

Carme B, Demar M, Ajzenberg D, Darde ML. Severe acquired toxoplasmosis caused by wild cycle of *Toxoplasma gondii*, French Guiana. Emerg Infect Dis. 2009;15:656–8. https://doi.org/10.3201/eid1504.081306.

Deng H, Dam-Deisz C, Luttikholt S, Maas M, Nielen M, Swart A, Vellema P, van der Giessen J, Opsteegh M. Risk factors related to *Toxoplasma gondii* seroprevalence in indoor-housed Dutch dairy goats. Prev Vet Med. 2016;124:45–51. https://doi.org/10.1016/j.prevetmed.2015.12.014.

Dietz HH, Henriksen P, Bille-Hansen V, Henriksen SA. Toxoplasmosis in a colony of new world monkeys. Vet Parasitol. 1997;68:299–304.

Dubey JP. Distribution of cysts and tachyzoites in calves and pregnant cows inoculated with *Toxoplasma gondii* oocysts. Vet Parasitol. 1983;13:199–211.

Dubey JP. A review of toxoplasmosis in pigs. Vet Parasitol. 1986;19:181–223.

Dubey JP. Strategies to reduce transmission of *Toxoplasma gondii* to animals and humans. Vet Parasitol. 1996;64:65–70.

Dubey JP. Advances in the life cycle of *Toxoplasma gondii*. Int J Parasitol. 1998a;28:1019–24.

Dubey JP. *Toxoplasma gondii* oocyst survival under defined temperatures. J Parasitol. 1998b;84:862–5.

Dubey JP. A review of toxoplasmosis in wild birds. Vet Parasitol. 2002;106:121–53.

Dubey JP. *Toxoplasma gondii* infections in chickens (*Gallus domesticus*): prevalence, clinical disease, diagnosis and public health significance. Zoonoses Public Health. 2010a;57:60–73. https://doi.org/10.1111/j.1863-2378.2009.01274.x.

Dubey JP. Toxoplasmosis of animals and humans. 2nd ed. Boca Raton, FL: CRC Press; 2010b.

Dubey JP, Carpenter JL. Histologically confirmed clinical toxoplasmosis in cats: 100 cases (1952-1990). J Am Vet Med Assoc. 1993;203:1556–66.

Dubey JP, Crutchley C. Toxoplasmosis in wallabies (*Macropus rufogriseus* and *Macropus eugenii*): blindness, treatment with atovaquone, and isolation of *Toxoplasma gondii*. J Parasitol. 2008;94:929–33. https://doi.org/10.1645/GE-1448.1.

Dubey JP, Emond JP, Desmonts G, Anderson WR. Serodiagnosis of postnatally and prenatally induced toxoplasmosis in sheep. Am J Vet Res. 1987;48:1239–43.

Dubey JP, Carpenter JL, Speer CA, Topper MJ, Uggla A. Newly recognized fatal protozoan disease of dogs. J Am Vet Med Assoc. 1988a;192:1269–85.

Dubey JP, Ott-Joslin J, Torgerson RW, Topper MJ, Sundberg JP. Toxoplasmosis in black-faced kangaroos (*Macropus fuliginosus melanops*). Vet Parasitol. 1988b;30:97–105.

Dubey JP, Lindsay DS, Speer CA. Structures of *Toxoplasma gondii* tachyzoites, bradyzoites, and sporozoites and biology and development of tissue cysts. Clin Microbiol Rev. 1998;11:267–99.

Dubey JP, et al. Prevalence of viable *Toxoplasma gondii* in beef, chicken, and pork from retail meat stores in the United States: risk assessment to consumers. J Parasitol. 2005;91:1082–93. https://doi.org/10.1645/GE-683.1.

Dubey JP, Webb DM, Sundar N, Velmurugan GV, Bandini LA, Kwok OC, Su C. Endemic avian toxoplasmosis on a farm in Illinois: clinical disease, diagnosis, biologic and genetic characteristics of *Toxoplasma gondii* isolates from chickens (*Gallus domesticus*), and a goose (Anser Anser). Vet Parasitol. 2007;148:207–12.

Dubey JP, Pas A, Rajendran C, Kwok OC, Ferreira LR, Martins J, Hebel C, Hammer S, Su C. Toxoplasmosis in sand cats (*Felis margarita*) and other animals in the breeding Centre for Endangered Arabian Wildlife in the United Arab Emirates and al Wabra wildlife preservation, the State of Qatar. Vet Parasitol. 2010;172:195–203. https://doi.org/10.1016/j.vetpar.2010.05.013.

Epe C, Ising-Volmer S, Stoye M. Parasitological fecal studies of equids, dogs, cats and hedgehogs during the years 1984-1991. Dtsch Tierarztl Wochenschr. 1993;100:426–8.

Faria EB, Gennari SM, Pena HF, Athayde AC, Silva ML, Azevedo SS. Prevalence of anti-*Toxoplasma gondii* and anti-*Neospora caninum* antibodies in goats slaughtered in the public slaughterhouse of Patos city, Paraiba state, northeast region of Brazil. Vet Parasitol. 2007;149:126–9. https://doi.org/10.1016/j.vetpar.2007.07.009.

Fentress SJ, et al. Phosphorylation of immunity-related GTPases by a *Toxoplasma gondii*-secreted kinase promotes macrophage survival and virulence. Cell Host Microbe. 2010;8:484–95. https://doi.org/10.1016/j.chom.2010.11.005.

Fleckenstein MC, Reese ML, Konen-Waisman S, Boothroyd JC, Howard JC, Steinfeldt T. A *Toxoplasma gondii* pseudokinase inhibits host IRG resistance proteins. PLoS Biol. 2012;10:e1001358. https://doi.org/10.1371/journal.pbio.1001358.

Gamble HR, Brady RC, Dubey JP. Prevalence of *Toxoplasma gondii* infection in domestic pigs in the New England states. Vet Parasitol. 1999;82:129–36.

Garcia-Bocanegra I, Cabezon O, Hernandez E, Martinez-Cruz MS, Martinez-Moreno A, Martinez-Moreno J. *Toxoplasma gondii* in ruminant species (cattle, sheep, and goats) from southern Spain. J Parasitol. 2013;99:438–40. https://doi.org/10.1645/12-27.1.

Gottstein B, Hentrich B, Wyss R, Thur B, Busato A, Stark KD, Muller N. Molecular and immunodiagnostic investigations on bovine neosporosis in Switzerland. Int J Parasitol. 1998;28:679–91.

Gyimesi ZS, Lappin MR, Dubey JP. Application of assays for the diagnosis of toxoplasmosis in a colony of woolly monkeys (*Lagothrix lagotricha*). J Zoo Wildl Med. 2006;37:276–80. https://doi.org/10.1638/05-018.1.

Hide G. Role of vertical transmission of *Toxoplasma gondii* in prevalence of infection. Expert Rev Anti Infect Ther. 2016;14:335–44. https://doi.org/10.1586/14787210.2016.1146131.

Howe DK, Sibley LD. *Toxoplasma gondii* comprises three clonal lineages: correlation of parasite genotype with human disease. J Infect Dis. 1995;172:1561–6.

Howe DK, Honore S, Derouin F, Sibley LD. Determination of genotypes of *Toxoplasma gondii* strains isolated from patients with toxoplasmosis. J Clin Microbiol. 1997;35:1411–4.

Iovu A, Gyorke A, Mircean V, Gavrea R, Cozma V. Seroprevalence of *Toxoplasma gondii* and *Neospora caninum* in dairy goats from Romania. Vet Parasitol. 2012;186:470–4. https://doi.org/10.1016/j.vetpar.2011.11.062.

Jones JL, Kruszon-Moran D, Wilson M, McQuillan G, Navin T, McAuley JB. T*oxoplasma gondii* infection in the United States: seroprevalence and risk factors. Am J Epidemiol. 2001a;154:357–65.

Jones JL, Lopez A, Wilson M, Schulkin J, Gibbs R. Congenital toxoplasmosis: a review. Obstet Gynecol Surv. 2001b;56:296–305.

Karsten V, Qi H, Beckers CJ, Reddy A, Dubremetz JF, Webster P, Joiner KA. The protozoan parasite *Toxoplasma gondii* targets proteins to dense granules and the vacuolar space using both conserved and unusual mechanisms. J Cell Biol. 1998;141:1323–33.

Labalette P, Delhaes L, Margaron F, Fortier B, Rouland JF. Ocular toxoplasmosis after the fifth decade. Am J Ophthalmol. 2002;133:506–15.

Lang C, Gross U, Luder CG. Subversion of innate and adaptive immune responses by *Toxoplasma gondii*. Parasitol Res. 2007;100:191–203. https://doi.org/10.1007/s00436-006-0306-9.

Lin MH, Chen TC, Kuo TT, Tseng CC, Tseng CP. Real-time PCR for quantitative detection of *Toxoplasma gondii*. J Clin Microbiol. 2000;38:4121–5.

Lindsay DS, Dubey JP. *Toxoplasma gondii*: the changing paradigm of congenital toxoplasmosis. Parasitology. 2011;138:1829–31. https://doi.org/10.1017/S0031182011001478.

Lopes AP, Dubey JP, Neto F, Rodrigues A, Martins T, Rodrigues M, Cardoso L. Seroprevalence of *Toxoplasma gondii* infection in cattle, sheep, goats and pigs from the north of Portugal for human consumption. Vet Parasitol. 2013;193:266–9. https://doi.org/10.1016/j.vetpar.2012.12.001.

Mancianti F, Nardoni S, D'Ascenzi C, Pedonese F, Mugnaini L, Franco F, Papini R. Seroprevalence, detection of DNA in blood and milk, and genotyping of *Toxoplasma gondii* in a goat population in Italy. Biomed Res Int. 2013;2013:905326. https://doi.org/10.1155/2013/905326.

McAllister MM, McGuire AM, Jolley WR, Lindsay DS, Trees AJ, Stobart RH. Experimental neosporosis in pregnant ewes and their offspring. Vet Pathol. 1996;33:647–55.

Melo MB, Jensen KD, Saeij JP. *Toxoplasma gondii* effectors are master regulators of the inflammatory response. Trends Parasitol. 2011;27:487–95. https://doi.org/10.1016/j.pt.2011.08.001.

Miller CM, Boulter NR, Ikin RJ, Smith NC. The immunobiology of the innate response to *Toxoplasma gondii*. Int J Parasitol. 2009;39:23–39. https://doi.org/10.1016/j.ijpara.2008.08.002.

Moré G, Basso W, Bacigalupe D, Venturini MC, Venturini L. Diagnosis of *Sarcocystis cruzi*, *Neospora caninum*, and *Toxoplasma gondii* infections in cattle. Parasitol Res. 2008;102:671–5. https://doi.org/10.1007/s00436-007-0810-6.

Moré G, Pardini L, Basso W, Machuca M, Bacigalupe D, Villanueva MC, Schares G, Venturini MC, Venturini L. Toxoplasmosis and genotyping of *Toxoplasma gondii* in *Macropus rufus* and *Macropus giganteus* in Argentina. Vet Parasitol. 2010;169:57–61. https://doi.org/10.1016/j.vetpar.2009.12.004.

Moré G, et al. *Toxoplasma gondii* infection in sentinel and free-range chickens from Argentina. Vet Parasitol. 2012;184:116–21. https://doi.org/10.1016/j.vetpar.2011.09.012.

Ortega-Mora L, Gottstein B, Conraths F, Buxton D. Protozoal abortificients in farm ruminants. Guidelines for diagnosis and control. Wallingford: CABI; 2007.

Pardini L, et al. Evaluation of an in-house TgSAG1 (P30) IgG ELISA for diagnosis of naturally acquired *Toxoplasma gondii* infection in pigs. Vet Parasitol. 2012;189:204–10. https://doi.org/10.1016/j.vetpar.2012.04.014.

Pardini L, et al. Isolation and molecular characterization of *Toxoplasma gondii* in a colony of captive black-capped squirrel monkeys (*Saimiri boliviensis*). Parasitol Int. 2015;64:587–90. https://doi.org/10.1016/j.parint.2015.08.009.

Pardini L, Moré G, Rudzinski M, Gos ML, Campero LM, Meyer A, Bernstein M, Unzaga JM, Venturini MC. *Toxoplasma gondii* isolates from chickens in an area with human toxoplasmic retinochoroiditis. Exp Parasitol. 2016;166:16–20. https://doi.org/10.1016/j.exppara.2016.03.006.

Pena HF, Gennari SM, Dubey JP, Su C. Population structure and mouse-virulence of *Toxoplasma gondii* in Brazil. Int J Parasitol. 2008;38:561–9. https://doi.org/10.1016/j.ijpara.2007.09.004.

Pollard AM, Onatolu KN, Hiller L, Haldar K, Knoll LJ. Highly polymorphic family of glycosylphosphatidylinositol-anchored surface antigens with evidence of developmental regulation in *Toxoplasma gondii*. Infect Immun. 2008;76:103–10. https://doi.org/10.1128/IAI.01170-07.

Saadatnia G, Haj Ghani H, Khoo BY, Maimunah A, Rahmah N. Optimization of *Toxoplasma gondii* cultivation in VERO cell line. Trop Biomed. 2010;27:125–30.

Saeij JP, Boyle JP, Coller S, Taylor S, Sibley LD, Brooke-Powell ET, Ajioka JW, Boothroyd JC. Polymorphic secreted kinases are key virulence factors in toxoplasmosis. Science. 2006;314:1780–3. https://doi.org/10.1126/science.1133690.

Schares G, Vrhovec MG, Pantchev N, Herrmann DC, Conraths FJ. Occurrence of *Toxoplasma gondii* and *Hammondia hammondi* oocysts in the faeces of cats from Germany and other European countries. Vet Parasitol. 2008;152:34–45. https://doi.org/10.1016/j.vetpar.2007.12.004.

Schares G, Herrmann DC, Maksimov P, Matzkeit B, Conraths FJ, Moré G, Preisinger R, Weigend S. Chicken line-dependent mortality after experimental infection with three type IIxIII recombinant *Toxoplasma gondii* clones. Exp Parasitol. 2016;180:101–11. https://doi.org/10.1016/j.exppara.2016.11.004.

Shwab EK, Zhu XQ, Majumdar D, Pena HF, Gennari SM, Dubey JP, Su C. Geographical patterns of *Toxoplasma gondii* genetic diversity revealed by multilocus PCR-RFLP genotyping. Parasitology. 2014;141:453–61. https://doi.org/10.1017/S0031182013001844.

Sibley LD, Boothroyd JC. Virulent strains of *Toxoplasma gondii* comprise a single clonal lineage. Nature. 1992;359:82–5. https://doi.org/10.1038/359082a0.

Spencer JA, Joiner KS, Hilton CD, Dubey JP, Toivio-Kinnucan M, Minc JK, Blagburn BL. Disseminated toxoplasmosis in a captive ring-tailed lemur (*Lemur catta*). J Parasitol. 2004;90:904–6. https://doi.org/10.1645/GE-249R.

Stormoen M, Tharaldsen J, Hopp P. Seroprevalence of *Toxoplasma gondii* infection in Norwegian dairy goats. Acta Vet Scand. 2012;54:75. https://doi.org/10.1186/1751-0147-54-75.

Su C, Zhang X, Dubey JP. Genotyping of *Toxoplasma gondii* by multilocus PCR-RFLP markers: a high resolution and simple method for identification of parasites. Int J Parasitol. 2006;36:841–8. https://doi.org/10.1016/j.ijpara.2006.03.003.

Su C, Shwab EK, Zhou P, Zhu XQ, Dubey JP. Moving towards an integrated approach to molecular detection and identification of *Toxoplasma gondii*. Parasitology. 2010;137:1–11. https://doi.org/10.1017/S0031182009991065.

Su C, et al. Globally diverse *Toxoplasma gondii* isolates comprise six major clades originating from a small number of distinct ancestral lineages. Proc Natl Acad Sci U S A. 2012;109:5844–9. https://doi.org/10.1073/pnas.1203190109.

Tait ED, Hunter CA. Advances in understanding immunity to *Toxoplasma gondii*. Mem Inst Oswaldo Cruz. 2009;104:201–10.

Tenter AM, Heckeroth AR, Weiss LM. *Toxoplasma gondii*: from animals to humans. Int J Parasitol. 2000;30:1217–58.

Tizard IR. Veterinary immunology and introduction. 8th ed. Barcelona: Elsevier; 2009.

Tzanidakis N, Maksimov P, Conraths FJ, Kiossis E, Brozos C, Sotiraki S, Schares G. *Toxoplasma gondii* in sheep and goats: seroprevalence and potential risk factors under dairy husbandry practices. Vet Parasitol. 2012;190:340–8. https://doi.org/10.1016/j.vetpar.2012.07.020.

Uggla A, Sjoland L, Dubey JP. Immunohistochemical diagnosis of toxoplasmosis in fetuses and fetal membranes of sheep. Am J Vet Res. 1987;48:348–51.

Unzaga JM, Moré G, Bacigalupe D, Rambeaud M, Pardini L, Dellarupe A, De Felice L, Gos ML, Venturini MC. *Toxoplasma gondii* and *Neospora caninum* infections in goat abortions from Argentina. Parasitol Int. 2014;63:865–7. https://doi.org/10.1016/j.parint.2014.07.009.

VanWormer E, Miller MA, Conrad PA, Grigg ME, Rejmanek D, Carpenter TE, Mazet JA. Using molecular epidemiology to track *Toxoplasma gondii* from terrestrial carnivores to marine hosts: implications for public health and conservation. PLoS Negl Trop Dis. 2014;8:e2852. https://doi.org/10.1371/journal.pntd.0002852.

Verma R, Khanna P. Development of *Toxoplasma gondii* vaccine: a global challenge. Hum Vaccin Immunother. 2013;9:291–3.

Wingstrand A, Lind P, Haugegaard J, Henriksen SA, Bille-Hansen V, Sorensen V. Clinical observations, pathology, bioassay in mice and serological response at slaughter in pigs experimentally infected with *Toxoplasma gondii*. Vet Parasitol. 1997;72:129–40.

Besnoitia

7

Heinz Mehlhorn

Abstract

The ten currently described *Besnoitia* species belong to those coccidians—*Apicomplexa*—that produce tissue cysts worldwide in domestic and wild ungulates, equids, rodents, lagomorphs, marsupials, and reptiles, which all act species-specifically as intermediate host. While the life cycles of those species, which use smaller animals such as rabbits, rodents, mice, rats, opossums, and lizards as intermediate hosts and cats as final hosts, are established, those life cycles involving cattle, goats, horses, donkeys, and related wild animals are still unknown. However, there are several indications that insects may play an important role in the life cycle, since they were found to be able to transmit cystozoites from the tissue cysts—which occur in the surface skin of their hosts—to wounds of other hosts thus initiating an infection, while other pathways of transmission, e.g., via final hosts producing infectious oocysts, are still unknown. This chapter summarizes the present knowledge of the species of this neglected but apparently spreading genus of coccidian parasites. Although the mortality rates in afflicted cattle herds rarely exceed 10%, they are of considerable economic importance.

7.1 Morphology, Life Cycle, and Host-Pathogen Interaction

7.1.1 History of the Genus *Besnoitia*

The first report of the cattle disease, later named besnoitiosis, was published by Cadéac in the year 1884, when he described the infected skin of cattle as

H. Mehlhorn
Department of Parasitology, Heinrich Heine University, D-40225 Düsseldorf, Germany
e-mail: mehlhorn@uni-duesseldorf.de

M. Florin-Christensen, L. Schnittger (eds.), *Parasitic Protozoa of Farm Animals and Pets*, https://doi.org/10.1007/978-3-319-70132-5_7

elephant-like without giving any indication of the origin of this aspect. In the year 1912, the French parasitologists Besnoit and Robin determined that this disease is caused by a parasite. They named the disease sarcosporidiosis, since they had noted similarities with *Miescher's tubes*—which had been detected several years before—but they noted also several differences to the typical sarcosporidians (Miescher 1843). Henry (1912), Marotel (1912), and Henry and Masson (1922) gave individual comments on the findings of Besnoit and Robin, while Franco and Borges (1916) reviewed the epidemiology of this cattle disease in Portugal and proposed the new genus name *Besnoitia* and *Besnoitia besnoiti* as species name. Although bovine besnoitiosis has been known for long in the sub-Saharan Africa (Bigalke 1960, 1967, 1968, 1970; Bigalke et al. 1967; Diesing et al. 1988; Bigalke and Prozesky 2004) and Asia—reviewed by Olias et al. 2011—main attention arose, when cases in Europe increased in numbers and intensity especially in Portugal, Spain, France, Switzerland, Italy, Germany, and Central-Eastern Europe (Alvarez-Garcia et al. 2013; Cortes et al. 2006a, b, 2007a, b, 2011, 2014; Fernandez-Garcia et al. 2008; Liénard et al. 2011; Mehlhorn et al. 2009; Kiehl et al. 2010; Rostaher et al. 2010; Gazzonis et al. 2014; Gentile et al. 2012; Gollnick et al. 2015; Hornok et al. 2014, 2015; Lesser et al. 2012; Basso et al. 2011; Jacquiet et al. 2010).

When *Besnoitia* infections of cattle were increasingly studied, also the knowledge on *Besnoitia* species in other hosts than cattle became more detailed leading to the description of the presently known ten species (Table 7.1). The life cycle of *Besnoitia* species, which lead to tissue cysts in rodents, rats, rabbits, opossums, and lizards, has been established, showing that cats are the final host excreting typical oocysts that look similar to those of *Toxoplasma gondii*, *Hammondia* species, and *Isospora* species. In contrast, the different phases and the final hosts of the life cycles of the *Besnoitia* species of cattle and horses are still unknown, Although

Table 7.1 Currently acknowledged *Besnoitia* sp. with their host species

Species	Intermediate host with tissue cysts	Final hosts	Size of oocysts (μm)	Pathogenicity (+/–)
B. akodoni	Rodents, grass mouse, gerbils	?	?	–
B. bennetti	Horses, donkeys, mules, zebras	?	?	–
B. besnoiti	Cattle, rodents?	Predator birds ?	?	+
B. caprae	Goats	?	?	+
B. darlingi	Opossums, lizards	Cats	11–13 × 10–13	–
B. jellisoni	Rodents, mice	?	?	+
B. neotomofelis	Wood rats	Cats	10–13	+
B. oryctofelisi	Rabbits	Cats	12 × 11	–
B. tarandi	Reindeers, caribou	?	?	+
B. wallacei	Rodents, rats	Cats	16–19 × 10–13	–

Peteshev et al. (1974a, b) and Peteshev and Polomoshnov (1976) claimed that cats are the final host of *Besnoitia besnoiti*, involved cats had been superinfected with an isosporan species and results have not yet been confirmed.

7.1.2 Morphology of *Besnoitia besnoiti* Stages

During the first 3–4 weeks of a not yet defined infection with *B. besnoiti*, tachyzoites found inside cattle host cells measure 6–9 μm as has been calculated from tissue cell experiments of several authors. This estimate corresponds to those reported from culture experiments with *B. jellisoni* (Sénaud and Mehlhorn 1978). Well-defined results, however, are available from measurements of stages inside tissue cysts. These stages—cystozoites or bradyzoites—reach a length of 7–9 μm and a width of approximately 2 μm (Fig. 7.5). They are situated inside tissue cysts with diameters of up to 3 mm and are mostly found in clusters of three to five specimens. In light microscopically studied semi-thin sections, it was seen that each tissue cyst was surrounded by a dense secondary cyst wall originating from different concentrically arranged laminae reaching about 0.2–0.3 mm in diameter. The laminae enclosed an enormously enlarged host cell including thousands of single parasites, which appeared spherical in cross sections and banana-shaped in longitudinal sections (Fig. 7.5). Electron micrographs showed that these cystozoites had their origin inside mother cells giving rise to two cystozoites in a process called endodyogeny, which was also seen in tissue cultures and is also characteristic for the species of the genera *Toxoplasma*, *Sarcocystis*, and *Frenkelia* (Mehlhorn and Heydorn 1978; Scholtyseck et al. 1973; Mehlhorn et al. 1974). The original host cell cytoplasm was reduced to just a small cytoplasmic layer containing several host cell nuclei (Figs. 7.4 and 7.5), while the internal large parasitophorous vacuole was closely filled with cyst merozoites (cystozoites or bradyzoites) and stages during endodyogeny (D'Haese et al. 1977; Dubey et al. 2013; Langenmayer et al. 2015; Njagi et al. 1998).

Cyst merozoites—bradyzoites—are limited by the typical three-layered pellicle, the outer of which surrounds the whole parasite, while the two inner ones are interrupted at the apical and posterior pole forming there polar rings. The anterior pole of these cyst merozoites shows the conoid protruding from the surrounding apical polar ring. Twenty-two subpellicular microtubules are attached to the polar ring and stretch below the lower side of the pellicle until the posterior region, but do not reach the terminal pole (D'Haese et al. 1977; Reis et al. 2006). The apical pole region of these cyst merozoites contains the ovoid micronemes and the club-shaped rhoptries. The anterior thin ductules of the rhoptries pass the interior of the conoid and reach the apical pole of the cell. In addition, the apical pole contains several spherical dense bodies and the apicoplast—formerly named double-walled body—which is limited by four closely attached membranes and represents the remnants of two former free-living organisms. Characteristic is also a tubular long-stretched mitochondrion and a Golgi apparatus just prior to the apical pole of the nucleus, which is mostly situated behind the middle region of these cysts stages—bradyzoites (Mehlhorn et al. 2009; D'Haese et al. 1977). The posterior pole of the cyst merozoites—bradyzoites—contained several granules of whitish-appearing amylopectin, ribosomes, and lacunes of the rough endoplasmic reticulum. The three-layered pellicle of the cyst merozoites contains a typical micropore used as a cytostome to ingest fluid material from the contents of the parasitophorous vacuole. These structures correspond to those of other *Besnoitia* species studied by several authors (Scholtyseck et al. 1973; Mehlhorn et al. 1974, 2009; Mehlhorn and Heydorn 1978; Dubey et al. 2003a; Dubey and Lindsay 2003; Mehlhorn 2016, 2017).

7.1.3 Transmission of *Besnoitia* Species

The life cycle and pathway of transmission of the *Besnoitia* species infecting smaller mammals—*B. darlingi*, *B. neotomofelis*, *B. wallacei*, and *B. oryctofelis*—has been demonstrated and showed that cats act as final hosts, producing the sexual stages and the infectious oocysts. In contrast, the transmission of *B. besnoiti*, *B. akodoni*, *B. caprae*, *B. jellisoni*, *B. tarandi*, and *B. benetti* is not fully understood (Smith and Frenkel 1977; Dubey and Yabsley 2010; Dubey et al. 2002, 2003a, b, c, 2005a, b). Since the cysts of these species lead to rather large and open wounds along the skin of infected animals, considerable economic losses may occur. Licking flies or bloodsucking insects take their food preferably at such places of infected animals. As vectors, they may play an important role in the transmission of tachyzoites and/or bradyzoites parasite stages, which can easily be transferred mechanically from one biting site to others. This aspect has been underestimated for a long time, but experimental transmission using and examining flies caught in nature shows high loads of agents of disease such as viruses, bacteria, and eggs of parasites (see Sect. 7.1.3.1) (Gestmann et al. 2012; Förster et al. 2012).

In the case of besnoitiosis (syn. besnoitiasis), such a mechanical transmission of agents of diseases by insects may be also accompanied by transmission of bradyzoites through skin-to-skin contacts of cattle when entering together narrow stable doors or when getting into contact during feeding in stables.

Infected members of herds of closely standing ruminants or groups of wandering sheep can easily transmit the infection through skin-to-skin contact. Thus, such groups/herds may reach high infection rates and many members of a group may have an unapparent infection. The existence of such a horizontal transmission has been experimentally shown (Bigalke 1960, 1968; Bigalke et al. 1967; Jellison et al. 1956; Pols 1954, 1960).

Importantly, all the outlined possibilities of the transmission of infection do not exclude the existence of a typical predator-prey life cycle, including free-living grass-feeding animals in nature. They would be infected as intermediate hosts by ingestion of oocysts excreted by meat-feeding carnivores or birds that had been infected when ingesting meat of living or dead intermediate hosts.

7.1.3.1 Potential Vectorial Transmission of Infectious Stages

Considering that on one hand several intense experimental transmission trials using *Besnoitia* species have failed when using meat-ingesting animals—such as carnivores or carrion eaters—and that on the other hand rather high infection rates have been diagnosed in cattle in many regions of several continents, it must be assumed that also other pathways of transmission are used. The existence of vectorial transmission pathways seems to be very probable especially when the parasites induce open wounds in the outer surface of the skin, where the licking or biting mouthparts of insects or maybe bats can become contaminated with cystozoites or tachyzoites and are finally transported to smaller wounds of other hosts. This scenery, however, does not exclude a not yet known typical coccidian pathway involving two hosts—predator/prey—in the close surroundings of cattle or other intermediate hosts. Flies are surely very important additional or even main vectors of *Besnoitia* species. This idea is strongly supported by experimental transmissions using many related insect species that are known as vectors of other agents of diseases. For example, biting midges (Ceratopogonidae), black flies (e.g., *Simulium* species), or tabanids (*Tabanus* sp.) show often bloody mouthparts when switching from one host to another (Adler and Crosskey 2011; Adler and McCreadie 2009). Gestmann et al. (2012) and Förster et al. (2012) showed in a broad spectrum of experiments that the mouthparts of germ-free flies could easily become contaminated by thousands of stages of bacteria and parasites, which they transmitted to cultures. Furthermore, fleas and other blood-sucking arthropods when imported to special regions by birds are suggested to introduce contained agents of diseases (Mehlhorn 2016a, b). Olias et al. (2011) discusses also the possibility that birds may be much more intensely involved in the transmission of agents of diseases such as besnoitiasis.

7.2 Clinical Effects and Diagnosis

The disease induced by infection with these protozoan—coccidian—parasites runs via two phases in different hosts:

7.2.1 Bovine Besnoitiasis or Besnoitiosis

The disease is called bovine besnoitiasis or besnoitiosis of cattle. It leads in 90% of cases to a nonfatal disease infecting practically all breeds worldwide except for Australia, where the parasite has not yet been found (Olias et al. 2011). However, the remaining mortality rate of about 10% and the destruction of the skin may be disastrous for some farmers in certain regions of the world. Depending on the intensity of the symptoms, the reduction of the productivity of affected animals reaches extreme high levels (Bigalke 1968; European Food Safety Authority 2010). Losses occur due to significant decrease of body weight, reduction of fertility, and milk production and due to the severe damage of the skin making it unusable for leather production. In severe cases, the skin becomes hairless and appears like that of elephants. Especially in France and Spain, but also in African countries, the disease is observed at start of spring time and reaches its peak in autumn. Most severe clinical cases have been observed mainly in young animals from 6 months of age until an age of 4–5 years.

Infected animals have been categorized by several authors into three main groups of disease manifestation:

- Inapparently infected, seropositive animals that appear to be weak: this group covers most of the parasitized animals.
- Animals showing scleroconjunctival cysts with related symptoms of disease.
- Animals with clear and dominant skin lesions and symptoms of weakness often cover 10–20% of the animals of an affected herd.

The disease itself can be divided into two main phases:

7.2.1.1 Anasarca Stage

The acute phase starts 1–3 weeks after infection and is characterized by the occurrence of tachyzoites. Tachyzoites reproduce themselves by repeated endodyogenies inside vacuoles of endothelial cells of blood vessels or inside vacuoles in fibroblasts and macrophages leading to high fever of 41–42 °C, edema, pyrexia, hyperemia, and orchitis. Transient or definitive infertility may occur in males, and abortions have been reported (Bigalke 1968; Cortes et al. 2014). The acute phase of the disease can be diagnosed most easily from far, when animals on a meadow show slow and stiff movements. Jacquiet et al. (2010) have described further common symptoms after intense internal investigations: increased heart and respiratory rates, arrest in rumination, increased nasal discharge, anorexia, epiphora, and/or photophobia. However, the latter symptoms may be confused with diseases such as bluetongue disease or malignant catarrhal fever (Alzieu 2007; Alzieu et al. 2007a, b; Mehlhorn et al. 2009; Gazzonis et al. 2014; Gutiérrez-Expósito et al. 2016, 2017).

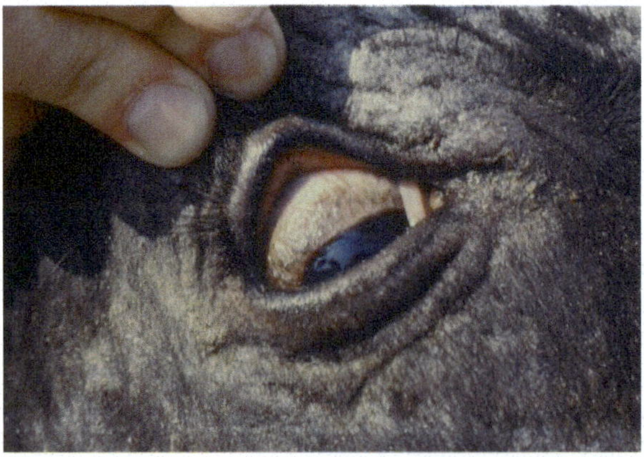

Fig. 7.1 Macrophoto of *Besnoitia besnoiti* cysts inside the eye of an infected cow

Fig. 7.2 Macrophoto showing the hairless skin of horny appearance of a cow infected with *B. besnoiti*

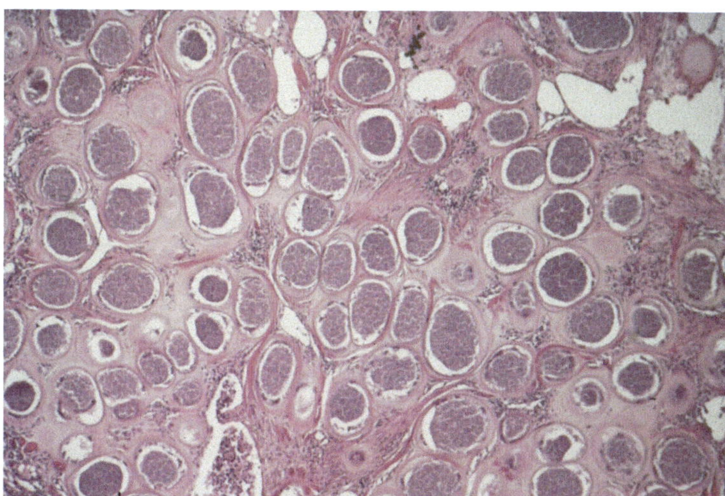

Fig. 7.3 Colored surface section through the hypodermis of an infected cow showing the density of tissue cysts of *B. besnoiti*

7.2.1.2 Chronic Stage

The chronic phase—sclerodermal stage—starts about 3–4 weeks after infection and is characterized by the formation of thick-walled tissue cysts (Figs. 7.1, 7.2, 7.3, 7.4 and 7.5). These cysts contain bradyzoites and are located in the cutis, subcutis, mucous layers, scleral conjunctiva, vestibulum vaginae, vagina, testes, and connective tissues and—though in lower numbers—in the muscles, spleen, liver, lung, and heart (Mehlhorn et al. 2009; Rostaher et al. 2010; Olias et al. 2011). The affected skin shows a strong reduction of hair and appears elephant-like.

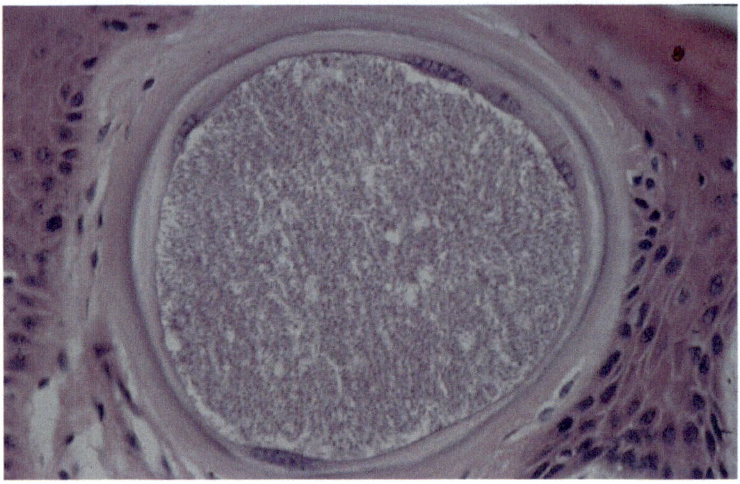

Fig. 7.4 High magnification of a section through a tissue cyst in the skin of a cow showing that the center is closely filled with cystozoites; note the enlarged nuclei in the small remnants of the host cell surrounded by a pale secondary cyst wall

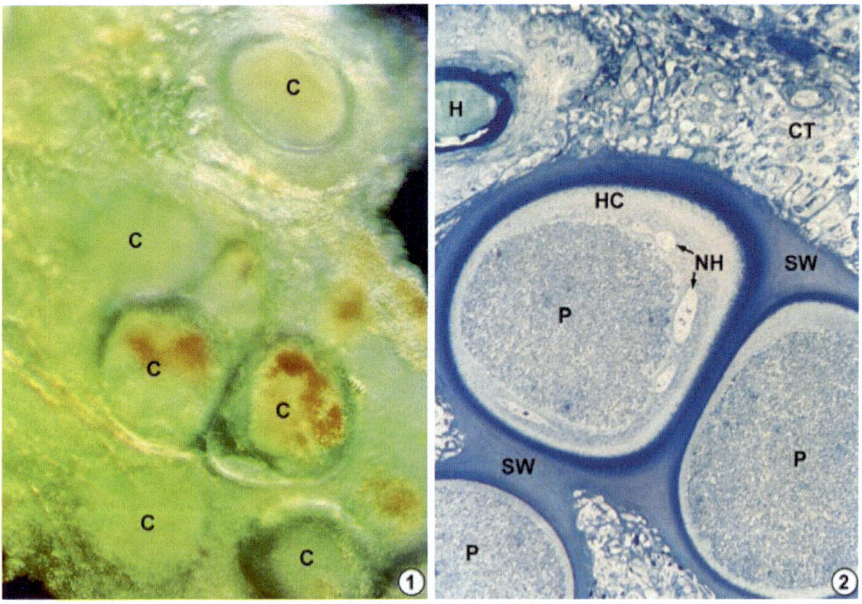

Fig. 7.5 Semi-thin sections through four tissue cysts of *Besnoitia besnoiti* showing the surface aspect (*1*) and a section through the midregion (*2*); the dense secondary cyst wall (SW); the host cells (HC), with their enlarged host cell nuclei (NH); and the large number of cystozoites (P) inside the tissue cyst surrounded by connective tissues (T) are designated

In enzootic regions, the seroprevalence rates in cattle may reach 50%, but in non-enzootic areas often seroprevalences of up to 80% are observed in herds (Fernandez-Garcia et al. 2010; Janitschke et al. 1984).

7.2.2 Besnoitiasis of Other Large Animals

Cysts of *Besnoitia* species may also occur in the tissues of other large mammals than cattle and have been described in goats, reindeers, caribous, antelopes, donkeys, and equids (Table 7.1). Their morphology appears very similar to that of *B. besnoiti*, and these species may infect the same organs (Ayroud et al. 1995; Diesing et al. 1988; Heydorn et al. 1984; Dubey et al. 2002, 2003a, b, c, 2004, 2005a, b; Oryan and Azizi 2008; Oryan et al. 2010; Ducrocq et al. 2009; Glover et al. 1990). However, the life cycles of these species have not yet been finally determined, since—as is the case in *B. besnoiti*—no final hosts have been demonstrated.

7.2.3 Besnoitiasis of Opossums, Lizards, and Rodents

The life cycles of *Besnoitia* species forming cysts in intermediate host species opossums, lizards, and rodents have been clarified. Cats have been shown to act as the final host and excrete the typical *Isospora*-type oocysts after ingestion of cyst-contaminated meat of the above-listed intermediate hosts (Frenkel 1953, 1977; Dubey et al. 2002; Dubey and Yabsley 2010, Dubey and Lindsay 2003; Venturini et al. 2002). The life cycle of these *Besnoitia* species seems to be clearly different from those infecting larger animals (see Sect. 7.2.2).

7.2.4 Molecular Discrimination of *Besnoitia* Species

By help of the ITS-1 region, species of the family of the *Sarcocystidae* may rather easily be differentiated (Tenter et al. 2002; Olias et al. 2011). In contrast, the chance of discrimination between *Besnoitia* species infecting small mammals is rather low, and it is totally absent for *Besnoitia* species infecting large mammals. The *p*-distances between the ITS-1 regions show clear differences between currently defined species (Table 7.2). However, the database is still too small to decide whether all claimed species are really valid. Thus, species determination remains doubtful as long as experimental transmission has not been demonstrated and suitable genetic markers are not available. Rabbits have been shown to be highly susceptible for the infection with specimens of the cattle-infecting species *B. besnoiti*, and they are also able to develop a high-graded and solid immunity (Bigalke 1968; Ellis et al. 2000; Kaggwa et al. 1979).

Table 7.2 Nucleotide distance values of the internal transcribed spacer (ITS-1) region of *Besnoitia* sp. using the Kimura-2 model (Olias et al. 2011)

	1	2	3	4	5	6	7	8	9
1. *B. besnoiti* Portugal AY833646									
2. *B. besnoiti* Israel DQ227420	0.0000000								
3. *B. besnoiti* S. Africa AF076859	0.0000000	0.0000000							
4. *B. caprae* Iran HM008988	0.0000000	0.0000000	0.0000000						
5. *B. bennetti* USA AY827839	0.0000000	0.0000000	0.0000000	0.0000000					
6. *B. tarandi* Finland AY665400	0.0000000	0.0000000	0.0000000	0.0000000	0.0000000				
7. *B. akodoni* Brazil AY545987	0.2550241	0.2550241	0.2550241	0.2550241	0.2550241	0.2550241			
8. *B. oryctofelisi* Argentina AY182000	0.2371550	0.2371550	0.2371550	0.2371550	0.2371550	0.2371550	0.0415611		
9. *B. darlingi* USA AF489696	0.2313065	0.2313065	0.2313065	0.2313065	0.2313065	0.2313065	0.0330173	0.0162828	
10. *B. jellisoni* USA AF076860	0.2313065	0.2313065	0.2313065	0.2313065	0.2313065	0.2313065	0.0590170	0.0458784	0.0415611

7.3 Epidemiology

7.3.1 Distribution of *Besnoitia* Species of Ungulates

7.3.1.1 *Besnoitia caprae*

This species is mainly enzootic in Kenya and Iran but single cases have also been found in Nigeria. In different regions of Iran, prevalence rates have been reported to range between 12 and 19%. Interestingly, the ITS-1 sequence of *B. caprae* differs only by one nucleotide from that of *B. besnoiti*. However, when looking at the fine structural aspects, *B. caprae* appears clearly different from *B. besnoiti*, and cannot be transmitted to rabbits, mice, guinea pigs, hamsters, sheep, and cattle (Ng'ang'a and Kasigazi 1994; Ng'ang'a et al. 1994; Mirzaie et al. 2007; Njenga et al. 1995). On the other hand, *B. caprae* from Kenya is transmittable to Balb/C mice (Oryan et al. 2010; Oryan and Azizi 2008).

7.3.1.2 *Besnoitia besnoiti*

This species is enzootic in the Pyrenean area of France and Spain and in Southern Portugal but is now also found in Southern and Central Spain, Italy, North and South France, and Germany (Olias et al. 2011; Jacquiet et al. 2010; Rostaher et al. 2010; Mehlhorn et al. 2009; Mutinelli et al. 2011; Schares et al. 2016). In addition, single cases have been reported from Israel and from South Africa, Nigeria, Kazakhstan, Uzbekistan, China, and South Korea (Pols 1954, 1960; Shkap et al. 1988, 1995, 2002; Bigalke and Prozesky 2004). However, details are scarce and these findings have to be confirmed. Cattle strains of *B. besnoiti* in South Africa have been experimentally transmitted to goats, merino sheep, and blue wildebeests (Bigalke 1967; Pols 1960). Although natural infections with this species have been found exclusively in cattle and antelopes, species specificity of the parasite seems to be rather low, since a variety of laboratory animals such as rabbits, gerbils, hamsters, mice, and rats have been experimentally infected (Basso et al. 2011; Olias et al. 2011).

7.3.1.3 *Besnoitia tarandi*

Widespread occurrence of *Besnoitia tarandi* reaching infection rates of 48.9% in reindeer and caribou—*Rangifer tarandus*—was shown in Alaska, Canada, Finland, and Russia, while no data exist from Greenland and Norway (Ducrocq et al. 2009). Gerbils and rabbits could not be experimentally infected with this parasite species. However, there are indications that transmission may occur between European reindeer and American caribou and even to mule deer (Glover et al. 1990).

The 18S rRNA gene—AY616163—and the ITS-1 region, AY665400, of *B. tarandi* are identical to strains of *B. besnoiti* from Germany, Portugal, Israel, and Spain, and only a single insert of 2 bp is found in the 28S rRNA gene, AY616164 (Olias et al. 2011). Thus it is presently very difficult to clearly differentiate by help of genetic markers between *B. besnoiti* and *B. tarandi*. Rabbits and gerbils were refractive to experimental infections with *B. tarandi*, and its bradyzoites showed clear differences in shape and number of rhoptries compared to *B. besnoiti* (Dubey et al. 2003b, 2004).

7.3.2 Besnoitiasis in Wild Ungulates

Besnoitia stages have also been diagnosed by McCully et al. (1966), Bigalke (1968), and Ellis et al. (2000) in wild ungulates such as antelopes, wild goats, musk ox, cervids, kudu (*Tragelaphus strepsiceros*), blue wildebeests (*Connochaetes taurinus*), and impala (*Aepyceros melampus*). The morphology of isolates of *Besnoitia* stages from cattle, impala, and wildebeest cannot be differentiated; however, results of isoenzyme electrophoresis point to the existence of different *Besnoitia* strains in these host species (Bigalke et al. 1967; Bigalke 1968).

7.3.3 Besnoitiasis in Equids

This disease has been diagnosed in donkeys, horses, mules, and zebras and was reported from the USA, South Africa, and Sudan (Bigalke 1970). In the USA, the infectious species is named *B. benetti* and was found only in donkeys (*Equus asinus*), while in Africa cases were known mainly from horses (*Equus caballus*) besides a single description from zebras (*Equus quagga*) and a few from donkeys and mules (Bigalke 1968). In Europe there is only a single report from horses, but symptoms may be easily missed during any inspection (Henry and Masson 1922). Molecular data reported from the USA show clear differences between *B. benetti* and *B. besnoiti*. South African isolates of *B. benetti* in contrast to *B. besnoiti* cannot be experimentally transmitted to guinea pigs, hamsters, white mice, rats, and rabbits (Dubey et al. 2003a, b).

7.3.4 *Besnoitia* Species in Rodents, Lagomorphs, Marsupials, and Reptiles

In the USA, *B. darlingi* is found in opossums—*Didelphis virginiana*—and uses cats as final hosts (Table 7.1) (Frenkel 1977). Cats have also shown to be the final host for *B. wallacei* (Wallace and Frenkel 1975). *B. jellisoni* has been described by Frenkel (1953) and occurs in the white-footed deer mice (*Peromyscus maniculatus*) and in kangaroo rats (*Dipodomys* sp.) yet its cycle is still unknown. Its genetics indicate, however, that this species is closely related to both, *B. neotomofelis* and *B. oryctofelisi*, which have also cats as final hosts (Dubey et al. 2003a). *B. akodoni* is another *Besnoitia* species closely related to *B. neotomofelis* and *B. oryctofelisi* which is found in the montane grass mouse—*Akodon montensis*—in South America (Dubey et al. 2003c). *B. neotomofelis* has been claimed to be closely related to *B. jellisoni* (Dubey and Yabsley 2010). These publications give an idea on how doubtful or valid species descriptions might be when based just on morphological criteria. An example of an erroneous description of a *Besnoitia* species based on the morphology of muscle cysts occurring in rodents and on oocysts found in snake feces after feeding such muscle cysts of rodents is Matuschka and Häfner's (1984). The authors claimed that snakes are final hosts. However, in the year 1987, the authors themselves show that the formerly observed stages belong to the newly established species *Sarcocystis hoarensis* that uses rodents as intermediate hosts and snakes as final hosts (Matuschka et al. 1987).

In contrast to other *Besnoitia* species, *B. wallacei* was first detected in the feces of naturally infected domestic cats in Hawaii, but the natural cycle has not yet been revealed until today (Wallace and Frenkel 1975). However, domestic mice and different rats are susceptible to experimental infections leading to typical *Besnoitia* cysts on serosa layers of viscera showing the existence of the typical two-host cycle, while hamsters and domestic rats were found not susceptible to experimental infections (Frenkel 1977). However, further results are either not available or have not been published in easily accessible journals. Venturini et al. (2002) described tissue cysts of a similar appearance in rabbits, but further details are lacking.

It can be concluded that the knowledge on the species concerning *Besnoitia* stages in target hosts such as rodents, lagomorphs, marsupials, and reptiles is rather scarce. Although many transmission trials had been started in many regions of the world, they were finally stopped due to the lack of financing and economic interest. Details are given in the review of Olias et al. (2011).

7.4 Treatment and Control of Besnoitiasis

Reports on successful treatment of besnoitiasis in cattle are scarce, and results are not always convincingly tested in larger amounts of infected animals. In South Africa, it has been recommended to use high dosages of long-term oxytetracyclines or sulfonamides and to separate high-graded infected animals from herds or flocks. In the case of infected goats, the parenteral application of an antimony solution—1%, 0.5 ml/kg bodyweight—led to healing of an infection, while thiazolides or arylimidamides controlled stages in vitro (Diesing et al. 1988; Cortes et al. 2006a, b, 2007a, b, 2011).

References

Adler PH, Crosskey RW. World blackflies (Diptera: Simuliidae): a fully revised edition of the taxonomic and geographical inventory. 2011. http://entweb.clemson.edu/biomia/pdfs/blackfly-inventory.pdf.

Adler PH, McCreadie JW. Black flies (Simuliidae). In: Mullen GR, Durden L, editors. Medical and veterinary entomology. San Diego: Academic Press; 2009. p. 189–206.

Alvarez-Garcia G, Frey C, Ortega-Mora LM, Schares G. A century of bovine besnoitiosis: an unknown disease re-emerging in Europe. Trends Parasitol. 2013;29:407–15.

Alzieu JP. Re-emerging cattle besnoitiosis (*Besnoitia besnoiti*) in France: update on clinical and epidemiological aspects. In: Proceedings of the 21st International Conference of the World Association for the Advancement of Veterinary Parasitology. Gent, Belgium, 2007. p. 222.

Alzieu JP, Cortes HC, Gottstein B, Jacquiet P, Dorchies P, Schelcher F, L'Hostis M. La besnoitiose bovine: actualités épidémiologiques et diagnostiques. Bull. GTV Hors-série parasitsme des bovins, 2007a. p. 41–9.

Alzieu J, Dorchies P, Schelcher F, Gottstein B. L'extension de la besnoitiose bovine en France. Le Point vétérinaire. 2007b;38:37–43.

Ayroud M, Leighton FA, Tessaro SV. The morphology and pathology of *Besnoitia* sp. in reindeer (*Rangifer tarandus tarandus*). J Wildl Dis. 1995;31:319–26.

Basso W, Schares G, Gollnick NS, Rutten M, Deplazes P. Exploring the life cycle of *Besnoitia besnoiti*—experimental infection of putative definitive and intermediate host species. Vet Parasitol. 2011;178:223–34.

Besnoit C, Robin V. Sarcosporidiose cutanée chez une vache. Rev Vet. 1912;37:649–63.

Bigalke RD. Preliminary observations on the mechanical transmission of cyst organisms of *Besnoitia besnoiti* from a chronically infected bull to rabbits by *Glossina brevipalpis*. J S Afr Vet Med Assoc. 1960;31:37–44.

Bigalke RD. The artificial transmission of *Besnoitia besnoiti* (Marotel, 1912) from chronically infected to susceptible cattle and rabbits. Onderstepoort J Vet Res. 1967;34:303–16.

Bigalke RD. New concepts on the epidemiological features of bovine besnoitiosis as determined by laboratory and field investigations. Onderstepoort J Vet Res. 1968;35:3–137.

Bigalke RD. Studies on equine besnoitiosis. J Parasitol. 1970;56:29.

Bigalke RD, Prozesky L. Besnoitiosis. In: Coetzer JAW, Tustin RC, editors. Infectious diseases of livestock, vol. 1. Cape Town: Oxford University Press; 2004. p. 331–59.

Bigalke RD, Van Niekerk JW, Basson PA, McCully RM. Studies on the relationship between *Besnoitia* of blue wildebeest and impala, and *Besnoitia besnoiti* of cattle. Onderstepoort J Vet Res. 1967;34:7–28.

Cadéac, C. Identit, de l, Elephantiasis et de l'anasarque du boeuf. Description de cette maladie. Revue Vétérinaire. 1884;521:521–40.

Cortes HC, Nunes S, Reis Y, Staubli D, Vidal R, Sager H, Leitao A, Gottstein B. Immunodiagnosis of *Besnoitia besnoiti* infection by Elisa and Western blot. Vet Parasitol. 2006a;141:216–25.

Cortes HCE, Reis Y, Waap H, Vidal R, Soares H, Marques I, Pereira da Fonseca I, Fazendeiro I, Ferreira ML, Caeiro V, et al. Isolation of *Besnoitia besnoiti* from infected cattle in Portugal. Vet Parasitol. 2006b;141:226–33.

Cortes HCE, Müller N, Esposito M, Leitao A, Naguleswaran A, Hemphill A. In vitro efficacy of nitro-and bromo-thiazolyl-salicylamide compounds (thiazolides) against *Besnoitia besnoiti* infection in vero cells. Parasitology. 2007a;134:975–85.

Cortes HCE, Reis Y, Gottstein B, Hemphill A, Leitao A, Muller N. Application of conventional and real-time fluorescent ITS1 rDNA PCR for detection of *Besnoitia besnoiti* infections in bovine skin biopsies. Vet Parasitol. 2007b;146:352–6.

Cortes HC, Muller N, Boykin D, Stephens CE, Hemphill A. In vitro effects of arylimidamides against *Besnoitia besnoiti* infection in Vero cells. Parasitology. 2011;138:583–92.

Cortes H, Leitão A, Gottstein B, Hemphill A. A review on bovine besnoitiosis: a disease with economic impact in herd health management, caused by *Besnoitia besnoiti* (Franco and Borges). Parasitology. 2014;141:1406–17.

D'Haese J, Mehlhorn H, Peters W. Comparative electron microscopic study of pellicular structures in coccidia (*Sarcocystis*, *Besnoitia*, *Eimeria*). Int J Parasitol. 1977;7:505–18.

Diesing L, Heydorn AO, Matuschka FR, Bauer C, Pipano E, de Waal DT, Potgieter FT. *Besnoitia besnoiti*: studies on the definitive host and experimental infections in cattle. Parasitol Res. 1988;75:114–7.

Dubey JP, Lindsay DS. Development and ultrastructure of *Besnoitia oryctofelisi* tachyzoites, tissue cysts, bradyzoites, schizonts and merozoites. Int J Parasitol. 2003;33:807–19.

Dubey JP, Yabsley MJ. *Besnoitia neotomofelis* n. sp. (Protozoa: Apicomplexa) from the Southern Plains Woodrat (*Neotoma micropus*). Parasitology. 2010;21:1–17.

Dubey JP, Lindsay DS, Rosenthal BM, Sreekumar C, Hill DE, Shen SK, Kwok OCH, Rickard LG, Black SS, Rashmir-Raven A. Establishment of *Besnoitia darlingi* from opossums (*Didelphis virginiana*) in experimental intermediate and definitive hosts, propagation in cell culture, and description of ultrastructural and genetic characteristics. Int J Parasitol. 2002;32:1053–64.

Dubey JP, Shkap V, Pipano E, Fish L, Fritz DL. Ultrastructure of *Besnoitia besnoiti* tissue cysts and bradyzoites. J Eukaryot Microbiol. 2003a;50:240–4.

Dubey JP, Sreekumar C, Lindsay DS, Hill D, Rosenthal BM, Venturini L, Venturini MC, Greiner EC. *Besnoitia oryctofelisi* n. sp. (Protozoa: Apicomplexa) from domestic rabbits. Parasitology. 2003b;126:526–39.

Dubey JP, Sreekumar C, Rosenthal BM, Lindsay DS, Grisard EC, Vitor RW. Biological and molecular characterization of *Besnoitia akodoni* n.sp. (Protozoa: Apicomplexa) from the rodent *Akodon montensis* in Brazil. Parassitologia. 2003c;45:61–70.

Dubey JP, Sreekumar C, Rosenthal BM, Vianna MCB, Nylund M, Nikander S, Oksanen A. Redescription of *Besnoitia tarandi* (Protozoa: Apicomplexa) from the reindeer (*Rangifer tarandus*). Int J Parasitol. 2004;34:1273–87.

Dubey JP, Sreekumar C, Donovan T, Rozmanec M, Rosenthal BM, Vianna MCB, Davis WP, Belden JS. Redescription of *Besnoitia bennetti* (Protozoa: Apicomplexa) from the donkey (*Equus asinus*). Int J Parasitol. 2005a;35:659–72.

Dubey JP, Sreekumar C, Donovan T, Rozmanee M, Rosenthal BM, Vianna MB, Davis WP, Belden JS. Cultivation and redescription of *Besnoitia bennetti* from the donkey (*Equus asinus*). Am J Vet Parasitol. 2005b;42:715–9.

Dubey JP, van Wilpe E, Blignaut DJC, Schares G, Williams JH. Development of early tissue cysts and associated pathology of *Besnoitia besnoiti* in a naturally infected bull (*Bos taurus*) from South Africa. J Parasitol. 2013;99:459–66.

Ducrocq J, Lair S, Beauchamp G, Kutz S. Prevalence and intensity of *Besnoitia tarandi* in caribou (*Rangifer tarandus*), associated risk factors and comparison between herds. Poster at the 6th Conference of the CircumArctic Rangifer Monitoring & Assessment Network, Vancouver, Canada. 2009. http://www.carmanetwork.com/display/public/Vancouver+2009+Conference.

Ellis JT, Holmdahl OJM, Ryce C, Njenga JM, Harper PAW, Morrison DA. Molecular phylogeny of *Besnoitia* and the genetic relationships among *Besnoitia* of cattle, wildebeest, and goats. Protist. 2000;151:329–36.

European Food Safety Authority. Scientific statement on bovine besnoitiosis. EPSA. 2010; 8:1–5.

Fernandez-Garcia A, Risco-Castillo V, Pedraza-Diaz S, Aguado-Martinez A, Alvarez-Garcia G, Gomez-Bautista M, Collantes-Fernandez E, Ortega-Mora LM. First isolation of *Besnoitia besnoiti* from a chronically infected cow in Spain. J Parasitol. 2010;95:474–6.

Förster M, et al. Flies as vectors of parasites potentially inducing severe diseases in humans and animals. In: Arthropods as vectors of emerging diseases, Parasitology research monographs, vol. 3. Heidelberg: Springer; 2012.

Franco EE, Borges I. Sur la sarcosporidiose bovine. Arquivos do Instituto Bacteriologico Câmara Pestana. 1916;4:269–89.

Frenkel JK. Infections with organisms resembling *Toxoplasma* together with the description of a new organism: *Besnoitia jellisoni*. Atti del VI Congresso Int Microbiol. 1953;5:426–34.

Frenkel JK. *Besnoitia wallacei* of cats and rodents: with a reclassification of other cyst-forming isosporoid coccidia. J Parasitol. 1977;63:611–28.

Gazzonis AL, Alvarez Garcia G, Zanzani SA, Garippa G, Rossi L, Maggiora M, Dini V, Invernizzi A, Luini M, Tranquillo VM, Ortega Mora L, Manfredi MT. *Besnoitia besnoiti* among cattle in insular and Northwestern Italy: endemic infection or isolated outbreaks? Parasit Vectors. 2014;7:585.

Gentile A, Militerno G, Schares G, Nanni A, Testoni S, Bassi P, Gollnick NS. Evidence for bovine besnoitiosis being endemic in Italy – first in vitro isolation of *Besnoitia besnoiti* from cattle born in Italy. Vet Parasitol. 2012;184:108–15.

Gestmann F, et al. Flies as vectors of microorganisms potentially inducing severe diseases. In: Arthropods as vectors of emerging diseases, Parasitology research monographs, vol. 3. Heidelberg: Springer; 2012.

Glover GJ, Swendrowski M, Cawthorn RJ. An epizootic of besnoitiosis in captive caribou (*Rangifer tarandus caribou*), reindeer (*Rangifer tarandus tarandus*) and mule deer (*Odocoileus hemionus hemionus*). J Wildl Dis. 1990;26:186–95.

Gollnick NS, Scharr JC, Schares G, Langenmayer MC. Natural *Besnoitia besnoiti* infections in cattle: chronology of disease progression. BMC Vet Res. 2015;11:35.

Gutiérrez-Expósito D, Arnal MC, Martínez-Durán D, Regidor-Cerrillo J, Revilla M, L Fernández de Luco D, Jiménez-Meléndez A, Calero-Bernal R, Habela MA, García-Bocanegra I, Arenas-Montes A, Ortega-Mora LM, Álvarez-García G. The role of wild ruminants as reservoirs of *Besnoitia besnoiti* infection in cattle. Vet Parasitol. 2016;223:7–13.

Gutiérrez-Expósito D, Ortega-Mora LM, Ara V, Marco I, Lavín S, Carvajal-Valilla J, Morales A, Álvarez-García G. Absence of antibodies specific to *Besnoitia* spp. in European sheep and goats from areas in Spain where bovine besnoitiosis is endemic. Parasitol Res. 2017;116:445–8.
Henry. Sarcosporidiose cutanée chez une Vache. Recueil de médecine vétérinaire XC, 327. 1912.
Henry A, Masson G. Cited by Neveu-Lemaire, M (1943) Traité de protozoologie médicale et vétérinaire. Paris: Vigot Frères; 1922.
Heydorn AO, Sénaud J, Mehlhorn H, Heinonen R. *Besnoitia* sp. from goats in Kenya. Parasitol Res. 1984;70:709–13.
Hornok S, Fedák A, Baska F, Hofmann-Lehmann R, Basso W. Bovine besnoitiosis emerging in Central-Eastern Europe, Hungary. Parasit Vectors. 2014;7:20.
Hornok S, Fedák A, Baska F, Basso W, Dencső L, Tóth G, Szeredi L, Abonyi T, Dénes B. Vector-borne transmission of *Besnoitia besnoiti* by blood-sucking and secretophagous flies: epidemiological and clinicopathological implications. Parasit Vectors. 2015;8:450.
Jacquiet P, Liénard E, Franc M. Bovine besnoitiosis: epidemiological and clinical aspects. Vet Parasitol. 2010;174:30–6.
Janitschke K, De Vos AJ, Bigalke RD. Serodiagnosis of bovine besnoitiosis by ELISA and immunofluorescence tests. Onderstepoort J Vet Res. 1984;51:239–43.
Jellison WL, Fullerton WJ, Parker H. Transmission of the protozoan *Besnoitia jellisoni* by ingestion. Ann N Y Acad Sci. 1956;64:271–4.
Kaggwa E, Weiland G, Rommel M. *Besnoitia besnoiti* and *Besnoitia jellisoni*: a comparison of the indirect immunofluorescence antibody test (IFAT) and the enzyme-linked immunosorbent assay (ELISA) in diagnosis of *Besnoitia* infections in rabbits and mice. Bull Anim Helath Prod Afr. 1979;27:127–37.
Kiehl H, Heydon AO, Schein E, Selmair J, Al-Quraishy S, Klimpel S, Mehlhorn H. Molecular biological comparison of *Besnoitia* species. Parasitol Res. 2010;106:889–94.
Langenmayer MC, Scharr JC, Sauter-Louis C, Schares G, Gollnick NS. Natural *Besnoitia besnoiti* infections in cattle: hematological alterations and changes in serum chemistry and enzyme activities. BMC Vet Res. 2015;11:32.
Lesser M, Braun U, Deplazes P, Gottstein B, Hilbe M, Basso W. Erste Fälle von Besnoitiose be Rindern in der Schweiz. Schweiz Arch Tierheilkd. 2012;154:469–74.
Liénard E, Salem A, Grisez C, Prévot F, Bergeaud JP, Franc M, Gottstein B, Alzieu JP, Lagalisse Y, Jacquiet P. A longitudinal study of *Besnoitia besnoiti* infections and seasonal abundance of *Stomoxys calcitrans* in a dairy cattle farm of southwest France. Vet Parasitol. 2011;177:20–7.
Marotel M. Discussion of paper by Besnoit and Robin. Bullet Mémoires de la Société de Sciences Veterinaires de Lyon et de la Société de Medicine Veterinaire de Lyon et du Sud-Est. 1912;15:196–217.
Matuschka FR, Häfner U. Cyclic transmission of an African *Besnoitia* species by snakes of the genus *Bitis* to several rodents. Parasitol Res. 1984;70:471–6.
Matuschka FR, Mehlhorn H, Abd-Al-Al Z. Replacement of *Besnoitia* sp. Matuschka and Häfner 1984 by *Sarcocystis hoarensis*. Parasitol Res. 1987;74:94–6.
McCully RM, Basson PA, Van Niekerk JW, Bigalke RD. Observations on *Besnoitia* cysts in the cardiovascular system of some wild antelopes and domestic cattle. Onderstepoort J Vet Res. 1966;33:245–76.
Mehlhorn H. Animal parasites. 7th ed. Heidelberg: Springer Spektrum; 2016a.
Mehlhorn H. Encyclopedia of parasitology. 4th ed. Heidelberg: Springer; 2016b.
Mehlhorn H, Heydorn AO. The Sarcosporidia (Protozoa, Sporozoa): life cycle and fine structure. Adv Parasitol. 1978;16:43–91.
Mehlhorn H, Sénaud J, Scholtyseck E. Etude ultrastructurale des coccidies formant des kystes: *Toxoplasma gondii*, *Sarcocystis tenella*, *Besnoitia jellisoni* et *Frenkelia* sp.: distribution de la phosphatase acide et des polysaccharides au niveau des ultrastructures chez les parasite et chez l'hôte. Protistologica. 1974;10:21–42.
Mehlhorn H, Klimpel S, Schein E, Heydorn AO, Al-Quraishy S, Selmair J. Another African disease in Central Europa: besnoitiosis of cattle. I. Light and electron microscopical study. Parasitol Res. 2009;104:861–8.
Miescher F. Über eigentümliche Schläuche in den Muskeln einer Hausmaus. Ber Verhandl Naturforsch Ges Basel. 1843;5:198–202.

Mirzaie M, Derakhshanfar A, Pourjafar M. Caprine besnoitiosis in the southwest of Iran. Vet Arhiv. 2007;77:435.

Mutinelli F, Schiavon E, Ceglie L, Fasolato M, Natale A, Rampin F, Carminato A. Bovine besnoitiosis in imported cattle in Italy. Vet Parasitol. 2011;178:198–204.

Ng'ang'a CJ, Kasigazi S. Caprine besnoitiosis: studies on the experimental intermediate hosts and the role of the domestic cat in transmission. Vet Parasitol. 1994;52:207–10.

Ng'ang'a CJ, Kanyari PW, Munyua WK. Isolation of *Besnoitia wallacei* in Kenya. Vet Parasitol. 1994;52:203–6.

Njagi ON, Ndarathi CM, Nyaga PN, Munga LK. An epidemic of besnoitiosis in cattle of Kenya. Onderstepoort J Vet Res. 1998;65:133–6.

Njenga J, Bwangamoi O, Kangethe EK, Mugera GM, Mutiga ER. Comparative ultrastructural studies on *Besnoitia besnoiti* and *Besnoitia caprae*. Vet Res Commun. 1995;19:295–308.

Olias P, Schade B, Mehlhorn H. Molecular pathology, taxonomy and epidemiology of *Besnoitia* species (Protozoa: Sarcocystidae). Infect Genet Evol. 2011;11:1564–76.

Oryan A, Azizi S. Ultrastructure and pathology of *Besnoitia caprae* in the naturally infected goats of Kerman, East of Iran. Parasitol Res. 2008;102:1171–6.

Oryan A, Namazi F, Namavari MM, Sharifiyazdi H, Moraveji M. Comparison of the pathogenesis of two isolates of *Besnoitia caprae* in inbred BALB/c mice. Vet Res Commun. 2010;34:423–34.

Peteshev VM, Polomoshnov AP. The distribution of *Besnoitia besnoiti*. Vet Protozool. 1976;83:84.

Peteshev VM, Galuzo IG, Polomoshnov AP. Cats-definitive hosts of *Besnoitia* (*Besnoitia besnoiti*). Akademiya Nauk Kazakhskoi SSR Izbestiya. Ser Biol. 1974a;1:33–8.

Peteshev VM, Galuzo IG, Polomoshnov AP. Koshki-definitivnye khozyaeva besnoiti. Jzv Akad Nauk Kaz, Ser Biol. 1974b;1:33–8.

Pols JW. The artificial transmission of *Globidium besnoiti* (Marotel, 1912) to cats and rabbits. J S Afr Vet Med Assoc. 1954;25:37–44.

Pols JW. Studies on bovine besnoitiosis with special reference to the aetiology. Onderstepoort J Vet Res. 1960;28:265–356.

Reis Y, Cortes H, Viseu-Melo L, Fazeindeiro I, Leitao A, Soares H. Microtubule cytoskeleton behaviour in the initial steps of host cell invasion by *Besnoitia besnoiti*. FEBS LeH. 2006;580:4673–82.

Rostaher A, Mueller RS, Majzoub M, Schares G, Gollnick NS. Bovine besnoitiosis in Germany. Vet Dermatol. 2010;21:329–34.

Schares G, Langenmayer MC, Majzoub-Altweck M, Scharr JC, Gentile A, Maksimov A, Schares S, Conraths FJ, Gollnick NS. Naturally acquired bovine besnoitiosis: differential distribution of parasites in the skin of chronically infected cattle. Vet Parasitol. 2016;216:101–7.

Scholtyseck E, Mehlhorn H, Müller BEG. Identification of merozoites of four cyst-forming Coccidia (*Sarcocystis*, *Toxoplasma*, *Besnoitia*, *Frenkelia*) based on ultrastructural criteria. Parasitol Res. 1973;42:185–206.

Sénaud J, Mehlhorn H. *Besnoitia jellsoni* en culture sur cellules de rein de chien. Protistologica. 1978;14:5–14.

Shkap V, Yakobson BA, Pipano E. Transmission and scanning electron microscopy of *Besnoitia besnoiti*. Int J Parasitol. 1988;18:761–6.

Shkap V, Matuschka FR, Yakobson B, Perl S, Frank M, Pipano E. Differentiating between *Besnoitia besnoiti* and *Sarcocystis hoarensis* from rodents. Parasitol Res. 1995;81:490–3.

Shkap V, Reske A, Pipano E, Fish L, Baszler T. Immunological relationship between *Neospora caninum* and *Besnoitia besnoiti*. Vet Parasitol. 2002;106:35–43.

Smith DD, Frenkel JK. *Besnoitia darlingi*. Cyclic transmission by cats. J Parasitol. 1977;63:1066–71.

Tenter AM, Barta JR, Beveridge I, Duszynski DW, Mehlhorn H, Morrison DA, Thompson RC, Conrad PA. The conceptual basis for a new classification of the coccidia. Int J Parasitol. 2002;32:595–616.

Venturini L, Petrucelli M, Piscopo M, Unzaga JM, Venturini MC, Baciagalupe D, Basso W, Dubey JP. Natural *Besnoitia* infection in domestic rabbits from Argentina. Vet Parasitol. 2002;107:273–8.

Wallace GD, Frenkel JK. *Besnoitia* species: recognition of cyclic transmission by cats. Science. 1975;188:369–71.

Theileria in Ruminants

8

Henry Kiara, Lucilla Steinaa, Vishvanath Nene, and Nicholas Svitek

Abstract

Theileria are important hemoprotozoan parasites of domestic and wild ruminants, transmitted by ixodid ticks leading to diseases which range from mild in apparent reactions to highly fatal diseases. Bovine-infecting *T. parva* and *T. annulata* and ovine-infecting *T. lestoquardi* are of major global economic importance, but other *Theileria* spp. are also briefly mentioned. Classification of *Theileria* has been a subject of great controversy without consensus on whether many taxons are different species, synonyms, or subspecies of the same parasite. However, with the development of new molecular tools, many of the outstanding difficulties could be resolved. *Theileria* have complex life cycles both in the vertebrate host and the tick vector, many of which are not clearly understood. One unique feature of some *Theileria* is their ability to transform infected host cells into a reversible cancer-like proliferation conferring to them an ability to proliferate without apoptosis. The transformation is not permanent because it can be reversed by treatment with anti-theilerial drugs. Understanding this mechanism could give insights into treatment of cancer. Control of the diseases caused by *Theileria* has largely relied on chemical drugs either to treat infected hosts or prevent infection by controlling the tick vectors. But resistance to chemicals by the parasites or vectors has led to the development of more sustainable control methods such as live vaccination against the three most pathogenic *Theileria* spp. of ruminants. Efforts are also under way to develop subunit vaccines against these parasites.

H. Kiara (✉) • L. Steinaa • V. Nene • N. Svitek
International Livestock Research Institute, Nairobi, Kenya
e-mail: h.kiara@gciar.org

M. Florin-Christensen, L. Schnittger (eds.), *Parasitic Protozoa of Farm Animals and Pets*, https://doi.org/10.1007/978-3-319-70132-5_8

8.1 Life Cycle, Morphology, and Host-Pathogen Interaction

8.1.1 Morphology

Being eukaryotes, *Theileria* species have the usual cellular organelles observed in other eukaryotes including mitochondria, ribosomes, vacuoles, and a nucleus. However, what makes *Theileria* sp. and most other apicomplexans unique is that they contain an apicoplast, a vestigial plastid-like organelle. The function of the apicoplast in Apicomplexa is not clearly understood, but enzymes involved in the metabolism of fatty acids have been found in this structure (Lim and McFadden 2010). Being a unique organelle—most likely of prokaryotic origin—the apicoplast provides an excellent target for developing anti-theilerial drugs that are not toxic for the mammalian host (Moore et al. 2008; Lizundia et al. 2009).

Most of the mammalian stages of *Theileria* sp. have similar morphologies, and it is difficult to distinguish the schizont life cycle stages of transforming *Theileria* species—i.e. *T. parva, T. annulata, T. taurotragi,* and *T. lestoquardi.* Generally, merozoites are pear-shaped, but piroplasms are variable in size and shape. Some have a coma, bacillary, ring, or oval shape. *T. verifera* piroplasms have a rectangular veil, while those of *T. taurotragi* have a characteristic bar. Piroplasms also come in various sizes between the different *Theileria* sp. The smallest of them being from *T. parva*, the middle sized are from *T. annulata* and *T. mutans*, and the largest are of the *T. taurotragi* species. Piroplasms seem to divide by binary fission.

8.1.2 *Theileria* Life Cycle

Most of the studies on the life cycle of *Theileria* species have been conducted on *Theileria parva* and *T. annulata*. Therefore, given the broader scientific publications on these species compared to available publications on the other *Theileria* species, most of the information contained in this chapter is based on morphological studies involving these two species reviewed in Francia and Striepen (2014), Norval et al. (1992), and Striepen et al. (2007).

Theileria parasites display a complex life cycle (Fig. 8.1). The complexity of their life cycle resides in their need to infect two different hosts to complete their life cycle, ticks and ruminants. *Theileria* can be classified as transforming or non-transforming, the former possessing a unique ability to cause infected leucocytes to behave like transformed cells (reviewed in Dobbelaere and Rottenberg 2003; Tretina et al. 2015). The transforming *Theileria* spp. include *T. parva, T. annulata, T. lestoquardi*, and *T. taurotragi*. The non-transforming *Theileria* are *T. mutans, T. orientalis,* and *T. velifera.*

The life cycle in the mammalian host starts when haploid sporozoites, generated in the tick salivary gland during sporogony, are released during tick feeding. The release of sporozoites occurs between 4 and 8 days post-tick attachment. During

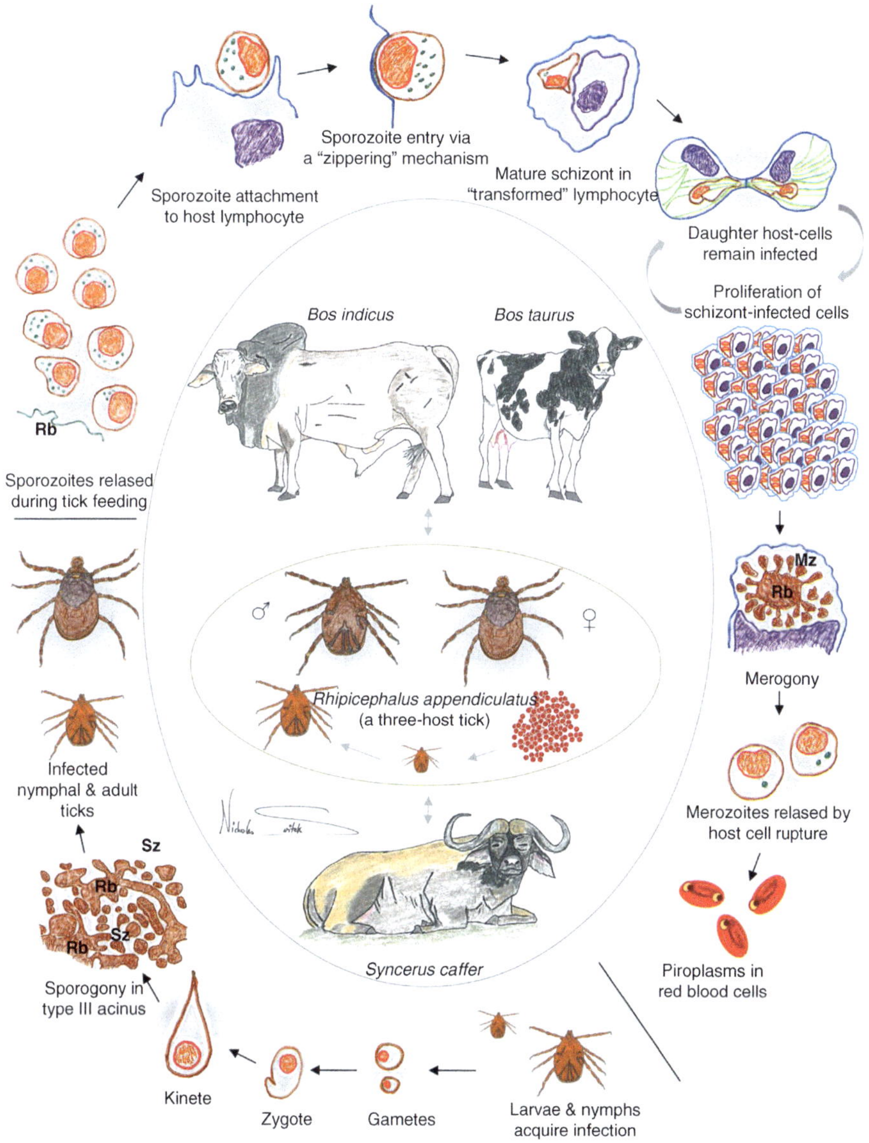

Fig. 8.1 Life cycle of *Theileria parva* in the cattle and tick hosts (drawings and artistic creation by Nicholas Svitek; used with permission from Nene et al. 2016)

this stage, sporozoites are released into the tick feeding site. Since *Theileria* sporozoites are non-motile, binding to target host cells is a random event. *T. parva* sporozoites preferentially establish infections in subsets of T lymphocytes and B lymphocytes. In the case of *T. annulata*, the main infected cells are the macrophages, dendritic cells, and B lymphocytes. Studies performed in vitro have shown that *T. parva* sporozoite host cell entry occurs through a zippering process of the sporozoite and host cell membrane. Secretory organelles, like rhoptries and microspheres, are released from sporozoites upon internalization in the mammalian cell. Sporozoite entry into lymphocytes can be inhibited by monoclonal antibodies specific toward the bovine leukocyte antigen (BoLA) class I molecules as well as the invariant chain beta 2 microglobulin. However, antibodies specific toward the BoLA class II molecules or toward pan-leukocyte surface antigen do not show any inhibitory activity toward sporozoites entry (Shaw et al. 1991). Upon internalization, sporozoites differentiate into a multinucleated macroschizont, a cellular entity having multiple nuclei that resides freely in the host cell cytoplasm without a parasitophorous vacuolar membrane (PVM) surrounding it. Lymphocytes can become infected by more than one sporozoite leading to the development of multiple schizonts inside a single cell (Fig. 8.2). Infected cells exhibit a cancer-like, but reversible, transformed phenotype. Division of the parasite occurs in synchrony with the host cell cycle and takes place during metaphase. In vitro, these infected lymphocytes can proliferate indefinitely, as long as the culture media is replenished. In some instances, the macroschizont will evolve into merozoites through a process called merogony. Each merozoite contains a single one nucleus and, once released from the lymphocytes, will infect red blood cells. The mechanism of entry into erythrocytes is merozoite-specific—as sporozoites do not enter erythrocytes—but similar in nature to sporozoite entry into lymphocytes. Inside red blood cells, merozoites continue their metamorphosis and develop into piroplasms that are residing freely in the host cell cytoplasm.

The tick stage of the parasite begins when piroplasms are released from red blood cells, in the tick gut. Inside the tick host, a proportion of *Theileria* piroplasms differentiate into *ray-bodies*, which are precursors of the gametes, and start to develop flagella-like protrusions leading to formation of microgametes. The *ray-bodies* also develop as spherical structures, which do not divide and form macrogametes. These structures have been observed for *T. parva*, *T. annulata*, *T. taurotragi*, *T. mutans,* and *T. velifera*. The gametocyte structures undergo syngamy—fusion—in the tick gut forming a diploid spherical zygote that invades the tick gut epithelium and matures into a motile kinete. Kinetes released into the tick hemolymph invade acinar cells in the tick salivary glands, where sporogony occurs. Both the zygote and kinete develop free in the host cell cytoplasm. Each infected acinus can produce between 30,000 and 50,000 sporozoites. In the case of *T. taurotragi*, sporogony generates up to 140,000 sporozoites. It is believed that it is because of this high number of sporozoites that *T. taurotragi* sporozoites appear due to the squeezing of the sporozoites as *frying-pan shaped*. The sporozoites of *T. mutans* are fewer in numbers but much bigger in size.

Transmission of *T. parva* by *R. appendiculatus* is transstadial—i.e., larval and nymphal stages acquire infections which are transmitted by the next tick stage. However, in the case of *T. annulata*, transovarial transmission has also been observed in *Hyalomma anatolicum excavatum* ticks (FAO 1983).

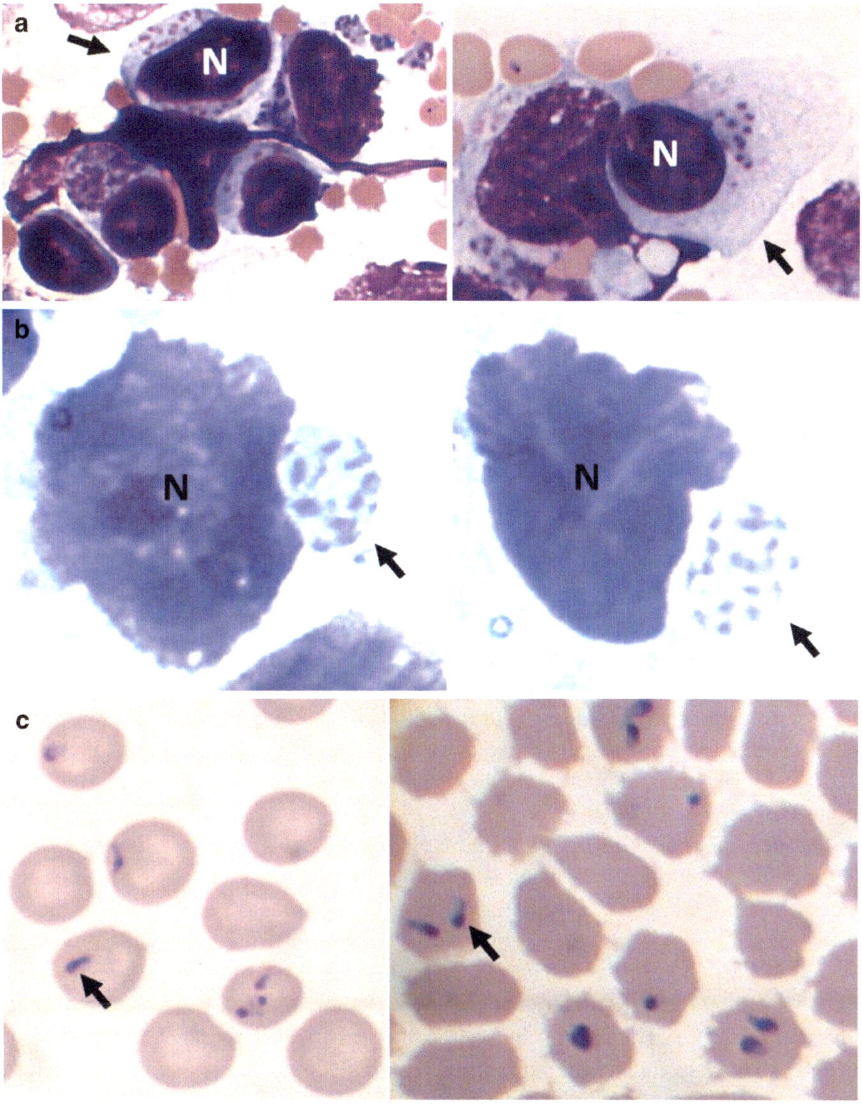

Fig. 8.2 *Theileria parva* schizonts and piroplasms: (**a**) multinucleated schizonts at 400× magnification; (**b**) multinucleated schizonts at 1000× magnification; (**c**) piroplasms in red blood cells at 1000× magnification; the *arrows* indicate the schizonts and the piroplasms inside the lymphocyte or red blood cell, respectively; N, cell nucleus; Giemsa staining of schizonts and piroplasms coloring the cell nucleus, schizont nuclei, and piroplasms in purple (photo courtesy of Thomas Njoroge and Robert Muriuki)

8.1.3 Host-Pathogen Interaction and Pathogenesis

As previously mentioned, a characteristic of the pathogenic *Theileria* spp. parasites is their ability to cause a lymphoproliferative disease in the mammalian host, a unique feature among eukaryotic parasites. *T. parva*, *T. annulata*, *T. taurotragi*, and *T. lestoquardi* transform cells into cancer-like cells. Importantly, *Theileria*-transformed cells are not permanently transformed—as in the case of chemically or virally induced transformation—but the process is reversible upon treatment with an anti-theilerial drug. Here again, most of studies have been conducted mainly on *T. parva* and *T. annulata*. With transforming *Theileria* spp. and particularly with *T. parva* and *T. annulata*, the schizont stage is associated with pathology due to invasion of several organs leading to tissue destruction. In *T. parva* there is little multiplication of the piroplasm stage, and anemia is usually not a feature of disease. This is not the case for *T. annulata*, where higher piroplasm parasitemias cause red blood cell associated pathology. In the case of *T. parva*, death is believed to be caused by pulmonary edema, and higher virulence has been associated with infected T cells rather than B cells (Morrison et al. 1996).

Several studies have investigated the mechanisms by which *Theileria* spp. transform cells. In general, *Theileria* spp. cause extensive rewiring of the cell signaling cascades and cell metabolism, which is known as the Warburg effect observed in cancer biology (Cheeseman and Weitzman 2015). In summary, research has demonstrated that *T. parva* has developed many and overlapping mechanisms to induce proliferation and prevent apoptosis of infected cells. A comprehensive list of experiments conducted on the subject is listed in the reviews of Dobbelaere and Rottenberg (2003) and Tretina et al. (2015), and only a few main mechanisms are highlighted in this chapter.

The ability of *Theileria* spp. to transform leukocytes resides mainly in their capacity to manipulate factors that control cell growth, prevent apoptosis, and encode homing receptors to other tissues. Among such factors is the nuclear factor (NF)-κB, a transcription factor known to induce pro-inflammatory cytokine genes as well as anti-apoptotic molecules, a factor known to be constitutively expressed with oncogenic viruses (Ivanov et al. 1989; Palmer et al. 1997; Hermann and Dobbelaere 2006). The NF-κB activation in parasitized lymphocytes was found recently to be activated by the schizont protein called the *T. parva* schizont-derived cytoskeleton-binding protein (TpSCOP) (Hayashida et al. 2010). The expression of this protein in a murine lymphocytic cell line has been shown to activate the NF-κB signaling cascade and resistance to apoptosis, indicating its important role in the transformation process. Other transcription factors that have been shown to be upregulated during *T. parva* infection includes c-Jun, JunD, c-Fos, and ATF-2 which are part of the AP-1 transcription factor complex known to be involved in cellular transformation (Botteron and Dobbelaere 1998; Chaussepied et al. 1998).

It was also observed that JNK-1 and JNK-2, members of the mitogen-activated protein (MAP) kinase family known to play a role in the regulation of cellular proliferation, are activated in a parasite-dependent manner (Galley et al. 1997; Seitzer et al. 2006). Other proto-oncogene molecules were discovered as participating in the transformation process of *T. parva*. Among these, the Src kinases and the Notch signaling cascade, known to be involved in cellular proliferation and oncogenesis, have been shown to be upregulated in *T. parva*-infected cells (Chaussepied et al. 2006; Fich et al. 1998).

Transforming *Theileria* have also the ability to upregulate expression and secretion of the cytokines interleukin (IL)-2, IL-10, and interferon (IFN)-γ, with the latter showing to increase the levels of infection by *T. parva* and therefore contributing to the transformed state of the cells (Dobbelaere et al. 1988; McKeever et al. 1997; DeMartini and Baldwin 1991). Another study showed the contribution of IL-2 and tick salivary gland extract in enhancing lymphocytes susceptibility to *T. parva* (Shaw et al. 1993). Some of these cytokines like IL-10 might contribute in the immunosuppression observed during *T. parva* infection. Pro-inflammatory cytokines such as IL-1β and IL-6 are also upregulated during a *Theileria* infection and might contribute to the virulence of disease (Graham et al. 2001). In the case of *T. annulata* infection, the cytokine transforming growth factor (TGF)-β 2 is associated with virulence, disease susceptibility, and invasiveness of *Theileria*-transformed macrophages (Chaussepied et al. 2010; Haidar et al. 2015a, b). *T. orientalis* also induce IL-10 and IFN-γ cytokines in Holstein cattle; however, a study suggested that IFN-γ might contribute to the resistance of *T. sergenti* infection (Hagiwara et al. 2005; Yamaguchi et al. 2010).

Like any successful pathogens, pathogenic *Theileria* sp. are, by definition, one step ahead of their hosts and have most probably developed mechanisms to interfere with the host innate immune responses. Not many studies have been conducted on the interaction of *Theileria* sp. with the innate immune response, and this deserves more attention. However, studies conducted on *T. annulata* showed that this parasite downregulates the interferon-stimulated gene (ISG) 15 which is known to stimulate $CD8^+$ antitumor activity as well as to play a role in antiviral immunity (Oura et al. 2006). The downregulation of ISG15 in a *T. annulata*-infected macrophage was found to be orchestrated by parasite polypeptides—TashAT family—located in the nucleus. It remains to be found if such mechanism is present in other *Theileria* species. Moreover, it is widely accepted that *T. parva* is more virulent in exotic breeds of cattle, *Bos taurus*, than in indigenous African breeds of cattle, *Bos indicus*, and this differential property of the parasite to cause pathogenesis could be due to the better ability of the parasite to control the innate immune response in susceptible cattle. Another interesting finding that deserves more attention is the observation that cattle infected with low pathogenic *Theileria* species are more resistant to pathogenic *Theileria* when infection occurs concomitantly (Woolhouse et al. 2015).

8.2 Diagnosis and Epidemiology

Theileria are important obligate intracellular protozoan parasites of both wild and domestic ruminants occurring throughout the world. They are transmitted by ixodid ticks and infect leucocytes and erythrocytes in mammalian host and develop in the gut epithelial cells and the salivary glands of the arthropod vectors. The subject has been reviewed by Dolan (1989), Bishop et al. (2004), Norval et al. (1992), and Uilenberg (1981).

8.2.1 Diagnosis

Theileria species are usually diagnosed, firstly by suspicion due to tick infestation and clinical symptoms such as fever and enlarged lymph nodes, which is followed up by diagnostic tests such as indirect fluorescence antibody test (IFAT) or by microscopy using Giemsa staining of smears from lymph nodes or blood. In order to diagnose for a particular parasite, serological tests or molecular tests are needed.

8.2.1.1 Microscopy

The parasites can be confirmed by Giemsa staining in smears from lymph nodes or blood, but differential diagnosis cannot be done because schizonts and piroplasms are morphologically similar. Two exceptions to this are *T. velifera,* which has a characteristic veil associated with the piroplasm, and *T. taurotragi,* which has bar-like structures in infected erythrocytes. The presence of multinucleate intracytoplasmic and free schizonts in leucocytes from lymph node biopsy smears is indicative for acute infections with *T. parva*, *T. annulata,* and *T. lestoquardi* among others. In cattle samples where schizonts are present in blood smears, lymph node impression smears or histological sections are diagnostic of ECF. Piroplasm in the erythrocytes can be present in later stages of the three parasites (OIE 2008).

8.2.1.2 Serological Tests

Serological tests are useful for diagnosis on a herd level rather than at individual animal basis and for epidemiological studies. This is because the antibody response needs time to develop, and animals will usually develop disease and pass the point of intervention before antibodies have developed. Furthermore, parasite-specific antibodies remain long in circulation after animals have recovered from disease. The most widely used test for *Theileria* species is IFAT in which both schizont and piroplasm antigens can be used. This test is sensitive, fairly specific, and relatively easy to perform but has limitations in areas where different *Theileria* species overlap because there are problems with cross-reactivities—e.g., between *T. parva* and *T. taurotragus* and between *T. parva* and *Theileria* sp. (buffalo). The use of IFAT is therefore not recommended if particular species need to be identified except if it is known that certain species are absent from the area. As an alternative, ELISA tests for detection of *T. parva* and *T. mutans* were developed. In the case of *T. parva*, the recombinant polymorphic immunodominant protein (PIM) from the parasite is used and for *T. mutans* the antigen p32 (Katende et al. 1990; Morzaria et al. 1999). These tests have shown a higher sensitivity of over 95% than the originally used IFAT. Similar tests have also been developed for *T. annulata* in which the recombinant merozoite surface 1 antigen—Tams1—and the *T. annulata* macroschizont protein—TaSP—were used (Gubbels et al. 2000; Seitzer et al. 2007; Renneker et al. 2008). TaSP was also used for developing a rapid lateral flow test for *T. annulata*, which compare well with other serological assays (Abdo et al. 2010). ELISA tests specific for *T. lestoquardi* has also been developed (Bakheit et al. 2006). However, none of these tests are commercially available.

8.2.1.3 Molecular DNA-Based Tests

Molecular tests can be used for the detection of the pathogen. The nature of these tests are very diverse spanning from conventional PCR and electrophoretic analysis, PCR-restriction fragment length polymorphism (RFLP)-based analysis methods, and nested PCR and PCR followed by dot blotting, capillary blotting, or slot blotting followed by hybridization with radio-isotope-labeled probes among others. Several molecular tests have been developed for detection of *T. parva*, *T. annulata,* and *T. lestoquardi*.

Reverse line blots (RLBs) have been developed for all three parasites. The principle of this test is that specific probes are blotted on a membrane in a line, and then samples—e.g., biotinylated amplicons—are added perpendicular to this. This allows probes to hybridize with sample amplicons and after washing, and color development by using streptavidin linked to, e.g., HRP, specific hybridization can be visualized. *Theileria*-specific tests have been based on amplification of various genes, the hypervariable V4 region of the small ribosomal RNA gene of *T. parva*, *T. annulata*, *T. lestoquardi*—but also *T. ovis* and others—and cytochrome b for *T. annulata* (Gubbels et al. 1999; Schnittger et al. 2004; Bilgic et al. 2010). RLB is capable of detecting very low level of parasites, but it requires sophisticated lab equipment as the controlled hybridization and reproducibility in different labs can be a problem.

Conventional PCR methods are also common tests for all parasites. Both conventional, blood spot PCR and nested PCRs were developed for *T. parva* based on p67 and p104 (Bishop et al. 1992; Skilton et al. 2002; Odongo et al. 2010). For *T. annulata*, the gene coding for the merozoite p30—Tams-1—and cytochrome b have been used (d'Oliveira et al. 1995; Martín-Sánchez et al. 1999; Kirvar et al. 2000; Santos et al. 2013). There are also PCRs for *T. ovis* which have been utilized for *T. lestoquardi* by adapting the primers (Imam and Taha 2015). Various multiplex PCRs have also been described for simultaneous detection of multiple pathogens (Bilgiç et al. 2012).

In addition to conventional PCRs, real-time PCRs have been developed. For *T. parva*, the 18S ribosomal RNA gene has been used and the cytochrome c oxidase (cox) III gene (Sibeko et al. 2008; Papli et al. 2011; Chaisi et al. 2013). These assays are more sensitive than the conventional PCR assay and should be able to reliably detect the parasite in carrier animals. A hybrid assay—a hybridization real-time PCR method—has also been developed for *T. parva* which should offer a more specific assay circumventing interference of simultaneous infection with *Theileria* sp. (buffalo) like parasites (Pienaar et al. 2011). For *T. annulata*, quantitative real-time PCRs have been set up using the 18S ribosomal RNA gene as well as Tams-1 (Ros-García et al. 2012). No quantitative PCR has been described for *T. lestoquardi* or *T. ovis*.

Loop-mediated isothermal amplification (LAMP) is a relatively new method in which DNA is amplified under isothermal conditions. This results in a high sensitivity and specificity, and the assay is rapid to perform. Several sets of primers are used for detection of target sequences. It can be performed on a drop of blood spotted on a filter paper. The mode of detection can vary from visual inspection of turbidity and fluorescent dye—SYBR Green—followed by gel electrophoresis. A LAMP assay has been developed for *T. parva* targeting PIM and p150, two for *T. annulata* based on GeneDB TA04795 as well as the 18S ribosomal RNA gene and internal transcribed spacer (ITS) sequences, and one for *T. lestoquardi* based on what has been described as the clone 5 sequence (Salih et al. 2008; Thekisoe et al. 2010; Liu et al. 2012; Salih et al. 2012).

Other molecular tools include Restriction Fragment Length Polymorphism analysis (RFLP). This methods digests DNA with restriction enzymes, and the resultant restriction fragment are run on a gel. Sizes of the fragment are used for identification of the pathogen. Such assays have been developed for *T. annulata* and *T. lestoquardi* based on the 18S rRNA gene (Spitalska et al. 2004; Heidarpour Bami et al. 2009). A multiplex oligonucleotide suspension microarray has been developed for bovine piroplasms, which included probes for various pathogens including *T. annulata* (Ros-García et al. 2011). It was shown to be more sensitive than RLB analysis.

8.2.2 Epidemiology

This section will concentrate mainly on two of the economically most important *Theileria* of cattle, *T. annulata* and *T. parva,* and briefly mention other *Theileria* that infect livestock and are of economic relevance (Uilenberg 1995).

8.2.2.1 Classification

Molecular phylogeny has been able to well define and delineate the group of *Theileria* parasites as a single monophyletic group that is, for convenience, sometimes referred to as *true Theileria*—corresponding to Clade V as defined in Schnittger et al. (2012). All ruminant-infecting *Theileria* covered in this chapter belong into this monophyletic clade of *true Theileria.* However, there are still piroplasmids referred to as *Theileria* that do not belong into this group and represent misnomers—e.g., *Theileria equi*, *Theileria youngi*, and *Theileria bicornis* (Schnittger et al. 2012). Within the *true Theileria*, the classification of some *Theileria* has been and still is a subject of great confusion at the species level (Irvin 1987). Since the discovery of *T. parva* in 1902 and its naming by Bettencourt in 1907—cited by Perry and Young 1993—the parasite has gone through multiple names. It has been previously named *T. lawrencei* and *T. bovis*, and a trinomial nomenclature was briefly used to create three subspecies of *T. parva*, namely, *T. parva parva*, cause of classical ECF; *T. parva lawrencei*, cause of corridor disease; and *T. parva bovis*, cause of January or Zimbabwean theileriosis (Uilenberg 1976; Lawrence 1979). The trinomial nomenclature was discarded for lack of biological justification for the subspeciation (Uilenberg 1981). It is now recognized that there are clinical and hematological differences of the disease caused when cattle are infected by ticks which have fed on buffalo and those which have fed on other cattle (Sitt et al. 2015). It is now widely accepted that there is a single species of *T. parva* but which can be buffalo-derived or cattle-derived. It is expected that as more sophisticated genomic tools are developed, such differences in classification and taxonomy will be resolved.

With respect to *T. buffeli/T. sergenti/T. orientalis*, there still is no consensus on whether these taxons are different species, synonyms, or subspecies of the same parasite (Chaisi et al. 2014; Gubbels et al. 2000; Dolan 1989). *T. orientalis* is considered by some to be identical to *T. buffeli*, but enough genotypic and phenotypic differences exist between *T. buffeli* and *T. sergenti* to regard the latter as different species. Gubbels et al. (2000) has proposed that the orientalis group should be named *T. buffeli* since they all are found in buffalo. Molecular characterization has not fully resolved the matter. Others believe enough differences exist that these species have to be regarded as different species. There is now consensus that *T. hirci* and *T. lestoquardi* are synonyms (Imam and Taha 2015).

For those interested in the taxonomy of *Theileria* there are excellent reviews on their classification by Dolan (1989), Norval et al. (1992), Levine et al. (1980), Mehlhorn and Schein (1984), and Irvin (1987). Table 8.1 lists the most common *Theileria* of livestock ruminants.

Table 8.1 Common *Theileria* in ruminant vectors, livestock hosts, disease, and geographic distribution (National Center for Biotechnology Information, Taxonomy (2017)

Theileria species	Tick vector	Hosts	Diseases	Region
T. annulata	*Hyalomma* sp.	Cattle, water buffalo	Tropical theileriosis or Mediterranean theileriosis	North Africa, Sudan, Middle East, Ethiopia, central, and W. Asia
T. parva	*Rhipicephalus appendiculatus/R. zambeziensis*	Cattle, African buffalo	East Coast fever, corridor disease, and Zimbabwean theileriosis	East, central, and southern Africa
T. mutans	*Amblyomma* sp.	Cattle, African buffalo	Benign theileriosis	Sub-Saharan Africa, Caribbean Islands
T. taurotragi	*Rhipicephalus* sp.	Cattle, eland	Benign	Eastern, southern, and Central Africa
T. verifera	*Amblyomma* sp.	Cattle, African buffalo	Benign	Sub-Saharan Africa
T. buffeli/T. sergenti/T. orientalis/T. sinensis	*Haemaphysalis* sp.	African buffalo, cattle, and water buffalo	Oriental theileriosis	Worldwide
T. Lestoquardi	*Hyalomma* sp.	Sheep, goats	Malignant theileriosis	Sudan, Middle East, India
T. Ovis	*Hyalomma* sp.	Sheep, goats	Benign	Worldwide
T. separata	*Rhipicephalus* sp./*Hyalomma* sp.	Sheep, antelope	ND	Southern and Eastern Europe, N. Africa
T. Uilenbergi	*Haemaphysalis quinhaiensis*	Sheep, goats, red deer	Theileriosis of sheep	China
T. Luwenshuni	*Haemaphysalis quinhaiensis*	Sheep, goats, red deer	Theileriosis of sheep	China

8.2.2.2 Geographic Distribution and Economic Importance

Theileria parva

Theileria parva is a parasite of cattle and the African buffalo—*Syncerus caffer*—and is transmitted by *Rhipicephalus appendiculatus*, *R. zambeziensis,* and *R. duttoni* ticks. Although other *Rhipicephalus* spp. and some *Hyalomma* species of ticks can transmit *T. parva* experimentally, their role in the epidemiology of the disease is doubtful. *T. parva* is thought to have evolved in the African buffalo, where it does not cause disease. In cattle *T. parva* causes a highly fatal lymphoproliferative disease known as East Coast fever (reviewed by Norval et al. 1992; Nene et al. 2016).

The disease follows very closely the distribution of its principal vectors. It is endemic in 12 countries in East, Central, and Southern Africa and in some Indian Ocean islands. These countries include South Sudan, Uganda, Democratic Republic of Congo, Rwanda, Burundi, Kenya, Tanzania (including Zanzibar), Malawi, Zambia, Zimbabwe, Mozambique, and the Comoro Islands (Fig. 8.3). The disease

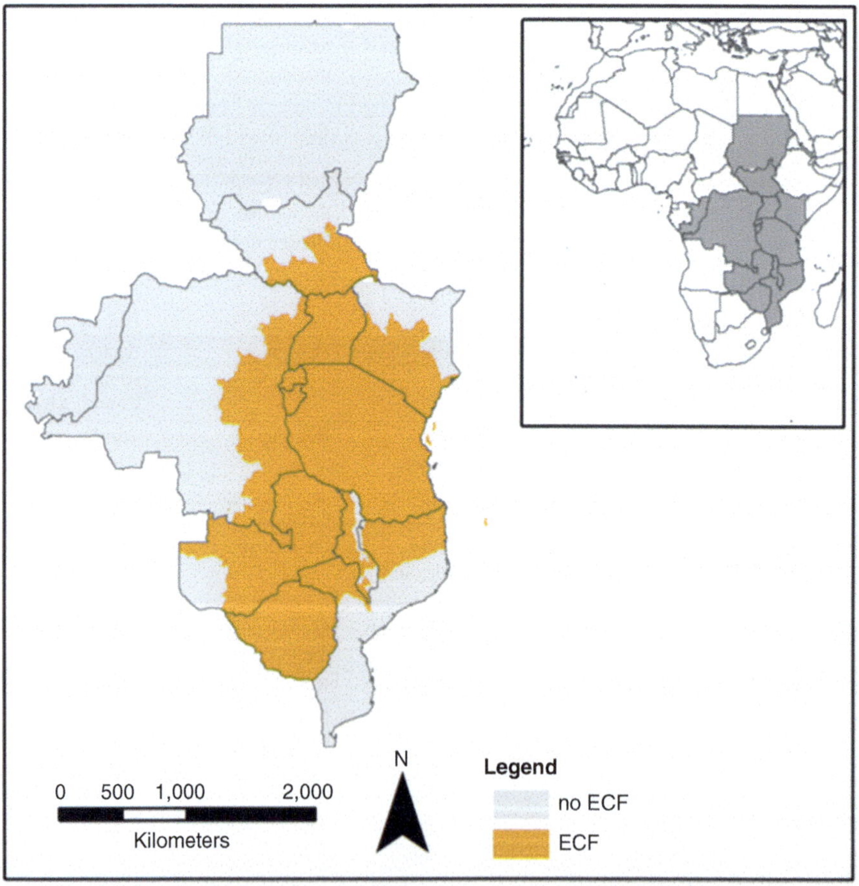

Fig. 8.3 Geographic distribution of *Theileria parva* (map created by Catherine Pfeifer)

spreads from the Indian Ocean coast west into the eastern rim of the Congo River Basin in the Democratic Republic of Congo. It is found as far north as Southern Sudan and extends southward to Northern Mozambique. In South Africa, a related disease commonly referred to as corridor disease in cattle is restricted to areas bordering national parks that contain buffalo. East Coast fever is transmitted by ticks that have fed on infected cattle, while corridor disease is transmitted by ticks that have fed on buffalo. Corridor disease is generally believed to be self-limiting in cattle because the disease is acute with very low levels of piroplasm parasitemia. However, *T. parva* infections have been reported in cattle outside of the corridor disease areas in South Africa (Thompson et al. 2008).

Given that the distribution of East Coast fever is closely linked to the distribution of the tick vector, which in turn is controlled by climatic factors and host availability, the current distribution of East Coast fever is less than the potential areas where the tick vector can survive (Norval et al. 1991). For some reasons that are as yet unclear, *R. appendiculatus* does not extend into Ethiopia, although the highlands are climatically suitable for the vector. Predictive models indicate that the vector can thrive in the Ethiopian Highlands (Norval et al. 1991). The absence of *R. appendiculatus* in Ethiopia has been explained by the absence of tick infestation of the highlands before human habitation and the absence of large herbivores in Ethiopia. It has also been argued that the Ethiopian Highlands are in the center of the country surrounded by arid lands to the east and south of the country and a tsetse barrier in the south western of the country excluding the presence of large numbers of suitable ruminant hosts.

It is also apparent from predictive models that large areas of the West African coast are climatically suitable for *R. appendiculatus*, but the vector there is currently absent. This has been explained because of the absence of large numbers of herbivores due to large forested areas and the presence of trypanosomiasis (Norval et al. 1992). But with large-scale deforestation and introduction of cattle accompanied by tsetse control, the situation might change. The recent introduction of ECF into Western Equatorial State of South Sudan, which borders Central African Republic, has increased the risk of introducing the tick and the disease into West Africa (Kivaria et al. 2012).

Inadvertent spread of ECF has been demonstrated in recent times when in 2002 animals immunized with the live ECF vaccine were imported into the Comoros from mainland Tanzania. In 2004 a severe outbreak of ECF was reported for the first time in that country (De Deken et al. 2007). Similarly, ECF has been spreading very rapidly northward in South Sudan causing huge losses when infected cattle came back into the country with returning refugees following the comprehensive peace agreement in 2005 (Malak et al. 2012; Kivaria et al. 2012; Wani et al. 2012).

Theileria parva infections occur in various epidemiological states (Norval et al. 1992). Endemic stability with regard to ECF has been reported in a few situations but is believed to be widespread in eastern Africa (Moll et al. 1986). Endemic stability is defined as a stable host-vector-parasite relation where almost all calves are infected very early in life but without clinical disease or death leading to a high degree of immunity in adult cattle and low incidence of clinical disease (Coleman et al. 2001). Because of changing breeds from indigenous to more *Bos taurus* breeds and various efforts at tick control, it is becoming more and more difficult to meet the conditions for endemic stability where nearly all calves become infected within a relatively short time after birth.

The disease is therefore found in a wide variety of syndromes ranging from subclinical to moderate to severe morbidity and mortality. There is evidence that *Bos taurus* cattle are more susceptible to ECF than *Bos indicus* cattle. There also appear to be breed differences in susceptibility to disease (Norval et al. 1992). However, these observations need to be taken cautiously because they might not reflect true genetic differences. The severity of ECF is parasite dose-dependent which in turn depends on the number of ticks infecting cattle and the infection rates of the ticks. It has been shown that individual cattle exhibit differences in the resistance to ticks (de Castro et al. 1991). There could also be differences in the threshold of infectivity in individual cattle.

Within the ECF endemic areas, about 50 million cattle are at risk from ECF. The annual economic loss due to ECF was estimated to be 168 million US$ in 1989 constituting both direct and indirect losses (Mukhebi et al. 1992). This figure was revised to 300 million in 1999 (McLeod and Kristjanson 1999). It must be pointed out that these are very crude estimates because of paucity of data for calculating the economic losses. Beyond the actual losses, ECF also limits the introduction of more productive exotic *Bos taurus* cattle in endemic areas, losses that are very difficult to quantify.

Several techniques have been developed to assess the genetic diversity of *T. parva* parasites. A number of loci and variable number of tandem repeat (VNTR) markers—micro- and minisatellites—derived from the *T. parva* genome sequence data have been used as markers to determine parasite diversity (Oura et al. 2005; Patel et al. 2016).

Results obtained so far indicate a high level of genetic diversity in populations of *T. parva*. Isolates from Kenya and Uganda were found to have higher levels of diversity compared to isolates from Zimbabwean and Zambia stocks (Oura et al. 2005). The evidence points to the fact that there is no obvious relationship between geographical origin of the parasite and level of genetic similarity between the parasite isolates. Distinct parasite genotypes have been isolated from the same farm, while genetically similar isolates have been found from geographically separated areas (Odongo et al. 2006; Oura et al. 2003). Older cattle have been shown to have a larger number of different parasite genotypes than younger ones, which tended to have a predominating genotype (Oura et al. 2005).

Some geographical sites exhibit a substructure in parasite populations, others did not, and some exhibit an epidemic structure characteristic of recent predominating infections (Oura et al. 2005). It has been explained that cattle movement, their interaction with buffalo, and parasite transmission rates are likely to play a major role in determining *T. parva* population structures. In general, buffalo-derived *T. parva* parasites are more diverse than cattle-derived *T. parva* parasites.

Theileria annulata

Theileria annulata is a parasite of cattle and domestic Asian buffalo—*Bubalus bubalis*. It is thought to have originated in the latter. It is transmitted by several species

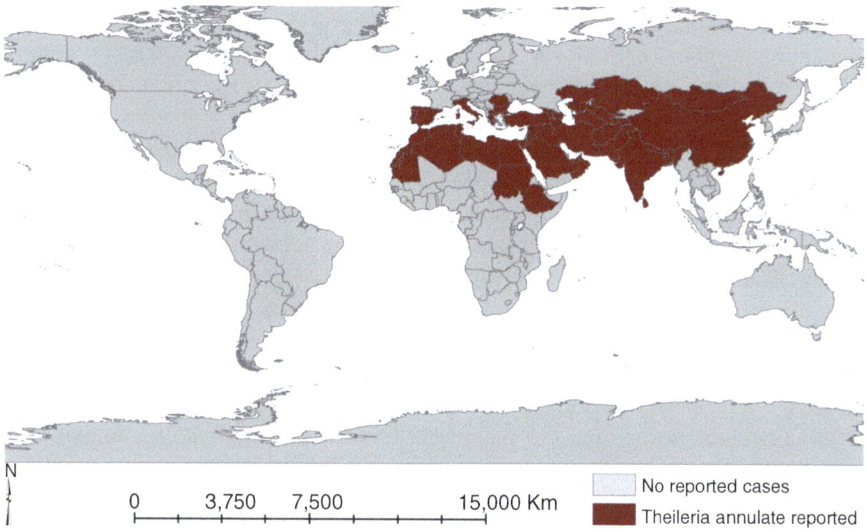

Fig. 8.4 Geographic distribution of *Theileria annulata* (map created by Catherine Pfeifer)

of the two-host *Hyalomma* ticks. It causes the diseases of cattle known as tropical or Mediterranean theileriosis. *T. annulata* is probably the most important of the *Theileria* of ruminants affecting an estimated 250 million cattle. The disease stretches from North Africa through Southern Europe and the Near and Middle East and into India and Central Asia. It extends from North Africa through Egypt and into South Sudan. The pathogen has recently been reported for the first time in Northern Ethiopia (Fig. 8.4) (Gebrekidan et al. 2014). Some reports indicate *T. annulata* is extending eastward from India to the Pacific coast including Southern China and all Far Eastern countries south of China and into Malaysia. But there could probably be some confusion with the *T. sergenti/T. orientalis* group (Uilenberg 1995).

The economic impact of *T. annulata* is not well documented. In India alone the economic cost was estimated at 384 million dollars (Minjauw and McLeod 2003). The impact though varies considerably depending on livestock production systems, ecological factors, and breeds of cattle. Although all cattle are affected, the disease causes higher morbidity and mortality in exotic European type breeds relative to indigenous cattle raised in endemic areas (Purnell 1977; Hashemi-Fesharki 1988). For instance, it has been shown that Sahiwal cattle—*Bos indicus*—which are native to Pakistan and the Kenana breed which is indigenous to the Sudan, are more resistance to *T. annulata* compared to nonindigenous susceptible breeds (Glass et al. 2005). The impact is also influenced by the epidemiological state, with regard to endemic stability or instability in which the disease exists. It has recently been shown that there is high level of genetic diversity with different isolates of *T. annulata* (Manuja et al. 2006; Weir et al. 2011). Most animals are infected by several genotypes at the same time. The importance of multiple genotype infections in the epidemiology of the disease is not well understood.

Other *Theileria*

Generally the other *Theileria* spp. including *T. mutans*, *T. taurotragi,* and *T. verifera* are nonpathogenic or cause asymptomatic transient infections in ruminants (Norval et al. 1992; Uilenberg 1981). However, pathogenic strains of *T. mutans* have been reported (Irvin et al. 1972). There are reports that *T. mutans* from buffalo are more pathogenic to cattle, but this has not been confirmed (Young 1981).

T. lestoquardi is the most pathogenic *Theileria* of sheep and goats. It is transmitted by ticks of the genus *Hyalomma* and has been reported in Iraq, Iran, India, Turkey, Sudan, Saudi Arabia, and Oman but probably has a worldwide distribution (Imam and Taha 2015). It causes malignant ovine theileriosis which is an acute disease of sheep. Clinical signs include fever, coughing, loss of condition, and enlargement of lymph nodes. The disease can also lead to abortions. Mortalities of up to 70% have been reported (Goh et al. 2016).

Oriental theileriosis caused by members of the *Theileria orientalis* complex—*T. buffeli*, *T. orientalis,* and *T. sergenti* and other unclassified *Theileria*—and transmitted by various species of *Haemaphysalis* ticks has a worldwide distribution. Because of the confusion in nomenclature, it is not clear if these are different species or subspecies or strains of the same parasite. Several biological differences among isolates have been observed—e.g., occurrence of macroschizonts, aberrant piroplasm morphology, and tick vector preferences. Although the disease is generally benign, in recent years clinical cases have been reported in Australia and the Asia-Pacific region (Kamau et al. 2011; Islam et al. 2011; Sugimoto and Fujisaki 2002). Oriental theileriosis is associated with high fever, anemia, jaundice, lethargy, weakness, abortion, and/or mortality. The disease is believed to be transmitted by ticks of the genus *Haemaphysalis* (Izzo et al. 2010). At present, the economic losses caused by oriental theileriosis are difficult to estimate. However, herds that have been clinically affected by this disease indicate a substantial loss.

Theileria sinensis has been reported in Central China as a new parasite of cattle and yaks transmitted by *Haemaphysalis qinghaiensis* (Liu et al. 2010). It is morphologically indistinguishable from the orientalis group, and some have doubted whether it is a new species or variant of *T. orientalis*. But it has been shown to be genetically distinct from T. *sergenti* to which it is most closely related (Gou et al. 2013).

Theileria uilenbergi and *T. luwenshuni* are pathogenic *Theileria* of sheep. They have been reported in China but probably have a much wider distribution (Liu et al. 2010).

Theileria ovis is a nonpathogenic *Theileria* of sheep reported in many countries but probably has a worldwide distribution.

8.3 Clinical Effects

8.3.1 *Theileria parva*

After a bite by a *T. parva*-infected tick, the parasite will localize and multiply in the lymph node that drains the site of the bite. This will often be the parotid lymph node as the ticks usually attach on the ears. The incubation period is 8–12 days with schizonts spreading to other lymph nodes. This will be associated with fever, sometimes up to 42 °C, and piroplasms appear in erythrocytes around day 10. Extensive leukopenia is a sign of severe disease where the white blood cells can disappear completely. Other clinical signs are anorexia, diarrhea, and soft coughs due to accumulated fluid in the lungs and thereby difficulties in breathing. Before death of the animal, the temperature usually falls, and a frothy nasal discharge will be visible due to the pulmonary edema (Fig. 8.5). Sometimes the parasite can invade the central nervous system which can result in nervous signs—turning sickness—and paralysis. The degree of pyrexia and pathogen load usually determines

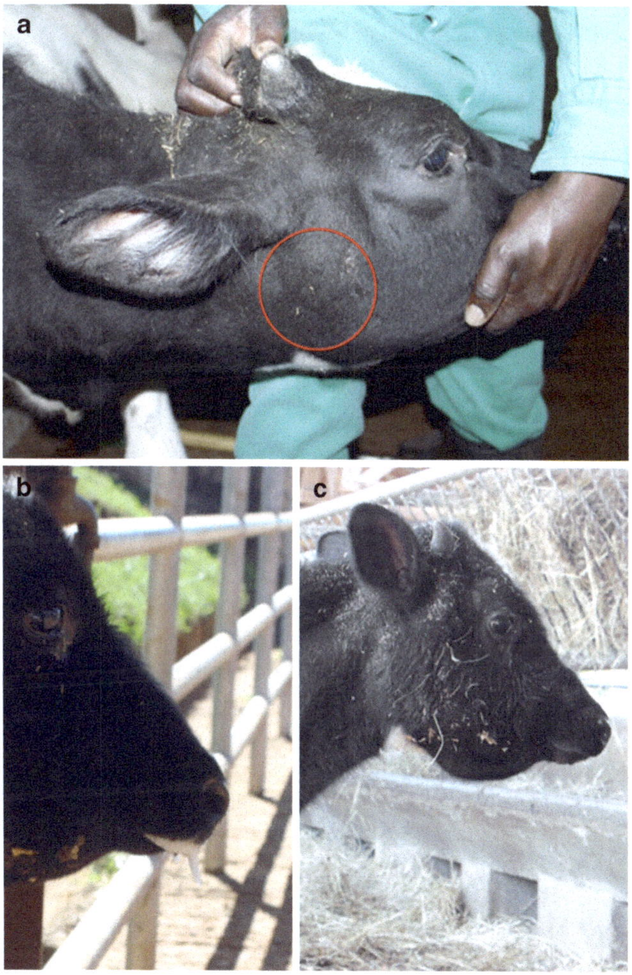

Fig. 8.5 Clinical signs of East Coast fever: (**a**) enlarged lymph node under the right ear; (**b**) frothing, lachrymation, and respiratory distress; (**c**) edema under the jaw, *bottle jaw* (pictures by Nicholas Svitek)

the severity of the disease. Young cattle appear to be more resistant than older animals, and Friesian-European breeds such as Friesian/Holstein appear to be more susceptible to the disease than indigenous cattle. Mortality can reach close to 100% on herd level if no interventions are implemented. In addition to the breed and age, the mortality also depends on the strain of the parasite and the amount of parasite exposure. ECF is a sporozoite dose-dependent disease (Irvin and Mwamachi 1983; OIE 2008).

8.3.2 *Theileria annulata*

Just as for *T. parva*, the *T. annulata* sporozoite localizes in the draining lymph node, which will enlarge, and the schizonts will proliferate and spread to other lymph nodes and nonlymphoid tissues. The incubation period for *T. annulata* is within 10–25 days a bit longer than for *T. parva*. Infected animals develop high fever—up to 42 ° C—and hemorrhages on visible mucus membranes and sometimes on the skin, mild to moderate anemia, and jaundice particularly in subacute and chronic forms of the disease. Other clinical signs are dyspnea, emaciation, and diarrhea. Ocular signs and masses may develop, and neurological and reproductive signs may develop in chronic diseases. The degree of pyrexia, pathogen load, and host susceptibility determines the severity of clinical signs at presentation. As is the case in *T. parva*, young cattle appear to be more resistant than older European breeds that are particularly susceptible to the disease. Mortality can vary between 30 to 90% depending on the parasite isolate and susceptibility of the animals (Gill et al. 1977).

8.3.3 *Theileria lestoquardi*

Clinical signs of *T. lestoquardi* infections in sheep and goats are very similar to the signs for *T. annulata* in cattle, and *T. annulata* is also the closest relative to *T. lestoquardi*. Natural infection of *T. lestoquardi* without clinical disease in cattle has been reported. The most prominent clinical signs are generalized enlargement of the superficial lymph nodes, high fever, anemia, anorexia, emaciation, diarrhea or constipation, and loss of condition. Initially, infected animals have a normal appetite, but with onset of fever they stop eating and become increasingly emaciated. The reason for the anemia is currently unknown. A marked fall in white blood cells (WBC) and packed cell volume (PVC) as well as hemoglobin (Hb) in blood are often reported. Disease appears to be associated with the schizont stage, where schizonts can be detected in the lymph nodes but also in internal organs. A high mortality of 46–100% is associated with the disease (Imam and Taha 2015).

8.4 Prevention and Treatment

8.4.1 Acaricides

Weekly treatments with acaricides are the most widespread method to prevent tick infestation and thereby also theileriosis, and other tick-borne diseases. However, there is a growing development of resistance to the available acaricides demanding more research in this area.

8.4.2 Live Vaccines

Live vaccines which prevent disease have been developed for *T. parva*, *T. annulata*, and lately also *T. lestoquardi*. Cattle can be vaccinated against *T. parva* by injecting live parasite concomitantly with a long-acting—i.e., 4–5 days—oxytetracycline; this procedure has been developed in the 1970s and is called the infection and treatment method (ITM) (Radley et al. 1975). The oxytetracycline slows down the parasite development, and this basically works as if the parasite was attenuated. The treatment gives a life-long immunity and an efficacy of close to 100% when tested against the strains in the vaccines. It is also broadly efficacious in the field except in areas where cattle get infected with *T. parva* strains circulating in buffalos. In the East African region, a vaccine referred to as the Muguga cocktail is used based on three different geographic isolates of the parasite. Vaccines based on the same principles are produced in other African countries but composed of local isolates. The sporozoites are injected close to the parotid or the prescapular lymph nodes and the oxytetracycline intramuscularly in the neck. These vaccines are produced by feeding clean ticks on cattle infected with a known strain of the parasites. After molting the infected ticks are pre-fed for 3 days to allow the sporozoites to mature before they are ground up to release the sporozoites from the salivary glands. The vaccine is then stored in liquid nitrogen until used (Patel et al. 2016). This requirement can be an obstacle of vaccine delivery in rural areas, and efforts are underway to improve this vaccine.

Vaccines have also been developed for *T. annulata* (Pipano and Shkap 2000). These are based on schizont-infected cell lines that have been attenuated by a prolonged in vitro passage. About 10^5–10^6 infected cells are sufficient for induction of immunity generating a robust immunity. As for *T. parva*, there is some degree of strain specificity making a vaccine based on one strain more efficient to this particular strain than to others. Such vaccines are made in a number of countries in Asia, Southern Europe, and Northern Africa where local strains are used. Similar techniques have been tried for *T. parva*, but for this parasite 10^8 to 10^9 cells are required for induction of immunity, which is not feasible for commercial production of a vaccine.

Similarly, cell-based vaccines have been developed for *T. lestoquardi* in Iraq, Iran, and Sudan where they have resulted in successful prophylaxis of malignant ovine theileriosis (Ahmed et al. 2013; Imam and Taha 2015).

The live ITM vaccine for *T. parva* is difficult to produce, and it needs a cold chain for distribution as do the cellular vaccines for *T. annulata* and *T. lestoquardi*. As a result they are relatively expensive.

8.4.3 Subunit Vaccines

The live vaccines described above come with various constraints and drawbacks which have encouraged engaging in the development of subunit vaccines especially for *T. parva* and *T. annulata* parasites. One approach has been to use sporozoite antigens for induction of antibodies which can neutralize sporozoite infectivity. The p67 sporozoite protein from *T. parva* and a shorter derivative of this, p67C, have been tested as vaccine candidates. These proteins have consistently resulted in 50% protection against ECF under experimental conditions (Musoke et al. 1992; Bishop et al. 2003). However, results from field trials led to protection levels of 25%. Currently, this vaccine antigen is being reevaluated, and attempts to increase its immunogenicity are underway.

The corresponding sporozoite protein from *T. annulata* SPAG-1 has also been evaluated as a vaccine antigen. SPAG-1 resulted in partial protection of immunized cattle (Boulter et al. 1999). It has been also shown that combining a cellular vaccine for *T. annulata* with the SPAG-1 resulted in a synergistic protective effect (Darghouth et al. 2006).

For *T. parva*, there have also been attempts to induce cytotoxic T lymphocytes (CTL) using a prime boost regimen with defined antigens cloned in plasmid DNA or canarypox followed by a boost with the antigens using an attenuated vaccinia virus vector. This resulted in some but a low level of protection (Graham et al. 2006). Currently, efforts are ongoing to evaluate various delivery systems for induction of CTLs and to combine these systems with p67C immunization to enhance the effect.

8.4.4 Treatment with Drugs

Parvaquone and buparvaquone are two important drugs for the treatment of *T. parva* and *T. annulata*. Buparvaquone has shown about 90% efficacy in field setting and is more efficient for treating *T. parva* and *T. annulata* than parvaquone. In severe cases an extra dose can be injected 48–72 h after the first dose (McHardy 1999). There are no major side effects of the drug, but there are reports showing parasite resistance to the drugs for *T. annulata*, which is of concern (Mhadhbi et al. 2010). A limitation with these drugs is that they have to be used relatively early in the course of the disease in order to be efficacious. Another drawback that may limit their use is that they are rather expensive and that the drug can be detected in muscles and milk for

Moore RB, Obornik M, Janouskovec J, Chrudinmsky T, Vancová M, Green DH, et al. A photosynthetic alveolate closely related to apicomplexan parasites. Nature. 2008;451:959–63.
Morrison WI, MacHugh ND, Lalor PA. Pathogenicity of *Theileria parva* is influenced by the host cell type infected by the parasite. Infect Immun. 1996;64:557–62.
Morzaria SP, Katende J, Musoke A, Nene V, Skilton R, Bishop R. Development of sero-diagnostic and molecular tools for the control of important tick-borne pathogens of cattle in Africa. Parassitologia. 1999;41(1):73–80. Review.
Mukhebi AW, Perry BD, Kruska RL. Estimated economics of theileriosis control in Africa. Prev Vet Med. 1992;12:73–85.
Musoke A, Morzaria S, Nkonge C, Jones E, Nene V. A recombinant sporozoite surface antigen of *Theileria parva* induces protection in cattle. Proc Natl Acad Sci U S A. 1992;89(2):514–8.
National Center for Biotechnology Information, Taxonomy. https://www.ncbi.nlm.nih.gov/taxonomy. 2017. Accessed 20 Aug 2017.
Nene V, Kiara H, Lacasta A, Pelle R, Svitek N, Steinaa L. The biology of *Theileria parva* and control of East Coast fever-current status and future trends. Ticks Tick Borne Dis. 2016;7:549–64.
Norval RAI, Perry BD, Gebreab F, Lessard P. East Coast fever: a problem of the future for the horn of Africa? Prev Vet Med. 1991;10:163–72.
Norval RAI, Perry BD, Young A. The epidemiology of Theileriosis in Africa. London: Academic; 1992.
Odongo DO, Oura CAL, Spooner PR, Kiara H, Mburu D, Hanotte OH, Bishop RB. Linkage disequilibrium between alles at highly polymorphic mini-and microsatellite loci of *Theileria parva* isolated from cattle in three regions of Kenya. Int J Parasitol. 2006;36:937–46.
Odongo DO, Sunter JD, Kiara HK, Skilton RA, Bishop RP. A nested PCR assay exhibits enhanced sensitivity for detection of *Theileria parva* infections in bovine blood samples from carrier animals. Parasitol Res. 2010;106(2):357–65. https://doi.org/10.1007/s00436-009-1670-z. Epub 2009 Nov 10.
OIE. Theileriosis. OIE Terrestrial manual, 2008.
Oura CAL, Odongo DO, Lubega GW, Spooner PR, Tait A, Bishop RP. A panel of microsatelite and minisatelite makers for the characterization of field isolates of *Theileria parva*. Int J parasitol. 2003;33:1641–53.
Oura CAL, Asiimwe BB, Weir W, Lubega GW, Tait A. Population genetic analysis and substructuring of *Theileria parva* in Uganda. Mol Biochem Parasitol. 2005;140:229–39.
Oura CA, McKellar S, Swan DG, Okan E, Shiels BR. Infection of bovine cells by the protozoan parasite *Theileria annulata* modulates expression of the ISGylation system. Cell Microbiol. 2006;8:276–88.
Palmer GH, Machado J Jr, Fernandez P, Heussler V, Perinat T, Dobbelaere DA. Parasite-mediated nuclear factor kappaB regulation in lymphoproliferation caused by *Theileria parva* infection. Proc Natl Acad Sci U S A. 1997;94:12527–32.
Papli N, Landt O, Fleischer C, Koekemoer JO, Mans BJ, Pienaar R, Josemans A, Zweygarth E, Potgieter F, Latif AA. Evaluation of a TaqMan real-time PCR for the detection of *Theileria parva* in buffalo and cattle. Vet Parasitol. 2011;175(3-4):356–9. https://doi.org/10.1016/j.vetpar.2010.10.038. Epub 2010 Oct 27.
Patel E, Mwaura S, Kiara H, Morzaria S, Peters A, Toye P. Production and dose determination of the ITM method (ITM) MC vaccine used to control East Coast fever in cattle. Ticks Tick Borne Dis. 2016;7(2):306–14.
Perry BD, Young AS. The naming game: the changing fortunes of East Coast fever and *Theileria parva*. Vet Rec. 1993;133:613–6.
Pienaar R, Potgieter FT, Latif AA, Thekisoe OM, Mans BJ. The hybrid II assay: a sensitive and specific real-time hybridization assay for the diagnosis of *Theileria parva* infection in cape buffalo (*Syncerus caffer*) and cattle. Parasitology. 2011;138(14):1935–44. https://doi.org/10.1017/S0031182011001454. Epub 2011 Sep 9.
Pipano E, Shkap V. Vaccination against tropical theileriosis. Ann N Y Acad Sci. 2000;916:484–500. Review.

Purnell RE. East Coast fever: some recent research in East Africa. Adv Parasitol. 1977;15:83–132.

Radley DE, Brown CGD, Cunningham MP, Kimber CD, Musisi FL, Payne RC, Purnell RE, Stagg DA, Young AS. East Coast fever. 3. Chemoprophylactic immunization of cattle using oxytetracycline and a combination of *Theileria* strains. Vet Parasitol. 1975;1:51–60.

Renneker S, Kullmann B, Gerber S, Dobschanski J, Bakheit MA, Geysen D, Shiels B, Tait A, Ahmed JS, Seitzer U. Development of a competitive ELISA for detection of *Theileria annulata* infection. Transbound Emerg Dis. 2008;55(5-6):249–56. https://doi.org/10.1111/j.1865-1682.2008.01036.x.

Ros-García A, Juste RA, Hurtado AA. Highly sensitive DNA bead-based suspension array for the detection and species identification of bovine piroplasms. Int J Parasitol. 2011;42(2):207–14. https://doi.org/10.1016/j.ijpara.2011.12.001. Epub 2011 Dec 27.

Ros-García A, Nicolás A, García-Pérez AL, Juste RA, Hurtado A. Development and evaluation of a real-time PCR assay for the quantitative detection of *Theileria annulata* in cattle. Parasit Vectors. 2012;5:171. https://doi.org/10.1186/1756-3305-5-171.

Salih DA, Liu Z, Bakheit MA, Ali AM, El Hussein AM, Unger H, Viljoen G, Seitzer U, Ahmed JS. Development and evaluation of a loop-mediated isothermal amplification method for diagnosis of tropical theileriosis. Transbound Emerg Dis. 2008;55(5-6):238–43. https://doi.org/10.1111/j.1865-1682.2008.01033.x.

Salih DA, Ali AM, Liu Z, Bakheit MA, Taha KM, El Imam AH, Kullmann B, El Hussein AM, Ahmed JS, Seitzer U. Development of a loop-mediated isothermal amplification method for detection of *Theileria lestoquardi*. Parasitol Res. 2012;110(2):533–8. https://doi.org/10.1007/s00436-011-2518-x.

Santos M, Soares R, Costa P, Amaro A, Inácio J, Gomes J. Revisiting the Tams1-encoding gene as a species-specific target for the molecular detection of *Theileria annulata* in bovine blood samples. Ticks Tick Borne Dis. 2013;4(1-2):72–7. https://doi.org/10.1016/j.ttbdis.2012.07.006. Epub 2012 Dec 11.

Schnittger L, Yin H, Qi B, Gubbels MJ, Beyer D, Niemann S, Jongejan F, Ahmed JS. Simultaneous detection and differentiation of *Theileria* and *Babesia* parasites infecting small ruminants by reverse line blotting. Parasitol Res. 2004;92(3):189–96. Epub 2003 Dec 3

Schnittger L, Rodriguez AE, Florin-Christensen M, Morrison DA. *Babesi*a: a world emerging. Infect Genet Evol. 2012;12:1788–809.

Seitzer U, Schnittger L, Boguslawski K, Ahmed JS. Investigation of MAP kinase activation in *Theileria*-infected cell lines. Ann N Y Acad Sci. 2006;1081:473–5.

Seitzer U, Bakheit MA, Salih DE, Ali A, Haller D, Yin H, Schnittger L, Ahmed J. From molecule to diagnostic tool: *Theileria annulata* surface protein TaSP. Parasitol Res. 2007;101(2):S217–23. Review.

Shaw MK, Tilney LG, Musoke AJ. The entry of *Theileria parva* sporozoites in bovine lymphocytes: evidence for MHC class I involvement. J Cell Biol. 1991;113:87–101.

Shaw MK, Tilney LG, McKeever DJ. Tick salivary gland extract and interleukin-2 stimulation enhance susceptibility of lymphocytes to infection by Theileria parva sporozoites. Infect Immun. 1993;61:1486–95.

Sibeko KP, Oosthuizen MC, Collins NE, Geysen D, Rambritch NE, Latif AA, Groeneveld HT, Potgieter FT, Coetzer JA. Development and evaluation of a real-time polymerase chain reaction test for the detection of *Theileria parva* infections in cape buffalo (Syncerus Caffer) and cattle. Vet Parasitol. 2008;155(1-2):37–48. https://doi.org/10.1016/j.vetpar.2008.03.033. Epub 2008 Jun 2.

Sitt T, Poole EJ, Ndambuki G, Mwaura S, Njoroge T, Omondi GP, Mutinda M, Mathenge J, Prettejohn G, Morrison IW, P(T. Exposure of vaccinated and naive cattle to natural challenge from buffalo-derived *Theileria parva*. Int J Parasitol Parasites Wildl. 2015;4:244–51.

Skilton RA, Bishop RP, Katende JM, Mwaura S, Morzaria SP. The persistence of *Theileria parva* infection in cattle immunized using two stocks which differ in their ability to induce a carrier state: analysis using a novel blood spot PCR assay. Parasitology. 2002;124(3):265–76.

Spitalska E, Torina A, Cannella V, Caracappa S, Sparagano O. Discrimination between *Theileria lestquardi* and *Theileria annulata* in their vectors and hosts by RFLP based on the 18S rRNA gene. Parasitol Res. 2004;94(4):318–20.

Striepen B, Jordan CN, Reiff S, van Dooren GG. Building the perfect parasite: cell division in Apicomplexa. PLoS Pathog. 2007;3(6):691–8.

Sugimoto C, Fujisaki K. Non-transforming Theileria parasites of ruminants. In: Dobbelaere D, McKeever D, editors. World class parasites, vol 3, *Theileria*. New York: Springer Science and Business Media LLC; 2002.

Thekisoe OM, Rambritch NE, Nakao R, Bazie RS, Mbati P, Namangala B, Malele I, Skilton RA, Jongejan F, Sugimoto C, Kawazu S, Inoue N. Loop-mediated isothermal amplification (LAMP) assays for detection of Theileria parva infections targeting the PIM and p150 genes. Int J Parasitol. 2010;40(1):55–61. https://doi.org/10.1016/j.ijpara.2009.07.004.

Thompson BE, Latif AA, Oosthuizen MC, Troskie M, Penzhorn BL. Occurrence of *Theileria parva* infection in cattle on a farm in the Ladysmith district, KwaZulu-Natal, South Africa. J S Afr Vet Assoc. 2008;79(1):31–5.

Tretina K, Gotia HT, Mann DJ, Silva JC. Theileria-transformed bovine leukocytes have cancer hallmarks. Trends Parasitol. 2015;31:306–14.

Uilenberg G. Tick-borne livestock diseases and their vectors. 2. Epizootiology of tick borne diseases. World Anim Rev. 1976;17:8–15.

Uilenberg G. Theileria species of domestic animals. In: Irvin AD, Cunningham MP, Young AS, editors. Advances in the control of Theileriosis. Proceedings of an International Conference. ILRAD, Nairobi, 9-13 February 1981. The Hague: Martinus Nijhoff; 1981. p. 227–37.

Uilenberg G. International collaborative research: significance of tick-borne hemoparasitic diseases to world animal health. Vet Parasitol. 1995;57:10–41.

Wani LM, Salih DA, Julla II, Rahim MA, Hussein E. Seroprevalence of East coast fever in Central Equatorial State, South Sudan. Vet Ital. 2012;48(4):379–85.

Weir W, Karagen T, Gharbi M, Simuunza M, Aypak S, Aysul N, Darghouth MA, Shiels B, Tait A. Population diversity and multiplicity of infection in *Theileria annulata*. Int J Parasitol. 2011;41:193–20.

Woolhouse MEJ, Thumbi SM, Jennings A, Chase-Topping M, Callaby R, Kiara H, Oosthuizen MC, Mbole-Kariuki MN, Conradie I, Handel IG, Poole EJ, Njiiri E, Collins NE, Murray G, Tapio M, Auguet OT, Weir W, Morrison WI, Kruuk LEB, Bronsvoort BMC, Hanottee O, Coetzer K, Toye PG. Co-infections determine patterns of mortality in a population exposed to parasite infection. Sci Adv. 2015;1:e1400026.

Yamaguchi T, Yamanaka M, Ikehara S, Kida K, Kuboki N, Mizuno D, Yokoyama N, Narimatsu H, Ikehara Y. Generation of IFN-gamma-producing cells that recognize the major piroplasm surface protein in *Theileria orientalis*-infected bovines. Vet Parasitol. 2010;171(3-4):207–15. https://doi.org/10.1016/j.vetpar.2010.03.038.

Young AS. The epidemiology of theileriosis in East Africa. In: Irvin AD, Cunningham MP, Young AS, editors. Advances in the control of Theileriosis. Proceedings of an International Conference. ILRAD, Nairobi, 9-13 February. The Hague: Martinus Nijhoff; 1981. p. 227–37.

Babesia in Domestic Ruminants

9

Sabrina Ganzinelli, Anabel Rodriguez,
Leonhard Schnittger, and Mónica Florin-Christensen

Abstract

Babesia spp. are intraerythrocytic tick-transmitted apicomplexan protozoans, considered to be the second most commonly found parasites in the blood of mammals after trypanosomes. Ever since the discovery of parasitic inclusions in erythrocytes of cattle by Victor Babes, a great number of *Babesia* species have been described, and thanks to the advances in microscopy, cell biology, and molecular biology techniques, our knowledge is continually expanding. Most *Babesia* species that affect domestic ruminants, including cattle, water buffalo, and small ruminants, are distributed in tropical and subtropical regions of the world. Acute *Babesia* infections of these animals are associated with fever, anemia, hemoglobinuria, and abortions and, in some cases, neurological symptoms, respiratory distress, and even death. Babesiosis outbreaks can be prevented by a combination of vaccination with living attenuated organisms and tick control programs. However, these control methods have numerous limitations, and, additionally, commercial live vaccines are only available for bovine babesiosis caused by *B. bovis* and *B. bigemina*. Our knowledge on relevant parasite molecules that act in the interface with the mammalian and tick hosts is rapidly increasing in the postgenomic era and will aid in the development of new and improvement of known immunotherapeutic interventions.

S. Ganzinelli • L. Schnittger • M. Florin-Christensen (✉)
Institute of Pathobiology, Center for Research on Veterinary and Agronomic Sciences (CICVyA), National Institute of Agricultural Technology (INTA-Argentina), Hurlingham, Province of Buenos Aires, Argentina

National Council of Scientific and Technological Research (CONICET), Buenos Aires, Argentina
e-mail: jacobsen.monica@inta.gob.ar

A. Rodriguez
Institute of Pathobiology, Center for Research on Veterinary and Agronomic Sciences (CICVyA), National Institute of Agricultural Technology (INTA-Argentina), Hurlingham, Province of Buenos Aires, Argentina

M. Florin-Christensen, L. Schnittger (eds.), *Parasitic Protozoa of Farm Animals and Pets*, https://doi.org/10.1007/978-3-319-70132-5_9

9.1 Morphology, Life Cycle, and Host-Pathogen Interactions

Victor Babes was the first to observe a microorganism—which he considered a bacterium—inside bovine erythrocytes of Rumanian cattle, and to connect this finding with hemoglobinuria or *red water*, a clinical sign that had long troubled cattle ranchers in many regions around the globe (Babes 1888). He later described a similar microorganism in sheep erythrocytes (Babes 1892). Not long after, in North America, Theobald Smith and Frederick Kilborne reported that *red water* in cattle was indeed associated with a tick-transmitted protozoon. Thus, *Babesia* was the first arthropod-transmitted pathogen ever described (Smith and Kilborne 1893). The parasites observed by Babes, and by Smith and Kilborne were later named *Babesia bovis*, *B. bigemina,* and *B. ovis*, and the diseases they cause in cattle and sheep, bovine and ovine babesiosis, respectively (Mihalca 2010). Since those early discoveries, a great number of *Babesia* species have been described, and thanks to the advances in microscopy, cell biology, and molecular biology techniques, our knowledge is continually expanding (Criado-Fornelio et al. 2004; Uilenberg 2006; Schnittger et al. 2012). *Babesia* spp. are considered the second most commonly found parasites in the blood of mammals after trypanosomes and have been reported to infect all kinds of domestic and wild animals, including marsupials and humans, as well as some birds. While infections are largely asymptomatic in wild animals, they are often associated with morbidity and mortality in domestic ones (Schnittger et al. 2012). This chapter will focus on *Babesia* species infecting domestic ruminants—bovines, water buffaloes, sheep, and goats—which are frequently associated with substantial economic losses (Table 9.1).

9.1.1 Morphology

Babesia spp. are unicellular eukaryotes, taxonomically classified as Piroplasmida —commonly referred to as piroplasmids—within the Apicomplexa group, together with *Theileria* spp. and *Cytauxzoon* spp. Piroplasmids have been defined as tick-transmitted, piriform, round, or rod-shaped parasites that lack conoids and flagella in all life stages; without oocysts; and with sexual stages associated with the formation of a large axopodium-like structure (Adl et al. 2005, 2012). As nonpigment-forming hemoparasites, piroplasmids can be distinguished from other erythrocyte-infecting genera, such as *Plasmodium* and *Haemoproteus*, which in contrast form pigment deposits—hemozoin—in the parasitized cell (Uilenberg 2006). All ruminant-infecting *Babesia* species belong to the sensu stricto group. This group is characterized by the lack of a schizont stage and the occurrence of transovarial transmission in the tick, two phenotypic features that clearly distinguish them from *Theileria* parasites (Uilenberg 2006; Schnittger et al. 2012).

Babesia spp. only invade erythrocytes within their vertebrate hosts. As will be described in Sect. 9.1.2, after entering an erythrocyte, the infectious sporozoite acquires a cyclic structure—a trophozoite ring—which then transforms into a replicative merozoite. Merozoites are tear-shaped and normally observed in pairs,

Table 9.1 *Babesia* parasites affecting domestic ruminants, their tick vectors and geographic distribution

Main host	*Babesia* species	Main tick vectors	Geographic distribution	References
Cattle Water buffalo	*B. bovis*	*Rhipicephalus microplus* *R. annulatus* *R. geigyi* *R. australis*	Africa America Asia Australia Europe	Bock et al. (2004), de Waal and Combrink (2006), Uilenberg (2006), Schnittger et al. (2012), Estrada-Peña et al. (2012)
	B. bigemina	*R. microplus* *R. annulatus* *R. geigyi* *R. decoloratus* *R. evertsi evertsi*		
Cattle	*B. divergens*	*Ixodes ricinus*, *I. persulcatus*	Europe Northern Africa	Zintl et al. (2003)
	B. ovata	*Haemaphysalis* sp.	Asia	Sivakumar et al. (2016)
	B. major	*Haemaphysalis* sp.	Asia Europe	Bock et al. (2004), Schnittger et al. (2012)
	B. beliceri	*Hyalomma* sp.	Russia	Uilenberg (2006)
	B. occultans	*Hy. marginatum*	Africa Europe	Blouin and van Rensburg (1988), Decaro et al. (2013)
	B. jakimovi	*Ixodes* sp.	Russia	Uilenberg (2006)
	B. sp. *Kashi*	*Hyalomma anatolicum*	Asia	Aktas et al. (2012)
Water buffalo	*B. orientalis*	*R. haemaphysaloides*	Asia	Liu et al. (2005, 2007a), He et al. (2012), Weerasooriya et al. (2016)
Sheep Goat	*B. ovis*	*R. bursa* *R. sturanicus*	Africa, Asia, Europe	Schnittger et al. (2003), Ranjbar-Bahadori et al. (2012)
	B. motasi	*Haemaphysalis* spp.	Africa, Asia, Europe	Schnittger et al. (2003), Niu et al. (2016b)
	B. crassa	*Haemaphysalis* spp.	Middle East Asia Europe	Schnittger et al. (2003), Hornok et al. (2015)
	B. sp. Xinjiang	*Hyalomma anatolicum*	China	Liu et al. (2007b), Niu et al. (2017)
	B. sp. BQ1	*Hy. anatolicum*	China	Niu et al. (2016a)

although Maltese cross structures can occasionally be found (Fig. 9.1). Based on merozoite morphology, *Babesia* spp. have been divided into two groups: *small babesias*, 1.0 to 2.5 μm long, and *large babesias*, 2.5 to 5.0 μm long, smaller and longer than the erythrocyte radius, respectively. Among the domestic ruminant-infecting species, the first group includes *B. bovis*, *B. divergens*, and *B. ovis*, and the second, *B. bigemina, B. major, B. ovata*, and *B. motasi*. *Babesia* pairs form an obtuse or acute angle, in the case of *small* or *large babesias*, respectively. This size division, however, has no clear genetic basis (Chauvin et al. 2009). Moreover, it is not useful for species identification, since changes of size and shape can take place

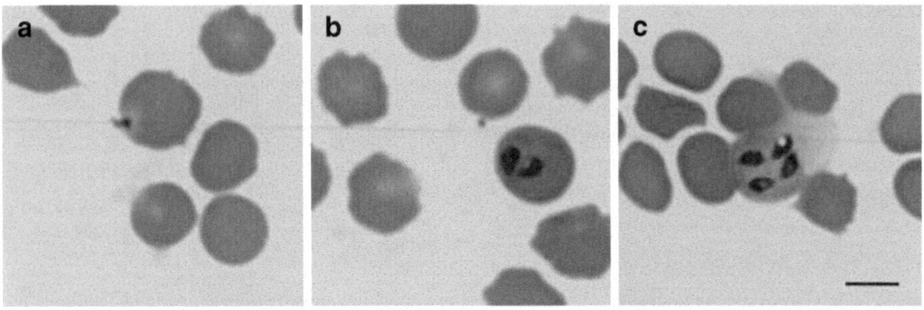

Fig. 9.1 Giemsa-stained smears of in vitro cultured *Babesia bovis* merozoites in bovine erythrocytes. *B. bovis* free merozoite attaching to the surface of an erythrocyte (**a**); typical appearance of a *B. bovis*-infected erythrocyte with a pair of tear-shaped merozoites forming a < 90° angle (**b**); bovine erythrocyte infected with two pairs of merozoites (**c**); magnification, 400×; bar, 5 μm

during different developmental stages within erythrocytes or when infecting different hosts (Homer et al. 2000).

As shown in Fig. 9.2, *Babesia* merozoites share with other eukaryotic cells the presence of membranous organelles, such as nucleus, endoplasmic reticulum (ER), Golgi apparatus, and mitochondria. In addition, as other Alveolata members, they are characterized by the presence of a cellular wrapping—the pellicle—which is composed of the cell membrane and two inner layers: a primary layer of vesicles or alveoli and a secondary layer of microtubules, involved in parasite motility and host cell invasion (Lew et al. 2002). As in other Apicomplexans, an assembly of invasion-specialized organelles—the apical complex—is present at the anterior apex. It is composed of a polar ring, micronemes, and rhoptries. In contrast, another typical apical structure—the conoid—is absent in piroplasmids (Blackman and Bannister 2001; Klinger et al. 2013). Spherical bodies are unique secretory organelles of *Babesia* spp. distributed in the parasite cytoplasm. They are considered to be homologous to the dense granules, present in other apicomplexans, and to participate in host-pathogen interactions during the intraerythrocytic stage. An apicoplast, a plastid with no photosynthetic activity, thought to have been acquired by secondary endosymbiosis from algae, also occurs in these cells. The apicoplast takes care of some metabolic activities, such as fatty acid and isoprenoid biosynthesis (Caballero et al. 2012). Due to its essential functions for parasite survival, this special plastid has gained attention as an attractive target for the development of parasiticidal drugs (Brayton et al. 2007; Huang et al. 2015).

Sexual stages in the tick gut were initially reported for *B. bigemina* by Robert Koch (1906). These forms are characterized by radial projections, for which they received the name of *Strahlenkorper—spiky-rayed stages* or *ray bodies*. Electron micrographs showed that *Strahlenkorper* contain a great number of micropores, abundant microtubules, and a large, circular, nonmembrane-bounded structure, named *bright body* because of its electron-lucid nature. The *spiky rays* consist of cytoplasmic protrusions that can vary in number from two to more than ten and are supported by a bundle of parallel microtubules (Weber and Friedhoff 1977). Differential expression of some proteins in sexual stages has recently been reported

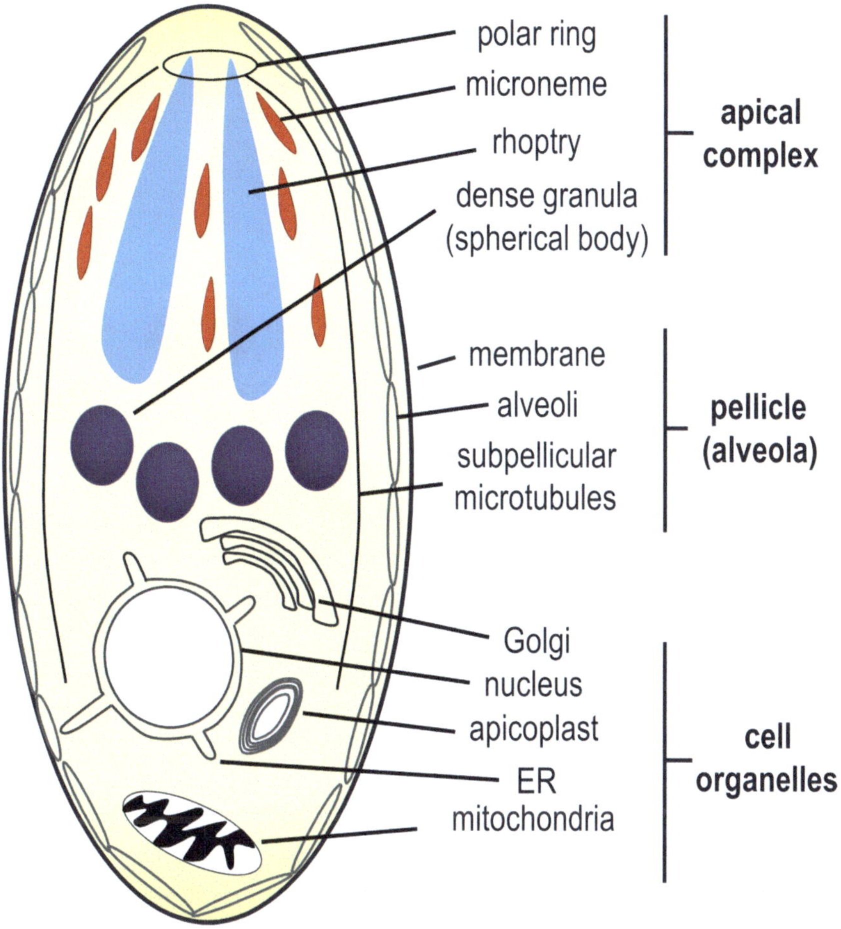

Fig. 9.2 Morphology of a *Babesia* merozoite. Organelles shared with other eukaryotes, other Alveolata and other Apicomplexa are shown. A conoid is not present in *Babesia* apical complex. Spherical bodies of this parasite are homologous to apicomplexan-typical dense granules. They do not co-localize with the apical complex but are functionally associated

for *B. bovis*. These proteins are likely to have essential roles in sexual reproduction mechanisms and might be applied in future studies to the development of transmission blocking vaccines (Bastos et al. 2013; Alzan et al. 2016).

9.1.2 Life Cycle

The life cycle of *Babesia* spp. involves two different hosts: an Ixodidae tick, where sexual reproduction takes place, and a vertebrate, where the parasite exclusively experiences asexual reproduction within erythrocytes (Fig. 9.3).

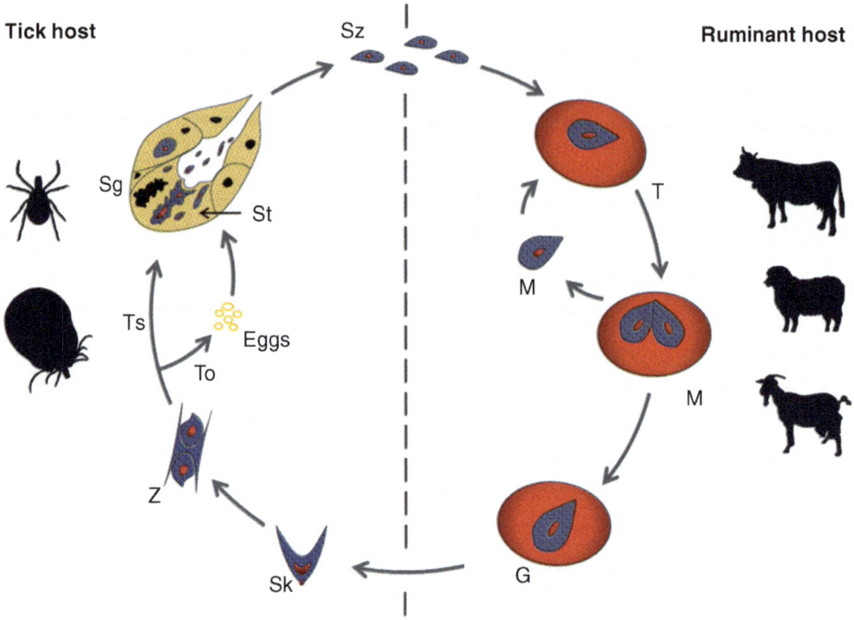

Fig. 9.3 *Babesia* sp. life cycle. Sporozoites (Sz) are injected into the bloodstream of a vertebrate host during the blood meal of an infected tick. After invading erythrocytes, they differentiate into trophozoites (T), which divide asexually into two or sometimes four merozoites (M). After erythrocyte lysis, each merozoite invades a new erythrocyte and successive merogonies occur. A few merozoites stop division and transform into gamonts or pregametocytes (G). Multiplication is asynchronous, and various divisional stages of the parasite can be seen in the bloodstream at the same time. Gamogony and sporogony take place in the tick. When gamonts are taken up by a tick feeding on an infected host, they differentiate in the gut into gametes, also known as *ray bodies* or *Strahlenkorper* (Sk) that fuse forming a diploid zygote (Z, gamogony). Zygotes undergo meiosis giving rise to motile haploid kinetes, which multiply by sporogony and access the hemolymph, invading and continuing their replication in several tick organs, including the salivary glands (Sg). Kinetes transform into sporozoites that will infect a vertebrate host after the tick has molded into the next stage (larvae to nymph or nymph to adult, transstadial transmission, Ts). In ruminant-infecting *Babesia* spp.—as in other *Babesia* sensu stricto species—kinetes also invade the tick ovaries and eggs, and infective sporozoites are formed in the salivary glands of the next-generation larvae (transovarial transmission, To)

Following uptake of intraerythrocytic parasites by an Ixodidae tick during a blood meal, parasites need to cross multiple cellular barriers in their migration through tick tissues and cavities, during which they undergo several—still only partially characterized—metamorphic changes (Kakoma and Mehlhorn 1994; Florin-Christensen and Schnittger 2009). Gametocytes, which are thought to be already present in the ingested blood, mature into dimorphic elongated gametes—*ray bodies*—that fuse, yielding diploid zygotes (Ribeiro and Patarroyo 1998). Zygotes adhere to and invade midgut epithelial cells, and eventually transform into motile kinetes, which are released into the tick hemocoel and invade multiple tissues, including granular acini of salivary glands. Here, parasites replicate asexually by sporogony, forming sporozoite colonies. *Babesia* sensu stricto parasites undergo transovarial transmission, meaning that kinetes also invade tick ovaries and eggs, and infective sporozoites are formed in the salivary glands of the next-generation larvae as well. Sporozoites are injected into a suitable host together with tick saliva during feeding and invade erythrocytes. They then convert into hemoglobin-feeding trophozoites, which in turn transform into merozoites. Merozoites multiply by binary fission—merogony—and, eventually, they lyse host cells and invade new ones, repeating this asexual propagation cycle (Kakoma and Mehlhorn 1994).

9.1.3 Host-Pathogen Interactions

Babesia spp. carry out an obligatory intracellular stage in the vertebrate host inside erythrocytes, where they get a nutritious medium and a shelter to escape from the dangers of the host immune system. An efficient invasion is guaranteed by the presence of erythrocyte receptors that are recognized by complementary ligand molecules on the parasite surface. Understanding the mechanisms and molecules involved in this event can greatly contribute to the development of vaccines and therapeutics against these parasites (Yokoyama et al. 2006; Rodriguez et al. 2013a; Florin-Christensen et al. 2014).

The mechanism of erythrocyte invasion by a *Babesia* spp. parasite is illustrated in Fig. 9.4. The first contact between parasite and host cell is apparently established by random collisions, followed by some kind of receptor-ligand interaction. Different surface proteins have been implicated in this process. Erythrocyte-binding assays, seroneutralization, and/or enzymatic cleavage experiments suggest that glycosylphosphatidylinositol (GPI)-anchored proteins are involved in erythrocyte recognition and attachment (Suarez et al. 2000; Wilkowsky et al. 2003; Delbecq et al. 2008; Dominguez et al. 2010; Rodriguez et al. 2014). In addition, protein-free GPI molecules, which are likely components of the parasite glycocalyx, might be involved in these first interactions with red blood cells (Rodriguez et al. 2010). *B. bovis* has several GPI-anchored antigens, defined by an amino-terminal hydrophobic signal sequence, a hydrophilic central region, and a conserved carboxy-terminal region containing a GPI-anchor signal sequence (Rodriguez et al. 2014). The best characterized of these antigens belong to the family of variable merozoite surface antigens (VMSA) and have been named MSA-1, MSA-2a_1, MSA-2a_2, MSA-2b, and MSA-2c (Suarez et al. 2000; Florin-Christensen et al. 2002). They contain neutralization-sensitive B-cell epitopes that are conserved among geographically distant isolates, as well as T-cell epitopes that induce lymphoproliferation and release of IFN-γ (Suarez et al. 2000; Mosqueda et al. 2002; Wilkowsky et al. 2003; Dominguez et al. 2010; Gimenez et al. 2016). MSA-2c is the most conserved member of this family (Florin-Christensen et al. 2002; Dominguez et al. 2010), and as other VMSA family members, it is immunodominant in *B. bovis* infections. MSA-2c has been applied successfully to the development of diagnostic tests and has also been proposed as a vaccine candidate (Wilkowsky et al. 2003; Kim et al. 2008; Alvarez et al. 2010; Dominguez et al. 2012; Rodriguez et al. 2013b; Gimenez et al. 2016; Jaramillo Ortiz et al. 2016).

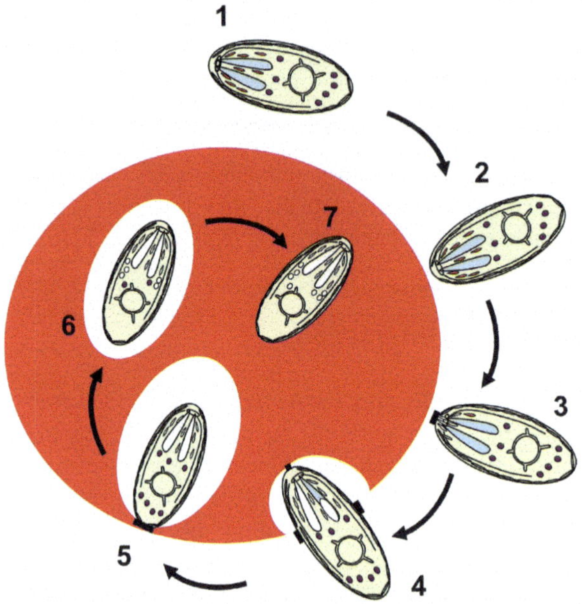

Fig. 9.4 *Babesia* sp. host cell invasion. The first contact between parasite and host red blood cells is established in random collisions (*1*), followed by the reorientation of the parasite apical end (*2*), leading to tight junction formation (*3*). This, in turn, leads to invagination of the erythrocyte cell membrane (*4*) and deeper engulfment of the merozoite inside a parasitophorous vacuole (*5*). The parasitophorous vacuole disintegrates soon after invasion (*6*), and the parasite membrane remains in direct contact with the erythrocyte cytoplasm during the rest of the intraerythrocytic life stage (*7*)

One of the host molecules thought to participate in the initial attachment of *Babesia* parasites is sialic acid present in erythrocyte glycolipids and/or glycoproteins, such as glycophorin A and B. Evidence in this direction comes from the observation that neuraminidase treatment of erythrocytes dramatically inhibits parasite invasion (Zintl et al. 2003; Gaffar et al. 2003; Lobo 2005; Cursino-Santos et al. 2014).

After initial recognition of the erythrocyte, the parasite reorients itself with its apical end perpendicular to the host membrane. Then, the erythrocyte membrane invaginates, and the merozoite is engulfed inside a parasitophorous vacuole (PV) (Lobo et al. 2012; Asada et al. 2012; Rodriguez et al. 2013a). Apical complex organelles—micronemes and rhoptries—participate in the invasion process through the sequential exocytosis of their contents at distinct stages (Gubbels and Duraisingh 2012). Microneme proteins (MICs) are secreted to the parasite surface in a calcium-dependent manner and, through interaction with host receptors, prompt the secretion of rhoptry proteins, which form complexes with host surface proteins. A dynamic interaction—known as moving junction—is established between these complexes and surface-exposed integral membrane MICs, which are connected to the actin-myosin engine of the pellicle. The moving junction starts at the parasite apical pole and proceeds toward the posterior end as the parasite invades the host cell, forming a PV (Besteiro et al. 2011).

Several MICs and rhoptry proteins have been described in *Babesia* spp. MICs include highly conserved proteins among apicomplexans, such as AMA-1, apical membrane antigen, and TRAP-1, thrombospondin-related adhesive protein, both of which are released to the extracellular medium, indicating a proteolytic cleavage after invasion (Gaffar et al. 2004a, b). This cleavage, necessary to allow the moving junction to proceed and to finally release the parasite into the PV, is likely carried out by *Babesia* rhomboid serine proteases, a family of transmembrane enzymes with their catalytic sites immersed in the cell membrane (Mesplet et al. 2011; Li et al. 2012). Among rhoptry proteins, the best characterized are the members of the rhoptry-associated protein-1 (RAP-1) family, which have been described in different *Babesia* species (Suarez et al. 1998, 2003; Niu et al. 2015, 2016a). Binding of *B. bovis* RAP-1 to the erythrocyte membrane appears essential for parasite invasion (Yokoyama et al. 2002). Other rhoptry proteins are believed to be involved in the modulation of protein function through phosphorylation, as well as in hydrolytic processes (Gubbels and Duraisingh 2012).

Different from what happens in *Toxoplasma* and *Plasmodium* zoites, *Babesia* PV disintegrates soon after invasion; thus the parasite membrane remains in direct contact with the erythrocyte cytoplasm at this life cycle stage (Asada et al. 2012).

Immediately after invasion and throughout the parasite life cycle, the content of spherical bodies is released. Spherical body proteins are localized in the erythrocyte cytoplasm or on the cytoplasmic side of the erythrocyte membrane and are believed to be responsible for host cell modifications (Dowling et al. 1996; Ruef et al. 2000; Blackman and Bannister 2001; Gubbels and Duraisingh 2012; Terkawi et al. 2011).

A special feature of *B. bovis* is that they cause infected erythrocytes to adhere to host endothelial cells, due to the appearance of protruding ridges on the erythrocyte surface. This results in the sequestration of infected erythrocytes inside capillaries. Parasites escape in this way the main circulation, avoiding destruction in the spleen. On the other hand, cytoadherence of infected erythrocytes to the bovine nervous microvasculature provokes ischemia and leads to the neurological signs often observed in acute *B. bovis* infections (Hutchings et al. 2007; Gohil et al. 2010). One of the proteins localized in the ridges and likely involved in cytoadherence is VESA1. It is encoded by a tightly regulated multigenic gene family of over 100 polymorphic genes. VESA1 undergoes antigenic variation during infection, which likely allows the parasite to escape the host immune response by expressing new variants without losing its cytoadherence properties (O'Connor and Allred 2000).

After *Babesia* merozoites have divided into two or four cells, they eventually lyse the erythrocyte and rapidly glide away to invade new ones (Asada et al. 2012). It is not thoroughly understood how lysis occurs, but in the case of *Plasmodium falciparum*, a papain-like cysteine protease—falcipain-2—has been implicated in erythrocyte egress, through the cleavage of erythrocytic cytoskeletal proteins. *P. falciparum* falcipain-2 homologues are present in *B. bovis*,

B. bigemina, and *B. ovis*—bovipain-2, babesipain, and ovipain-2, respectively. They are secreted to the erythrocyte cytoplasm, which suggests a similar function to falcipain-2 in the egress process, as well as their participation in the digestion of erythrocyte cytoplasmic proteins, such as hemoglobin (Mesplet et al. 2010; Martins et al. 2012; Carletti et al. 2016). These cysteine proteases are part of the parasite degradome, which in the case of *B. bovis* is composed of 66 cysteine, serine, aspartic, threonine, and metalloproteases. They are likely to fulfill essential roles in parasite nutrition and differentiation, as well as in invasion and egress mechanisms, and await further characterization (Mesplet et al. 2011).

In *Bos taurus*, the etiology of bovine babesiosis varies according to the infecting species, parasite strain, and the age of the animal. Clinical cases are typically observed in naïve adults, while animals younger than 9–10 months usually remain asymptomatic upon infection. Contrary to what was initially thought, this inverse age resistance is not based on the transfer of maternal colostral antibodies but rather on innate immune mechanisms (Zintl et al. 2005; Rodriguez et al. 2013a). The innate response starts with the non-specific recognition of pathogen-associated molecular patterns (PAMPs), such as *B. bovis* unmethylated CpG DNA, whole lipid extracts, or GPIs, by Toll-like receptors present on effector cells of the immune system, particularly in the spleen (Brown et al. 1999; Shoda et al. 2000). This stimulation leads to the secretion of cytokines, such as IFN-γ and TNF-α, which promote a type 1 cellular response with production of babesicidal oxygen reactive species. The importance of the spleen in this initial protection mechanism is highlighted by the observation that splenectomy provokes severe clinical disease in calves upon *B. bovis* experimental infection (Goff et al. 2010). *B. divergens*, on the other hand, seems to be resistant to oxidative radicals, and this mechanism may rather contribute to the observed immunopathological side effects of infection (Zintl et al. 2005). Innate and adaptive responses are linked via the natural killer cell-dependent activation, maturation, antigen-processing, and migration of dendritic cells to eventually present parasite antigens to CD4+ T cells. After immunization or secondary pathogen exposure, memory T cells are responsible for the immediate kick-off of protective cellular and humoral effector mechanisms, including production of IFN-γ and opsonizing IgG2 antibodies (Brown et al. 2006; Goff et al. 2010). The immune response mechanisms elicited by infections with other ruminant *Babesia* spp. remain to be elucidated.

9.2 Diagnostics and Epidemiology

9.2.1 Diagnostics

Numerous tests have been developed for diagnostic as well as epidemiological purposes. Bovine and ovine babesiosis can be confirmed by microscopic examination of blood or tissue fluids stained with Giemsa. However, this method is laborious, has low sensitivity, and does not differentiate between morphologically similar organisms (Maharana et al. 2016). Yet, parasite detection can be improved by using a fluorescent dye, such as acridine orange, and observation under an epifluorescence microscope (OIE 2014).

With the advent of molecular techniques, several direct, nested, and quantitative PCR assays have been described for the detection of ruminant *Babesia* spp. (Figueroa et al. 1993; Aktaş et al. 2005; Buling et al. 2007; Liu et al. 2007a, Liu et al. 2007b; Criado-Fornelio et al. 2009; He et al. 2011; Ramos et al. 2011; Horta et al. 2014; Romero-Salas et al. 2016; Erster et al. 2016). Nucleic acid-based assays are highly sensitive, can be designed to be specific for the pathogen of interest, and can be applied to both mammals and ticks (Maharana et al. 2016). Molecular diagnosis is a rapidly evolving technological field. As an example, qPCR has been recently applied in a high-throughput format to the screening of thousands of *Ixodes ricinus* ticks in Europe, simultaneously detecting 37 different tick-transmitted pathogens (Michelet et al. 2014).

Among molecular targets, the most commonly used is the 18S rRNA gene, because of its adequate species-related degree of polymorphism and repetitive arrangement within the genome (Lempereur et al. 2017). In addition, other genes have also proved to be adequate diagnostic targets, such as those encoding species-specific surface proteins, or the mitochondrial cytochrome b gene, which is also a member of a multigenic family.

In spite of their clear advantages, current PCR assays cannot be implemented in every diagnostic laboratory due to the need of specialized equipment (OIE 2014). As an attractive alternative, protocols based on LAMP—loop-mediated isothermal amplification—have been developed for the molecular diagnosis of babesiosis in cattle, buffalo, and small ruminants (Guan et al. 2008; He et al. 2009; Yang et al. 2016). LAMP shows similar, or in some cases higher, sensitivity than PCR and is cost-effective, simple, and fast, which makes it suitable for pen-side diagnostics. Amplification is carried out at constant temperature in a water bath, and results can be observed by the naked eye, greatly simplifying the processing of large amounts of samples (OIE 2014).

Reverse line blot hybridization (RLB) tests simultaneously detect DNA from different piroplasmids in blood or tick samples, with high sensitivity and specificity (Gubbels et al. 1999; Schnittger et al. 2004). The first step in this test consists in PCR amplification of a polymorphic 18S rRNA gene fragment, flanked by piroplasmid-conserved regions, where designed oligonucleotide primers will bind. In a second step, after transferring the amplification products to a nylon membrane, hybridization with species-specific oligonucleotide probes in a macroarray format allows the identification of the piroplasmid species present in each sample. Although technically elaborate, RLB has been applied to a great number of epidemiological studies in cattle, small ruminants, and ticks around the world (Hurtado 2015; Schötta et al. 2017).

Serological tests allow assessing vaccine efficacy, evaluating the immunological status of particular herds, and confirming diagnosis, and are highly useful epidemiology tools. A number of indirect and competitive enzyme-linked immunosorbent assays (ELISAs) have been developed for bovine and ovine babesiosis, using *B. bovis*, *B. bigemina*, *B. divergens*, and *B. ovis* soluble parasite extracts, culture supernatants, or recombinant proteins as antigens (Duzgun et al. 1991; Boonchit et al. 2004, 2006; Bono et al. 2008; Dominguez et al. 2012; Gabrielli et al. 2012; OIE 2014; Chung et al. 2017). ELISAs have a higher processing efficiency and objectivity in interpretation of results than the indirect fluorescent antibody test (IFAT). However, IFAT continues being the method of choice in numerous veterinary laboratories and is considered the gold standard for serological diagnosis (OIE 2014). Immunochromatographic tests (ICT) using recombinant *Babesia* sp. merozoite proteins display increased practical applicability in different settings and have been used in epidemiological surveys of bovine babesiosis (Guswanto et al. 2017).

9.2.2 Epidemiology

Ticks are becoming a growing problem worldwide due to climate changes that prolong their period of activity or alter their distribution. Domestic ruminant-infecting ticks provoke huge economic losses through reductions in productivity and fertility. These losses are further enhanced by mortality and morbidity caused by tick-transmitted pathogens. In addition, control measures, losses of potential production in tick-infested areas, and cattle trade restrictions add to the economic burden imposed by ticks and tick-borne pathogens on ruminant production around the world (Schnittger et al. 2012).

The global distribution of *Babesia* spp. strictly follows that of their tick vectors, which is mainly influenced by temperature and humidity (de Waal and Combrink 2006). *Babesia* species known to infect domestic ruminants as well as their tick vectors and distribution are shown in Table 9.1. Morphological descriptions exist for most of these parasites, while some have only been described based on the DNA sequence information of isolates found in different ruminants and/or ticks. Importantly, recent epidemiological studies followed by phylogenetic analysis have led to the discovery of new *Babesia* species that affect these animals, so it can be expected that this list of parasites will increase in the near future.

From an economic point of view, bovine babesiosis is the most important arthropod-transmitted pathogenic disease of cattle (de Waal and Combrink 2006). Its deleterious effects on livestock production in the USA prompted a 37-year tick eradication effort that finally allowed declaring the country free of the disease in 1943. Since then, around three billion dollars have been annually saved by the livestock industry. However, the occurrence of sporadic bovine babesiosis outbreaks in Southern US states indicates the need of permanent surveillance and the planning of long-term interventions (Peréz de León et al. 2010).

Most bovine babesiosis cases around the world are caused by *B. bovis* and *B. bigemina*, which are mainly transmitted by the tick vectors *Rhipicephalus microplus*—previously known as *Boophilus microplus*–*R. annulatus*, and *R. geigyi*. These ticks have been described in tropical and subtropical regions of the world, including

Africa, Asia, the Americas, and Australia (Gohil et al. 2013). Recently, however, the common Australian cattle tick has been redescribed as a new species, *R. australis*, instead of *R. microplus* (Estrada-Peña et al. 2012). *B. bigemina* can also be transmitted by *R. decoloratus* and *R. evertsi*, which further extends the distribution of this parasite into the African continent and makes it the most widespread bovine *Babesia* species (Rodriguez et al. 2013a).

Another important causative agent of bovine babesiosis is *B. divergens* that infects cattle in Europe—from Scandinavia to the Mediterranean Sea—and Northern Africa and whose only known vector is *Ixodes ricinus*. This tick only grows in moisture-saturated microhabitats and tolerates a large range of temperatures, which determines its distribution in various climatic regions (Zintl et al. 2003; de Waal and Combrink 2006). *B. divergens* infections of humans have also been reported, mostly corresponding to immunocompromised farmers or people vacationing in rural areas (Zintl et al. 2003).

In addition to the infection of European cattle (*Bos taurus*), which can result in high morbidity and mortality, *B. bovis* and *B. bigemina* reportedly cause subclinical infections in zebu cattle (*Bos indicus*) and water buffalo (*Bubalus bubalis*). Absent or attenuated clinical signs upon infection of these ruminants might have resulted from long coevolutionary adaptations with *Rhipicephalus* ticks and *Babesia* parasites (Ferreri et al. 2008; Florin-Christensen et al. 2014).

B. bovis and *B. bigemina* infections of deer have also been detected by molecular and serological methods (Cantu et al. 2007; Ramos et al. 2010; da Silveira et al. 2011; Holman et al. 2011). A role of deer as reservoir for these parasites would have important implications for the epidemiology and control of bovine babesiosis. However, since attempts to experimentally infect deer with *B. bovis* have so far been unsuccessful, their relevance in *Babesia* spp. transmission remains uncertain (Ueti et al. 2015).

Contrary to their resistance to develop clinical babesiosis upon *B. bovis* or *B. bigemina* infections, water buffaloes are highly susceptible to *B. orientalis*, which is transmitted by *R. haemaphysaloides* in Asia (Ferreri et al. 2008; Gohil et al. 2013; He et al. 2014).

Babesia infections of small ruminants—sheep and goats—have great economic importance in Southern Europe, Middle East, and some African and Asian countries (Schnittger et al. 2003; Ranjbar-Bahadori et al. 2012; Haghi et al. 2017; Zhou et al. 2017). The main infectious species include *B. ovis*, transmitted by *R. bursa* and *R. turanicus*, and *B. motasi*, transmitted by *Haemaphysalis* spp. ticks. In addition, *B. crassa* was described to infect small ruminants exclusively in the Middle East; however, it was also recently found by molecular methods in ticks in Hungary, suggesting that the distribution of this parasite is wider than initially expected (Schnittger et al. 2003; Hornok et al. 2015). In addition, a *Babesia* parasite transmitted by *H. longicornis* ticks has recently been reported to infect sheep in China and has been temporarily assigned the name *Babesia sp.* Xinjiang. *B. ovis* and *B. sp.* Xinjiang are highly pathogenic, while *B. motasi* shows moderate virulence (Uilenberg 2006; Schnittger et al. 2012; Carletti et al. 2016). Other *Babesia* isolates, such as *Babesia* sp. BQ1, have been recently detected in sheep in China and are considered to be novel species as determined by phylogenetic analysis (Niu et al. 2016a).

Babesiosis epidemiology relies on several parameters such as availability of adequate hosts, presence of tick vectors, existence of parasites within vectors, and host and environmental conditions. The ability of the pathogen to infect both vertebrate and invertebrate hosts is essential for the spread of disease, and this is determined by a great number of epidemiological factors (Fig. 9.5). In bovine babesiosis caused by *B. bovis* and *B. bigemina*, a state of enzootic or endemic stability corresponds to a situation where the relationship between host, parasite, vector, and environment remains in such a way that clinical disease rarely or not at all occurs, and thus no control measures are needed. Animals become naturally infected by tick bites at an early age, when normally no clinical signs are elicited, and develop into parasite carriers with strong acquired immunity and resistance to disease. However, endemic stability is a rare condition that can be easily broken by variations in climate, host genotypes, and management strategies (Mahoney 1974; Florin-Christensen et al. 2014).

Variability among *Babesia* populations is likely always high, as has been demonstrated for *B. bovis* and *B. bigemina* using micro- and minisatellite markers and multilocus sequence typing (Perez-Llaneza et al. 2010; Simuunza et al. 2011; Flores et al. 2013; Guillemi et al. 2013). Micro- and minisatellite analysis of different *B. bovis* isolates showed evidence of genetic recombination, association between genetic diversity and geographical location, and a highly structured population (Flores et al. 2013). Similar studies for other ruminant-infecting *Babesia* parasites are still pending.

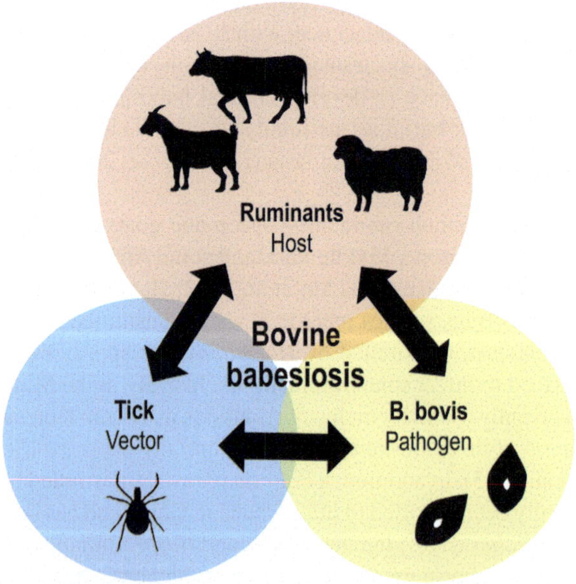

Fig. 9.5 Interrelation between *Babesia* pathogen, vertebrate host, and tick vector in bovine babesiosis epidemiology. The complex interactions between *Babesia* spp., their tick vectors, and their ruminant hosts need to be taken into account in the development of control strategies

9.3 Clinical Effects, Prevention, and Treatment

9.3.1 Clinical Effects

Clinical signs observed in bovine babesiosis depend on virulence and pathogenic effects of a particular *Babesia* species and host factors associated with disease, which include age, breed, and immune status. Among *Babesia* species, *B. bovis* is considered the most pathogenic, followed by *B. bigemina* and *B. divergens*. Other *Babesia* parasites of limited geographic distribution, such as *B. ovata*, have usually low pathogenicity; however, immunocompromised animals or those coinfected with *Theileria orientalis* can develop clinical disease (Sivakumar et al. 2016).

As mentioned in Sect. 9.1.3, clinical cases are typically observed in adults, while younger animals remain asymptomatic. Infections with *B. bovis* result in earlier more complicated and serious reactions, as compared to other bovine *Babesia* spp., associated with the induction of cytokines and other pharmacologically active agents (Brown et al. 1999). In addition, sequestration of *B. bovis*-infected erythrocytes in endothelial tissues can lead to cerebral babesiosis, respiratory distress, and multiorgan failure (Gohil et al. 2013). Bovine babesiosis caused by *B. bigemina* and *B. divergens* results in high parasitemias, and pathogenicity is mainly associated with massive erythrocyte destruction, leading to severe anemia, which may cause behavioral changes due to brain anoxia. Hemoglobinuria is observed at the peak of the hemolytic crisis in *B. bigemina* and *B. divergens* infections and is observed in the latter stages of babesiosis caused by *B. bovis*, hence the common name of *red water* for this disease. Fever is frequent in all three infections and can lead to abortions in pregnant cattle or temporary reduction of fertility in bulls. Other frequent signs are depression, anorexia, icterus, hemoglobinemia, and dramatic decreases in hematocrit values (Zintl et al. 2003; Florin-Christensen et al. 2014).

B. ovis infections of naïve sheep cause fever, hemolytic anemia, hemoglobinuria, and ictericia, while parasitemia is usually low (Habela et al. 1990). The disease is fatal if untreated, and outbreaks can produce significant mortality in sheep herds (Hurtado et al. 2015).

Buffalo babesiosis caused by *B. orientalis* is one of the most important diseases that affect these ruminants in China. Clinical signs include fever, anemia, icterus, hemoglobinuria, and frequent mortality (Liu et al. 2005).

9.3.2 Prevention

Vaccines consisting of live-attenuated strains of *B. bovis* and *B. bigemina* are produced in several countries either from the blood of infected donor animals or by in vitro culture. These vaccines are recommended in cases of enzootic instability or when animals are transported from a tick-free to a tick-infested area. Attenuated *B. bovis* parasites are derived by 20 to 30 serial and rapid passages of a virulent strain in splenectomized steers. Attenuation of *B. bigemina* is achieved by a similar procedure, but in this case slow successive passages among spleen-intact calves are used (Bock et al. 2004; de Waal and Combrink 2006; Florin-Christensen et al. 2014).

Attenuation of *B. divergens* was only possible by in vitro cultivation, but this methodology was not applied to vaccine formulation (Zintl et al. 2003). For this parasite, a live vaccine based on virulent field isolates was produced for a while in Ireland, either in splenectomized calves or intact gerbils. Treatment with babesicides was needed after vaccination to avoid appearance of clinical signs. However, this vaccine was discontinued over 10 years ago (Zintl et al. 2003; Bock et al. 2004).

The mechanisms behind attenuation of parasites through serial blood passages among steers remain unknown. It has been hypothesized that, during the process, parasites lose their capacity to express certain virulence-mediating genes and/or that a subpopulation of parasites with a mild pathogenicity phenotype, already present in the original pathogenic field isolate, is selected (Lau et al. 2011; Baravalle et al. 2012; Combrink et al. 2014). However, the attenuation scenario is likely more complex than just a selection procedure since, on one hand, an attenuated strain can be composed of virulent and avirulent subpopulations and, on the other hand, an avirulent clone can reverse its phenotype to a virulent one upon passage through a spleen-intact bovine. Future studies are needed to unravel which types of gene regulation mechanisms are involved in attenuation (Pedroni et al. 2013; Florin-Christensen et al. 2014).

Most *B. bovis* and *B. bigemina* live vaccines are produced in government-supported production facilities as a service to the livestock industries, in particular in Australia, Argentina, South Africa, and Israel. They are prepared as a bivalent formula, in which erythrocytes infected by each of these two parasites are mixed, or as a trivalent formula. The latter contains erythrocytes infected with *Anaplasma centrale* and provides protection against *A. marginale*, an intraerythrocytic tick-borne rickettsia with wide distribution in tropical and temperate regions (OIE 2014). Refrigerated and frozen forms of these vaccines exist. The former have limited shelf life, lasting 4–7 days at 2–8 °C; while when frozen in liquid nitrogen, vaccines can be stored for much longer periods of time. However, once thawed and reconstituted, they need to be immediately used, since parasites die within 8 h. Long-term storage of frozen vaccines allows postproduction control of each batch before commercialization. The risk of contamination with other microorganisms makes thorough quality control essential, although the high cost of these tests can hamper their routine application (de Waal and Combrink 2006; Florin-Christensen et al. 2014).

Live vaccines are usually safely administered to 4–10-month-old calves and only a transient clinical response to vaccination can sometimes take place. Older vaccinated animals, on the other hand, can develop clinical disease and should be held under surveillance and treated with a babesicide if adverse effects are observed. Normally, lifelong protective immunity develops in 3–4 weeks. Vaccine failures occasionally take place due to incorrect handling or storage, administration of chemotherapeutics immediately before or during vaccination, stress, concomitant infections, pathogenicity reversion of the attenuated vaccine strains, and/or changes in the parasite population that can lead to lack of protection (Bock and de Vos 2001; OIE 2014; de Waal and Combrink 2006). A further concern is associated with tick transmissibility. Argentine *B. bigemina* and *B. bovis* vaccine strains are non-transmissible, while the Dixie strain of *B. bovis* in the current Australian vaccine remains transmissible by ticks. This poses a potential threat if the vaccine is used in situations where the tick vector is present but *B. bovis* is not (Mangold et al. 1996). Thus, in spite of the general efficiency of current vaccines in preventing clinical cases of bovine babesiosis, there is considerable interest in the development of improved subunit vaccines that provide protection against disease and are safer and easier to handle and produce. Characterization of ruminant-infecting *Babesia* antigens as potential vaccine candidates is an active field of research that has been greatly facilitated by whole genome sequencing projects (Chap. 16). However, an important bottleneck in the development of subunit vaccines is represented by the costs and difficulties of experimental vaccination-challenge experiments in the natural ruminant hosts. Availability of an animal model—the gerbil—is so far restricted to bovine babesiosis caused by *B. divergens*. In this model, an experimental subunit vaccine based on a recombinant form of a *B. divergens* GPI-anchored surface protein provided significant protection upon challenge with a virulent parasite strain (Hadj-Kaddour et al. 2007). Until now, no subunit vaccine against any ruminant *Babesia* spp. has been successful to confer protective immunity in the natural hosts. Yet, given the copious amounts of knowledge that is rapidly accumulating about these and related parasites—including their pathogenicity mechanisms and molecules that articulate host-pathogen interactions—subunit vaccines might be an achievable goal in the not too distant future (Florin-Christensen et al. 2014).

Acaricides have long been applied by farmers around the world to reduce the deleterious impact of ticks and tick-borne diseases on health and productivity of domestic ruminants. In spite of their powerful action, chemical acaricides bear important drawbacks, such as contamination of the environment and animal products. In addition, the progressive emergence of ticks resistant to multiple acaricides calls for the development of new effective drugs, which is a lengthy and costly process (Florin-Christensen et al. 2014). Anti-tick vaccines offer a more affordable and environmental-friendly solution. Indeed, inoculation of cattle with a recombinant form of the concealed *R. microplus* tick antigen Bm86 produces partial but significant protection against ticks of the same species as well as against *R. annulatus* (Willadsen 2006). Other tick antigens and improved anti-tick vaccine formulations are actively sought for (de la Fuente 2012). Concerted measures that involve tick control and vaccination, as well as the use of resistant breeds of livestock, are likely to produce the most satisfactory results in the prevention of these diseases.

9.3.3 Treatment

Successful treatment of babesiosis relies on early diagnosis, followed by prompt administration of chemotherapeutic drugs. Chemotherapy of babesiosis is important not only to control field disease cases but also to control artificially induced infections. Currently, the only two anti-*Babesia* drugs available for treatment of babesiosis in ruminants are diminazene aceturate and imidocarb dipropionate (de Waal and Combrink 2006; Gohil et al. 2013). Imidocarb became the product of choice in those countries that licensed it, because in addition to its therapeutic utility, it also proved to be an effective prophylactic at twice the therapeutic dose. Besides, it is the only babesicide that consistently clears the host of parasites (Zintl et al. 2003; Mosqueda et al. 2012). However, the concern of drug residues in meat and dairy products following prolonged treatment has led to the withdrawal of its use in many European countries (de Waal and Combrink 2006; Mosqueda et al. 2012; Gohil et al. 2013). Several pharmacological compounds have been recently developed and evaluated, offering new options to control the disease. *Babesia* research is now well into a post-genomic era which will bring a new light on the development of new chemotherapy targets (Mosqueda et al. 2012; Rodriguez et al. 2013a).

References

Adl SM, Simpson AG, Farmer MA, Andersen RA, Anderson OR, Barta JR, Bowser SS, Brugerolle G, Fensome RA, Fredericq S, James TY, Karpov S, Kugrens P, Krug J, Lane CE, Lewis LA, Lodge J, Lynn DH, Mann DG, Mc Court RM, Mendoza L, Moestrup O, Mozley-Standridge SE, Nerad TA, Shearer CA, Smirnov AV, Spiegel FW, Taylor MF. The new higher level classification of eukaryotes with emphasis on the taxonomy of protists. J Eukaryot Microbiol. 2005;52:399–451.

Adl SM, Simpson AG, Lane CE, Lukeš J, Bass D, Bowser SS, Brown MW, Burki F, Dunthorn M, Hampl V, Heiss A, Hoppenrath M, Lara E, Le Gall L, Lynn DH, McManus H, Mitchell EA, Mozley-Stanridge SE, Parfrey LW, Pawlowski J, Rueckert S, Shadwick RS, Schoch CL, Smirnov A, Spiegel FW. The revised classification of eukaryotes. J Eukaryot Microbiol. 2012;59:429–93.

Aktaş M, Altay K, Dumanli N. Development of a polymerase chain reaction method for diagnosis of *Babesia ovis* infection in sheep and goats. Vet Parasitol. 2005;133(4):277–81.

Aktas M, Altay K, Ozubek S, Dumanli N. A survey of ixodid ticks feeding on cattle and prevalence of tick-borne pathogens in the Black Sea region of Turkey. Vet Parasitol. 2012;187(3–4): 567–71.

Alvarez J, Lopez U, Rojas C, Borgonio VM, Sanchez V, Castañeda R, Vargas P, Figueroa JV. Immunization of *Bos taurus* steers with *Babesia bov*is recombinant antigens MSA-1, MSA-2c and 12D3. Transbound Emerg Dis. 2010;57(1–2):87–90.

Alzan HF, Lau AO, Knowles DP, Herndon DR, Ueti MW, Scoles GA, Kappmeyer LS, Suarez CE. Expression of 6-Cys gene superfamily defines *Babesia bov*is sexual stage development within *Rhipicephalus microplus*. PLoS One. 2016;11(9):e0163791.

Asada M, Goto Y, Yahata K, Yokoyama N, Kawai S, Inoue N, Kaneko O, Kawazu S. Gliding motility of *Babesia bovis* merozoites visualized by time-lapse video microscopy. PLoS One. 2012;7(4):e35227.

Babes V. Sur l'hemoglobinurie bacterienne du boeuf. Comptesren dus hebdomadaires des seances de l'Academie des Sciences Paris. 1888;107:692–4.

Babes V. L'etiologie d'une enzootie des moutons, denommee Carceag en Roumanie. (the etiology of an enzootic disease of sheep, called Carceag, in Romania) (in French). CR Hebd Acad Sci. 1892;115:359–61.

Baravalle ME, Thompson C, Valentini B, Ferreira M, Torioni de Echaide S, Christensen MF, Echaide I. *Babesia bovis* biological clones and the inter-strain allelic diversity of the Bv80 gene support subpopulation selection as a mechanism involved in the attenuation of two virulent isolates. Vet Parasitol. 2012;190(3–4):391–400.

Bastos RG, Suarez CE, Laughery JM, Johnson WC, Ueti MW, Knowles DP. Differential expression of three members of the multidomain adhesion CCp family in *Babesia bigemina*, *Babesia bovis* and *Theileria equi*. PLoS One. 2013;8(7):e67765.

Besteiro S, Dubremetz JF, Lebrun M. The moving junction of apicomplexan parasites: a key structure for invasion. Cell Microbiology. 2011;13:797–805.

Blackman MJ, Bannister LH. Apical organelles of Apicomplexa: biology and isolation by subcellular fractionation. Mol Biochem Parasitol. 2001;117(1):11–25.

Blouin EF, van Rensburg L. An ultrastructural study of the development of Babesia occultans in the salivary glands of adult *Hyalomma marginatum* rufipes. Onderstepoort J Vet Res. 1988;55(2):93–100.

Bock RE, de Vos AJ. Immunity following use of Australian tick fever vaccine: a review of the evidence. Aust Vet J. 2001;79(12):832–9.

Bock R, Jackson L, de Vos A, Jorgensen W. Babesiosis of cattle. Parasitology. 2004;129(S):247–69.

Bono MF, Mangold AJ, Baravalle ME, Valentini BS, Thompson CS, Wilkowsky SE, Echaide IE, Farber MD, Torioni de Echaide SM. Efficiency of a recombinant MSA-2c-based ELISA to establish the persistence of antibodies in cattle vaccinated with Babesia bovis. Vet Parasitol. 2008;157(3–4):203–10.

Boonchit S, Xuan X, Yokoyama N, Goff WL, Waghela SD, Wagner G, Igarashi I. Improved enzyme-linked immunosorbent assay using C-terminal truncated recombinant antigens of *Babesia bovis* rhoptry-associated protein-1 for detection of specific antibodies. J Clin Microbiol. 2004;42(4):1601–4.

Boonchit S, Alhassan A, Chan B, Xuan X, Yokoyama N, Ooshiro M, Goff WL, Waghela SD, Wagner G, Igarashi I. Expression of C-terminal truncated and full-length Babesia bigemina rhoptry-associated protein 1 and their potential use in enzyme-linked immunosorbent assay. Vet Parasitol. 2006;137(1–2):28–35.

Brayton KA, Lau AO, Herndon DR, Hannick L, Kappmeyer LS, Berens SJ, Bidwell SL, Brown WC, Crabtree J, Fadrosh D, Feldblum T, Forberger HA, Haas BJ, Howell JM, Khouri H, Koo H, Mann DJ, Norimine J, Paulsen IT, Radune D, Ren Q, Smith RK Jr, Suarez CE, White O, Wortman JR, Knowles DP Jr, McElwain TF, Nene VM. Genome sequence of *Babesia bovis* and comparative analysis of apicomplexan hemoprotozoa. PLoS Pathog. 2007;3(10):1401–13.

Brown WC, Suarez CE, Shoda LK, Estes DM. Modulation of host immune responses by protozoal DNA. Vet Immunol Immunopathol. 1999;72(1–2):87–94.

Brown WC, Norimine J, Knowles DP, Goff WL. Immune control of *Babesia bovis* infection. Vet Parasitol. 2006;138:75–87.

Buling A, Criado-Fornelio A, Asenzo G, Benitez D, Barba-Carretero JC, Florin-Christensen M. A quantitative PCR assay for the detection and quantification of *Babesia bovis* and *B. bigemina*. Vet Parasitol. 2007;147(1–2):16–25.

Caballero MC, Pedroni MJ, Palmer GH, Suarez CE, Davitt C, Lau AO. Characterization of acyl carrier protein and LytB in Babesia bovis apicoplast. Mol Biochem Parasitol. 2012;181(2):125–33.

Cantu A, Ortega-S JA, Mosqueda J, Garcia-Vazquez Z, Henke SE, George JE. Immunologic and molecular identification of *Babesia bovis* and *Babesia bigemina* in free-ranging white-tailed deer in northern Mexico. J Wildl Dis. 2007;43(3):504–7.

Carletti T, Barreto C, Mesplet M, Mira A, Weir W, Shiels B, Oliva AG, Schnittger L, Florin-Christensen M. Characterization of a papain-like cysteine protease essential for the survival of *Babesia ovis* merozoites. Ticks Tick Borne Dis. 2016;7(1):85–93.

Chauvin A, Moreau E, Bonnet S, Plantard O, Malandrin L. Babesia and its hosts: adaptation to long-lasting interactions as a way to achieve efficient transmission. Vet Res. 2009;40(2):37.

Chung CJ, Suarez CE, Bandaranayaka-Mudiyanselage CL, Bandaranayaka-Mudiyanselage CB, Rzepka J, Heiniger TJ, Chung G, Lee SS, Adams E, Yun G, Waldron SJ. A novel modified-indirect ELISA based on spherical body protein 4 for detecting antibody during acute and long-term infections with diverse *Babesia bovis* strains. Parasit Vectors. 2017;10(1):77.

Combrink MP, Troskie PC, Pienaar R, Latif AA, Mans BJ. Genotypic diversity in Babesia bovis field isolates and vaccine strains from South Africa. Vet Parasitol. 2014;199(3–4):144–52.

Criado-Fornelio A, Gónzalez-del-Río MA, Buling-Saraña A, Barba-Carretero JC. The "expanding universe" of piroplasms. Vet Parasitol. 2004;119(4):337–45.

Criado-Fornelio A, Buling A, Asenzo G, Benitez D, Florin-Christensen M, Gonzalez-Oliva A, Henriques G, Silva M, Alongi A, Agnone A, Torina A, Madruga CR. Development of fluorogenic probe-based PCR assays for the detection and quantification of bovine piroplasmids. Vet Parasitol. 2009;162(3–4):200–6.

Cursino-Santos JR, Halverson G, Rodriguez M, Narla M, Lobo CA. Identification of binding domains on red blood cell glycophorins for *Babesia divergens*. Transfusion. 2014;54(4):982–9.

da Silveira JA, Rabelo EM, Ribeiro MF. Detection of Theileria and *Babesia* in brown brocket deer (Mazamagoua zoubira) and marsh deer (Blastocerus Dichotomus) in the state of Minas Gerais, Brazil. Vet Parasitol. 2011;177(1–2):61–6.

de la Fuente J. Vaccines for vector control: exciting possibilities for the future. Vet Journal. 2012;194:139–40.

de Waal DT, Combrink MP. Live vaccines against bovine babesiosis. Vet Parasitol. 2006;138(1–2):88–96.

Decaro N, Larocca V, Parisi A, Losurdo M, Lia RP, Greco MF, Miccolis A, Ventrella G, Otranto D, Buonavoglia C. Clinical bovine piroplasmosis caused by *Babesia occultans* in Italy. J Clin Microbiol. 2013;51(7):2432–4.

Delbecq S, Auguin D, Yang YS, Löhr F, Arold S, Schetters T, Précigout E, Gorenflot A, Roumestand C. The solution structure of the adhesion protein Bd37 from *Babesia divergens* reveals structural homology with eukaryotic proteins involved in membrane trafficking. J Mol Biol. 2008;375(2):409–24.

Dominguez M, Echaide I, Echaide ST, Mosqueda J, Cetrá B, Suarez CE, Florin-Christensen M. In silico predicted conserved B-cell epitopes in the merozoite surface antigen-2 family of *B. bovis* are neutralization sensitive. Vet Parasitol. 2010;167(2–4):216–26.

Dominguez M, Echaide I, de Echaide ST, Wilkowsky S, Zabal O, Mosqueda JJ, Schnittger L, Florin-Christensen M. Validation and field evaluation of a competitive enzyme-linked immunosorbent assay for diagnosis of *Babesia bovis* infections in Argentina. Clin Vaccine Immunol. 2012;19(6):924–8.

Dowling SC, Perryman LE, Jasmer DP. A *Babesia bovis* 225-kilodalton spherical-body protein: localization to the cytoplasmic face of infected erythrocytes after merozoite invasion. Infect Immun. 1996;64(7):2618–26.

Duzgun A, Wright IG, Waltisbuhl DJ, Gale KR, Goodger BV, Dargie JD, Alabay M, Cerci H. An ELISA for the diagnosis of *Babesia ovis* infection utilizing a synthetic, *Babesia bovis*-derived antigen. Vet Parasitol. 1991;39(3–4):225–31.

Erster O, Roth A, Wollkomirsky R, Leibovich B, Savitzky I, Zamir S, Molad T, Shkap V. Quantitative analysis of *Babesia ovis* infection in sheep and ticks. Vet Parasitol. 2016;221:39–45.

Estrada-Peña A, Venzal JM, Nava S, Mangold A, Guglielmone AA, Labruna MB, de la Fuente J. Reinstatement of *Rhipicephalus* (*Boophilus*) *australis* (Acari: Ixodidae) with redescription of the adult and larval stages. J Med Entomol. 2012;49(4):794–802.

Ferreri L, Benitez D, Dominguez M, Rodriguez A, Asenzo G, Mesplet M, Florin-Christensen M, Schnittger L. Water buffalos as carriers of *Babesia bovis* in Argentina. Ann N Y Acad Sci. 2008;1149:149–51.

Figueroa JV, Chieves LP, Johnson GS, Buening GM. Multiplex polymerase chain reaction based assay for the detection of *Babesia bigemina*, *Babesia bovis* and *Anaplasma marginale* DNA in bovine blood. Vet Parasitol. 1993;50(1–2):69–81.

Flores DA, Minichiello Y, Araujo FR, Shkap V, Benítez D, Echaide I, Rolls P, Mosqueda J, Pacheco GM, Petterson M, Florin-Christensen M, Schnittger L. Evidence for extensive genetic diversity and substructuring of the *Babesia bovis* metapopulation. Transbound Emerg Dis. 2013;60(Suppl 2):131–6.

Florin-Christensen M, Schnittger L. Piroplasmids and ticks: a long-lasting intimate relationship. Front Biosci. 2009;14:3064–73.

Florin-Christensen M, Suarez CE, Hines SA, Palmer GH, Brown WC, McElwain TF. The *Babesia bovis* merozoite surface antigen 2 locus contains four tandemly arranged and expressed genes encoding immunologically distinct proteins. Infect Immun. 2002;70(7):3566–75.

Florin-Christensen M, Suarez CE, Rodriguez AE, Flores DA, Schnittger L. Vaccines against bovine babesiosis: where we are now and possible roads ahead. Parasitology. 2014;141:1563–92.

Gabrielli S, Galuppi R, Marcer F, Marini C, Tampieri MP, Moretti A, Pietrobelli M, Cancrini G. Development of culture-based serological assays to diagnose *Babesia divergens* infections. Vector Borne Zoonotic Dis. 2012;12(2):106–10.

Gaffar FR, Franssen FF, de Vries E. Babesia bovis merozoites invade human, ovine, equine, porcine and caprine erythrocytes by a sialic acid-dependent mechanism followed by developmental arrest after a single round of cell fission. Int J Parasitol. 2003;33(14):1595–603.

Gaffar FR, Yatsuda AP, Franssen FF, de Vries E. Erythrocyte invasion by Babesia bovis merozoites is inhibited by polyclonal antisera directed against peptides derived from a homologue of *Plasmodium falciparum* apical membrane antigen 1. Infect Immun. 2004a;72: 2947–295.

Gaffar FR, Yatsuda AP, Franssen FF, de Vries E. A Babesia bovis merozoite protein with a domain architecture highly similar to the thrombospondin-related anonymous protein (TRAP) present in *Plasmodium* sporozoites. Mol Biochem Parasitol. 2004b;136:25–34.

Gimenez AM, Françoso KS, Ersching J, Icimoto MY, Oliveira V, Rodriguez AE, Schnittger L, Florin-Christensen M, Rodrigues MM, Soares IS. A recombinant multi-antigen vaccine formulation containing *Babesia bovis* merozoite surface antigens MSA-2a(1), MSA-2b and MSA-2c elicits invasion-inhibitory antibodies and IFN-γ producing cells. Parasit Vectors. 2016;9(1):577.

Goff WL, Bastos RG, Brown WC, Johnson WC, Schneider DA. The bovine spleen: interactions among splenic cell populations in the innate immunologic control of hemoparasitic infections. Vet Immunol Immunopathol. 2010;138(1–2):1–14.

Gohil S, Kats LM, Sturm A, Cooke BM. Recent insights into alteration of red blood cells by *Babesia bovis*: moovin' forward. Trends Parasitol. 2010;26(12):591–9.

Gohil S, Herrmann S, Günther S, Cooke BM. Bovine babesiosis in the 21st century: advances in biology and functional genomics. Int J Parasitol. 2013;43(2):125–32.

Guan G, Chauvin A, Luo J, Inoue N, Moreau E, Liu Z, Gao J, Thekisoe OM, Ma M, Liu A, Dang Z, Liu J, Ren Q, Jin Y, Sugimoto C, Yin H. The development and evaluation of a loop-mediated isothermal amplification (LAMP) method for detection of Babesia spp. infective to sheep and goats in China. Exp Parasitol. 2008;120(1):39–44.

Gubbels MJ, Duraisingh MT. Evolution of apicomplexan secretory organelles. Int J Parasitol. 2012;42(12):1071–81.

Gubbels JM, de Vos AP, van der Weide M, Viseras J, Schouls LM, de Vries E, Jongejan F. Simultaneous detection of bovine *Theileria* and *Babesia* species by reverse line blot hybridization. J Clin Microbiol. 1999;37(6):1782–9.

Guillemi E, Ruybal P, Lia V, González S, Farber M, Wilkowsky SE. Multi-locus typing scheme for *Babesia bovis* and *Babesia bigemina* reveals high levels of genetic variability in strains from northern Argentina. Infect Genet Evol. 2013;14:214–22.

Guswanto A, Allamanda P, Mariamah ES, Munkjargal T, Tuvshintulga B, Takemae H, Sivakumar T, AbouLaila M, Terkawi MA, Ichikawa-Seki M, Nishikawa Y, Yokoyama N, Igarashi I. Evaluation of immunochromatographic test (ICT) strips for the serological detection of *Babesia bovis* and *Babesia bigemina* infection in cattle from western java, Indonesia. Vet Parasitol. 2017;239:76–9.

Habela M, Reina D, Nieto C, Navarrete IK. Antibody response and duration of latent infection in sheep following experimental infection with *Babesia ovis*. Vet Parasitol. 1990;35:1–10.

Hadj-Kaddour K, Carcy B, Vallet A, Randazzo S, Delbecq S, Kleuskens J, Schetters T, Gorenflot A, Precigout E. Recombinant protein Bd37 protected gerbils against heterologous challenges with isolates of *Babesia divergens* polymorphic for the bd37 gene. Parasitology. 2007;134(Pt 2):187–96.

Haghi MM, Etemadifar F, Fakhar M, Teshnizi SH, Soosaraei M, Shokri A, Hajihasani A, Mashhadi H. Status of babesiosis among domestic herbivores in Iran: a systematic review and meta-analysis. Parasitol Res. 2017;116(4):1101–9.

He L, Zhou YQ, Oosthuizen MC, Zhao JL. Loop-mediated isothermal amplification (LAMP) detection of *Babesia orientalis* in water buffalo (*Bubalus babalis*, Linnaeus, 1758) in China. Vet Parasitol. 2009;165(1–2):36–40.

He L, Feng HH, Zhang QL, Zhang WJ, Khan MK, Hu M, Zhou YQ, Zhao JL. Development and evaluation of real-time PCR assay for the detection of *Babesia orientalis* in water buffalo (Bubalus Bubalis, Linnaeus, 1758). J Parasitol. 2011;97(6):1166–9.

He L, Feng HH, Zhang WJ, Zhang QL, Fang R, Wang LX, Tu P, Zhou YQ, Zhao JL, Oosthuizen MC. Occurrence of *Theileria* and *Babesia* species in water buffalo (Bubalus babalis, Linnaeus, 1758) in the Hubei province, South China. Vet Parasitol. 2012;186(3–4):490–6.

He L, Zhang Y, Zhang QL, Zhang WJ, Feng HH, Khan MK, Hu M, Zhou YQ, Zhao JL. Mitochondrial genome of *Babesia orientalis*, apicomplexan parasite of water buffalo (Bubalus Bubalis, Linnaeus, 1758) endemic in China. Parasit Vectors. 2014;7:82.

Holman PJ, Carroll JE, Pugh R, Davis DS. Molecular detection of Babesia bovis and *Babesia bigemina* in white-tailed deer (*Odocoileus virginianus*) from tom Green County in central Texas. Vet Parasitol. 2011;177(3–4):298–304.

Homer MJ, Aguilar-Delfin I, Telford SR III, Krause PJ, Persing DH. Babesiosis. Clin Microbiol Rev. 2000;13:451–69.

Hornok S, Takács N, Kontschán J, György Z, Micsutka A, Iceton S, Flaisz B, Farkas R, Hofmann-Lehmann R. Diversity of *Haemaphysalis*-associated piroplasms of ruminants in Central-Eastern Europe, Hungary. Parasit Vectors. 2015;8:627.

Horta S, Barreto MC, Pepe A, Campos J, Oliva A. Highly sensitive method for diagnosis of subclinical *Babesia ovis* infection. Ticks Tick Borne Dis. 2014;5(6):902–6.

Huang Y, He L, Hu J, He P, He J, Yu L, Malobi N, Zhou Y, Shen B, Zhao J. Characterization and annotation of *Babesia orientalis* apicoplast genome. Parasit Vec. 2015;8:543.

Hurtado A. Reverse line blot hybridization with species-specific oligonucleotide probes: application to piroplasm detection. Methods Mol Biol. 2015;1247:183–92.

Hurtado A, Barandika JF, Oporto B, Minguijón E, Povedano I, García-Pérez AL. Risks of suffering tick-borne diseases in sheep translocated to a tick infested area: a laboratory approach for the investigation of an outbreak. Ticks Tick Borne Dis. 2015;6(1):31–7.

Hutchings CL, Li A, Fernandez KM, Fletcher T, Jackson LA, Molloy JB, Jorgensen WK, Lim CT, Cooke BM. New insights into the altered adhesive and mechanical properties of red blood cells parasitized by *Babesia bovis*. Mol Microbiol. 2007;65(4):1092–110.

Jaramillo Ortiz JM, Molinari MP, Gravisaco MJ, Paoletta MS, Montenegro VN, Wilkowsky SE. Evaluation of different heterologous prime-boost immunization strategies against *Babesia bovis* using viral vectored and protein-adjuvant vaccines based on a chimeric multi-antigen. Vaccine. 2016;34(33):3913–9.

Kakoma I, Mehlhorn H. Babesia of domestic ruminants. In: Kreier JP, Baker JR, editors. Parasitic protozoa. 2nd ed. New York: Academic; 1994. p. 141–216.

Kim CM, Blanco LB, Alhassan A, Iseki H, Yokoyama N, Xuan X, Igarashi I. Development of a rapid immunochromatographic test for simultaneous serodiagnosis of bovine babesioses caused by *Babesia bovis* and *Babesia bigemina*. Am J Trop Med Hyg. 2008;78(1):117–21.

Klinger CM, Nisbet RE, Ouologuem DT, Roos DS, Dacks JB. Cryptic organelle homology in apicomplexan parasites: insights from evolutionary cell biology. Curr Opin Microbiol. 2013;16(4):424–31.

Koch R. Beitrage zur Entwicklungsgeschichte der Piroplasmen. Z Hyg Infekt-Kr. 1906;54:1–9.

Lau AO, Kalyanaraman A, Echaide I, Palmer GH, Bock R, Pedroni MJ, Rameshkumar M, Ferreira MB, Fletcher TI, McElwain TF. Attenuation of virulence in an apicomplexan hemoparasite results in reduced genome diversity at the population level. BMC Genomics. 2011;12:410.

Lempereur L, Beck R, Fonseca I, Marques C, Duarte A, Santos M, Zúquete S, Gomes J, Walder G, Domingos A, Antunes S, Baneth G, Silaghi C, Holman P, Zintl A. Guidelines for the detection of *Babesia* and *Theileria* parasites. Vector Borne Zoonotic Dis. 2017;17(1):51–65.

Lew AE, Dluzewski AR, Johnson AM, Pinder JC. Myosins of *Babesia bovis*: molecular characterisation, erythrocyte invasion, and phylogeny. Cell Motil Cytoskeleton. 2002;52:202–20.

Li H, Child MA, Bogyo M. Proteases as regulators of pathogenesis: examples from the Apicomplexa. Biochim Biophys Acta. 2012;1824(1):177–85.

Liu Q, Zhao JL, Zhou YQ, Liu EY, Yao BA, Fu Y. Study on some molecular characterization of *Babesia orientalis*. Vet Parasitol. 2005;130(3–4):191–8.

Liu Q, Zhou YQ, Zhou DN, Liu EY, Du K, Chen SG, Yao BA, Zhao JL. Semi-nested PCR detection of *Babesia orientalis* in its natural hosts *Rhipicephalus haemaphysaloides* and buffalo. Vet Parasitol. 2007a;143(3–4):260.

Liu AH, Yin H, Guan GQ, Schnittger L, Liu ZJ, Ma ML, Dang ZS, Liu JL, Ren QY, Bai Q, Ahmed JS, Luo JX. At least two genetically distinct large *Babesia* species infective to sheep and goats in China. Vet Parasitol. 2007b;147(3–4):246–51.

Lobo CA. *Babesia divergens* and *Plasmodium falciparum* use common receptors glycophorins a and B, to invade the human red blood cell. Infect Immun. 2005;73(1):649–51.

Lobo CA, Rodriguez M, Cursino-Santos JR. *Babesia* and red cell invasion. Curr Opin Hematol. 2012;19(3):170–5.

Maharana BR, Tewari AK, Saravanan BC, Sudhakar NR. Important hemoprotozoan diseases of livestock: challenges in current diagnostics and therapeutics: an update. Vet World. 2016;9(5):487–95.

Mahoney DF. The application of epizootiological principals in the control of babesiosis in cattle. Bull Off Int Epizoot. 1974;81:123–38.

Mangold AJ, Vanzini VR, Echaide IE, de Eschaide ST, Volpogni MM, Guglielmone AA. Viability after thawing and dilution of simultaneously cryopreserved vaccinal *Babesia bovis* and *Babesia bigemina* strains cultured in vitro. Vet Parasitol. 1996;61:345–8.

Martins TM, do Rosário VE, Domingos A. Expression and characterization of the *Babesia bigemina* cysteine protease BbiCPL1. Acta Trop. 2012;121(1):1–5.

Mesplet M, Echaide I, Dominguez M, Mosqueda JJ, Suarez CE, Schnittger L, Florin-Christensen M. Bovipain-2, the falcipain-2 ortholog, is expressed in intraerythrocytic stages of the tick-transmitted hemoparasite *Babesia bovis*. Parasit Vectors. 2010;3:113.

Mesplet M, Palmer GH, Pedroni MJ, Echaide I, Florin-Christensen M, Schnittger L, Lau AO. Genome-wide analysis of peptidase content and expression in a virulent and attenuated *Babesia bovis* strain pair. Mol Biochem Parasitol. 2011;179(2):111–3.

Michelet L, Delannoy S, Devillers E, Umhang G, Aspan A, Juremalm M, Chirico J, van der Wal FJ, Sprong H, Boye Pihl TP, Klitgaard K, Bødker R, Fach P, Moutailler S. High-throughput screening of tick-borne pathogens in Europe. Front Cell Infect Microbiol. 2014;4:103.

Mihalca AD. The quest for piroplasms: from babes to smith to molecules. Sci Parasitol. 2010;11:14–9.

Mosqueda J, McElwain TF, Palmer GH. *Babesia bovis* merozoite surface antigen 2 proteins are expressed on the merozoite and sporozoite surface, and specific antibodies inhibit attachment and invasion of erythrocytes. Infect Immun. 2002;70(11):6448–55.

Mosqueda J, Olvera-Ramirez A, Aguilar-Tipacamu G, Canto GJ. Current advances in detection and treatment of babesiosis. Curr Med Chem. 2012;19(10):1504–18.

Niu Q, Marchand J, Yang C, Bonsergent C, Guan G, Yin H, Malandrin L. Rhoptry-associated protein (rap-1) genes in the sheep pathogen *Babesia* sp. Xinjiang: multiple transcribed copies differing by 3′ end repeated sequences. Vet Parasitol. 2015;211(3–4):158–69.

Niu Q, Liu Z, Yang J, Yu P, Pan Y, Zhai B, Luo J, Moreau E, Guan G, Yin H. Expression analysis and biological characterization of *Babesia* sp. BQ1 (Lintan) (Babesia motasi-like) rhoptry-associated protein 1 and its potential use in serodiagnosis via ELISA. Parasit Vectors. 2016a;9(1):313.

Niu Q, Liu Z, Yang J, Yu P, Pan Y, Zhai B, Luo J, Yin H. Genetic diversity and molecular characterization of *Babesia motasi*-like in small ruminants and ixodid ticks from China. Infect Genet Evol. 2016b;41:8–15.

Niu Q, Liu Z, Yang J, Gao S, Pan Y, Guan G, Luo J, Yin H. Genetic characterization and molecular survey of *Babesia* sp. Xinjiang infection in small ruminants and ixodid ticks in China. Infect Genet Evol. 2017;49:330–5.

O'Connor RM, Allred DR. Selection of *Babesia bovis* infected erythrocytes for adhesion to endothelial cells co-selects for altered variant erythrocyte surface antigen isoforms. J Immunol. 2000;164:2037–45.

OIE World Organization for Animal Health Terrestrial Manual. Bovine Babesiosis. Chapter 2.4.2; 2014. http://www.oie.int/fileadmin/Home/eng/Health_standards/tahm/2.04.02_BOVINE_BABESIOSIS.pdf.

Pedroni MJ, Sondgeroth KS, Gallego-Lopez GM, Echaide I, Lau AO. Comparative transcriptome analysis of geographically distinct virulent and attenuated *Babesia bovis* strains reveals similar gene expression changes through attenuation. BMC Genomics. 2013;14:763.

Pérez de León AA, Strickman DA, Knowles DP, Fish D, Thacker E, de la Fuente J, Krause PJ, Wikel SK, Miller RS, Wagner GG, Almazán C, Hillman R, Messenger MT, Ugstad PO, Duhaime RA, Teel PD, Ortega-Santos A, Hewitt DG, Bowers EJ, Bent SJ, Cochran MH, TF ME, Scoles GA, Suarez CE, Davey R, Howell Freeman JM, Lohmeyer K, Li AY, Guerrero FD, Kammlah DM, Phillips P, Pound JM, Group for Emerging Babesioses and One Health Research and Development in the U.S. One health approach to identify research needs in bovine and human babesioses: workshop report. Parasit Vectors. 2010;3(1):36.

Perez-Llaneza A, Caballero M, Baravalle E, Mesplet M, Mosqueda J, Suarez CE, Echaide I, Katzer F, Pacheco GM, Florin-Christensen M, Schnittger L. Development of a tandem repeat-based multilocus typing system distinguishing *Babesia bovis* geographic isolates. Vet Parasitol. 2010;167(2–4):196–204.

Ramos CM, Cooper SM, Holman PJ. Molecular and serologic evidence for *Babesia bovis*-like parasites in white-tailed deer (*Odocoileus virginianus*) in south Texas. Vet Parasitol. 2010;172(3–4):214–20.

Ramos CA, Araújo FR, Souza II, Bacanelli G, Luiz HL, Russi LS, Oliveira RH, Soares CO, Rosinha GM, Alves LC. Real-time polymerase chain reaction based on msa2c gene for detection of *Babesia bovis*. Vet Parasitol. 2011;176(1):79–83.

Ranjbar-Bahadori S, Eckert B, Omidian Z, Shirazi NS, Shayan P. *Babesia ovis* as the main causative agent of sheep babesiosis in Iran. Parasitol Res. 2012;110(4):1531–6.

Ribeiro MFB, Patarroyo JHS. Ultrastructure of *Babesia bigemina* gametes obtained in "in vitro" erythrocyte cultures. Vet Parasitol. 1998;76:19–25.

Rodriguez AE, Couto A, Echaide I, Schnittger L, Florin-Christensen M. *Babesia bovis* contains an abundant parasite-specific protein-free glycerophosphatidylinositol and the genes predicted for its assembly. Vet Parasitol. 2010;167(2–4):227–35.

Rodriguez AE, Schnittger L, Tomazic ML, Florin-Christensen M. Current and prospective tools for the control of cattle-infecting *Babesia* parasites. In: Castillo V, Harris R, editors. Protozoa: biology, classification and role in disease. Hauppauge, NY: Nova; 2013a. p. 1–44.

Rodriguez AE, Zamorano P, Wilkowsky S, Torrá F, Ferreri L, Dominguez M, Florin-Christensen M. Delivery of recombinant vaccines against bovine herpes virus type 1 gD and *Babesia bovis* MSA-2c to mice using liposomes derived from egg yolk lipids. Vet J. 2013b;196(3):550–1.

Rodriguez AE, Florin-Christensen M, Flores DA, Echaide I, Suarez CE, Schnittger L. The glycosylphosphatidylinositol-anchored protein repertoire of *Babesia bovis* and its significance for erythrocyte invasion. Ticks Tick Borne Dis. 2014;5(3):343–8.

Romero-Salas D, Mira A, Mosqueda J, García-Vázquez Z, Hidalgo-Ruiz M, Vela NA, de León AA, Florin-Christensen M, Schnittger L. Molecular and serological detection of *Babesia bovis*- and *Babesia bigemina*-infection in bovines and water buffaloes raised jointly in an endemic field. Vet Parasitol. 2016;217:101–7.

Ruef BJ, Dowling SC, Conley PG, Perryman LE, Brown WC, Jasmer DP, Rice-Ficht AC. A unique *Babesia bovis* spherical body protein is conserved among geographic isolates and localizes to the infected erythrocyte membrane. Mol Biochem Parasitol. 2000;105(1):1–12.

Schnittger L, Yin H, Gubbels MJ, Beyer D, Niemann S, Jongejan F, Ahmed JS. Phylogeny of sheep and goat *Theileria* and *Babesia* parasites. Parasitology Res. 2003;91:398–406.

Schnittger L, Yin H, Qi B, Gubbels MJ, Beyer D, Niemann S, Jongejan F, Ahmed JS. Simultaneous detection and differentiation of *Theileria* and *Babesia* parasites infecting small ruminants by reverse line blotting. Parasitol Res. 2004;92(3):189–96.

Schnittger L, Rodriguez AE, Florin-Christensen M, Morrison DA. Babesia: a world emerging. Infect Genet Evol. 2012;12(8):1788–809.

Schötta AM, Wijnveld M, Stockinger H, Stanek G. Reverse line blot-based detection approaches of microbial pathogens in *Ixodes ricinus* ticks collected in Austria and impact of the chosen method. Appl Environ Microbiol. 2017; pii: AEM.00489-17.

Shoda LK, Palmer GH, Florin-Christensen J, Florin-Christensen M, Godson DL, Brown WC. *Babesia bovis*-stimulated macrophages express interleukin-1beta, interleukin-12, tumor necrosis factor alpha, and nitric oxide and inhibit parasite replication in vitro. Infect Immun. 2000;68:5139–45.

Simuunza M, Bilgic H, Karagenc T, Syakalima M, Shiels B, Tait A, Weir W. Population genetic analysis and sub-structuring in *Babesia bovis*. Mol Biochem Parasitol. 2011;177(2):106–15.

Sivakumar T, Igarashi I, Yokoyama N. *Babesia ovata*: taxonomy, phylogeny and epidemiology. Vet Parasitol. 2016;229:99–106.

Smith T, Kilborne FL. Investigations into the nature, causation and prevention of southern cattle fever. In: Ninth annual report of the Bureau of Animal Industry for the year 1892. Washington, DC: Government Printing Office; 1893. p. 177–304.

Suarez CE, Palmer GH, Hötzel I, McElwain TF. Structure, sequence, and transcriptional analysis of the *Babesia bovis* rap-1 multigene locus. Mol Biochem Parasitol. 1998;93(2):215–24.

Suarez CE, Florin-Christensen M, Hines SA, Palmer GH, Brown WC, McElwain TF. Characterization of allelic variation in the *Babesia bovis* merozoite surface antigen 1 (MSA-1) locus and identification of a cross-reactive inhibition-sensitive MSA-1 epitope. Infect Immun. 2000;68(12):6865–70.

Suarez CE, Palmer GH, Florin-Christensen M, Hines SA, Hötzel I, McElwain TF, Organization, transcription, and expression of rhoptry associated protein genes in the *Babesia bigemina* rap-1 locus. Mol Biochem Parasitol. 2003;127(2):101–12.

Terkawi MA, Seuseu FJ, Eko-Wibowo P, Huyen NX, Minoda Y, Abou Laila M, Kawai S, Yokoyama N, Xuan X, Igarashi I. Secretion of a new spherical body protein of *Babesia bovis* into the cytoplasm of infected erythrocytes. Mol Biochem Parasitol. 2011;178(1–2):40–5.

Ueti MW, Olafson PU, Freeman JM, Johnson WC, Scoles GA. A virulent *Babesia bovis* strain failed to infect white-tailed deer (*Odocoileus virginianus*). PLoS One. 2015;10(6):e0131018.

Uilenberg G. Babesia—a historical overview. Vet Parasitol. 2006;138:3–10.

Weber G, Friedhoff KT. Preliminary observations on the ultrastructure of supposed sexual stages of *Babesia bigemina* (Piroplasmea). Z Parasitenkund. 1977;53:83–92.

Weerasooriya G, Sivakumar T, Lan DT, Long PT, Takemae H, Igarashi I, Inoue N, Yokoyama N. Epidemiology of bovine hemoprotozoa parasites in cattle and water buffalo in Vietnam. J Vet Med Sci. 2016;78(8):1361–7.

Wilkowsky SE, Farber M, Echaide I, Torioni de Echaide S, Zamorano PI, Dominguez M, Suarez CE, Florin-Christensen M. *Babesia bovis* merozoite surface protein-2c (MSA-2c) contains highly immunogenic, conserved B-cell epitopes that elicit neutralization-sensitive antibodies in cattle. Mol Biochem Parasitol. 2003;127(2):133–41.

Willadsen P. Tick control: thoughts on a research agenda. Vet Parasitol. 2006;138:161–8.

Yang Y, Li Q, Wang S, Chen X, Du A. Rapid and sensitive detection of *Babesia bovis* and *Babesia bigemina* by loop-mediated isothermal amplification combined with a lateral flow dipstick. Vet Parasitol. 2016;219:71–6.

Yokoyama N, Suthisak B, Hirata H, Matsuo T, Inoue N, Sugimoto C, Igarashi I. Cellular localization of *Babesia bovis* merozoite rhoptry-associated protein 1 andits erythrocyte-binding activity. Infect Immun. 2002;70(10):5822–6.

Yokoyama N, Okamura M, Igarashi I. Erythrocyte invasion by *Babesia* parasites: current advances in the elucidation of the molecular interactions between the protozoan ligands and host receptors in the invasion stage. Vet Parasitol. 2006;138(1–2):22–32.

Zhou M, Cao S, Sevinc F, Sevinc M, Ceylan O, Ekici S, Jirapattharasate C, Moumouni PF, Liu M, Wang G, Iguchi A, Vudriko P, Suzuki H, Xuan X. Molecular detection and genetic characterization of *Babesia*, *Theileria* and *Anaplasma* amongst apparently healthy sheep and goats in the central region of Turkey. Ticks Tick Borne Dis. 2017;8(2):246–52.

Zintl A, Mulcahy G, Skerrett HE, Taylor SM, Gray JS. *Babesia divergens*, a bovine blood parasite of veterinary and zoonotic importance. Clin Microbiol Rev. 2003;16(4):622–36.

Zintl A, Gray JS, Skerrett HE, Mulcahy G. Possible mechanisms underlying age-related resistance to bovine babesiosis. Parasite Immunol. 2005;27:115–20.

Babesia of Domestic Dogs

10

Gad Baneth

Abstract

Canine babesiosis caused by different *Babesia* species is a protozoal tick-borne disease with worldwide distribution and global significance. Historically, *Babesia* infection in dogs was identified based on the morphologic appearance of the parasite in the erythrocyte. All large forms of *Babesia* were designated *B. canis*, whereas all small forms of *Babesia* were considered to be *B. gibsoni*. However, the development of molecular methods has demonstrated that additional *Babesia* species infect dogs and cause distinct diseases. The geographical distribution of canine *Babesia* species and thus the occurrence of babesiosis are largely dependent on the habitat of relevant tick vector species, with the exception of *B. gibsoni* where evidence for dog-to-dog transmission indicates that infection can be transmitted among fighting dog breeds independently of the limitations of vector tick infestation. Knowledge of the prevalence and clinicopathological aspects of *Babesia* species infecting dogs around the world is of epidemiological and medical interest. *Babesia* infection causes a disease with clinical manifestations that may vary considerably with the different species and strains involved and with factors that determine the host response to infection such as age, individual immune status, and the presence of concurrent infections or other diseases. Hemolytic anemia with systemic inflammatory responses may lead to tissue hypoxia and organ dysfunction, which account for the clinical signs observed in severe canine babesiosis. Babesiosis caused by large *Babesia* species is treated with imidocarb dipropionate or diminazene aceturate, while small *Babesia* species are more resistant to anti-babesial therapy and often require treatment with combinations of other drugs such as atovaquone, azithromycin, and clindamycin. Accurate detection and species recognition are important for the selection of the correct therapy and predicting the course of disease.

G. Baneth
Koret School of Veterinary Medicine, Hebrew University of Jerusalem, Rehovot, Israel
e-mail: gad.baneth@mail.huji.ac.il

M. Florin-Christensen, L. Schnittger (eds.), *Parasitic Protozoa of Farm Animals and Pets*, https://doi.org/10.1007/978-3-319-70132-5_10

10.1 Morphology, Life Cycle, and Host-Pathogen Interactions

10.1.1 Morphology

Babesia are tick-borne protozoan parasites that belong to the phylum Apicomplexa, class Piroplasmea, and order Piroplasmida and infect erythrocytes of domestic and wild animals and humans. *Babesia* belong to the Aconoidasida as they lack a conoid structure in their apical complex in all of their life stages except for the ookinete stage—in contrast to apicomplexans of the Conoidasida which have a conoid in all life stages (Mehlhorn et al. 1980). The babesial species that infect dogs are divided into those that present with relatively large merozoite forms in erythrocytes (5 × 2 μm) termed *large* canine *Babesia* species and those species which have distinctly smaller merozoite stages (0.3 × 3 μm) termed *small* canine *Babesia* species (Table 10.1). Historically, *Babesia* infection in dogs was identified based on the morphologic appearance of the parasite in the erythrocyte, and all large forms of *Babesia* were designated *B. canis*, whereas all small forms of *Babesia* were considered to be

Table 10.1 Distribution, vectors, and size of the main *Babesia* species that infect dogs

Species	Geographical distribution	Potential or confirmed vectors	Size (μm)	Typical form of parasites in blood
B. rossi	Southern Africa, Nigeria, Sudan	*Haemaphysalis elliptica* *Haemaphysalis leachi*	2 × 5 (large forms)	Two or more pyriform merozoites
B. canis	Europe	*Dermacentor reticulatus*	2 × 5 (large forms)	Two pyriform merozoites
B. vogeli	Africa, Asia, Europe, north, central and South America, Australia	*R. sanguineus* s.l.	2.5 × 4.5 (large forms)	Double or single pyriform merozoites
Babesia sp. (Coco)	Eastern United States	Unknown	2 × 6 (large forms)	Ameboid forms, two pyriform merozoites
B. gibsoni	Southeast Asia, United States, Australia, Europe	*H. longicornis* *H. bispinosa*? *R. sanguineus*?	1 × 3 (small forms)	Frequently individual forms
B. conradae	United States (California)	*R. sanguineus* s.l.?	0.3 × 3 (small forms)	Annular, in tetrads, amoeboid forms
B. vulpes (*Babesia microti*-like; *Theileria annae*)	Europe, North America	*Dermacentor reticulatus*? *Ixodes hexagonus*? *I. ricinus*? *I. canisuga*? *R. sanguineus* s.l.?	1 × 2.5 (small forms)	Frequently individual forms

B. gibsoni. However, the development of molecular methods has demonstrated that other genetically distinct *Babesia* species exist. This division into large and small species of canine *Babesia* is not supported by phylogenetic studies. The large *Babesia* species *Babesia canis*, *Babesia vogeli*, *Babesia rossi*, and *Babesia* sp. (Coco) as well as the small *Babesia* species *B. gibsoni* have been placed into a single monophyletic group, which is sometimes for convenience referred to as *true Babesia* or *Babesia* sensu stricto and corresponds to Clade V as defined in Schnittger et al. (2012). In contrast, other small *Babesia* species, *Babesia vulpes* and *Babesia conradae*, are sometimes for convenience referred to as *Babesia* sensu lato as they place into two different unrelated monophyletic groups—*Babesia vulpes* into Clade I as defined by Schnittger et al. (2012) and Baneth et al. (2015) and *Babesia conradae* into Clade II (Schnittger et al. 2012). Lack et al. (2012) and Schreeg et al. (2016) have confirmed this phylogenetic classification. Thus, altogether seven canine *Babesia* species are currently known to infect canines of which one—*Babesia* sp. (Coco)—has not yet been named.

The morphology of the different life stages of canine babesial parasites has been described in studies using light microscopy and electron microscopy. Most of these studies focused on *Babesia canis*—i.e., before this taxon has been later divided into the large species *Babesia vogeli*, *Babesia rossi*, and *Babesia canis* as known today—and on *Babesia gibsoni*, which represents a small *Babesia* species. The morphology of the red blood cell stages has been reported in more detail than the morphology of the tick stages of some canine *Babesia* species (Kjemtrup et al. 2006).

The life cycle of *Babesia* species in canines generally includes a trophozoite stage, which develops immediately after infection of the erythrocyte, and a merozoite stage, which is the result of the asexual division of the trophozoite and the formation of two pyriform bodies that may further divide within the erythrocyte. Parasites taken up during the blood meal of the feeding vector tick develop in the tick intestine into the gamete stage—termed *ray body* because of its thin projections. Zygotes formed by fusion of two *ray body* gametes develop into sporokinetes which are motile and infect the salivary glands of the tick to form the infective sporozoite stage or—in *Babesia* sensu stricto with transovarial transmission—also infect the ovaries of the tick and are transmitted via the eggs transovarially (Mehlhorn et al. 1994).

Intraerythrocytic *B. canis* merozoites have an outer pellicle layer consisting of three membranes, have no conoid, but possess apical and posterior polar rings, mitochondria, rhoptries, micronemes, subpellicular microtubules, and a membrane-bound nucleus (Mehlhorn and Schein 1984). *Babesia canis* merozoites maintain their structure and size also when grown in culture (Walter et al. 2002). In contrast, merozoites of small *Babesia* species such as *B. gibsoni* and *B. conradae* may take on variable forms that consist of four to six shapes (Walter et al. 2002; Radi et al. 2004; Kjemtrup et al. 2006). *B. gibsoni* may develop in culture differently than in the host, and its morphology as seen in the blood of infected dogs changes into larger merozoites that almost fill the entire erythrocyte (Walter et al. 2002).

The *ray bodies* of *B. canis* initially develop within erythrocytes in the tick gut and are spherical, polymorphic, or pyramidal with diameters of about 4–7 μm. They are characterized by short thornlike projections, which measure about 1.0–1.2 μm in length (Mehlhorn and Schein 1984). More developed *ray bodies* leave the decaying erythrocytes and possess several raylike projections before they fuse with another *ray body* to develop the zygote and sporokinete.

The sporozoites of *B. canis* measure about 2.5 μm in length. They have a broad apical pole and a pointed posterior pole and are bound by a pellicle composed of an outer cell membrane and an inner layer consisting of two membranes. They contain microtubules and several rhoptries that attach to the apical complex, which discharges proteolytic enzymes that enable the invasion of the parasite into host erythrocytes upon their transmission by the tick saliva (Mehlhorn and Schein 1984).

10.1.2 Life Cycle

10.1.2.1 Transmission by the Vector Tick

Dogs are infected following a tick bite when *Babesia* sporozoites are injected with saliva into their skin during the blood meal. In the canine host, parasites attach to erythrocyte membranes and invade the cell where they form ring-shaped or ameboid trophozoites. Merogony with asexual replication takes place when the parasite replicates by binary fission within the erythrocyte and forms merozoites observed as pairs of attached pear-shaped parasites—also termed pyriform bodies—in some *Babesia* species (Fig. 10.1). Merozoites may further divide forming eight or more parasites in

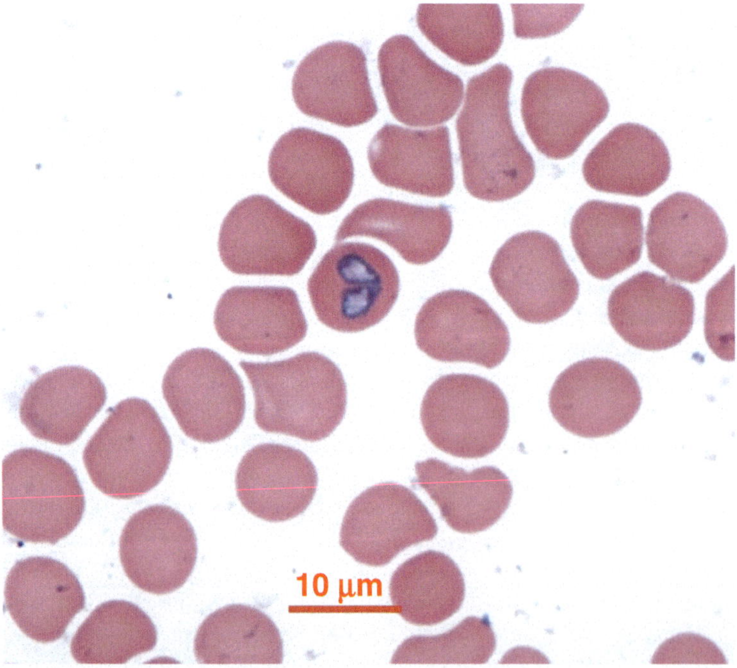

Fig. 10.1 *Babesia vogeli* merozoites in a canine erythrocyte from the blood of a naturally infected dog; blood smear stained by May Grunwald Giemsa

the same host erythrocyte, and eventually the cell disintegrates, and merozoites are released in the blood to invade new cells. Some merozoites develop to spheroid forms in erythrocytes considered as gamonts (Mehlhorn and Schein 1984).

Natural transmission of *Babesia* species usually occurs during an infected tick bite; however, congenital transplacental infection with passage of the parasite through the placenta has been documented in some species but is considered rarer than transmission by tick bite (Fukumoto et al. 2005; Konishi et al. 2008).

The sexual part of the babesial life cycle takes place in the tick. Ticks feeding on infected blood take up intraerythrocytic parasites, and while most merozoites are destroyed in the tick, spheroid forms differentiate further in the tick gut into gametes with discrete projections—*ray bodies*—which associate in couples and fuse to form zygotes. Zygotes develop into motile sporokinetes, which cross the tick gut into the hemocoele and migrate to the salivary glands where the infective sporozoites are produced by sporogony (Chauvin et al. 2009). Some species of *Babesia* including *B. canis* have been shown to be infective to animals only up to 2–3 days after tick attachment (Schein et al. 1979; Mehlhorn and Schein 1984). It is presumed that the change in temperature or the presence of a blood meal in the tick gut acts as an activation stimulus for the maturation of the infective sporozoites.

The life cycle of *Babesia* sp. in the mammalian host takes place exclusively in erythrocytes, whereas *Theileria* sp. have a preerythrocytic life stage in leukocytes—the schizont (Chauvin et al. 2009; Uilenberg 2006). Some *Babesia* species—*Babesia* sensu lato—are transmitted transstadially from one developmental tick stage to another after feeding on the vertebrate host, whereas other *Babesia* species, *Babesia* sensu stricto, are transmitted transovarial through the tick eggs and may therefore be passed through to the next generation without having to feed on an infected host (Uilenberg 2006; Chauvin et al. 2009; Schnittger et al. 2012). Furthermore, all *Theileria* sensu stricto are transmitted exclusively transstadially.

10.1.2.2 Transmission by Blood Transfusion or Dog-to-Dog Bite

Canine *Babesia* infection can also be transmitted via transfusion of blood products from an infected blood donor to a recipient dog. This has been shown for several canine babesial species including *B. gibsoni* and *B. canis* (Freeman et al. 1994; Stegeman et al. 2003). *Babesia* species are among the most important pathogens listed for testing in the blood of canine blood donors by the American college of Veterinary Internal Medicine and the Association of Veterinary Hematology and Transfusion Medicine (Wardrop et al. 2005, 2016). In addition to transmission by blood transfusion, studies have provided evidence that *B. gibsoni* is likely transmitted directly from dog to dog by bites. This has been especially observed for fighting dog breeds such as the Pit Bull Terrier and the Tosa (Birkenheuer et al. 2005; Jefferies et al. 2007; Lee et al. 2009; Yeagley et al. 2009).

10.1.3 Host-Pathogen Interactions

Babesia infection causes a disease with clinical manifestations that may vary considerably between the different species and strains involved as they may have different degrees of virulence. Important for the course of infection are also factors that determine the host immune response to the infection such as age, individual immune status, and the presence of concurrent infections or other diseases (Irwin 2009). Hemolytic anemia with massive erythrocyte destruction and a systemic inflammatory response, which may lead to multiple-organ dysfunction syndrome, account for most of the clinical signs observed in canine babesiosis. The disease onset is often acute with affected dogs suffering from fever and lethargy.

Acute canine babesiosis is characterized by the induction and increased activity of a multitude of inflammatory mediators including cytokines, chemokines, and acute phase proteins. The interaction between these inflammatory mediators, the infective organism, and the canine host have been studied in both experimental and natural infection settings with different species of *Babesia* including *B. canis*, *B. gibsoni*, and *B. rossi* (Zygner et al. 2014; Brown et al. 2015; Goddard et al. 2016). In some situations, these interactions may be beneficial to the host, but in others, they may prove detrimental. A study on dogs naturally infected with *B. canis* from Poland found increases of serum tumor necrosis factor alpha (TNF-α) concentration during canine babesiosis, which were associated with renal failure. The authors concluded that the increased TNF-α concentration influenced the development of hypotension and renal failure probably via TNF-α-mediated production of nitric oxide and induction of vasodilation and hypotension, leading to renal ischemia and hypoxia (Zygner et al. 2014). Another study that evaluated an experimental *B. gibsoni* infection of beagles found that it was associated with marked increases in the concentration of several cytokines such as TNF-α, granulocyte-macrophage colony-stimulating factor (GM-CSF), IL-2, IL-6, IL7, and IL-18, which showed a delayed onset that followed the infection and occurred subsequent to the detection of peripheral parasitemia. Increases in the levels of the acute phase protein c-reactive protein (CRP) occurred a few days prior to the detection of parasitemia (Brown et al. 2015). A study on *B. rossi* from South Africa investigated cytokines in naturally infected dogs and evaluated whether their levels were associated with disease outcome. Ninety-seven dogs naturally infected with *B. rossi* were studied, and 15 healthy dogs were included as controls. IL-10 and monocyte chemotactic protein 1 (MCP-1) concentrations were significantly elevated for the *Babesia*-infected dogs compared to the healthy controls. In contrast, the IL-8 concentration was significantly decreased in the *Babesia*-infected dogs. Concentrations of IL-6 and MCP-1 were increased in the non-survivor dogs compared to the survivors. Concentrations of IL-2, IL-6, IL-18, and GM-CSF were significantly higher in those cases that presented during the more acute stage of the disease. The result demonstrates a mixed cytokine response in *B. rossi* infection, and the authors suggested that an excessive pro-inflammatory response might result in a poor outcome (Goddard et al. 2016).

Hemolytic anemia in canine babesiosis can be intravascular—within blood vessels—or extravascular, in parenchymal organs of which the spleen plays a major role. It often presents with a combination of erythrocyte destruction in both locations. Hemolytic anemia in canine babesiosis is multifactorial and associated with several mechanisms. It can occur due to direct red blood cell lyses produced by replicating intracellular parasites or due to the binding of antibodies to the erythrocyte cell membrane displaying parasite antigens leading to complement activation (Adachi et al. 1994, 1995; Carli et al. 2009). Furthermore, hemolytic anemia can be caused by the production of serum hemolytic factors, the oxidative damage of erythrocytes, an increased red blood cell phagocytosis, the creation of spherocytes, and a decrease in the osmotic fragility of red blood cells (Onishi et al. 1990; Makinde and Bobade 1994; Murase et al. 1996; Otsuka et al. 2001, 2002). Antibodies against red blood cells have been documented in dogs infected with *B. gibsoni* and *B. vogeli* (Adachi et al. 1994; Carli et al. 2009). Intense hemolysis results in hemoglobinemia, hemoglobinuria, bilirubinemia, and bilirubinuria.

Thrombocytopenia is frequently found in canine babesiosis and may be caused by immune-mediated mechanisms such as antibodies coating platelets, splenic sequestration, or coagulatory consumption of platelets from hemolytic or vascular injury. Thrombocytopenia has been demonstrated in experimental canine babesiosis caused by *B. gibsoni* as well as in natural infections with other *Babesia* species (Wilkerson et al. 2001; Sikorski et al. 2010; Brown et al. 2015; Eichenberger et al. 2016).

Particular virulence factors encoded by *B. rossi* have been identified and associated with the high pathogenic phenotype of this species. A polymorphic phosphoprotein localized on the cytoplasmic surface of *B. rossi*-infected red blood cells that has been named *Babesia rossi* erythrocyte membrane antigen 1 (Br*EMA1*) is suspected to be a virulence factor in *B. rossi* canine babesiosis (Matjila et al. 2009). The gene that encodes this protein is not found in other species of *Babesia* infecting dogs such as *B. canis* and *B. vogeli*. A preliminary study suggests that there are also clinically important differences in the virulence between various *B. rossi* genotypes/strains (Matjila et al. 2009). Different proportions of the prevalence of genotypes of the BC28 gene—encoding the major merozoite surface antigens of *B. canis*—have been found in different regions of Europe. These differences may be related to variable biologic properties of parasite strains within the same *Babesia* species (Carcy et al. 2015).

Tissue hypoxia is a severe consequence of canine babesiosis associated with multiple-organ damage. It has been studied in depth in *B. rossi* and *B. canis* infections (Leisewitz et al. 2001; Jacobson 2006; Mathe et al. 2007). The causes of hypoxia include anemia, hypotensive shock, vascular stasis by sludging of erythrocytes, excessive endogenous production of carbon monoxide, parasitic damage to hemoglobin, and decreased ability of hemoglobin to offload oxygen in *Babesia*-infected dogs (Ayoob et al. 2010). The central nervous system, kidney, and muscle are the organs most affected by the resultant tissue hypoxia (Jacobson 2006). The histological changes observed in the kidneys in naturally acquired *B. canis* infections included vacuolar-hydropic degeneration, necrosis, and detachment of renal tubular epithelial cells in the proximal convoluted tubules, while no significant histological changes were seen in the glomeruli (Mathe et al. 2007). Tubular hemoglobin casts and hemoglobin droplets in the renal tubular epithelial cells were more rarely observed (Mathe et al. 2007).

Tissue hypoxia, hypotensive shock, multiple-organ dysfunction, and high mortality rates have been documented mostly in association with *B. rossi* infection (Jacobson 2006; Reyers et al. 1998). Infection with this species may present acutely or even as a peracute and fatal syndrome with massive hemolysis, renal failure, and acid-base abnormalities (Leisewitz et al. 2001; Jacobson 2006). Free oxygen radical release and mechanisms associated with harmful cytokine effects have been associated with endothelial damage and increased vascular permeability in canine babesiosis. These may result in non-cardiogenic pulmonary edema (Jacobson 2006). *B. rossi* infection of dogs can lead to cerebral babesiosis and mortality due to a more severe consumptive coagulopathy compared to that found in dogs surviving clinical disease (Jacobson 2006; Goddard et al. 2013).

The spleen has an important function in controlling babesiosis (Homer et al. 2000). Splenectomized and immune-compromised dogs are more susceptible to infection with *Babesia* spp. Splenectomized dogs that are experimentally infected with *Babesia* spp. rapidly develop parasitemia and clinical disease and may reach high parasitemia levels (Vercammen et al. 1995). Accordingly, splenectomy is an important risk factor for the development of natural and potentially fatal babesiosis in humans and has been documented to be associated with clinical natural canine babesiosis (Rosner et al. 1984; Camacho et al. 2002; Sikorski et al. 2010).

10.2 Diagnosis and Epidemiology

10.2.1 Diagnosis

The diagnosis of canine babesiosis and detection of *Babesia* infection are carried out by several diagnostic techniques ranging from simple microscopy of blood that can be carried out in field conditions to advanced and very sensitive molecular techniques (Lempereur et al. 2017). Although observation of stained blood smear can usually distinguish between infection with large or small canine *Babesia* species, the morphology of parasites in the blood as observed by light microscopy is not sufficient in separating between the different large canine babesial species and between the varieties of small species. Another consideration when selecting an appropriate diagnostic method is that subclinical infection will be often associated with extremely low parasitemia, and therefore molecular techniques for the detection of parasite DNA should be preferred over the relatively less sensitive technique of blood smear microscopy. Canine blood donors or blood transfusion products should be tested by sensitive PCR detection protocols as parasitemia, if present, is expected to be low in apparently healthy donors (Solano-Gallego and Baneth 2011).

10.2.1.1 Detection by Light Microscopy

Observation of large or small species of *Babesia* in stained blood smears has been the standard for diagnosis for many years. This method is deemed reliable when a moderate to high parasitemia is present; however, there is not always a correlation between the level of parasitemia and the severity of clinical signs. Furthermore, the diagnosis of chronically infected and carrier dogs remains a diagnostic challenge due to low and often intermittent parasitemia that is frequently difficult to observe by microscopic evaluation. In those cases, the use of molecular diagnostic assays is

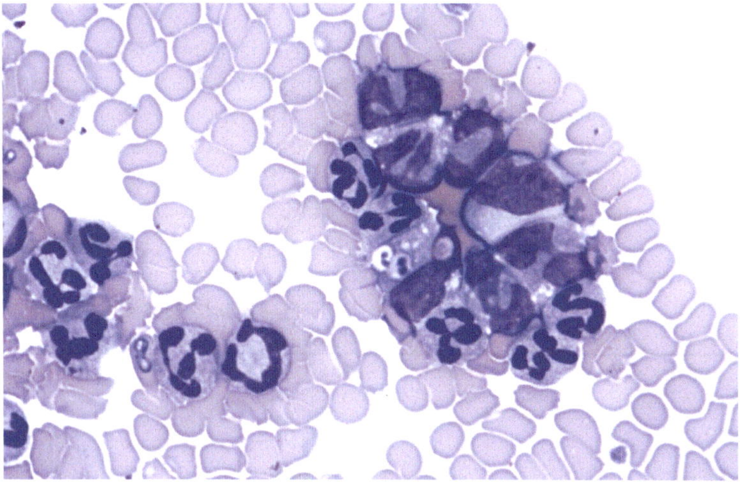

Fig. 10.2 *Babesia vogeli* merozoites in dog erythrocytes phagocytosed by a monocyte in the blood of a naturally infected dog; note also merozoites in a free erythrocyte; blood smear stained by May Grunwald Giemsa

strongly recommended. Smears made from ear tip or toe nail capillary blood may be beneficial in exhibiting large form *Babesia* parasites versus blood from a central vein (Bohm et al. 2006). A fresh smear is recommended for the accurate diagnosis of infection. Parasites may be detected in phagocytosed erythrocytes within macrophages (Fig. 10.2). Distinguishing between small and between large canine *Babesia* species based solely on morphology is not possible, and molecular analysis is required for speciation.

10.2.1.2 PCR-Based Molecular Detection

PCR is a sensitive and specific diagnostic technique frequently used for the detection of canine babesiosis and particularly useful for low parasitemia levels and for the determination of parasite species. A large number of PCR assays and protocols using a variety of gene targets have been described. Several PCR assays and additional procedures have been developed and used for the detection of canine babesiosis. Real-time PCR techniques have been developed to detect and quantify *Babesia* infection in canine blood (Qurollo et al. 2017). PCR-restriction fragment length polymorphism is also used to separate between canine *Babesia* species (Carret et al. 1999). A reverse line blotting (RLB) technique in which PCR products are hybridized to a membrane containing specific probes for the known babesial species and possibly for other pathogens has been developed for simultaneous detection of piroplasm species and coinfections. The RLB confirmed the presence of *B. vogeli* in addition to *B. rossi* in dogs from South Africa (Matjila et al. 2004). In addition, a high-resolution melting curve quantitative fluorescence resonance energy transfer-PCR has been developed to discriminate between species based on melting curve analysis (Wang et al. 2010).

10.2.1.3 Detection by Serology

Measurement of antibodies reactive with *Babesia* antigen indicates a past or present infection. The indirect fluorescent antibody test (IFAT) is the most commonly used test for canine babesiosis (Vercammen et al. 1995); however, cross-reactivity between different *Babesia* species and with other protozoan parasites occurs (Vercammen et al. 1995; Yamane et al. 1993). Enzyme-linked immunosorbent assays (ELISA) have been used in research and epidemiologic surveys (Schetters et al. 1996). The use of recombinant proteins such as the thrombospondin-related adhesive protein (TRAP) of *B. gibsoni* has been employed as an alternative for whole-parasite antigen with good sensitivities and specificities (Goo et al. 2008). False-negative results are possible in peracute or acute infection. In these cases, the use of convalescent antibody titers is strongly recommended to confirm acute infection.

10.2.2 Epidemiology

The geographical distribution of the causative agents and thus the occurrence of babesiosis are largely dependent on the habitat of their tick vector species, with the exception of *B. gibsoni*, where evidence for dog-to-dog transmission between fighting dog breeds independent of the vector tick has been presented (Birkenheuer et al. 2005; Jefferies et al. 2007; Yeagley et al. 2009). *Babesia vogeli* and *B. gibsoni* have a worldwide distribution, whereas *B. rossi* and *B. canis* have been mostly restricted to Africa and Europe, respectively. The unnamed large *Babesia* species most closely related to *B. bigemina* and *B. conradae* have been reported only from North America, whereas *B. vulpes* has been reported in Europe, Asia, and North America (Solano-Gallego and Baneth 2011).

10.2.2.1 *Babesia rossi*

Babesia rossi—(*Babesia canis rossi*) is a large form *Babesia* species, which has been described from South Africa as well as from other sub-Saharan African countries including Sudan and Nigeria (Adamu et al. 2014; Oyamada et al. 2005). Its tick vectors are *Haemaphysalis elliptica* and *Haemaphysalis leachi*, which were thought to be the same species in the past (Apanaskevich et al. 2007). It is considered the most virulent large *Babesia* species that affects dogs.

10.2.2.2 *Babesia canis*

Babesia canis—(*B. canis canis*) is a large *Babesia* mostly prevalent in central and northern Europe and transmitted by the tick *Dermacentor reticulatus*. It causes a moderate to severe disease that often has an acute onset (Zygner et al. 2014; Eichenberger et al. 2016).

10.2.2.3 *Babesia vogeli*

Babesia vogeli—(*B. canis vogeli*) is a large *Babesia* with a very wide distribution. It is transmitted by *Rhipicephalus sanguineus* sensu lato ticks and found mostly in tropical and subtropical regions including the Mediterranean basin, the Middle East, large areas of Asia, Australia, and South, Central, and North America (Solano-Gallego and Baneth 2011). It can cause subclinical infection or mild to moderate disease. It may cause a severe illness in young puppies and in greyhounds. *Babesia vogeli*-infected dogs often present with coinfections, frequently with *Ehrlichia canis* or *Hepatozoon canis* which are also transmitted by the same tick vector (Singla et al. 2016).

10.2.2.4 *Babesia gibsoni*

Babesia gibsoni is a small form *Babesia* that is endemic in Southeast Asia and appears to have spread from there to other continents including North and South America, Australia, and Europe. It is a common and often subclinical cause of infection in pit bull terriers but can also inflict a severe disease in this as well as in other dog breeds. It is transmitted by *Haemaphysalis longicornis* and possibly by *H. bispinosa* and *R. sanguineus* s.l. There is also evidence for its direct transmission by dog bites (Birkenheuer et al. 2005; Jefferies et al. 2007; Yeagley et al. 2009).

10.2.2.5 *Babesia conradae*

Babesia conradae is a small *Babesia*, which has mostly been reported in dogs from California and appears not to be prevalent outside North America. It was initially thought to be a strain of *B. gibsoni* but later found to be a genetically distinct species causing a severe and potentially fatal disease in dogs. The tick vector of *B. conradae* has not been described to date (Kjemtrup et al. 2006).

10.2.2.6 *Babesia vulpes*

Babesia vulpes described initially as *Theileria annae* and also termed *Babesia microti*-like and *Babesia* cf. *microti* is a third small species of *Babesia* that infects dogs (Zahler et al. 2000; Baneth et al. 2015). It was initially described from a dog that originated from Northern Spain and has subsequently been found in other European countries and in North America (Yeagley et al. 2009; Miró et al. 2015). *Babesia vulpes* is a common cause of infection of wild red foxes—i.e., *Vulpes vulpes*. Although it has been speculated that it is transmitted by the Hedgehog tick *Ixodes hexagonus* and/or by *Dermacentor reticulatus*, there is currently no definitive evidence of its transmission by any of these or by other tick species (Camacho et al. 2003; Hodžić et al. 2017).

10.2.2.7 *Babesia* sp. (Coco)

A currently unnamed large form *Babesia* species was detected for the first time in North Carolina in a dog under chemotherapy for lymphoma (Birkenheuer et al. 2004). Seven dogs infected with this pathogen were subsequently reported in eastern United States. All the dogs presented with immunocompromised conditions such as splenectomy or chemotherapy due to neoplasia (Sikorski et al. 2010). Analyses of the 18S rRNA gene of the unnamed large *Babesia* have revealed a unique sequence that shared a 93.9% identity with *B. bigemina* (Birkenheuer et al. 2004).

10.2.2.8 *Rangelia vitalii*

Rangelia vitalii is a piroplasm described in dogs in the southeast of Brazil, Uruguay, and Northern Argentina. This infection causes a disease referred to as *Nambiuvú*—i.e., bloody ears. The life cycle of *R. vitalii* is different from *Babesia* species because it has a tissue stage in the cytoplasm of endothelial cells as well as a developmental phase in blood cells. Piroplasm-like intracellular organisms of *R. vitalii* have been described in erythrocytes, monocytes, and neutrophils by observation of stained blood smears, and it has been reported to be transmitted by the tick *Amblyomma aureolatum*. Clinical manifestations are associated with fever, anemia, jaundice, splenomegaly, lymphadenomegaly, hemorrhage in the gastrointestinal tract, and persistent bleeding from the nose, oral cavity, and the ear pinna (Franca et al. 2010; Da Silva et al. 2011; Eiras et al. 2014). It also has been described as a cause of disease in wild canids in South America (Fredo et al. 2015).

10.3 Clinical Effects, Prevention, and Treatment

10.3.1 Clinical Effects

The clinical manifestations of babesiosis are mainly dependent on the infecting species and host-related factors. The main effects of babesiosis are related to anemia, tissue anoxia, and effect of toxins and inflammatory mediators produced during infection. Babesiosis due to *B. rossi* can present with clinical signs similar to those described for other babesial species such as fever, dehydration, lethargy, pale mucous membranes, and anorexia or with complicated severe clinical disorders including acute renal failure with anuria, icterus, hypotension, acute respiratory distress syndrome (ARDS), vomiting, diarrhea, pancreatitis, myalgia, rhabdomyolysis, ascites, pulmonary edema, encephalomyelitis, and peracute shock. Dogs with *B. rossi* infection may present with mild to severe anemia, thrombocytopenia, leucocytosis, bilirubinemia, pigmenturia, bilirubinuria, hypoglycaemia, acid-base imbalances, azotemia, and hyperlactemia.

The main clinical and clinicopathological findings reported in dogs suffering from *B. canis* infection include dehydration, lethargy, anorexia, fever, lethargy and dehydration with mild to severe thrombocytopenia, hyperfibrinogenemia, mild to severe anemia, hemolysis, and neutropenia (Adaszek et al. 2009; Eichenberger et al. 2016). Hemoglobinuria has also been described in urinalysis of naturally infected dogs. *Babesia vogeli* causes a mild to moderate clinical disease, which often accompanies other concomitant diseases or immunosuppressive conditions or affects splenectomized dogs. Severe fatal hemolytic anemia has been reported in puppies. The most common laboratory findings are regenerative anemia and thrombocytopenia (Solano-Gallego et al. 2008). The clinicopathological findings in *B. gibsoni* infection include fever, regenerative anemia, thrombocytopenia, splenomegaly, lymphadenomegaly, hepatomegaly, and lethargy. *Babesia conradae* infection has been described as more virulent than *B. gibsoni* infection resulting in higher parasitemia, more pronounced anemia, and higher rate of mortality. The clinicopathological findings resemble those reported in *B. gibsoni* infections. The most common clinical findings reported in dogs infected with *B. vulpes* from the north west of Spain include fever, lethargy, weakness, azotemia and pigmenturia. Infected dogs had moderate to severe regenerative anemia and thrombocytopenia (Guitián et al. 2003; Miró et al. 2015).

10.3.2 Prevention

Prevention of babesiosis is based on acaricidal treatments administered topically, systematically, orally as chewable tablets, or spread in the environment. Acaricidal treatments are aimed at reducing the exposure to vector ticks, their bites, and the transmission of the pathogen to dogs. A variety of products based on commercially available acaricidal chemicals have been tested for efficacy in the prevention of babesial infections under experimental or field conditions, and have been licensed for use in dogs (Solano-Gallego et al. 2016). As *Babesia* species are transmitted by blood product transfusions, it is recommended to screen canine blood donors for *Babesia* infection on a regular basis (Wardrop et al. 2016). Non-vectorial dog-to-dog transmission of babesia by fighting can be responsible for the spread of babesiosis into previously non-endemic areas. Vaccines against canine *Babesia* species are commercially available in some countries in Europe.

10.3.3 Treatment

The differences between *Babesia* species that infect dogs are also reflected in their susceptibility to drugs. Accurate detection and species recognition are important for the selection of the correct therapy and predicting the course of disease. While large form *Babesia* species of dogs are usually susceptible to certain drugs, small form *Babesia* are often resistant to these drugs, and treatment of their infections requires the use of other drugs and combinations of drugs. Large *Babesia* species infections of dogs are commonly treated with one dose of imidocarb dipropionate at 5–6 mg per kilogram dog weight intramuscular (IM) or subcutaneously (SC) with good clinical response and a repeated dose 14 days later. Large *Babesia* species and *Rangelia vitalii* have been reported to respond to diminazene aceturate treatment at 3.5 mg/kg IM once.

Babesia gibsoni and *B. conradae* infections are often resistant to imidocarb dipropionate and diminazene aceturate. The treatment of choice for these small *Babesia* species is the combination of the antimalarial atovaquone and the macrolide azithromycin. The most commonly used dose of atovaquone is 13.5 mg/kg, administered orally (PO) every 8 h with fatty food to maximize drug absorption, in combination with azithromycin at 10 mg/kg PO for 10 days. Buparvaquone at 5 mg/kg IM, 2 days apart in combination with azithromycin at 10 mg/kg PO for 10 days, has also been studied for the treatment of dogs infected with *B. vulpes* and found less effective than the combination of atovaquone and azithromycin (Checa et al.

2017). Resistance to atovaquone associated with irresponsiveness to treatment has been described in *B. gibsoni* in dogs from Japan and Taiwan and results from mutations in the parasite cytochrome b gene (Sakuma et al. 2009; Iguchi et al. 2012; Liu et al. 2016).

Parasitological cure with complete elimination of the parasite is commonly not achieved by treatment of small *Babesia* spp. infections in dogs, and clinical relapses frequently occur. Medical management of infection may require supportive treatments including blood transfusions, intravenous fluids, and the use of anti-inflammatory drugs. The prognosis of dogs infected with large forms of *Babesia* species and treated with effective drugs is generally good in uncomplicated disease. Canine disease with small *Babesia* spp. may be more resistant to treatment and carry a poorer prognosis. Currently, there are no zoonotic canine *Babesia* species known.

References

Adachi K, Tateishi M, Horii Y, Nagatomo H, Shimizu T, Makimura S. Elevated erythrocyte-bound IgG value in dogs with clinical *Babesia gibsoni* infection. J Vet Med Sci. 1994;56:757–9.

Adachi K, Tateishi M, Horii Y, Nagatomo H, Shimizu T, Makimura S. Immunologic characteristics of anti-erythrocyte membrane antibody produced in dogs during *Babesia gibsoni* infection. J Vet Med Sci. 1995;57:121–3.

Adamu M, Troskie M, Oshadu DO, Malatji DP, Penzhorn BL, Matjila PT. Occurrence of tick-transmitted pathogens in dogs in Jos, Plateau State, Nigeria. Parasit Vectors. 2014;7:119.

Adaszek Ł, Winiarczyk S, Skrzypczak M. The clinical course of babesiosis in 76 dogs infected with protozoan parasites *Babesia canis canis*. Pol J Vet Sci. 2009;12:81–7.

Apanaskevich DA, Horak IG, Camicas JL. Redescription of *Haemaphysalis* (*Rhipistoma*) *elliptica* (Koch, 1844), an old taxon of the *Haemaphysalis* (*Rhipistoma*) *leachi* group from east and southern Africa, and of *Haemaphysalis* (*Rhipistoma*) *leachi* (Audouin, 1826) (Ixodida, Ixodidae). Onderstepoort J Vet Res. 2007;74:181–208.

Ayoob AL, Hackner SG, Prittie J. Clinical management of canine babesiosis. J Vet Emerg Crit Care (San Antonio). 2010;20:77–89.

Baneth G, Florin-Christensen M, Cardoso L, Schnittger L. Reclassification of *Theileria annae* as *Babesia vulpes* sp. nov. Parasit Vectors. 2015;8:207.

Birkenheuer AJ, Neel J, Ruslander D, Levy MG, Breitschwerdt EB. Detection and molecular characterization of a novel large *Babesia* species in a dog. Vet Parasitol. 2004;124:151–60.

Birkenheuer AJ, Correa MT, Levy MG, Breitschwerdt EB. Geographic distribution of babesiosis among dogs in the United States and association with dog bites: 150 cases (2000–2003). J Am Vet Med Assoc. 2005;227:942–7.

Bohm M, Leisewitz AL, Thompson PN, Schoeman JP. Capillary and venous *Babesia canis rossi* parasitaemias and their association with outcome of infection and circulatory compromise. Vet Parasitol. 2006;141:18–29.

Brown AL, Shiel RE, Irwin PJ. Clinical, haematological, cytokine and acute phase protein changes during experimental *Babesia gibsoni* infection of beagle puppies. Exp Parasitol. 2015;157:185–96.

Camacho AT, Pallas E, Gestal JJ, Guitian FJ, Olmeda AS. *Babesia canis* infection in a splenectomized dog. Bull Soc Pathol Exot. 2002;95:17–9.

Camacho AT, Pallas E, Gestal JJ, Guitián FJ, Olmeda AS, Telford SR, Spielman A. *Ixodes hexagonus* is the main candidate as vector of *Theileria annae* in northwest Spain. Vet Parasitol. 2003;112:157–63.

Carcy B, Randazzo S, Depoix D, Adaszek L, Cardoso L, Baneth G, Gorenflot A, Schetters TP. Classification of *Babesia canis* strains in Europe based on polymorphism of the Bc28.1-gene from the *Babesia canis* Bc28 multigene family. Vet Parasitol. 2015; 211(3–4):111–23.

Carli E, Tasca S, Trotta M, Furlanello T, Caldin M, Solano-Gallego L. Detection of erythrocyte binding IgM and IgG by flow cytometry in sick dogs with *Babesia canis canis* or *Babesia canis vogeli* infection. Vet Parasitol. 2009;162:51–7.

Carret C, Walas F, Carcy B, Grande N, Precigout E, Moubri K, Schetters TP, Gorenflot A. *Babesia canis canis, Babesia canis vogeli, Babesia canis rossi*: differentiation of the three subspecies by a restriction fragment length polymorphism analysis on amplified small subunit ribosomal RNA genes. J Eukaryot Microbiol. 1999;46:298–303.

Chauvin A, Moreau E, Bonnet S, Plantard O, Malandrin L. *Babesia* and its hosts: adaptation to long-lasting interactions as a way to achieve efficient transmission. Vet Res. 2009;40:37.

Checa R, Montoya A, Ortega N, González-Fraga JL, Bartolomé A, Gálvez R, Marino V, Miró G. Efficacy, safety and tolerance of imidocarb dipropionate versus atovaquone or buparvaquone plus azithromycin used to treat sick dogs naturally infected with the *Babesia microti*-like piroplasm. Parasit Vectors. 2017;10:14.

Da Silva AS, Franca RT, Costa MM, Paim CB, Paim FC, Dornelles GL, Soares JF, Labruna MB, Mazzanti CM, Monteiro SG, Lopes ST. Experimental infection with *Rangelia vitalii* in dogs: acute phase, parasitemia, biological cycle, clinical-pathological aspects and treatment. Exp Parasitol. 2011;128:347–52.

Eichenberger RM, Riond B, Willi B, Hofmann-Lehmann R, Deplazes P. Prognostic markers in acute *Babesia canis* infections. J Vet Intern Med. 2016;30:174–82.

Eiras DF, Craviotto MB, Baneth G, Moré G. First report of *Rangelia vitalii* infection (canine rangeliosis) in Argentina. Parasitol Int. 2014;63:729–34.

Franca RT, Silva AS, Paim FC, Costa MM, Soares JF, Mazzanti CM, Lopes STA. *Rangelia vitalii* in dogs in southern Brazil. Comp Clin Path. 2010;19:383–7.

Fredo G, Bianchi MV, De Andrade CP, De Souza SO, Leite-Filho RV, Bandinelli MB, Amorim DB, Driemeier D, Sonne L. Natural infection of wild canids (*Cerdocyon thous* and *Lycalopex gymnocercus*) with the Intraendothelial Piroplasm *Rangelia vitalii* in southern Brazil. J Wildl Dis. 2015;51:880–4.

Freeman MJ, Kirby BM, Panciera DL, Henik RA, Rosin E, Sullivan LJ. Hypotensive shock syndrome associated with acute *Babesia canis* infection in a dog. J Am Vet Med Assoc. 1994;204:94–6.

Fukumoto S, Suzuki H, Igarashi I, Xuan X. Fatal experimental transplacental *Babesia gibsoni* infections in dogs. Int J Parasitol. 2005;35:1031–5.

Goddard A, Wiinberg B, Schoeman JP, Kristensen AT, Kjelgaard-Hansen M. Mortality in virulent canine babesiosis is associated with a consumptive coagulopathy. Vet J. 2013;196:213–7.

Goddard A, Leisewitz AL, Kjelgaard-Hansen M, Kristensen AT, Schoeman JP. Excessive pro-inflammatory serum cytokine concentrations in virulent canine babesiosis. PLoS One. 2016;11:e0150113.

Goo YK, Jia H, Aboge GO, Terkawi MA, Kuriki K, Nakamura C, Kumagai A, Zhou J, Lee EG, Nishikawa Y, Igarashi I, Fujisaki K, Xuan X. *Babesia gibsoni*: serodiagnosis of infection in dogs by an enzyme-linked immunosorbent assay with recombinant BgTRAP. Exp Parasitol. 2008;118:555–60.

Guitián FJ, Camacho AT, Telford SR III. Case-control study of canine infection by a newly recognised *Babesia microti*-like piroplasm. Prev Vet Med. 2003;61:137–45.

Hodžić A, Zörer J, Duscher GG. *Dermacentor reticulatus*, a putative vector of *Babesia* cf. *microti* (syn. *Theileria annae*) piroplasm. Parasitol Res. 2017;116:1075–7.

Homer MJ, Aguilar-Delfin I, Telford SR III, Krause PJ, Persing DH. Babesiosis. Clin Microbiol Rev. 2000;13:451–69.

Iguchi A, Matsuu A, Ikadai H, Talukder MH, Hikasa Y. Development of in vitro atovaquone-resistant *Babesia gibsoni* with a single-nucleotide polymorphism in cytb. Vet Parasitol. 2012;185:145–50.

Irwin PJ. Canine babesiosis: from molecular taxonomy to control. Parasit Vectors. 2009;26(2 Suppl 1):S4.

Jacobson LS. The south African form of severe and complicated Canine babesiosis: clinical advances 1994-2004. Vet Parasitol. 2006;138:126–39.

Jefferies R, Ryan UM, Jardine J, Broughton DK, Robertson ID, Irwin PJ. Blood, Bull Terriers and babesiosis: further evidence for direct transmission of *Babesia gibsoni* in dogs. Aust Vet J. 2007;85:459–63.

Kjemtrup AM, Wainwright K, Miller M, Penzhorn BL, Carreno RA. *Babesia conradae*, sp. nov., a small canine *Babesia* identified in California. Vet Parasitol. 2006;138:103–11.

Konishi K, Sakata Y, Miyazaki N, Jia H, Goo YK, Xuan X, Inokuma H. Epidemiological survey of *Babesia gibsoni* infection in dogs in Japan by enzyme-linked immunosorbent assay using *B. gibsoni* thrombospondin-related adhesive protein antigen. Vet Parasitol. 2008;155:204–8.

Lack JB, Reichard MV, van den Bussche RA. Phylogeny and evolution of the piroplasmida as inferred from 18S rRNA sequences. Int J Parasitol. 2012;42:353–63.

Lee MJ, DH Y, Yoon JS, Li YH, Lee JH, Chae JS, Park J. Epidemiologic and clinical surveys in dogs infected with *Babesia gibsoni* in South Korea. Vector Borne Zoonotic Dis. 2009;9:681–6.

Leisewitz AL, Jacobson LS, de Morais HS, Reyers F. The mixed acid-base disturbances of severe canine babesiosis. J Vet Intern Med. 2001;15:445–52.

Lempereur L, Beck R, Fonseca I, Marques C, Duarte A, Santos M, Zúquete S, Gomes J, Walder G, Domingos A, Antunes S, Baneth G, Silaghi C, Holman P, Zintl A. Guidelines for the detection of *Babesia* and *Theileria* parasites. Vector Borne Zoonotic Dis. 2017;17:51–65.

Liu PC, Lin YL, Lin CN, BL S. A SimpleProbe(®) real-time PCR assay for differentiating the cytochrome b M121I mutation in clinical specimens from dogs infected with *Babesia gibsoni*. Ticks Tick Borne Dis. 2016;7:639–43.

Makinde MO, Bobade PA. Osmotic fragility of erythrocytes in clinically normal dogs and dogs infected with parasites. Res Vet Sci. 1994;57:343–8.

Mathe A, Dobos-Kovacs M, Voros K. Histological and ultrastructural studies of renal lesions in *Babesia canis* infected dogs treated with imidocarb. Acta Vet Hung. 2007;55:511–23.

Matjila PT, Penzhorn BL, Bekker CP, Nijhof AM, Jongejan F. Confirmation of occurrence of *Babesia canis vogeli* in domestic dogs in South Africa. Vet Parasitol. 2004;122:119–25.

Matjila PT, Carcy B, Leisewitz AL, Schetters T, Jongejan F, Gorenflot A, Penzhorn BL. Preliminary evaluation of the BrEMA1 gene as a tool for associating *Babesia rossi* genotypes and clinical manifestation of canine babesiosis. J Clin Microbiol. 2009;47:3586–92.

Mehlhorn H, Schein E. The piroplasms: life cycle and sexual stages. Adv Parasitol. 1984;23:37–103.

Mehlhorn H, Peters W, Haberkorn A. The formation of kinetes and oocysts in *Plasmodium gallinaceum* and considerations on phylogenetic relationships between Haemosporidia, Piroplasmida, and other Coccidia. Protistologica. 1980;16:135–54.

Mehlhorn H, Schein E, Ahmed JS. Theileria. In: Kreier JP, editor. Parasitic protozoa, vol. 7. San Diego: Academic; 1994. p. 217–304.

Miró G, Checa R, Paparini A, Ortega N, González-Fraga JL, Gofton A, Bartolomé A, Montoya A, Gálvez R, Mayo PP, Irwin P. *Theileria annae* (syn. *Babesia microti*-like) infection in dogs in NW Spain detected using direct and indirect diagnostic techniques: clinical report of 75 cases. Parasit Vectors. 2015;8:217.

Murase T, Ueda T, Yamato O, Tajima M, Maede Y. Oxidative damage and enhanced erythrophagocytosis in canine erythrocytes infected with *Babesia gibsoni*. J Vet Med Sci. 1996;58:259–61.

Onishi T, Ueda K, Horie M, Kajikawa T, Ohishi I. Serum hemolytic activity in dogs infected with *Babesia gibsoni*. J Parasitol. 1990;76:564–7.

Otsuka Y, Yamasaki M, Yamato O, Maede Y. Increased generation of superoxide in erythrocytes infected with *Babesia gibsoni*. J Vet Med Sci. 2001;63:1077–81.

Otsuka Y, Yamasaki M, Yamato O, Maed Y. The effect of macrophages on the erythrocyte oxidative damage and the pathogenesis of anemia in *Babesia gibsoni*-infected dogs with low parasitemia. J Vet Med Sci. 2002;64:221–6.

Oyamada M, Davoust B, Boni M, Dereure J, Bucheton B, Hammad A, Itamoto K, Okuda M, Inokuma H. Detection of *Babesia canis rossi*, *B. canis vogeli*, and *Hepatozoon canis* in dogs in a village of eastern Sudan by using a screening PCR and sequencing methodologies. Clin Diagn Lab Immunol. 2005;12:1343–6.

Qurollo BA, Archer NR, Schreeg ME, Marr HS, Birkenheuer AJ, Haney KN, ThomasBS BEB. Improved molecular detection of *Babesia* infections in animals using a novel quantitative real-time PCR diagnostic assay targeting mitochondrial DNA. Parasit Vectors. 2017;10:128.

Radi ZA, Styer EL, Frazier KS. Electron microscopic study of canine *Babesia gibsoni* infection. J Vet Diagn Investig. 2004;16:229–33.

Reyers F, Leisewitz AL, Lobetti RG, Milner RJ, Jacobson LS, van Zyl M. Canine babesiosis in South Africa: more than one disease. Does this serve as a model for falciparum malaria? Ann Trop Med Parasitol. 1998;92:503–11.

Rosner F, Zarrabi MH, Benach JL, Habicht GS. Babesiosis in splenectomized adults. Review of 22 reported cases. Am J Med. 1984;76:696–701.

Sakuma M, Setoguchi A, Endo Y. Possible emergence of drug-resistant variants of *Babesia gibsoni* in clinical cases treated with atovaquone and azithromycin. J Vet Intern Med. 2009;23:493–8.

Schein E, Mehlhorn H, Voigt WP. Electron microscopical studies on the development of *Babesia canis* (Sporozoa) in the salivary glands of the vector tick *Dermacentor reticulatus*. Acta Trop. 1979;36:229–41.

Schetters TP, Scholte NC, Kleuskens JA, Bos HJ. Not peripheral parasitaemia but the level of soluble parasite antigen in plasma correlates with vaccine efficacy against *Babesia canis*. Parasite Immunol. 1996;18:1–6.

Schnittger L, Rodriguez AE, Florin-Christensen M, Morrison DA. *Babesia*: a world emerging. Infect Genet Evol. 2012;12:1788–809.

Schreeg ME, Marr HS, Tarigo JL, Cohn LA, Bird DM, Scholl EH, Levy MG, Wiegmann BM, Birkenheuer AJ. Mitochondrial genome sequences and structures aid in the resolution of Piroplasmida phylogeny. PLoS One. 2016;11:e0165702.

Sikorski LE, Birkenheuer AJ, Holowaychuk MK, McCleary-Wheeler AL, Davis JM, Littman MP. Babesiosis caused by a large *Babesia* species in 7 immunocompromised dogs. J Vet Intern Med. 2010;24:127–31.

Singla LD, Sumbria D, Mandhotra A, Bal MS, Kaur P. Critical analysis of vector-borne infections in dogs: *Babesia vogeli*, *Babesia gibsoni*, *Ehrlichia canis* and *Hepatozoon canis* in Punjab, India. Acta Parasitol. 2016;61:697–706.

Solano-Gallego L, Baneth G. Babesiosis in dogs and cats—expanding parasitological and clinical spectra. Vet Parasitol. 2011;181:48–60.

Solano-Gallego L, Trotta M, Carli E, Carcy B, Caldin M, Furlanello T. *Babesia canis canis* and *Babesia canis vogeli* clinicopathological findings and DNA detection by means of PCR-RFLP in blood from Italian dogs suspected of tick-borne disease. Vet Parasitol. 2008;157:211–21.

Solano-Gallego L, Sainz Á, Roura X, Estrada-Peña A, Miró G. A review of canine babesiosis: the European perspective. Parasit Vectors. 2016;9:336.

Stegeman JR, Birkenheuer AJ, Kruger JM, Breitschwerdt EB. Transfusion-associated *Babesia gibsoni* infection in a dog. J Am Vet Med Assoc. 2003;222:959–63.

Uilenberg G. *Babesia*—a historical overview. Vet Parasitol. 2006;138:3–10.

Vercammen F, De Deken R, Maes L. Clinical and serological observations on experimental infections with *Babesia canis* and its diagnosis using the IFAT. Parasite. 1995;2:407–10.

Walter S, Mehlhorn H, Zweygarth E, Schein E. Electron microscopic investigations on stages of dog piroplasms cultured in vitro: Asian isolates of *Babesia gibsoni* and strains of *B. canis* from France and Hungary. Parasitol Res. 2002;88:32–7.

Wang C, Ahluwalia SK, Li Y, Gao D, Poudel A, Chowdhury E, Boudreaux MK, Kaltenboeck B. Frequency and therapy monitoring of canine *Babesia* spp. infection by high-resolution melting curve quantitative FRET-PCR. Vet Parasitol. 2010;168:11–8.

Wardrop KJ, Reine N, Birkenheuer A, Hale A, Hohenhaus A, Crawford C, Lappin MR. Canine and feline blood donor screening for infectious disease. J Vet Intern Med. 2005;19:135–42.

Wardrop KJ, Birkenheuer A, Blais MC, Callan MB, Kohn B, Lappin MR, Sykes J. Update on canine and feline blood donor screening for blood-borne pathogens. J Vet Intern Med. 2016;30:15–35.

Wilkerson MJ, Shuman W, Swist S, Harkin K, Meinkoth J, Kocan AA. Platelet size, platelet surface-associated IgG, and reticulated platelets in dogs with immune-mediated thrombocytopenia. Vet Clin Pathol. 2001;30:141–9.

Yamane I, Thomford JW, Gardner IA, Dubey JP, Levy M, Conrad PA. Evaluation of the indirect fluorescent antibody test for diagnosis of *Babesia gibsoni* infections in dogs. Am J Vet Res. 1993;54:1579–84.

Yeagley TJ, Reichard MV, Hempstead JE, Allen KE, Parsons LM, White MA, Little SE, Meinkoth JH. Detection of *Babesia gibsoni* and the canine small *Babesia* 'Spanish isolate' in blood samples obtained from dogs confiscated from dog fighting operations. J Am Vet Med Assoc. 2009;235:535–9.

Zahler M, Rinder H, Schein E, Gothe R. Detection of a new pathogenic *Babesia microti*-like species in dogs. Vet Parasitol. 2000;89:241–8.

Zygner W, Gójska-Zygner O, Bąska P, Długosz E. Increased concentration of serum TNF alpha and its correlations with arterial blood pressure and indices of renal damage in dogs infected with *Babesia canis*. Parasitol Res. 2014;113:1499–503.

Equine Piroplasmids

11

Massaro W. Ueti and Donald P. Knowles

Abstract

Equine piroplasmosis (EP), caused by the tick-borne protozoan parasites *Theileria equi* or *Babesia caballi*, is an infectious disease that affects equids worldwide. The disease is of global economic importance due to morbidity, mortality, costs associated with treatment, diagnosis, and impact on the international movement of horses for commerce and competition. Infected horses remain lifelong carriers of *T. equi* or *B. caballi*, although horses appear able to occasionally eliminate *B. caballi* infection. Persistently infected equids are a continuous source for transmission of the protozoal pathogens by tick vectors or iatrogenic transfer. The presence of ticks capable of transmitting these pathogens on all continents increases the need for global surveillance to prevent EP dissemination. This chapter presents life cycle differences between *T. equi* and *B. caballi*, the world distribution of competent tick vectors, efficacy of drug treatment to eliminate persistent infection, acaricide treatments to lower tick burden, and improvement of diagnostic assays to prevent dissemination of protozoan parasite strains throughout the world.

M.W. Ueti (✉) • D.P. Knowles
Animal Diseases Research Unit, Agricultural Research Service, U.S. Department of Agriculture, Pullman, WA, USA

Program in Vector-borne Diseases, Department of Veterinary Microbiology and Pathology, Washington State University, Pullman, WA, USA

School for Global Animal Health, Washington State University, Pullman, WA, USA
e-mail: massaro@vetmed.wsu.edu

M. Florin-Christensen, L. Schnittger (eds.), *Parasitic Protozoa of Farm Animals and Pets*, https://doi.org/10.1007/978-3-319-70132-5_11

11.1 Morphology, Life Cycle, and Host-Pathogen Interactions

Theileria equi and *Babesia caballi* are both members of the phylum Apicomplexa. *Theileria equi* has an initial lymphocytic stage and then forms a tetrad configuration in erythrocytes, termed a Maltese cross formation (Fig. 11.1a) (Ramsay et al. 2013; Schein et al. 1981; Holbrook 1969). *Babesia caballi* has the characteristics of a true *Babesia* sp. in that it only infects erythrocytes, forming two large pear-shaped piroplasms (Fig. 11.1b) (Scoles and Ueti 2015). Parasites in mammalian host cells are haploid and undergo asexual reproduction (Uilenberg 2006). During replication of *T. equi* in erythrocytes, parasites express immunodominant surface proteins equine merozoite antigen-1 and equine merozoite antigen-2 (EMA-1 and EMA-2) (Kappmeyer et al. 1993). It has been postulated that these proteins are important for parasite attachment to erythrocytes. In *B. caballi*, parasites express rhoptry-associated protein-1 (RAP-1), a highly conserved immunogenic protein. It has been postulated that RAP-1 is critical for erythrocyte invasion (Ikadai et al. 1999; Kappmeyer et al. 1999).

When competent tick vectors ingest infected erythrocytes, the intra-erythrocytic parasite stages begin a transformation within the lumen of the tick midgut (Ueti et al. 2003). Initially, the parasite undergoes gametogenesis. Gamonts begin nuclear division followed by formation of protrusions and then ray bodies (Fig. 11.2a, b). These multinuclear forms divide into two to four mononuclear stages called ray bodies (Zapf and Schein 1994a). This stage elongates and forms microgametes. Concurrently, macrogametes are formed with little morphologic change from merozoites. Microgametes and macrogametes fuse to form diploid zygotes that attach to and invade tick midgut epithelial cells and, subsequently, develop into kinetes. *Theileria equi* kinetes migrate from the midgut to infect tick salivary glands (Guimaraes et al. 1998a, b; Ueti et al. 2008, Zapf and Schein 1994b). *Babesia caballi* kinetes infect tick ovaries resulting in vertical transmission with infection of eggs and subsequent larvae (Fig. 11.2a, b) (Schwint et al. 2008; Ueti et al. 2008).

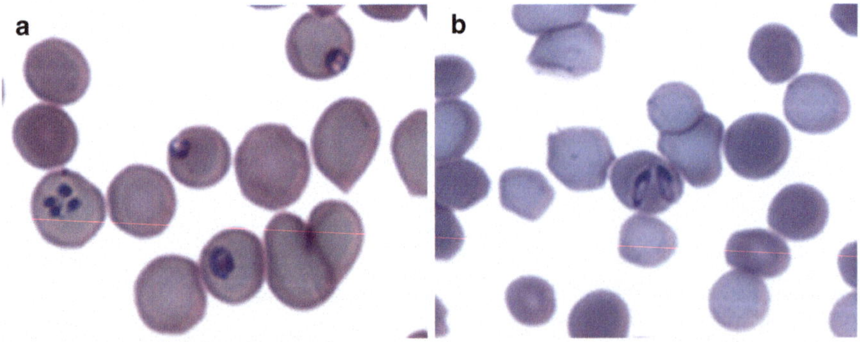

Fig. 11.1 Giemsa blood smear from infected horses during acute infection. *Theileria equi* (**a**) and *Babesia caballi* (**b**)

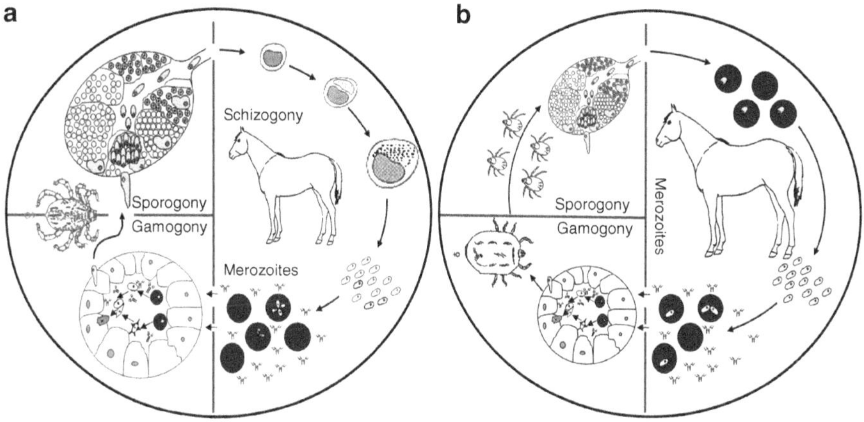

Fig. 11.2 Life cycle comparison between *Theileria equi* and *Babesia caballi* during infection in the horse and tick vector

Within tick salivary glands, sporozoite maturation occurs during tick acquisition of a blood meal. Tick vectors inoculate mature sporozoites into mammalian hosts via tick saliva. *Theileria equi* sporozoites invade peripheral blood mononuclear cells and transform into schizonts prior to invading erythrocytes. In contrast, *B. caballi* sporozoites directly invade erythrocytes. Infected horses are lifelong reservoirs for tick infection (Ueti et al. 2008).

11.1.1 Model of Tick-Borne Transmission of *Theileria equi* and *Babesia caballi*

Horses infected with *T. equi* or *B. caballi* are found throughout the world and are sources for tick infection. Both parasites are efficiently transmitted by tick vectors. However, the model of transmission between the two parasites is distinct. *Theileria equi* is commonly transmitted transstadially or intrastadially as opposed to *B. caballi* that is commonly transmitted transovarially (Fig. 11.3). Transstadial transmission occurs when nymphal ticks acquire *T. equi* infections during blood meal on either acutely or chronically infected horses and, after molting and movement to a new host, successfully infect naïve horses (Ueti et al. 2005). Intrastadial transmission occurs when adult ticks acquire *T. equi* infection during blood meal on either acutely or chronically infected horses and, after movement to a new host, successfully transmit *T. equi* parasites (Ueti et al. 2008). In contrast, transovarial transmission of *B. caballi* occurs when female ticks acquire the parasite during blood meal and vertically transmit the parasite to the next generation (Fig. 11.3). Infected ticks transmit *B. caballi* to naïve horses during acquisition of a blood meal. Furthermore *B. caballi* parasites can persist in the tick vectors over several generations in the absence of reinfection from an infected host (Schwint et al. 2008; Ueti et al. 2008). Therefore, unlike the case for *T. equi*, tick vectors are a reservoir host for *B. caballi*.

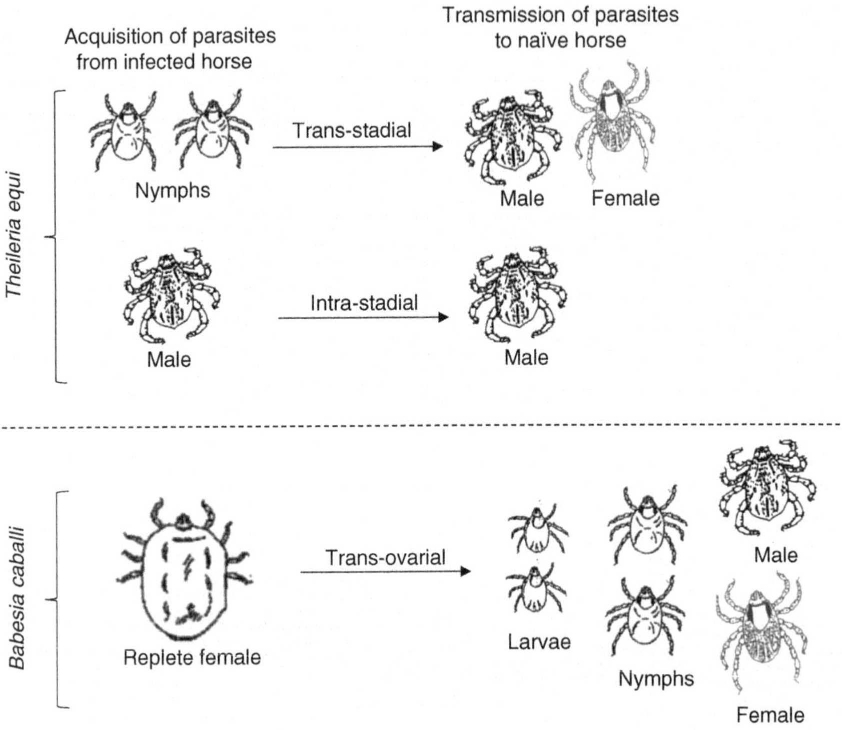

Fig. 11.3 Model of tick-borne transmission of protozoan parasites that cause equine piroplasmosis

11.2 Diagnosis and Epidemiology

11.2.1 Diagnosis

Diagnostic assays have been used to control the movement of infected horses and prevent dissemination of the parasites (Table 11.1). The methods vary in their specificity and sensitivity. Therefore, several methods need to be considered to prevent the movement of infected horses. Definitive diagnosis of *T. equi* and *B. caballi* can be made by microscopic examination of Giemsa-stained blood smears during the acute phase of infection (Fig. 11.1). However, in the chronic phase of infection, the parasitemia in peripheral blood may be too low for reliable microscopic diagnosis. For this reason, nucleic acid- and serologic-based techniques have been developed for the detection of parasite DNA and specific antibody, respectively.

Prior to DNA-based detection methods, the standard approach for confirming a carrier status was blood subinoculation into immunocompromised recipient horses. This approach is costly and time-consuming. Therefore, an alternative means of parasite detection was desired. DNA probes were developed as one alternative method to replace the use of horses for confirmation of infection. Polymerase chain

Table 11.1 Diagnostic assays to detect *Theileria equi* or *Babesia caballi* infection

Assay	Infection status	
	Acute infection	Persistent infection
Giemsa-stained blood smear	D	ND
PCR	D	D
IFA	D	D
CFT	D	ND
cELISA	D	D

D test detects stage of infection, *ND* test is unable to detect stage of infection

reaction (PCR) showed promise in providing a means to detect *T. equi* and *B. caballi* carrier infections with improved sensitivity (Bashiruddin et al. 1999; Posnett and Ambrosio 1991). The most sensitive detection method for *T. equi* involves nested PCR using oligonucleotides associated with *T. equi* merozoite antigen-1 gene (*ema*-1) (Ueti et al. 2012).

Serology remains the diagnostic approach of choice primarily due to the ease of protocols and higher throughput of the assays. A number of serologic assays have been used to detect specific *T. equi* and *B. caballi* antibody from equine serum. Although the indirect immunofluorescence (IFA) assay has greater sensitivity than the complement fixation test (CFT), the CFT has been used for many years as an official test by Australia, Canada, Japan, and the USA for screening imported horses (Kuttler et al. 1988). The CFT is limited in several respects including problems with anticomplementary activity and/or anti-erythrocytic autoantibody in certain sera. The CFT has poor sensitivity due to the inability of immunoglobulin isotype IgG5, an isotype present during chronic infection, to fix complement (Knowles et al. 1991). For these reasons, efforts to improve serological assays for better surveillance have resulted in the development of competitive enzyme linked immunosorbent assays (cELISAs) for both *T. equi*- and *B. caballi*-specific serum antibody. This approach was based on the hypothesis that immunodominant conserved epitopes for antibody existed and that epitope-specific antibody from infected equine serum would inhibit the binding of related epitope-specific monoclonal antibody (Kappmeyer et al. 1999; Knowles et al. 1991, 1992).

Equi merozoite antigen-1 (EMA-1) from *T. equi*, a 34 kDa surface protein, and rhoptry-associated protein-1 (RAP-1) from *B. caballi*, a 60 kDa protein, contain immunodominant and geographically conserved epitopes defined by monoclonal antibodies 36/133.97 and 79/17.18.5, respectively (Kappmeyer et al. 1999; Knowles et al. 1991). Recombinant peptides bearing these epitopes have been used in each specific cELISA, and the assays applied to detect specific antibodies in serum from experimentally infected horses and from serum obtained from every continent (Kappmeyer et al. 1999; Knowles et al. 1991). The *T. equi* EMA-1 associated epitope is conformationally dependent, while the *B. caballi* RAP-1-associated epitope is linear and found in the C-terminal repeat region (Cunha et al. 2002; Kappmeyer et al. 1999). A major attribute of the cELISA is the possibility to standardize the assay. This attribute is based on the use of recombinant antigen and monoclonal antibody. Recombinant EMA-1 used in the cELISA is a consistent, standardized source of antigen, unlike CFT antigen that may vary significantly from laboratory to laboratory. The lack of standardization of the CFT antigen makes comparison with the cELISA challenging, and concordance values must be interpreted with caution. Despite the low throughput and subjective nature of the IFA, it has proven to be a valuable assay, and concordance between cELISA and IFA for both *T. equi* and *B. caballi* is very high (Rhalem et al. 2001).

11.2.2 Epidemiology

EP is globally distributed, and the majority of horse populations live in endemic areas (Friedhoff 1982; Kerber et al. 2009, Potgieter et al. 1992). Distribution of *T. equi* and *B. caballi* coincides with the presence of tick vectors capable of transmitting these parasites to horses (Table 11.2). With few exceptions, three genera including *Hyalomma*, *Dermacentor*, and *Rhipicephalus* serve as biological vectors for both hemoprotozoan parasites (Scoles and Ueti 2015; Stiller et al. 2002). Few countries in the world are considered free of EP. The disease has not been reported in Japan, and despite the occurrence of potential vector ticks, it has not been reported in northern Europe, including Great Britain (Joyner et al. 1981; Ikadai et al. 2002). *Theileria equi* was introduced into Australia but did not spread due to the lack of suitable vectors, although it is now known that *T. equi* can be transmitted by *R. microplus* that was considered to be prevalent in Australia. Recently, however, the common cattle tick from Australia was redescribed as a new species, *R. australis* (Mahoney et al. 1977; Guimaraes et al. 1998a; Ueti et al. 2003, 2005, 2008;

Table 11.2 Global distribution of tick vectors of equine piroplasmosis

Continent	Tick vectors	EP status
Africa	*D. marginatus*[a,b], *H. anatolicum*[a,b], *H. detritum*[a,b], *H. dromedarii*[a], *H. excavatum*[a],*H. lusitanicum*[a], *H. marginatum*[a,b], *H. truncatum*[b], *R. annulatus*[a], *R. bursa*[a,b], *R. evertsi*[a,b], *R. microplus*[a], *R. sanguineus*[a,b]	Endemic
Australia	*R. microplus*	Not present
Asia	*H. anatolicum*[a,b], *H. detritum*[a,b], *H. dromedarii*[a], *H. marginatum*[a,b], *H. uralense*[a], *R. microplus*[a]	Endemic. Except Japan
Europe	*D. marginatus*[a,b], *D. nuttalli*[b], *D. silvarum*[b], *D. reticulatus*[a,b], *H. anatolicum*[a,b], *H. detritum*[a,b], *H. dromedarii*[a], *H. lusitanicum*[a], *H. marginatum*[a,b], *R. bursa*[a,b]	Endemic. Except northern Europe
North America	*A. mixtum*[a], *D. albipictus*[b], *D. nitens*[b], *D. variabilis*[a,b], *R. annulatus*[a], *R. microplus*[a]	Endemic. Except Canada and the USA
South America	*A. mixtum*[a], *D. nitens*[b], *R. annulatus*[a], *R. microplus*[a]	Endemic

A Amblyomma, *D* Dermacentor, *H* Hyalomma, *R* Rhipicephalus
[a]*Theileria equi*
[b]*Babesia caballi*

Estrada-Pena et al. 2012). The genetic diversity between *R. microplus* and *R. australis* might contribute to explain the inability of the latter to transmit the parasites. Both *T. equi* and *B. caballi* were introduced into the USA but were contained through strict control measures (Short et al. 2012; Ueti et al. 2003, 2005, 2008, 2012). However, a few isolated outbreaks have been detected in the USA. In an outbreak in Florida, transmission occurred iatrogenically, while in the Texas outbreak, transmission occurred by tick vectors (Scoles et al. 2011; Scoles and Ueti 2013; Short et al. 2012). These outbreaks were controlled by euthanizing infected horses or eliminating the infection using drug therapy (Short et al. 2012; Ueti et al. 2012).

11.3 Clinical Signs, Prevention, and Treatment

11.3.1 Clinical Signs

Clinical signs of EP develop approximately 15 days after tick transmission (Ueti et al. 2005). The acute disease is characterized by fever, reduced appetite and malaise, increased pulse and respiration rates, anorexia, constipation followed by diarrhea, tachycardia, petechiae, splenomegaly, anemia, hemoglobinuria, and hemoglobinemia (Taylor et al. 1969). Central nervous system involvement has been reported from *B. caballi* infections, but not *T. equi* (Taylor et al. 1969). Chronic infection may be characterized by weight loss, poor condition, and performance (Laus et al. 2015).

It is common to detect varying complications from the disease including pneumonia and enteritis (De Waal 1992) and abortion in mares due to severe disease and pyrexia of the mare or to intrauterine infection of the fetus (Erbsloh 1975).

Anemia is a consequence of hemolysis of infected erythrocytes and is generally characterized initially as normocytic followed by macrocytic with an increase in reticulocytes (Mahoney et al. 1977). Decreased blood iron, phosphorous, and plasma fibrinogen have been reported as well as an increase in bilirubin (De Waal et al. 1987).

Gross pathology includes edema of the subcutaneous and subserosal tissues, hepatomegaly and splenomegaly, epicardial and endocardial hemorrhages, enlargement of kidneys and lymph nodes, and lung congestion. Histopathology includes necrosis of the liver with bile stasis, nephrosis, and edema of the lungs. In addition, there is proliferation of reticuloendothelial cells in the liver, lungs, kidneys, and lymph nodes (Mahoney et al. 1977).

11.3.2 Prevention

Control of tick-borne protozoan parasites affecting equids remains a major concern and challenge in tropical and subtropical regions of the world. Several means of controlling tick-borne pathogens have been attempted including tick control and drug treatment. However, the efficacies of both strategies are limited. Due to the lack of effective means for controlling protozoan parasites and ticks, EP is still one of the most important threats to global equine health.

Acaricides have been used extensively for controlling ticks with a subsequent reduction in tick-borne infection transmission. Acaricide classes such as organophosphates, amidines, and pyrethroids have served as the primary means of tick control and are considered crucial as a short-term solution to control tick infestations on horses. However, the long-term effectiveness of acaricide tick control is limited due to emerging resistance (Taylor 2001). Over the past several years, *Rhipicephalus* ticks have developed resistance to all of the major classes of acaricides (Foil et al. 2004). Long-term effects of acaricide resistance along with other environmental factors may lead to new areas of tick-borne disease endemicity or create areas of enzootic instability where previous transmission rates resulted in stable herd immunity (Guerrero et al. 2012). *Dermacentor nitens* can acquire infection from horses with low levels of *B. caballi* in the peripheral blood. Furthermore, as few as one *T. equi* infected tick has been reported to be sufficient to transmit the parasite to a naïve horse (Ueti et al. 2005). This illustrates the high risk of introducing chronically infected horses into a non-endemic area possessing competent vectors that may have, or will develop, acaricide resistance. In the case of *Rhipicephalus* ticks, acaricide resistance is well documented (George et al. 2004).

11.3.3 Treatment

Few drugs have shown efficacy in eliminating *T. equi* or *B. caballi* infection (Knowles 1996). To date, treatment with imidocarb (Imizol®, Schering Plough Animal Health) has eliminated both parasites during chronic infections. A relatively high-dose regimen of imidocarb dipropionate, 4.0 mg per kg of body weight delivered four times at 72 h intervals, showed efficacy in parasite elimination from the peripheral blood. Subinoculation of blood or attempted tick transmission from treated horses failed to transmit parasites to susceptible naïve animals (Grause et al. 2013; Schwint et al. 2009; Ueti et al. 2012). Significantly, treated horses returned to seronegative status (Knowles 1996; Schwint et al. 2009; Ueti et al. 2012). The results indicated that imidocarb dipropionate eliminates both *T. equi* and *B. caballi* infections. However, other drug treatments have been attempted with varied efficacy in parasite control. Other drugs most commonly used for treatment of acute EP are shown below. These drugs have not been shown to eliminate infection of either *T. equi* or *B. caballi*.

Amicarbalide (Diampron, May and Baker, England), an aromatic diamidine used as an antiprotozoal in the treatment of babesiosis at 10 mg/kg body weight via intramuscular injection is effective at moderating the acute disease associated with both *T. equi* and *B. caballi* (De Waal et al. 1987; De Waal 1992). Diminazene (Berenil, Hoechst, Germany) at 11 mg/kg body weight delivered intramuscularly is effective against acute disease associated with both parasites, although two additional treatments at 24 h intervals may be necessary for *T. equi* infection (Wise et al. 2013).

Finally, parvaquone (Clexon) and buparvaquone (Butalex) (Burroughs Wellcome & Co., England) at 20 mg/kg and 5 mg/kg body weight via intramuscular injection, respectively, are effective against acute *T. equi* infection (Zaugg and Lane 1989).

As mentioned before, the introduction of horses chronically infected with either *T. equi* or *B. caballi* into non-endemic regions that contain suitable ticks for transmission can jeopardize the horse industry. Therefore, surveillance and restriction of international movement of equids are required with the goal of preventing introduction of chronically infected horses that maintain parasitemia levels at or above the minimal threshold for tick transmission. In addition, small numbers of infected ticks are capable of transmission, increasing the risk of an EP outbreak. Hence, vigilance utilizing improved diagnostics and quarantine will be necessary to prevent the dissemination of *T. equi* and/or *B. caballi* throughout the world.

Acknowledgment This work was supported by USDA-ARS CRIS project number 2090-32000-039-00D.

References

Bashiruddin JB, Camma C, et al. Molecular detection of *Babesia equi* and *Babesia caballi* in horse blood by PCR amplification of part of the 16S rRNA gene. Vet Parasitol. 1999;84(1–2):75–83.

Cunha CW, Kappmeyer LS, et al. Conformational dependence and conservation of an immunodominant epitope within the *Babesia equi* erythrocyte-stage surface protein equi merozoite antigen 1. Clin Diagn Lab Immunol. 2002;9(6):1301–6.

De Waal DT. Equine piroplasmosis: a review. Br Vet J. 1992;148(1):6–14.

De Waal DT, Van HJ, et al. An investigation into the clinical pathological changes and serological response in horses experimentally infected with *Babesia equi* and *Babesia caballi*. Onderstepoort J Vet Res. 1987;54(4):561–8.

Erbsloh JK. Babesiosis in the newborn foal. J Reprod Fertil Suppl. 1975;23:725–6.

Estrada-Pena A, Venzal JM, et al. Reinstatement of *Rhipicephalus* (*Boophilus*) *australis* (Acari: Ixodidae) with redescription of the adult and larval stages. J Med Entomol. 2012;49(4):794–802.

Foil LD, Coleman P, et al. Factors that influence the prevalence of acaricide resistance and tick-borne diseases. Vet Parasitol. 2004;125(1–2):163–81.

Friedhoff KT. Piroplasmas of horses—impact on the international horse trade. Berl Munch Tierarztlc Wochenschr. 1982;95(19):368–74.

George JE, Pound JM, et al. Chemical control of ticks on cattle and the resistance of these parasites to acaricides. Parasitology. 2004;129(Suppl):S353–66.

Grause JF, Ueti MW, et al. Efficacy of imidocarb dipropionate in eliminating Theileria equi from experimentally infected horses. Vet J. 2013;196(3):541–6.

Guerrero FD, Lovis L, et al. Acaricide resistance mechanisms in *Rhipicephalus* (*Boophilus*) *microplus*. Rev Bras Parasitol Vet. 2012;21(1):1–6.

Guimaraes AM, Lima JD, et al. Sporogony and experimental transmission of *Babesia equi* by *Boophilus microplus*. Parasitol Res. 1998a;84(4):323–7.

Guimaraes AM, Lima JD, et al. Ultrastructure of sporogony in *Babesia equi* in salivary glands of adult female *Boophilus microplus* ticks. Parasitol Res. 1998b;84(1):69–74.

Holbrook AA. Biology of equine piroplasmosis. J Am Vet Med Assoc. 1969;155(2):453–4.

Ikadai H, Tamaki Y, et al. Inhibitory effect of monoclonal antibodies on the growth of *Babesia caballi*. Int J Parasitol. 1999;29(11):1785–91.

Ikadai H, Nagai A, et al. Seroepidemiologic studies on *Babesia caballi* and *Babesia equi* infections in Japan. J Vet Med Sci. 2002;64(4):325–8.

Joyner LP, Donnelly J, et al. Complement fixation tests for equine piroplasmosis (*Babesia equi* and B caballi) performed in the UK during 1976 to 1979. Equine Vet J. 1981;13(2):103–6.

Kappmeyer LS, Perryman LE, et al. A *Babesia equi* gene encodes a surface protein with homology to *Theileria* species. Mol Biochem Parasitol. 1993;62(1):121–4.

Kappmeyer LS, Perryman LE, et al. Detection of equine antibodies to babesia caballi by recombinant *B. caballi* rhoptry-associated protein 1 in a competitive-inhibition enzyme-linked immunosorbent assay. J Clin Microbiol. 1999;37(7):2285–90.

Kerber CE, Labruna MB, et al. Prevalence of equine Piroplasmosis and its association with tick infestation in the state of Sao Paulo, Brazil. Rev Bras Parasitol Vet. 2009;18(4):1–8.

Knowles DP Jr. Control of *Babesia equi* parasitemia. Parasitol Today. 1996;12(5):195–8.

Knowles DP Jr, Perryman LE, et al. Detection of equine antibody to *Babesia equi* merozoite proteins by a monoclonal antibody-based competitive inhibition enzyme-linked immunosorbent assay. J Clin Microbiol. 1991;29(9):2056–8.

Knowles DP Jr, Kappmeyer LS, et al. Antibody to a recombinant merozoite protein epitope identifies horses infected with *Babesia equi*. J Clin Microbiol. 1992;30(12):3122–6.

Kuttler KL, Goff WL, et al. Serologic response of *Babesia equi*-infected horses as measured by complement-fixation and indirect fluorescent antibody tests. Vet Parasitol. 1988;26(3–4):199–205.

Laus F, Spaterna A, et al. Clinical investigation on *Theileria equi* and *Babesia caballi* infections in Italian donkeys. BMC Vet Res. 2015;11:100.

Mahoney DF, Wright IG, et al. The identification of *Babesia equi* in Australia. Aust Vet J. 1977;53(10):461–4.

Posnett ES, Ambrosio RE. DNA probes for the detection of *Babesia caballi*. Parasitology. 1991;102(Pt 3):357–65.

Potgieter FT, De Waal DT, et al. Transmission and diagnosis of equine babesiosis in South Africa. Mem Inst Oswaldo Cruz. 1992;87(Suppl 3):139–42.

Ramsay JD, Ueti MW, et al. Lymphocytes and macrophages are infected by *Theileria equi*, but T cells and B cells are not required to establish infection *in vivo*. PLoS One. 2013;8(10):e76996.

Rhalem A, Sahibi H, et al. Validation of a competitive enzyme-linked immunosorbent assay for diagnosing *Babesia equi* infections of Moroccan origin and its use in determining the seroprevalence of *B. equi* in Morocco. J Vet Diagn Invest. 2001;13(3):249–51.

Schein E, Rehbein G, et al. *Babesia equi* (Laveran 1901) 1. Development in horses and in lymphocyte culture. Tropenmed Parasitol. 1981;32(4):223–7.

Schwint ON, Knowles DP, et al. Transmission of *Babesia caballi* by *Dermacentor nitens* (Acari: Ixodidae) is restricted to one generation in the absence of alimentary reinfection on a susceptible equine host. J Med Entomol. 2008;45(6):1152–5.

Schwint ON, Ueti MW, et al. Imidocarb dipropionate clears persistent *Babesia caballi* infection with elimination of transmission potential. Antimicrob Agents Chemother. 2009;53(10):4327–32.

Scoles GA, Ueti MW. *Amblyomma cajennense* is an intrastadial biological vector of *Theileria equi*. Parasit Vectors. 2013;6(1):306.

Scoles GA, Ueti MW. Vector ecology of equine piroplasmosis. Annu Rev Entomol. 2015;60:561–80.

Scoles GA, Hutcheson HJ, et al. Equine piroplasmosis associated with *Amblyomma cajennense* ticks, Texas, USA. Emerg Infect Dis. 2011;17(10):1903–5.

Short MA, Clark CK, et al. Outbreak of equine piroplasmosis in Florida. J Am Vet Med Assoc. 2012;240(5):588–95.

Stiller D, Goff WL, et al. *Dermacentor variabilis* and *Boophilus microplus* (Acari: Ixodidae): experimental vectors of *Babesia equi* to equids. J Med Entomol. 2002;39(4):667–70.

Taylor MA. Recent developments in ectoparasiticides. Vet J. 2001;161(3):253–68.

Taylor WM, Bryant JE, et al. Equine piroplasmosis in the United States-a review. J Am Vet Med Assoc. 1969;155(6):915–9.

Ueti MW, Palmer GH, et al. Expression of equi merozoite antigen 2 during development of *Babesia equi* in the midgut and salivary gland of the vector tick *Boophilus microplus*. J Clin Microbiol. 2003;41(12):5803–9.

Ueti MW, Palmer GH, et al. Ability of the vector tick *Boophilus microplus* to acquire and transmit *Babesia equi* following feeding on chronically infected horses with low-level parasitemia. J Clin Microbiol. 2005;43(8):3755–9.
Ueti MW, Palmer GH, et al. Persistently infected horses are reservoirs for intrastadial tick-borne transmission of the apicomplexan parasite *Babesia equi*. Infect Immun. 2008;76(8):3525–9.
Ueti MW, Mealey RH, et al. Re-emergence of the apicomplexan *Theileria equi* in the United States: elimination of persistent infection and transmission risk. PLoS One. 2012;7(9):e44713.
Uilenberg G. Babesia—a historical overview. Vet Parasitol. 2006;138(1–2):3–10.
Wise LN, Kappmeyer LS, et al. Review of equine piroplasmosis. J Vet Intern Med. 2013;27(6):1334–46.
Zapf F, Schein E. The development of *Babesia* (*Theileria*) *equi* (Laveran, 1901) in the gut and the haemolymph of the vector ticks, *Hyalomma* species. Parasitol Res. 1994a;80(4):297–302.
Zapf F, Schein E. New findings in the development of *Babesia* (*Theileria*) *equi* (Laveran, 1901) in the salivary glands of the vector ticks, *Hyalomma* species. Parasitol Res. 1994b;80(7):543–8.
Zaugg JL, Lane VM. Evaluations of buparvaquone as a treatment for equine babesiosis (*Babesia equi*). Am J Vet Res. 1989;50(5):782–5.

Trypanosoma 12

Silvina E. Wilkowsky

Abstract

Pathogenic trypanosomes affecting domestic animals are a major cause of ill-health and death. These pathogens represent a major constraint to economic development in Africa and their negative impact is increasing in South America and Asia. The key issues surrounding the main economically important animal trypanosome species and the diseases they cause in farm animals and pets are presented in this chapter including biology, structure, host-parasite interaction and immune evasion mechanisms. In spite of the fact that many biochemical and molecular aspects of these pathogens have been clearly established, effective control of the diseases they cause has proven unsatisfactory. Control of trypanosomiasis in animals involves disease monitoring and the use of curative and prophylactic trypanocidal drugs although drug resistance is becoming increasingly common. The available and environmentally acceptable vector control tactics are expensive, and reinfestation usually occurs. While vaccines against animal trypanosomes are not available; trypanotolerance, the innate ability of certain livestock breeds to tolerate African trypanosomes and remain productive, has been described as an economical and sustainable option for combating these parasites. Effective control of the great economic burden of trypanosomiasis to developing countries will be accomplished only by the coordinated international support toward better vector control programs and new and safer chemotherapeutic drugs.

S.E. Wilkowsky
Consejo Nacional de Investigación Científica y Tecnológica (CONICET),
Buenos Aires, Argentina

Instituto Nacional de Tecnología Agropecuaria (INTA),
Instituto de Biotecnología, Buenos Aires, Argentina
e-mail: wilkowsky.silvina@inta.gob.ar

M. Florin-Christensen, L. Schnittger (eds.), *Parasitic Protozoa of Farm Animals and Pets*, https://doi.org/10.1007/978-3-319-70132-5_12

12.1 Morphology, Life Cycle, and Host-Pathogen Interactions

12.1.1 Morphology

Trypanosomes are ubiquitous protozoan parasites of insects, plants, birds, fish, amphibians and mammals including humans. They are principally transmitted from one vertebrate to another by blood-sucking arthropods or leeches. They belong to the Trypanosomatidae family that includes several monogenetic and digenetic taxa which infect plants, invertebrates and vertebrates (Vickerman 1994). In mammals, the main pathogenic trypanosomes include genera whose species can cause severe diseases in humans like *Trypanosoma cruzi* and *T. brucei*, while other species like *T. vivax* and *T. congolense* or subspecies like *T. brucei brucei* are responsible of cattle diseases of great economic importance (Yaro et al. 2016). There are also a large number of nonpathogenic *Trypanosoma* species that infect domestic animals, such as *T. theileri* of bovines, *T. melophagium* of sheep and *T. caninum* of dogs that must be differentiated for diagnosis purposes (Mansfield 1977; Madeira et al. 2014). However, it should be noted that the nonpathogenic state can change whenever the host immunological status is affected. In this chapter, only species of pathogenic trypanosomes affecting farm animals and pets will be covered. The list is described in Table 12.1.

According to the mechanism of transmission by the insect vector, Hoare and Losos divided the genera *Trypanosoma* into two sections (Hoare 1972; Losos 1986): Stercoraria, which develop in the posterior part of the digestive tract and are transmitted by contamination with the feces, such as *T. cruzi*, and Salivaria in which parasites develop in the anterior part of the digestive tract and infection occurs via inoculation, such as the main African livestock pathogenic trypanosomes. Most of the pathogenic trypanosomes of domestic animals fall into the Salivaria section; however, dogs and cats play key roles as amplifying hosts and sources of Stercoraria *T. cruzi* in many domestic or peri-domestic transmission cycles (Enriquez et al. 2014; Gürtler and Cardinal 2015).

Table 12.1 General characteristics of trypanosomes affecting domestic animals

Subgenus	Parasite	Domestic hosts	Size (μm)	Vector
Trypanozoon	*Trypanosoma evansi*	Bovine, equine	15–36	Tabanids
Trypanozoon	*T. equiperdum*	Equine, donkey	15–31	None
Trypanozoon	*T. brucei brucei*	Bovine, sheep, goat	11–42	*Glossina* sp.
Nannomonas	*T. simiae*	Pigs	8–24	*Glossina* sp.
Nannomonas	*T. congolense*	Bovine, sheep, goat	8–24	*Glossina* sp.
Duttonella	*T. vivax*	Bovine, sheep, goat	21–25	*Glossina* sp. Tabanidae *Stomoxys*
Pycnomonas	*Trypanosoma suis*	Pigs	14–19	*Glossina* sp.
Schizotrypanum	*Trypanosoma cruzi*	Dogs, cats	16–21	Triatomine

Size corresponds to the mammalian stage

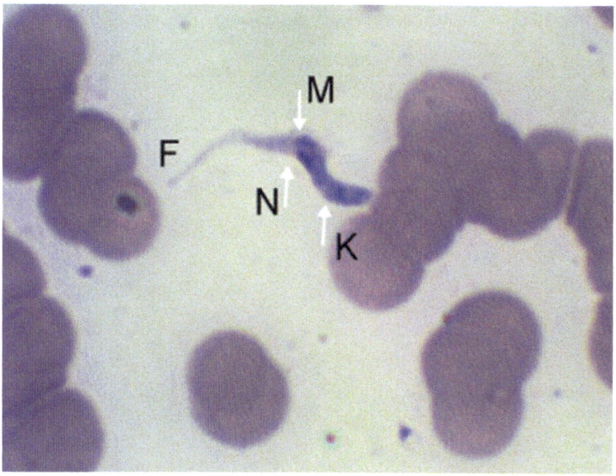

Fig. 12.1 Giemsa-stained slides of a trypomastigote form of *Trypanosoma evansi* in the blood of an infected horse. *Arrows* indicates nucleus (N), kinetoplast (K), undulating membrane (M) and the flagellum (F). (Courtesy of Fernando Dubois, CIT Formosa, Argentina)

Trypanosomes have a typical elongated shape and single flagellum (Fig. 12.1). Size ranges between 8 and 40 μm. They are motile and possess a writhing movement among red blood cells of vertebrates when observed in wet blood smears. The direction of motion of trypanosomes is toward the end of the flagellum. This part of the parasite is commonly called anterior end and has a pointed shape. On the contrary, the posterior end has a different shape, either rhomboid or rounded.

In addition to the flagellum, trypanosomes have a characteristic organelle called kinetoplastid (Simpson et al. 2002) located opposite to the flagellar basal body. The kinetoplast is an enlarged region of the mitochondrion which contains condensed mitochondrial DNA forming a network of interlocked circular molecules of different size: maxicircles and minicircles. Some trypanosomes like *Trypanosoma evansi* have lost the maxicircles of kinetoplastic mitochondrial DNA, and they are no longer able to undergo their cycle in *Glossina* vectors (Desquesnes et al. 2013).

Other features found exclusively in Kinetoplastida are the presence of multiple peroxisome-like organelles, called glycosomes, that harbor the enzymes of a major part of the glycolytic/gluconeogenic pathways (Haanstra et al. 2016), and a particular genomic organization, which arranges genes in long polycistronic transcription units. These units of up to 100 open reading frames are transcribed and processed by trans-splicing and polyadenylation. In trans-splicing, which is essential for the processing of each mRNA, an exon known as the spliced leader (SL) is added to all mRNAs (Preußer et al. 2012).

12.1.2 Life Cycle

As mentioned before, the life cycle of most trypanosomes involves an insect vector and a mammalian host. The two major patterns of life cycles are related to whether the trypanosome belongs to the Salivarian or Stercorarian section (Table 12.2). The Salivarian pattern (Fig. 12.2b) includes the parasites that cause African animal (AAT) and human trypanosomiasis, such as *T. brucei* and *T. congolense*. In this

Table 12.2 Classification of trypanosomes according to their transmission

Section	Subgenus	Species
Salivaria	*Duttonella*	*Trypanosoma vivax*
	Nannomonas	*T. congolense*, *T. simiae*
	Trypanozoon	*T. brucei*, *T. evansi*, *T. equiperdum*
	Pycnomonas	*T. suis*
Stercoraria	*Schizotrypanum*	*T. cruzi*

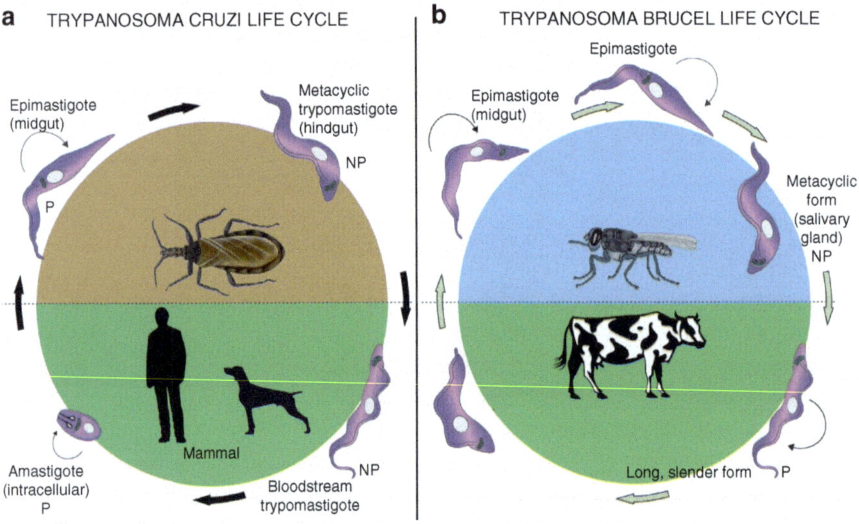

Fig. 12.2 Life cycles of (**a**) *Trypanosoma cruzi* and (**b**) *Trypanosoma brucei* as examples of the Stercoraria and Salivaria sections, respectively. *P* proliferative form, *NP* nonproliferative form

group, extracellular trypomastigote forms circulate within the blood of a vertebrate host. When a blood-feeding vector, such as the tsetse fly, ingests trypomastigotes along with its blood meal, parasites continually divide by fission within the midgut over a period of 1–2 weeks. Over the next week, they migrate anteriorly in the gut, eventually entering the salivary gland and accumulate there. Within the salivary gland, the parasite transforms into a replicative noninfective epimastigote form and continues to divide. Some epimastigotes differentiate into infectious metacyclic trypomastigotes. After inoculation into the bloodstream of a new vertebrate host, the cycle starts again, and the parasite continues to divide in this form.

Parasites from the Stercorarian section (Fig. 12.2a) include the causative agent of American trypanosomiasis or Chagas' disease (*T. cruzi*). In this group, extracellular trypomastigote forms also circulate within the blood of a mammalian vertebrate host. When the blood-feeding vector (triatomine bugs) ingests trypomastigotes along with its blood meal, these parasites differentiate into epimastigotes within the insect midgut. After about 1 week, they differentiate again in the hindgut to metacyclic trypomastigotes. When the insect host feeds, it often defecates at the same time, releasing metacyclic trypanosomes in the feces which may contaminate the wound or mucosa.

Once in the vertebrate blood, trypomastigotes are phagocytosed by cells of the host immune system, or they enter other cells, such as cardiac muscle cells. During this intracellular phase of existence, trypomastigotes differentiate into rounded amastigotes and begin multiplication. When large numbers of amastigotes accumulate intracellularly, they form pseudocysts. Eventually, these amastigotes may transform again into trypomastigotes, escape the host cell by lysis, and enter into circulation. This latter stage is the one that may be ingested by a vector or infect a new host cell, again becoming amastigotes and repeating the cycle.

Trypanosoma equiperdum, the causative agent of a sexually transmitted disease of equids called "dourine," is the only trypanosome that is not transmitted by an invertebrate vector, and it differs from other trypanosomes in that it is primarily a tissue parasite that rarely invades the blood (Claes et al. 2005).

12.1.3 Host-Pathogen Interactions

Trypanosomes live in the bloodstream and tissue spaces of a wide range of mammalian hosts, and within this environment, they challenge the immune system and therefore induce a host response. This host-parasite relationship is the most important factor in determining whether the parasitic infection will succeed or it will be resolved by the host immune system.

The different trypanosomes have developed successful means of evading the consequences of the immune system activation, and several mechanisms are involved in this complex interaction. Except for *Trypanosoma cruzi* that invades host cells and is thus an intracellular pathogen, the rest of trypanosomes that infect farm animals and pets are extracellular.

The model organism *T. brucei*, but also *T. vivax*, *T. equiperdum*, and *T. congolense*, has a repertoire of developmentally regulated genes that code for a protective cell coat of antigens throughout the mammalian infectious cycle and mediate immune evasion (Jackson et al. 2012). These proteins, called VSG (variant surface glycoproteins), are polymorphic along much of their length; they are located on the trypanosome surface and are decorated with multiple sugar residues (Horn 2014). The evasion strategy is the clone-specific singular expression of a dense VSG coat responsible for antigenic variation and infection relapses. This clone-specific expression combined with switching from one VSG to another results in that when the host antibody titers increase, the vast majority of parasites are eliminated, and only cells with distinct VSG coats survive (Horn 2014) (Fig. 12.3).

The VSG coat itself does not protect the parasite from the host immune system; in fact, it activates a lytic antibody response via the complement system. This mechanism allows opsonization and cell lysis, but only of those parasite clones carrying the VSG variant against which the response was made. However, a small proportion of the parasite population switches VSG coats, which stimulates a new antibody response to the new prevalent VSG species, and this process continues until the host immune system fails (Hovel-Miner et al. 2015). Soluble as well as membrane-fixed VSGs have been shown to contribute to macrophage over-activation and uncontrolled production

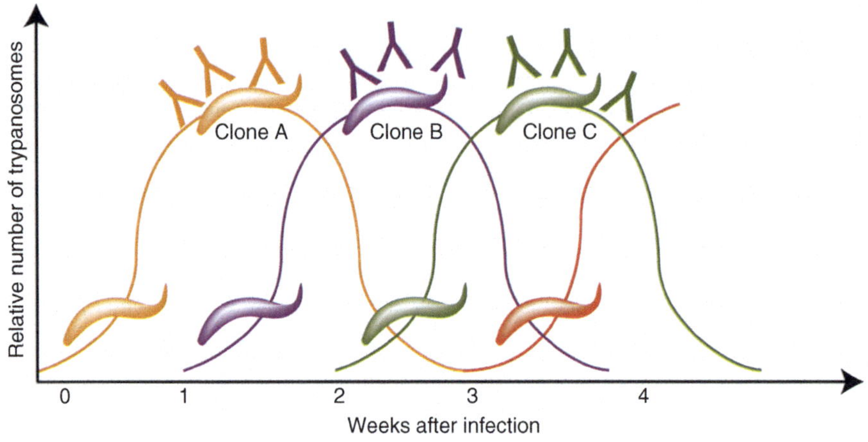

Fig. 12.3 VSG expression and switching in African trypanosomes: once antibody titers against a certain VSG variant increase, the vast majority of parasites are eliminated and only those with a distinct VSG coat survive

of harmful substances, such as tumor necrosis factor (TNF), which induces immune suppression, anemia, organ lesions, and cachexia (Antoine-Moussiaux et al. 2009).

Genome analysis of African trypanosomes has shown that a repertoire of hundreds of VSG genes under the control of a single transcription site is located at the end of the chromosomes, near the telomeres. Only one of these sites is transcriptionally active at any given time. The parasite can switch the expression to an antigenically distinct VSG by two mechanisms: (1) transcriptional VSG switching which occurs by the simultaneous activation of the transcription of a new VSG and inactivation of the previously expressed site or (2) copying a VSG from a separate expression site into the active transcription site (recombinational switch) (Hovel-Miner et al. 2015). With a repertoire of ~2000 distinct VSG genes, the parasite has an enormous potential to elude existing host antibodies by remodeling its surface coat.

This coat of VSG is also regulated during parasite differentiation. VSG expression is activated in the tsetse fly salivary gland (before infecting the mammalian host), is actively shed and renewed in bloodstream stages, and is inactivated upon return to the tsetse fly midgut (Ponte-Sucre 2016).

Although VSG are the main molecules involved in immune evasion of African trypanosomes, other molecules also play a role in this mechanism. This is the case of cysteine peptidases, which have been associated to anemia and immunosuppression, and oligopeptidases related with reduction of plasmatic atrial natriuretic factor. Parasite sialidases are other virulence factors that have been identified in trypanosomes. In infections caused by *T. congolense*, *T. vivax*, and potentially *T. evansi*, direct alteration of red blood cells seems to be an important mechanism leading to subsequent phagocytosis, in which sialidases could play a central role (Antoine-Moussiaux et al. 2009).

For intracellular trypanosomes like *T. cruzi*, the first step in the interaction process is the binding of the parasite to the host cell membrane. Parasites have a variety of surface and secreted molecules to attach and invade host cells. These molecules, in turn, activate many signal transduction pathways to ensure successful entry and intracellular survival.

In *T. cruzi*, which can invade a wide variety of phagocytic and non-phagocytic mammalian cell types, host cell invasion is linked to the activation of signal transduction pathways that lead to an increase in cytosolic calcium concentration in both the parasite and the host cell (Moreno and Docampo 2003). The parasite must transit the host cell lysosomal compartment to establish a productive intracellular infection. This pathway can be undertaken directly by inducing Ca^{2+}-dependent lysosome exocytosis or indirectly by intersecting with the endocytic or autophagic pathway. This mechanism implies that the acidic environment of lysosomes provides the appropriate conditions for the parasite-mediated disruption of the parasitophorous vacuole and release into the host cell cytosol, where replication of intracellular amastigotes occurs (Caradonna and Burleigh 2011).

With respect to *T. cruzi* surface antigens, the best characterized are members of the polymorphic gp85/transialidase superfamily that are expressed by mammalian-infective stages (Alves and Colli 2007). Transsialidase (TS) is an enzyme that transfers sialic acid from glycoconjugates to molecules of the parasite, mainly mucins. TS, as well as other members of Gp85/TS superfamily, and together with mucins and some proteases are shed into the milieu by *T. cruzi* and are related to the evasion from the parasitophorous vacuole. It was claimed that TS induces cell apoptosis which seems to be an essential event to explain the pathogenic role of *T. cruzi* in cardiac damage. Gp82, another member of the Gp85/TS superfamily, and mucin-like molecules (Gp35/50) induce signaling pathways with Gp82 eliciting Ca^{2+} mobilization in both parasite and host cell and tyrosine phosphorylation of a *T. cruzi* substrate (Yoshida 2006).

Proteases appear to be of main importance for the infection of the vertebrate host by *T. cruzi*, similarly to other pathogens or metastatic transformed cells. Cruzipain is the most abundant protease expressed by this parasite, and its role in infection has been associated with its ability to act on bradykinin receptors. Other proteases described in *T. cruzi* are the hemolysin TcTox, which helps trypomastigotes escape from the parasitophorous vacuole; POPPTc80, a serine protease that mediates hydrolysis of native collagen and fibronectin; and oligopeptidase B, a serine protease implicated in cell invasion by generating a Ca^{2+} agonist necessary for recruitment and fusion of host lysosomes at the site of parasite attachment (Alves and Colli 2007).

12.2 Diagnosis and Epidemiology

12.2.1 Diagnosis

Diagnosis of trypanosomes in farm and domestic animals can be achieved by parasitological, immunological, and molecular methods. The choice of a particular test will be guided by economic principles and the availability of expertise but especially by the diagnostic requirement. For example, different degrees of sensitivity and specificity are applied to the confirmation of the infection in an individual animal as compared to the detection of infection at a herd level (OIE World Organization for Animal Health 2013).

Parasitological methods are based on demonstrating the presence of trypanosomes. This can be achieved by inoculation of susceptible animals and by the microscopic examination of parasites in blood or lymph fluid. Animal inoculation methods appear to be more sensitive but are laborious and not useful for immediate diagnosis. The simplest microscopic methods are wet and thick or thin films of fresh blood, but sensitivity depends on the blood volume used and the operator expertise.

The hematocrit centrifuge technique (Woo 1970) is one of the most widely used parasitological methods for the detection of blood trypanosomes, in which a capillary tube is filled with blood and centrifuged. Motile trypanosomes can be viewed under a microscope between the leukocyte layer and the plasma. Another concentration technique is the Murray method (Murray et al. 1977) in which the capillary tube is cut 0.5 mm below the buffy coat, and this coat and the uppermost layer of red blood cells are extruded on to a clean microscope slide. Although relatively simple, all parasitological methods have low sensitivity when applied during the chronic phase of the disease. Differentiation of species is also difficult with these methods.

Immunological methods that detect parasite antigens such as sandwich ELISA (also called capture or antigen ELISA) or indirect ELISA for antibody detection have the benefit of allowing high-throughput screening, but both depend on suitable detection reagents (monoclonal antibodies and crude or recombinant antigens, respectively). Antigen-detection ELISAs are not recommended by the OIE because sensitivity and specificity are not suitable for the diagnosis of trypanosomiasis (OIE World Organization for Animal Health 2013).

Detection of antibodies against the parasites is usually performed by indirect fluorescent antibody test (IFAT) that uses whole parasites fixed on a slide, or indirect ELISA, which uses trypanosome lysates as target. Both antibody detection tests have high sensitivity and genus specificity, but the species specificity is low. A serious limitation of both techniques is the dependence of propagation methods (culture or animal propagation of parasites) and the standardization of whole trypanosome lysates.

Many ELISA tests using recombinant trypanosome species-specific antigens have been developed, but they remain unproven (Boulangé et al. 2002; Pillay et al. 2013; Sengupta et al. 2014). Furthermore, due to the homology between *T. congolense*, *T. vivax*, and *T. brucei* subspecies, species-specific diagnosis is difficult, although possible through analyzing the level of cross-reactions on all three species concurrently (Desquesnes et al. 2001).

For *T. evansi*, the most relevant tests recommended by the OIE are IFAT, an indirect ELISA, and a card agglutination test (CATT/T). The CATT/T test makes use of fixed and stained trypanosomes that take part in an agglutination reaction. When serum and antigens are mixed, agglutination takes place, and blue granular deposits reveal a positive reaction visible to the naked eye.

The indirect ELISA is based on soluble antigens of *T. evansi* from whole cell lysates (OIE 2008) and has been adapted to other species such as buffaloes, cattle, and pigs (Kocher et al. 2015). A recombinant antigen of *T. evansi* called GM6 which consisted of four repeat domains was tested in indirect ELISA and in an immunochromatographic test (ICT) and used in a recent epidemiological study in South Africa (Nguyen et al. 2015). For field use, CATT/T and ICT can be applied. ELISA for detecting IgG is more likely to correctly diagnose uninfected animals, while the CATT is more likely to correctly diagnose truly infected animals. ELISA would thus be suitable for verifying the disease-free status of animals prior to movement or during quarantine. CATT can be used to target individual animals for treatment with trypanocidal drugs. In areas where *T. cruzi*, *T. equiperdum*, or tsetse-transmitted trypanosomes are present, cross-reactions may occur with any so far available serological test.

It is extremely difficult to detect *T. equiperdum* in the body fluids of infected horses; therefore, diagnosis is based on serological methods like the complement fixation test. This method does not distinguish *T. equiperdum* from *T. evansi* or *T. b. brucei*, and possible cross-reactions might occur (Claes et al. 2005). These observations, along with phylogenetic analysis of total internal transcribed sequences within the three members of the subgenus *Trypanozoon* (*Trypanosoma brucei*, *Trypanosoma evansi*, and *Trypanosoma equiperdum*), showed a high degree of genetic similarities providing evidence to postulate that some *T. equiperdum* strains are actually *T. b. brucei* or members of a subspecies of *T. brucei* (*Trypanosoma brucei equiperdum*) and that other *T. equiperdum* strains are misidentified and are, in fact, *T. evansi* (Wen et al. 2016).

Molecular techniques for trypanosome detection and differentiation in livestock, humans, and tsetse flies are becoming available, but they generally require well-equipped laboratories. In conventional PCR, a large number of samples can be processed at one time, making it potentially suitable for large-scale surveys. However, at the moment, the cost of PCR analyses is prohibitive for the routine use of these tests. Specific repetitive sequences of *T. vivax* and of the three types of *T. congolense* are used, and a common primer set is also available for detection of the three *T. brucei* subspecies (Desquesnes and Davila 2002). PCR restriction fragment length polymorphism (RFLP) assays and ITS1 of ribosomal DNA amplification allow the identification of all *Trypanosoma* species as single or mixed infections using one single test (Delespaux et al. 2003); however, these tests are not yet suitable for routine diagnosis.

For *T. cruzi* diagnosis in dogs and cats, molecular and serological methods are available and were tested in naturally infected populations (Enriquez et al. 2014).

Loop-mediated isothermal amplification tests have been developed for many trypanosomes affecting domestic animals (Thekisoe et al. 2010; Njiru et al. 2010, 2011; Chaouch et al. 2013).

12.2.2 Epidemiology

In vector-borne trypanosomiasis, as in any other parasitic disease, infection occurrence depends on the interaction of three elements that need to be present within a particular environment (Fig. 12.4): (1) the mammalian reservoirs of parasites that can also be hosts suffering from the disease, (2) the insect vector which is critical for transmission and fully dependent on environmental factors, and (3) the pathogenic parasite, the trypanosome itself. Because of these interactions, transmission of trypanosomiasis occurs in spatially limited areas called "foci," which are located mainly in remote rural areas.

In Africa, tsetse flies are present in 36 countries across sub-Saharan regions, between the latitudes of 14°N and 29° (Moloo 1993). The wide distribution of the *Glossina* vector is one of the greatest obstacles to livestock development and has prevented the establishment of sustainable agricultural systems in many areas of enormous potential. In addition, climate change, deforestation, and loss of native species

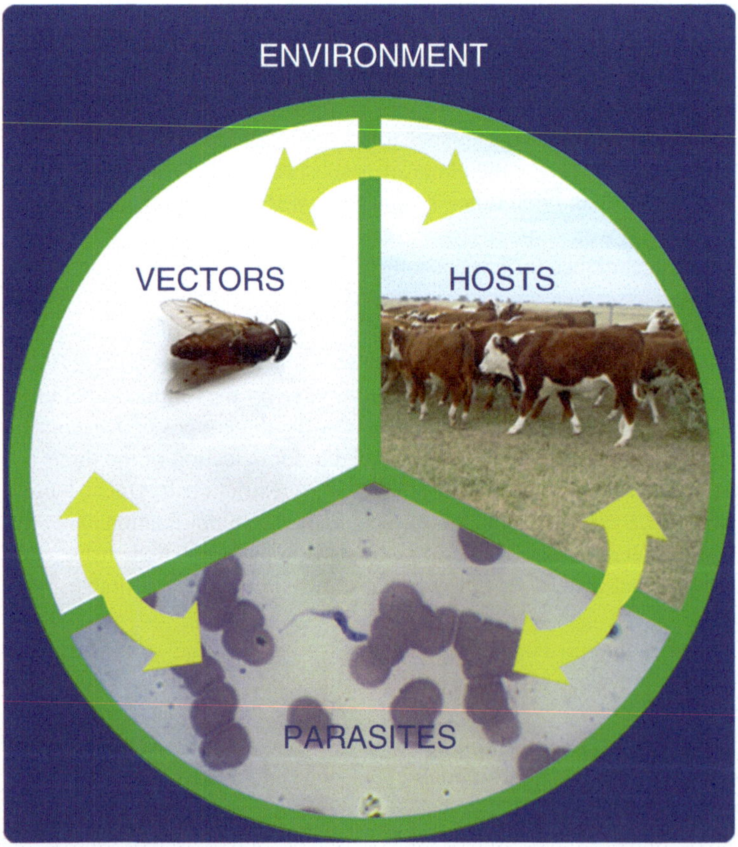

Fig. 12.4 Effect of environmental changes on vector-host-parasite relationships in tsetse-transmitted trypanosomiasis and their repercussions on the epidemiology of the disease

as blood meal for vectors are influencing the distribution of the disease, and domestic animals and particularly livestock are becoming the main source of food for tsetse flies (Van den Bossche et al. 2010). It is estimated that trypanosomiasis reduces the production of meat and milk from cattle by at least 50% in tsetse-infested areas (Swallow 1999). According to FAO, AAT causes about 3 million deaths in cattle per year. The economic losses in cattle production alone are in the range of US$ 1.0–1.2 billion. An estimated evaluation extrapolated for the total tsetse-infested lands calculates total losses, in terms of agricultural gross domestic product, at US$ 4.75 billion per year (FAO 2016).

Cyclical transmission (Fig. 12.2) occurs with *T. congolense*, *T. vivax*, *T. simiae*, *T. suis*, and *T. brucei* and the trypanosomes responsible for human sleeping sickness, *T. gambiense* and *T. rhodesiense*. This transmission is achieved by *Glossina* spp. flies, which are strictly blood-feeding insects living exclusively in tropical Africa. There are about 30 species or subspecies, classified in three groups depending on the environment they inhabit. The subgenus *Nemorhina*, also known as the palpalis group, is found in Western and Central Africa and lives in vegetation close to a water source, such as forests, gallery forests, riverbanks and lakes. The subgenus *Glossina* sensu stricto, the morsitans group, occurs in woodland Savannah mainly in Eastern Africa and is linked to the presence of wild fauna and cattle. Finally, the subgenus *Austenina*, the fusca group, lives in primary forest belts, including rainforests, savannah, and coastal forests (Franco et al. 2005).

Mechanical transmission is effected by various blood-sucking insects, such as flies of the family Tabanidae—horseflies—and *Stomoxys* spp. Tsetse flies themselves can also act as mechanical vectors. During interrupted blood feeding on an infected animal, the biting insect passes trypomastigotes to another healthy host. The time between the two feeds is crucial for effective transmission because the trypanosomes die when blood dries. This form of transmission is the rule for *T. evansi* but may also occur with trypanosomes habitually transmitted cyclically by *Glossina*, particularly *T. vivax* which may therefore be found in regions far from the *Glossina* distribution area, such as South and Central America.

In general, natural transmission of *T. equiperdum* occurs only during copulation of horses. However, experimental infections inoculating parasites by the intravenous or intraperitoneal route indicate that mechanical transmission by blood-feeding flies cannot be excluded as a possible route.

American trypanosomiasis (AT) or Chagas' disease (CD) caused by *T. cruzi* follows the pattern of other vector-borne diseases, in that parasite distribution correlates with eco-epidemiological conditions that favor the presence of its triatomine vector. Besides classical vector-borne transmission, congenital, transfusion-associated, and oral/foodborne transmission is also possible. The disease occurs in North, Central, and South America where the parasite is mainly transmitted via infected feces from numerous different species of the genera *Triatoma*, *Rhodnius*, and *Panstrongylus*.

T. cruzi is quite an unrestricted parasite and can infect approximately 180 species of mammals (WHO 2002). Dogs and cats are major reservoir hosts of the parasite in the domestic environment throughout the Americas and display much higher infectiousness to triatomine bugs than humans (Enriquez et al. 2014). Studies demonstrated that vectors prefer feeding on mammals and dogs over birds or cats and that only a small fraction of the infected dogs and cats are highly infectious to triatomine bugs and could be considered superspreaders (Gürtler and Cardinal 2015). Household dogs are estimated to be at a threefold higher risk for transmission than cats in regions of endemicity (Gürtler and Cardinal 2015). Oral transmission of *T. cruzi* in sylvatic and domestic animals was also described due to insectivorous habits (Esch and Petersen 2013).

12.3 Clinical Effects, Prevention, and Treatment

12.3.1 Clinical Effects

The features of a trypanosomal infection vary considerably according to the species of trypanosome involved and a number of host factors. Depending on the affected species and the genetic background of the animal or breed, the outcome will fluctuate in either susceptibility or relative resistance to infection, which has been termed "trypanotolerance" (Naessens 2006).

Within the parasite species, genetic variation results in different clinical outcomes and relevance to disease in domestic animals. In cattle, for example, *T. congolense* Savannah has a greater pathogenicity compared with *T. congolense* forest or Kilifi strains (Morrison et al. 2016).

However, general characteristics of animal trypanosomiasis are intermittent fever and presence of parasites in the blood, anemia, loss of body condition, reduced productivity, and, often, high mortality. Hemolytic anemia is the main sign of trypanosomiasis, and hemolysis is immune-mediated and results from the deposition of immunocomplexes on the erythrocyte surface, triggering erythrophagocytosis. Nonspecific phagocytosis of erythrocytes by means of a hyperactivated phagocytic system or specific phagocytosis of damaged cells may contribute to anemia (Osório et al. 2008). Cattle usually have a chronic course of disease with high mortality, especially if there is poor nutrition or other stress factors are present. Ruminants die, usually of congestive heart failure, or gradually recover if the level of challenge is low, while stress results in relapse. The strong immune response to trypanosomes and the formation of immune complexes cause inflammation and contribute to the signs and lesions of the disease. Immunosuppression is another key characteristic of trypanosome infections which increases the risk of concurrent infections (Osório et al. 2008).

In some *Trypanosoma* species, such as *T. congolense* and *T. vivax* and to a minor extent *T. brucei*, an immune reaction to the entrance of metacyclic parasites at the biting site causes a huge swelling in the skin and enlargement of the local draining lymph node. Once metacyclics differentiate into bloodstream forms and migrate to the blood, they cause a systemic infection. Both *T. congolense* and *T. vivax* are intravascular parasites, although in the latter, some strains can be found extravascularly in lymph nodes, eyes, choroid plexus, and cerebrospinal fluid (Osório et al. 2008). For this reason, *T. vivax* can directly damage tissues, as when ocular tissues are involved, with ensuing keratitis, or when the cardiac muscle exhibits lesions characterized by hemorrhage and mononuclear infiltration (Osório et al. 2008).

T. congolense shows a preference for microvascular sites where it can potentially bind to the endothelium. Parasites provoke dilation of the microvasculature, compromising capillary circulation and impairing nutrient and metabolite exchange. No cerebral infections have been observed with monospecific infections with *T. congolense* or *T. vivax* in cattle, but a high frequency of central nervous system involvement was observed in concurrent infections (Ikede and Losos 1972).

T. brucei and *T. evansi* can leave the blood vessels and invade solid tissues (Ikede and Losos 1972). Trypanosomes can be found in peritoneal, pericardic, as well as cerebrospinal fluids and in the aqueous and vitreous humors of the eye. *T. b. brucei* has been reported to cause central nervous system abnormalities (Ikede and Losos 1972).

Both domestic and wild pigs can become infected with various species of tsetse-transmitted trypanosomes but infrequently show symptoms or disease pathology unless coinfected with *T. simiae*, which is highly pathogenic for domestic pigs (Hamill et al. 2013). According to these authors, domestic pigs are acting as a significant reservoir of infection mainly for *T. vivax* and *T. congolense* in cattle, sheep, and goats.

Equines are considered to be the only natural host of *T. equiperdum*, which causes a chronic disease in horses. The first stage of the disease is characterized by edema, tumefaction and damage of the genitalia and begins 1–2 weeks after infection. The second stage is pathognomonic for dourine, where periodical, typical cutaneous plaques or skin thicknesses can occur, with sizes ranging from extremely small to hand-sized. The third phase of dourine is characterized by progressive anemia, disorders of the nervous system mainly paralysis of the hind legs, paraplegia and, finally, death (Claes et al. 2005).

Infection of dogs with *T. cruzi* resembles Chagas' disease in humans, with the three characteristic phases: acute, indeterminate or latent, and chronic. The disease usually manifests with myocarditis, although many infected dogs remain asymptomatic for life. The cause of the myocarditis is thought to be due to cell damage (when trypomastigotes rupture from myocardiocytes) and the resulting inflammation. During peak parasitemia, clinical signs of generalized lymphadenopathy and acute myocarditis may appear. Lethargy, lymphadenopathy, pale mucous membranes, and, in some cases, splenomegaly and hepatomegaly are the main presenting signs in young puppies. In older dogs, clinical signs are often much less severe and sometimes not apparent at all (Barr 2009).

12.3.2 Prevention and Treatment

Control of trypanosomiasis in livestock must be carried out through wide geographic regions. For instance, African animal trypanosomes remain endemic in 37 of 54 African countries, and the disease has a very significant combined economic and health burden in the sub-Saharan African region. These negative effects are also increasing in South America and Southeast Asia where economic consequences have not been satisfactorily estimated. Control measures involve disease monitoring, the use of curative and prophylactic trypanocidal drugs, and area-wide tsetse control. Vaccines against animal trypanosomes are not available.

For *T. equiperdum*, there are no officially approved drugs to treat horses suffering from dourine, and international regulations currently impose the slaughtering of serologically positive horses. The situation regarding the use of chemotherapeutic drugs is also very unsatisfactory. There are only three commonly used drugs, and they were all developed over 50 years ago, they have a narrow therapeutic index, and drug resistance is becoming increasingly common (Holmes 2013). In 2008, 17 sub-Saharan countries reported trypanocidal resistance problems, and by early 2015, this number had risen to 21 countries (Tsegaye et al. 2015). The drugs currently in use are diminazene aceturate, which is purely curative in action, homidium bromide and chloride, which are curative with some prophylactic activity, and isometamidium which is curative but also has a strong prophylactic action. As with human trypanosomiasis, there is an urgent need for safer new drugs for the treatment and control of trypanosomiasis in domestic animals.

Trypanotolerance, the innate ability of certain livestock breeds to tolerate African trypanosomes and remain productive, has been described as an economical and sustainable option for combating trypanosomiasis. The use of taurine bovine breeds such as N'Dama as a control strategy could have a major positive effect on long-term food security of the affected regions (Naessens 2006; Yaro et al. 2016), and is the only option available for many farmers.

Several approaches to control the transmitting tsetse fly vector have been tested, including the sterile insect release technique, the destruction of fly habitat, the use of insect traps, treatment of livestock with insecticide and mass spraying of insecticide. These interventions yielded limited positive outcomes in the first half of the last century along with negative environmental consequences such as pollution of water bodies and deforestation (Yaro et al. 2016). Vector control by habitat destruction or elimination of wildlife hosts is no longer considered acceptable and the ground spraying of known tsetse resting sites with persistent insecticide formulations is only allowed under exceptional situations (Holmes 2013).

Control methods for the prevention of *T. cruzi* infection in dogs include limiting or complete avoidance of vectors and possible reservoir hosts to block domestic or peri-urban transmission; use of insecticide collars, sprays, pour-ons, and treatment. Euthanasia of *T.* cruzi-infected dogs is neither feasible as a public health measure nor warranted. Enhanced prevention may be achieved through targeted residual insecticide spraying of kennels or other dog resting sites and use of insecticide-impregnated collars with repellent effects. Treatment of infected dogs with the two available parasiticidal drugs (benznidazole and nifurtimox) is feasible and moderately effective, especially during the acute phase (Barr 2009; Gurtler and Cardinal 2015).

References

Alves MJM, Colli W. Trypanosoma cruzi: adhesion to the host cell and intracellular survival. IUBMB Life. 2007;59:274–9. https://doi.org/10.1080/15216540701200084.

Antoine-Moussiaux N, Biischer P, Desmecht D. Host-parasite interactions in trypanosomiasis: on the way to an antidisease strategy. Infect Immun. 2009;77:1276–84. https://doi.org/10.1128/IAI.01185-08.

Barr SC. Canine Chagas' disease (American Trypanosomiasis) in North America. Vet Clin North Am Small Anim Pract. 2009;39:1055–64. https://doi.org/10.1016/j.cvsm.2009.06.004.

Boulangé A, Katende J, Authié E. Trypanosoma congolense: expression of a heat shock protein 70 and initial evaluation as a diagnostic antigen for bovine trypanosomosis. Exp Parasitol. 2002;100:6–11. https://doi.org/10.1006/expr.2001.4667.

Caradonna KL, Burleigh BA. Mechanisms of host cell invasion by trypanosoma cruzi. 1st ed. Amsterdam: Elsevier; 2011.

Chaouch M, Mhadhbi M, Adams ER, et al. Development and evaluation of a loop-mediated isothermal amplification assay for rapid detection of Leishmania infantum in canine leishmaniasis based on cysteine protease B genes. Vet Parasitol. 2013;198:78–84. https://doi.org/10.1016/j.vetpar.2013.07.038.

Claes F, Büscher P, Touratier L, Goddeeris BM. Trypanosoma equiperdum: master of disguise or historical mistake? Trends Parasitol. 2005;21:316–21. https://doi.org/10.1016/j.pt.2005.05.010.

Delespaux V, Ayral F, Geysen D, Geerts S. PCR-RFLP using Ssu-rDNA amplification: applicability for the diagnosis of mixed infections with different trypanosome species in cattle. Vet Parasitol. 2003;117:185–93. https://doi.org/10.1016/j.vetpar.2003.08.004.

Desquesnes M, Bengaly Z, Millogo L, et al. The analysis of the cross-reactions occurring in antibody-ELISA for the detection of trypanosomes can improve identification of the parasite species involved. Ann Trop Med Parasitol. 2001;95:141–55. doi: 1080/00034980120050251.

Desquesnes M, Dávila AMR. Applications of PCR-based tools for detection and identification of animal trypanosomes: A review and perspectives. Vet Parasitol. 2002;109:213–31. https://doi.org/10.1016/S0304-4017(02)00270-4.

Desquesnes M, Holzmuller P, Lai DH, et al. Trypanosoma evansi and surra: a review and perspectives on origin, history, distribution, taxonomy, morphology, hosts, and pathogenic effects. Biomed Res Int. 2013;2013:194176. https://doi.org/10.1155/2013/194176.

Enriquez GF, Bua J, Orozco MM, et al. High levels of Trypanosoma cruzi DNA determined by qPCR and infectiousness to Triatoma infestans support dogs and cats are major sources of parasites for domestic transmission. Infect Genet Evol. 2014;25:36–43. https://doi.org/10.1016/j.meegid.2014.04.002.

Esch KJ, Petersen CA. Transmission and epidemiology of zoonotic protozoal diseases of companion animals. Clin Microbiol Rev. 2013;26:58–85. https://doi.org/10.1128/CMR.00067-12.

FAO Programme against African trypanosomiasis. 2016. http://www.fao.org/ag/againfo/programmes/en/paat/disease.html.

Franco JR, Simarro PP, Diarra AJJ. Epidemiology of human African trypanosomiasis. Clin Epidemiol. 2005;6:257–75. https://doi.org/10.2147/CLEP.S39728.

Gürtler RE, Cardinal MV. Reservoir host competence and the role of domestic and commensal hosts in the transmission of Trypanosoma cruzi. Acta Trop. 2015;151:32–50. https://doi.org/10.1016/j.actatropica.2015.05.029.

Gurtler RE, Cardinal MV. Reservoir host competence and the role of domestic and commensal hosts in the transmission of Trypanosoma cruzi. Acta Trop. 2015;151:32–50. https://doi.org/10.1016/j.actatropica.2015.05.029.

Haanstra JR, González-Marcano EB, Gualdrón-López M, Michels PAM. Biogenesis, maintenance and dynamics of glycosomes in trypanosomatid parasites. Biochim Biophys Acta. 2016;1863:1038–48. https://doi.org/10.1016/j.bbamcr.2015.09.015.

Hamill LC, Kaare MT, Welburn SC, Picozzi K. Domestic pigs as potential reservoirs of human and animal trypanosomiasis in Northern Tanzania. Parasit Vectors. 2013;6:322. https://doi.org/10.1186/1756-3305-6-322.

Hoare C. The trypanosomes of mammals: a zoological monograph. Oxford: Blackwell Scientific; 1972.

Holmes P. Tsetse-transmitted trypanosomes—their biology, disease impact and control. J Invertebr Pathol. 2013;112:S11–4. https://doi.org/10.1016/j.jip.2012.07.014.

Horn D. Antigenic variation in African trypanosomes. Mol Biochem Parasitol. 2014;195:123–9. https://doi.org/10.1016/j.molbiopara.2014.05.001.

Hovel-Miner G, Mugnier M, Papavasiliou FN, et al. A host-pathogen interaction reduced to first principles: antigenic variation in T. brucei. Results Probl Cell Differ. 2015;57:23–46. https://doi.org/10.1007/978-3-319-20819-0_2.

Ikede BO, Losos GJ. Pathological changes in cattle infected with Trypanosoma brucei. Vet Pathol. 1972;9:272–7. https://doi.org/10.1177/030098587200900407.

Jackson AP, Berry A, Aslett M, et al. Antigenic diversity is generated by distinct evolutionary mechanisms in African trypanosome species. Proc Natl Acad Sci. 2012;109:3416–21. https://doi.org/10.1073/pnas.1117313109.

Kocher A, Desquesnes M, Kamyingkird K, et al. Evaluation of an indirect-ELISA test for Trypanosoma evansi infection (Surra) in buffaloes and its application to a serological survey in Thailand. Biomed Res Int. 2015;2015:1–8. https://doi.org/10.1155/2015/361037.

Losos J. Infectious tropical diseases of domestic animals. Harlow Essex: Longman Scientific and Technical; 1986.

Madeira MF, Almeida ABPF, Barros JHS, et al. Trypanosoma caninum, a new parasite described in dogs in Brazil: aspects of natural infection. J Parasitol. 2014;100:231–4. https://doi.org/10.1645/13-297.1.

Mansfield JM. Nonpathogenic trypanosomes of mammals. In: Kreier JP, editor. Parasitic protozoa, Taxonomy, kinetoplastids, and flagellates of fish, vol. 1. London: Academic; 1977. p. 297–327.

Moloo SK. The distribution of *Glossina* species in Africa and their natural hosts. Int J Trop Insect Sci. 1993;14:511–27. https://doi.org/10.1017/S1742758400014211.

Moreno SNJ, Docampo R. Calcium regulation in protozoan parasites. Curr Opin Microbiol. 2003;6:359–64. https://doi.org/10.1016/S1369-5274(03)00091-2.

Morrison LJ, Vezza L, Rowan T, Hope JC. Animal African trypanosomiasis: time to increase focus on clinically relevant parasite and host species. Amsterdam: Elsevier; 2016.

Murray M, Murray PK, McIntyre WI. An improved parasitological technique for the diagnosis of African trypanosomiasis. Trans R Soc Trop Med Hyg. 1977;71:325–6.

Naessens J. Bovine trypanotolerance: a natural ability to prevent severe anaemia and haemophagocytic syndrome? Int J Parasitol. 2006;36:521–8. https://doi.org/10.1016/j.ijpara.2006.02.012.

Nguyen T-T, Motsiri MS, Taioe MO, et al. Application of crude and recombinant ELISAs and immunochromatographic test for serodiagnosis of animal trypanosomosis in the Umkhanyakude district of KwaZulu-Natal province, South Africa. J Vet Med Sci. 2015;77:217–20. https://doi.org/10.1292/jvms.14-0330.

Njiru ZK, Ouma JO, Enyaru JC, Dargantes AP. Loop-mediated Isothermal Amplification (LAMP) test for detection of Trypanosoma evansi strain B. Exp Parasitol. 2010;125:196–201. https://doi.org/10.1016/j.exppara.2010.01.017.

Njiru ZK, Ouma JO, Bateta R, et al. Loop-mediated isothermal amplification test for Trypanosoma vivax based on satellite repeat DNA. Vet Parasitol. 2011;180:358–62. https://doi.org/10.1016/j.vetpar.2011.03.021.

OIE World Organisation for Animal Health. Trypanosoma evansi infections (including surra). In: OIE manual of diagnostic tests and vaccines (mammals, birds and bees). 7th ed. Paris: Office International des Épizooties; 2008. p. 66.

OIE World Organization for Animal Health. Trypanosomosis (tse-tse transmitted). In: Manual of diagnostic tests and vaccines for terrestrial animals. 2013.

Osório ALAR, Madruga CR, Desquesnes M, et al. Trypanosoma (Duttonella) vivax: its biology, epidemiology, pathogenesis, and introduction in the New World—a review. Mem Inst Oswaldo Cruz. 2008;103:1–13.

Pillay D, Izotte J, Fikru R, et al. Trypanosoma vivax GM6 antigen: a candidate antigen for diagnosis of African animal trypanosomosis in cattle. PLoS One. 2013;8:1–10. https://doi.org/10.1371/journal.pone.0078565.

Ponte-Sucre A. An overview of trypanosoma brucei infections: an intense host-parasite interaction. Front Microbiol. 2016;7:2126. https://doi.org/10.3389/fmicb.2016.02126.

Preußer C, Jaé N, Bindereif A. MRNA splicing in trypanosomes. Int J Med Microbiol. 2012;302:221–4. https://doi.org/10.1016/j.ijmm.2012.07.004.

Sengupta PP, Rudramurthy GR, Ligi M, et al. Sero-diagnosis of surra exploiting recombinant VSG antigen based ELISA for surveillance. Vet Parasitol. 2014;205:490–8. https://doi.org/10.1016/j.vetpar.2014.08.012.
Simpson AGB, Lukes J, Roger AJ. The evolutionary history of kinetoplastids and their kinetoplasts. Mol Biol Evol. 2002;19:2071–83. https://doi.org/10.1093/oxfordjournals.molbev.a004032.
Swallow B. Impacts of trypanosomiasis on African agriculture. Nairobi: Int Livest Res Institute; 1999. p. 1–46.
Thekisoe OMM, Coronel-Servian AM, Fukumoto S, et al. Detection of Trypanosoma cruzi and T. rangeli infections from Rhodnius pallescens bugs by loop-mediated isothermal amplification (LAMP). Am J Trop Med Hyg. 2010;82:855–60. https://doi.org/10.4269/ajtmh.2010.09-0533.
Tsegaye B, Dagnachew S, Terefe G. Review on drug resistant animal trypanosomes in Africa and overseas. African J Basic Appl Sci. 2015;7:73–83. https://doi.org/10.5829/idosi.ajbas.2015.7.2.9370.
Van den Bossche P, de La Rocque S, Hendrickx G, Bouyer J. A changing environment and the epidemiology of tsetse-transmitted livestock trypanosomiasis. Trends Parasitol. 2010;26:236–43. https://doi.org/10.1016/j.pt.2010.02.010.
Vickerman K. The evolutionary expansion of the trypanosomatid flagellates. Int J Parasitol. 1994;24:1317–31.
Wen Y-Z, Lun Z-R, Zhu X-Q, et al. Further evidence from SSCP and ITS DNA sequencing support Trypanosoma evansi and Trypanosoma equiperdum as subspecies or even strains of Trypanosoma brucei. Infect Genet Evol. 2016;41:56–62. https://doi.org/10.1016/j.meegid.2016.03.022.
Woo PT. The haematocrit centrifuge technique for the diagnosis of African trypanosomiasis. Acta Trop. 1970;27:384–6.
World Health Organization. Control of chagas disease: Second report of the WHO Expert Committee. Geneva: WHO; 2002.
Yaro M, Munyard KA, Stear MJ, Groth DM. Combatting African Animal Trypanosomiasis (AAT) in livestock: the potential role of trypanotolerance. Vet Parasitol. 2016;225:43–52. https://doi.org/10.1016/j.vetpar.2016.05.003.
Yoshida N. Molecular basis of mammalian cell invasion by Trypanosoma cruzi. An Acad Bras Cienc. 2006;78:87–111. https://doi.org/10.1590/S0001-37652006000100010.

Leishmania

13

Anabel Elisa Rodriguez, José Octavio Estévez,
María Cecilia Nevot, Alejandra Barrios,
and Monica Florin-Christensen

Abstract

Leishmania spp. are kinetoplastid protozoan parasites that infect numerous mammalian hosts, including humans, and are transmitted by the bite of female phlebotomine sand flies. Their distribution area has been broadly subdivided into the "New World"–the Americas, and the "Old World"–Africa, Asia, and Europe. The disease complex they cause, known as leishmaniosis or leishmaniasis, is endemic in large areas of the tropics, subtropics, and the Mediterranean basin, affecting more than 98 countries. More than 23 species of *Leishmania* have been described, most of which are zoonotic. The most important *Leishmania* parasite that infects domestic animals is *L. infantum*, also known as *L. chagasi* in Latin America. Dogs are very susceptible to this parasite and act as reservoirs. They

A.E. Rodriguez (✉)
Institute of Pathobiology, Center for Research on Veterinary and Agronomic Sciences (CICVyA), National Institute of Agricultural Technology (INTA-Argentina), Hurlingham, Province of Buenos Aires, Argentina
e-mail: rodriguez.anabel@inta.gob.ar

J.O. Estévez • M.C. Nevot
Veterinaria del Oeste, Posadas, Misiones, Argentina

A. Barrios
Microbiology and Parasitology Department, School of Health Sciences, National University of Salta, Salta, Argentina

M. Florin-Christensen
Institute of Pathobiology, Center for Research on Veterinary and Agronomic Sciences (CICVyA), National Institute of Agricultural Technology (INTA-Argentina), Hurlingham, Province of Buenos Aires, Argentina

National Council of Scientific and Technological Research (CONICET), Buenos Aires, Argentina

School of Exact, Chemical and Natural Sciences (FCEQN), University of Moron, Moron, Province of Buenos Aires, Argentina

M. Florin-Christensen, L. Schnittger (eds.), *Parasitic Protozoa of Farm Animals and Pets*, https://doi.org/10.1007/978-3-319-70132-5_13

may suffer from a complex and deadly syndrome, canine leishmaniosis, though many course asymptomatic infections. Cats and horses can also be infected, with milder clinical manifestations. Several serological and molecular diagnostic methods have been developed, but the gold standard is still the demonstration of parasites in stained tissue smears. Control strategies are largely limited to destruction of animal reservoirs, treatment of infected patients, and management of sand fly populations. Development of an effective vaccine against leishmaniosis is an active field of research.

13.1 Morphology, Life Cycle, and Host-Pathogen Interactions

13.1.1 Morphology

Leishmania parasites present two main morphologically distinguishable forms: amastigotes and promastigotes (Fig. 13.1).

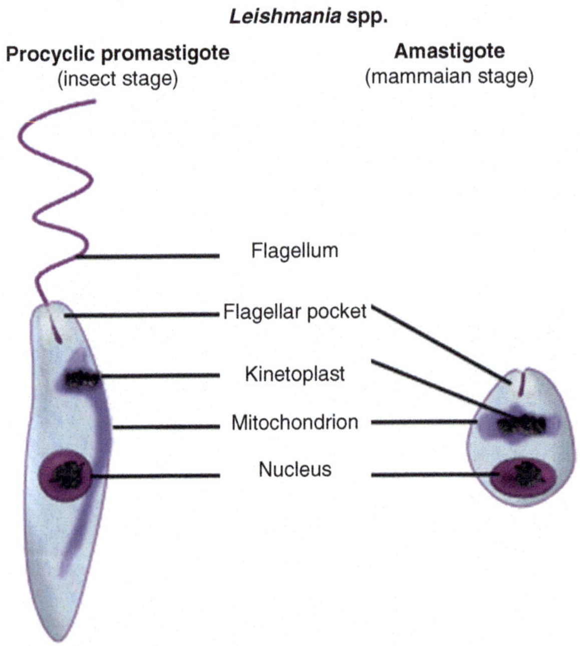

Fig. 13.1 Schematic representation of a procyclic promastigote and an amastigote of *Leishmania* sp. The promastigote is characterized by the presence of a free flagellum with no undulating membrane that emerges from the "anterior end", which receives this name because swimming takes place in this direction. The kinetoplast is located toward the anterior end. The amastigote is spherical in shape and has no free flagellum, although a basal body and the base of the flagellum are still present. The kinetoplast is usually detectable as a darkly stained body near the nucleus. This form is a non-motile intracellular stage

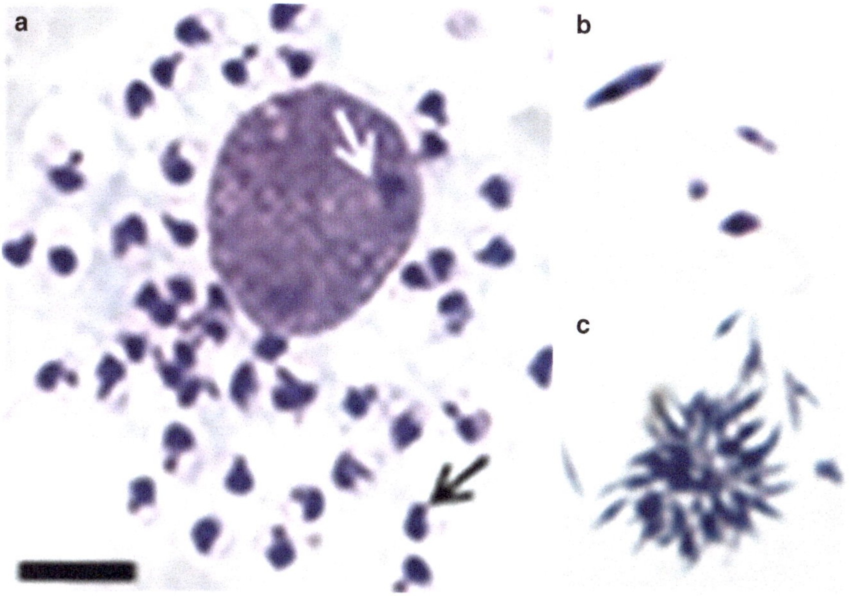

Fig. 13.2 Giemsa-stained *Leishmania* sp. amastigotes and promastigotes. (**a**) Intracellular and extracellular oval-shaped *L. infantum* amastigotes from a bone marrow smear of an infected dog. A nucleus and a small rod-shaped kinetoplast are stained in each amastigote (*black arrow*). An amastigote inside a macrophage is shown (*white arrow*). (**b** and **c**) In vitro cultured promastigotes of *L. braziliensis*, reference strain M2903, free (**b**) or forming a rosette (**c**). Magnification 1000×. Bar: 10 μm

Amastigotes are replicative forms found in mammalian hosts. They thrive within cells of the mononuclear phagocytic system, including macrophages, monocytes, and Langerhans cells. The presence of amastigotes inside neutrophils and fibroblasts in skin lesions has also been reported (Laskay et al. 2003; Ritter et al. 2009). They have oval or rounded shape, with dimensions ranging from 2.5 to 6.5 μm in length and 1.5 to 3 μm in width. There is no free flagellum. The nucleus is relatively large and generally eccentric, and a kinetoplast, consisting of a rodlike mitochondria and unique DNA, is closely associated to the flagellar pocket and the basal body of the flagellum (OIE 2014). They can be observed upon microscopic examination of smears of tissue suspensions or biopsies of infected individuals after Giemsa staining (Figs. 13.1 and 13.2a).

Promastigotes are found in the digestive system of sand flies and can be in vitro cultured. They are elongated, fusiform, extracellular, and motile. Larger than amastigotes, they measure between 15 to 20 μm in length and 1.5 to 3.5 μm in width. The nucleus is oval and central, and the kinetoplast appears as an electrodense granular band within the extension of the mitochondria, located at 1–2 μm from the anterior end of the parasite. From this end, a single flagellum, measuring between 15 and 28 μm, emerges. The axoneme, the shaft within a flagellum, contains 20 microtubules arranged as 9 doublets and 2 singlets. It originates in the

Morpholological category	Body length (μm)	Body width (μm)	Flagellum (F) vs Body(B) length	Illustration
Amastigote	Ovoid body form, 2.5 to 6.5 um in length and 1.5 to 3 um in width, nofreeflagellum			
Procyclic promastigote	6.5 to 11.5	>1.5	F < B	
Nectomonard promastigote	>12	>1.5	Variable flagellar length	
Leptomonard promastigote	6.5 to 11.5	>1.5	F > B	
Haptomonard promastigote	Variable body form		Variable flagellar length	
Metacyclic promastigote	6.5 to 11.5	<1.5	F > B	

Fig. 13.3 Morphological characteristics of *Leishmania* sp. life cycle stages

basal body and is contained within the flagellar pocket, a small invagination of the plasma membrane where the flagellum emerges from the cell body. In vitro cultured promastigotes are morphologically identical to developmental stages found in infected phlebotomine sand flies (Figs. 13.1 and 13.2b) (OIE 2014). According to their morphology and function, different categories of promastigotes have been described in *Leishmania mexicana* and *L. infantum*: (1) *procyclic*, (2) *nectomonad*, (3) *haptomonad*, (4) *leptomonad*, and (5) *metacyclic promastigotes*. Their morphological differences can be observed in Fig. 13.3. While a similar body length of <11.5 μm is characteristic of three of these forms—procyclic, leptomonad, and metacyclic promastigotes—body width and the relative length of the flagellum with respect to the body allow to discriminate between them (Rogers et al. 2002; Gossage et al. 2003). Although free swimming individual promastigotes are easily observed in culture, aggregation of promastigotes, sometimes forming rosettes, is frequent (Fig. 13.2c).

13.1.2 Life Cycle

Leishmania is a two-host parasite, alternating between a mammalian host and a vector insect during its life cycle. Transmission to the mammalian hosts—which include humans and domestic animals such as canines, equines, and felines—is effected by the bite of female sand flies of the genus *Phlebotomus* in the Old World

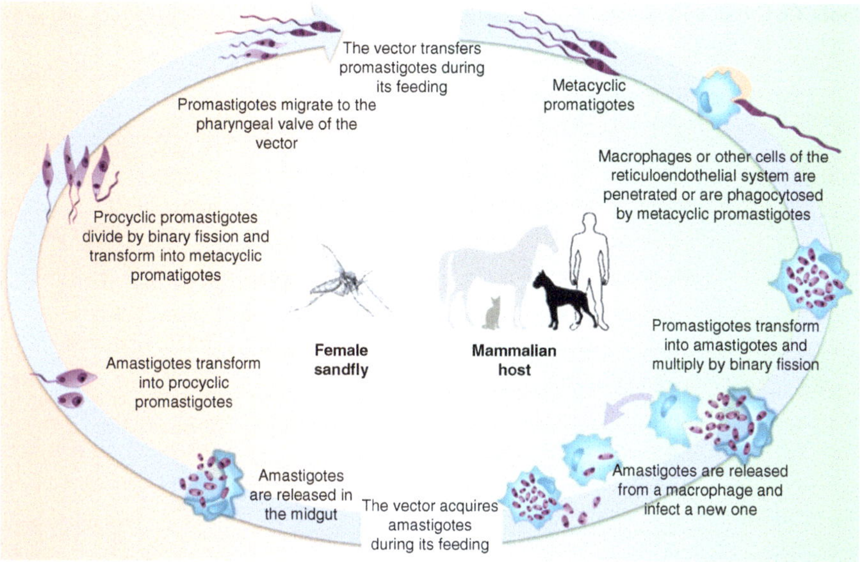

Fig. 13.4 Schematic diagram of the life cycle of *Leishmania* spp. Metacyclic promastigotes are transferred to the vertebrate host as the sand fly takes a blood meal. Promastigotes are phagocytosed by macrophages and transform into amastigotes, which undergo repeated cycles of intracellular binary fission. Amastigotes are then released from the infected macrophage and reinitiate the replicative cycle in new macrophages. If taken up by a sand fly, they transform to promastigotes in the insect midgut and undergo binary fission. They adhere to the gut epithelium and thus avoid being excreted, stop dividing, and undergo a terminal differentiation into metacyclic promastigotes, which are infective for the vertebrate host

or *Lutzomyia* in the New World (Fig. 13.4, Table 13.1). Infection of mammals starts when metacyclic, extracellular promastigotes, present in the proboscis of sand flies, are introduced into the dermis of a susceptible host during a blood meal. Promastigotes are penetrated or are phagocytosed by cells of the mononuclear phagocytic system, resulting in the formation of a parasitophorous vacuole (PV). Inside the PV, promastigotes are transformed into amastigotes that actively multiply by binary fission. Uncontrolled proliferation of the parasites leads to the bursting of the macrophage, setting the parasites free. Free amastigotes interact with other macrophages or spread through the circulatory or lymphatic pathway and infect organs rich in these immune cells such as the liver, spleen, or bone marrow (Kaye and Scott 2011; Gallego Berenguer 2014).

Female sand flies acquire *Leishmania* parasites when feeding from an infected mammalian host. They insert their saw-mouth parts into the skin and agitate them to produce a small wound in which blood flows from the superficial capillaries. This damaged tissue associated with wound creation promotes the release of free or intracellular amastigotes, present in macrophages and other phagocytes, from the skin of the infected mammal. Once taken up by the sand fly, amastigotes are

Table 13.1 Causative agents of canine, feline, and equine leishmaniosis, their vectors, and geographical distribution

	Leishmania species	Sand fly species	Geographical distribution	References
Canine leishmaniosis	*L. infantum*	*Lu. longipalpis* *Lu.evansi* *Lu. olmeca*	America	Desjeux (2001), Schwartz et al. (2006), Bates (2007), Solano-Gallego et al. (2009)
		P. ariasi *P. perniciosus* *P. perfiliewi* *P. neglectus* *P. tobbi*	Mediterranean basin, Central and West Asia	Bates (2007), Solano-Gallego et al. (2009), Gramiccia (2011)
		P. chinensis *P. alexandri*	Central Asia, China	Solano-Gallego et al. (2009)
		P. kandelakii	Southern Asia, Iran, Armenia, Afghanistan	Solano-Gallego et al. (2009)
		P. perfiliewi *P. neglectus* *P. langeroni* *P. tobbi*	Middle East	Solano-Gallego et al. (2009)
	L. tropica	*P. sergenti* *P. arabicus*	Middle East, Central Asia, Africa	Gramiccia and Gradoni (2005), Solano-Gallego et al. (2009)
	L. donovani	*P. orientalis* *P. martini*		
	L. braziliensis	*Lu. wellcomei* *Lu. spinicrassa* *Lu. whitmani* *Lu. yucumensis* *Lu. carrerai carrerai* *Lu. llanosmartinsi* *Lu. ovallesi* *Lu. intermedia* *Lu. gomezi* *Lu. trapidoi* *Lu. ylephiletor* *Lu. umbratilis*	Central to South America	Solano-Gallego et al. (2009)
	L. panamanensis	*Lu. trapidoi* *Lu. ylephiletor* *Lu. gomezi* *Lu. panamensis* *Lu. hartmanni*	Central America	Solano-Gallego et al. (2009), Gramiccia (2011)
	L. peruviana	*Lu. peruensis* *Lu. verrucarum* *Lu. Ayacuchensis*	Peruvian Andes	Solano-Gallego et al. (2009)
	L. amazonensis	ND		Gramiccia (2011)

Table 13.1 (continued)

Feline leishmaniosis	*L. mexicana*	ND	North and Central America	Pennisi (2015)
	L. venezuelensis	ND	South America	Pennisi (2015)
	L.braziliensis	ND	Central and South America	Solano-Gallego et al. (2009)
	L. amazonensis	ND	South America	
	L. infantum	ND	China, Middle East, Mediterranean basin Central and South America	Pennisi (2015)
Equine leishmaniosis	*L. brazilensis*	ND	Latin America	Bonfante-Garrido et al. (1981), Gramiccia (2011)
	L. infantum	ND	Europe	Müller et al. (2009), Gramiccia (2011)
	L. siamensis	ND	North America	Reuss et al. (2012)

transformed into procyclic promastigotes, a replicative form that multiplies in the ingested blood. This initial phase is limited by the peritrophic matrix—PM—a non-cellular, semipermeable mesh of chitin microfibrils and proteins embedded in a proteoglycan matrix, which is secreted by the midgut epithelium and surrounds the food bolus. Procyclic promastigotes eventually differentiate into elongated and strongly motile nectomonad promastigotes, which migrate to the anterior midgut, where some adhere to the microvilli and others to the stomodeal valve, present at the junction between the midgut and the foregut. This adherence of parasites prevents their expulsion during defecation. After reaching the stomodeal valve, parasites transform into leptomonad promastigotes, shorter forms that resume replication. Some of the nectomonad and/or leptomonad promastigotes adhere to each other at the surface of the valves, differentiating into haptomonad promastigotes, with no clear function or destiny. Finally, some of the leptomonad promastigotes differentiate into infective metacyclic promastigotes and are introduced into the skin of a mammalian host when the fly swallows blood again, leading to the transmission of the parasite (Killick Kendrick and Rioux 1991; Rogers et al. 2002; Gossage et al. 2003).

Some differences in sand fly stages are observed between the two subgenera of the *Leishmania* genus, *Leishmania* and *Viannia*. While the subgenus *Leishmania* develops exclusively in the midgut and in the anterior gut of their vectors, *Viannia* also includes a developmental stage in the posterior intestine (Rogers et al. 2002; Gossage et al. 2003).

A mechanism used by *Leishmania* parasites to ensure transmission is the creation of a promastigote gel in the anterior midgut. This consists of mucin-like filamentous proteophosphoglycan molecules, forming a plug that hampers blood intake by the sand fly. The plug is regurgitated, carrying metacyclic promastigotes and depositing them on the mammalian host skin (Stierhof et al. 1999). Other possible routes of transmission include parasite damage to the stomodeal valve, which remains permanently opened, infection of salivary glands and delivery with the saliva, and excretion of parasites from the anus of infected sand flies (Schlein et al. 1992; Volf et al. 2004).

Evidence of a sexual stage with mating and recombination in the midgut of the sand fly has been obtained (Akopyants et al. 2009; Sadlova et al. 2011). However, these events appear to be nonobligatory for the completion of the life cycle (Calvo-Álvarez et al. 2014).

13.1.3 Host-Pathogen Interactions

An intricate net of interactions takes place between *Leishmania* parasites and their mammalian and insect hosts, a few of which are described below.

During the blood meal of a sand fly on a mammal, the skin is damaged, and a pool, containing blood, different cells, and dermis extracellular matrix—ECM—components, is formed at the bite site. If the sand fly is infected, *Leishmania* promastigotes present in its saliva are deposited in this milieu, where the first critical interactions of the parasites with the mammalian host take place. Promastigotes are able to attach to collagen I, the major dermis ECM component, and degrade scaffolds of this protein with secreted specific proteases (Lira et al. 1997; de Menezes et al. 2016). Moreover, *Leishmania* experimental infections elicit a collagen shift from the more rigid collagen I to the softer collagen III. These interactions with collagen expedite parasite migration and tissue invasion (Fatoux-Ardore et al. 2014).

Another ECM component, the glycoprotein fibronectin—FN—binds to surface proteins of promastigotes and amastigotes, facilitating parasite uptake by monocytes through a FN-recognition pathway. On the other hand, peptides generated by parasite protease cleavage of FN inhibit macrophage production of antimicrobial oxygen reactive species, such as nitric oxide and hydrogen peroxide (Kulkarni et al. 2008). Interestingly, the same type of inhibitory activity is displayed by some sand fly salivary components. Sand fly saliva also aids *Leishmania* parasites to thrive in the mammalian host environment by its anticoagulant activity and its modulation of cytokines, promoting the expression of some that enhance parasite survival while downregulating others that display protective activity (de Menezes et al. 2016). Recruitment of neutrophils and macrophages, stimulated by parasite-secreted proteophosphoglycans, takes place at the site of the sand fly bite (Rogers et al. 2010). *Leishmania*-parasitized neutrophils can be observed in early stages, while macrophages are the main host cells in later stages of the infection. Neutrophils infected with metacyclic promastigotes undergo apoptosis and are, in turn, phagocytosed by macrophages through receptor-mediated pathways, without triggering defensive responses. In this way, infected neutrophils have been suggested to act as "Trojan horses", safely delivering live promastigotes into the macrophage phagosome, where parasites can differentiate and multiply (Laskay et al. 2003). On the other hand, neutrophils exert a defensive role against infections through a process known as NETosis, in which their

DNA, associated to proteins and granulocytic material, is released to the extracellular medium. A mesh-like structure, known as Neutrophil Extracellular Net (NET), is formed and leads to the trapping and killing of microorganisms (Brinkmann et al. 2004). NETosis is induced by the presence of promastigotes and also amastigotes, but parasites have developed mechanisms to escape NETs, such as the production and release of nucleases (Guimarães-Costa et al. 2014). Notably, an endonuclease—*Lutzomyia* NET destroying protein or Lundep—described in the saliva of *Lutzomyia longipalpis* sand flies, allows parasites to escape NET toxic activity (Chagas et al. 2014). Given the diverse actions described for neutrophils, their importance in the establishment and progression of *Leishmania* infections in vivo awaits to be experimentally confirmed (Kaye and Scott 2011).

Invasion of macrophages is essential for the establishment of *Leishmania* infections. Internalization is initiated with interaction between parasites and macrophage molecules, which, in addition to the abovementioned FN receptors, involves receptors for complement, mannose, and Fcγ, each one eliciting a different course of infection. Parasites are then internalized in a process that may involve cholesterol-rich membrane lipid microdomains and/or the actin cytoskeleton. A parasite-containing phagosome is formed out of membranes from different macrophage compartments and contributes to the formation of the parasitophorous vacuole—PV— where the parasite lives, transforms to the amastigote form, and multiplies by binary fission. The main stimulus that induces the expression of amastigote-specific genes appears to be the shift of temperature associated with the passage from a poikilothermic invertebrate vector into a homoeothermic mammalian host (Alcolea et al. 2010). Several parasite-mediated mechanisms allow them to survive in the phagosome, such as delaying lysosome fusion and preventing acidification. A *Leishmania* iron transporter is activated and pumps iron inside the phagosome, competing with the macrophage transporter which pumps iron from the phagosome into the cytosol. Iron depletion in the macrophage cytosol leads to an increased production of transferrin and transferrin-mediated iron uptake (Kaye and Scott 2011; Gupta et al. 2013; Podinovskaia and Descoteaux 2015; de Menezes et al. 2016).

An array of innate and acquired immune responses is mounted upon infection by *Leishmania* parasites. Protective immunity is associated with a T helper-1—Th-1—response, with production of intracellular nitric oxide within macrophages that elicits amastigote apoptotic cell death (Holzmuller et al. 2006). Conversely, susceptibility to the infection is associated with a Th-2 response. This type of response favors the production of *Leishmania*-specific antibodies, which are ineffective to stop the infection. In the absence of an adequate cellular response, high parasite burdens appear in the skin, bone marrow, liver, spleen, and lymph nodes. In addition, immune complexes are deposited at the glomerular basal membrane, resulting in glomerulonephritis and renal failure. Host immunogenetic factors, nutritional status, concomitant infections, and previous exposure to the parasite influence the type of prevailing immune response and, therefore, the clinical outcome of an infection (Baneth et al. 2008; Reis et al. 2010). In addition, different *Leishmania* molecules interact with host cell signaling pathways involved in the immune response. Among them, lipophosphoglycans—LPG—the most abundant surface glycolipids of promastigotes, stimulate macrophage Toll-like receptors 2, leading to the production of cytokines that favor parasite survival (Kaye and Scott 2011). Also GP63, a parasite metalloprotease anchored to its surface through a glycosylphosphatidylinositol anchor, cleaves a number of host proteins inside the macrophage. As examples, GP63 has been reported to cleave transcription factors, translation regulators, micro-RNA processing proteins, and enzymes involved in apoptotic mechanisms. The array of intricate mechanisms developed by *Leishmania* parasites to manipulate host cell functioning and responses is under active research, as well as how to use this knowledge to develop novel therapeutic approaches (Kaye and Scott 2011; Podinovskaia and Descoteaux 2015).

After *Leishmania* amastigotes are ingested by a sand fly during its blood meal, the parasites need to thrive, transform, and multiply in the insect midgut and finally be regurgitated on the skin of a mammalian host to continue their life cycle. One of the first hazards they encounter in the sand fly gut milieu consists of blood-degrading proteases with parasiticide effects that provoke significant parasite losses (Dostálová and Volf 2012). *Leishmania* protection against proteolytic attack is likely exerted by their secreted protease inhibitors and by modulation of the expression of sand fly peptidases (Ramalho-Ortigao et al. 2010). Promastigotes appear to be more resistant to proteolytic cleavage than amastigotes, an effect suggested to be connected with the surface expression of phosphoglycan molecules (Secundino et al. 2010).

As mentioned in Sect. 13.1.2, blood and parasites ingested by a sand fly are contained inside the midgut in a PM. On one hand, this matrix exerts a protective effect on the protease-susceptible amastigotes from the attack of digestive enzymes at the first stages of digestion. On the other, it could interfere with the continuation of parasite life cycle by hampering promastigote migration through the sand fly gut, once they have been formed. It has been observed that promastigote egress from the PM is facilitated by a secreted parasite chitinase activity, although a synergistic action of parasite and sand fly chitinases has also been suggested (Pimenta et al. 1997; Ramalho-Ortigao et al. 2010). Once outside the PM, parasites migrate to the anterior region of the midgut and attach to the epithelium, mainly through their flagellum. This event, which prevents them to be excreted with the feces, appears to be mediated by the interaction between LPG at the parasite surface and specific midgut receptors (Dobson et al. 2010).

Several signals from the midgut environment acting synergistically can trigger metacyclogenesis, including acidic pH, absence of hemoglobin or oxygen, decreased levels of tetrahydrobiopterin, and exposure to sand fly saliva (Ramalho-Ortigao et al. 2010). In addition, it has been recently shown that the midgut microbiota critically influences this process. Indeed, the diversity of gut microorganisms decreases in infected sand flies, as *Leishmania* parasites multiply and develop to their metacyclic stage, and antibiotic treatment impedes parasite growth and differentiation (Kelly et al. 2017).

Only some parasite-sand fly combinations successfully lead to the metacyclic stage, such as *L. donovani-Phlebotomus argentipes*, *L. major-P. papatasi*, and *L. tropica-P. sergenti*. It has been postulated that this restricted vectorial capacity is connected with parasite interspecies LPG heterogeneity and sand fly midgut receptor specificity. Likewise, permissive sand fly varieties that allow metacyclogenesis of different *Leishmania* species might express more than one LPG receptor in their midguts (Ramalho-Ortigao et al. 2010).

The final stages of the life cycle within the sand fly involve attaching to the insect stomodeal valve. Parasite-secreted chitinases also act at this point, damaging this valve and thus facilitating regurgitation of parasites embedded in the promastigote gel with a backflow of ingested blood (Schlein et al. 1992; Volf et al. 2004). Moreover, parasitized sand flies tend to feed more frequently and on different hosts, which promotes parasite transmission, illustrating an adaptive manipulation of sand fly behavior by the parasite (Rogers and Bates 2007).

Finally, it is important to note that parasites are exposed to components of the sand fly innate immune system during their passage through the midgut. Indeed, expression of defensins—cationic peptides with antimicrobial activity—is enhanced in response to *Leishmania* infections. Defensins, as well as other innate immune molecules, such as serpins, pattern recognition proteins, and antioxidant enzymes that have been described in the sand fly midgut, are likely to impact the outcome of *Leishmania* infections in this vector (Dostálová and Volf 2012).

13.2 Diagnosis and Epidemiology

13.2.1 Diagnosis

Diagnostic methods for leishmaniosis in company animals were developed for dogs but can be applied to cats and horses, as well (Gramiccia 2011). It is important to distinguish if diagnosis is meant from an epidemiological or a clinical point of view. Different diagnosis strategies can be applied to detect clinically healthy dogs that live in endemic regions, as well as symptomatic animals in endemic or non-endemic regions; to avoid transmission from subclinical carriers by blood transfusions; regulate the import of infected dogs from endemic into non-endemic regions; evaluate the response to treatment; and analyze the prevalence of the disease in a certain region. When diagnosis is focused on a particular animal, it needs to be accompanied by a complete clinical examination. Diagnosis of canine leishmaniosis is complex, since the range of pathological anomalies, clinical signs, and biochemical parameters can be wide and non-specific. In addition, the potential presence of concomitant infectious or noninfectious diseases in the same animal calls for diverse differential diagnostic procedures. The following are different aspects and methodologies related to the diagnosis of leishmaniosis in company animals.

13.2.1.1 Indirect Diagnosis

The origin of the patient needs to be recorded. Given the large variety of clinical presentations of the disease, any animal from an endemic region can be potentially sick of leishmaniosis, which needs to be considered during differential diagnosis. On the other hand, in a non-endemic area, where vectors are not present, animals can be considered suspicious of having the disease only if they have characteristic signs. Relevant information includes whether the animal could have been exposed to the vector in the past or if there is the chance of non-vectorial routes of infection, such as vertical transmission, sexual contact, or blood transfusion (da Silva et al. 2009; Naucke and Lorentz 2012). In endemic regions, larger animals that spend most of their time outdoors are more exposed to sand fly bites and are thus more prone to be infected.

Clinical signs with diagnostic value associated to leishmaniosis will be described in Sect. 13.3.1.

Biochemical parameters associated to leishmaniosis are non specific. Hemograms frequently show different degrees of anemia, transaminase increases, and high urea and creatinine blood levels, indicative of renal damage. Probably the most frequently observed biochemical sign is hyperglobulinemia—in particular an increase of gamma and beta globulins—which can result in higher total protein values and an alteration in albumin/globulin relationships. In urinalysis tests, proteinuria and other altered parameters can be observed, according to the magnitude of renal damage.

Complementary studies using ultrasound can reveal splenomegaly and renal changes produced by glomerulonephritis, while X-rays frequently show alterations of bones and/or articulations. Ocular damage is common, so ophthalmological studies are also recommended.

Differential diagnosis should include detection of other cutaneous and systemic diseases, including dermatitis, seborrhea, atopies, and immunological diseases, such as lupus, demodicosis, sarcoptic mange, cryptococcosis, pyoderma, malassezia dermatitis, ehrlichiosis, tuberculosis, hepatozoonosis, and neoplasms—lymphoma and squamous cell carcinoma. Comorbidity with demodicosis, malassezia dermatitits, and lymphoma is frequent.

13.2.1.2 Specific Diagnosis

Cytology consists in the identification of amastigotes in cytological smears of cutaneous lesions and in aspirates of lymphatic nodes, bone marrow, spleen, and cutaneous or subcutaneous nodes. Amastigotes stain purple with Giemsa or fast stains and can be visualized free or inside macrophages under 400 to 1000× magnification (Figs. 13.2a and 13.5). A pattern of mixed inflammation—presence of neutrophils, eosinophils, macrophages, and plasmocytes—or granulomatous inflammation is frequently observed. Ganglia present reactive hyperplasia. Cytology is a highly specific technique, since detection of a single amastigote has diagnostic value. Its sensitivity, however, is variable and depends on the experience of the observer, the quality of the sample, and its origin. Highest sensitivity is achieved with spleen and bone marrow samples (Cowell et al. 2000). In addition, it is fast, can be noninvasive—according to the sample source—and allows to identify the presence of other pathogens as well, such as *Ehrlichia*, *Hepatozoon*, *Anaplasma*, or *Babesia*. Immunohistochemical techniques applied to sections of the skin or other organs increase the sensitivity of parasite detection, which is particularly useful in cases of low infection levels (Tafuri et al. 2004).

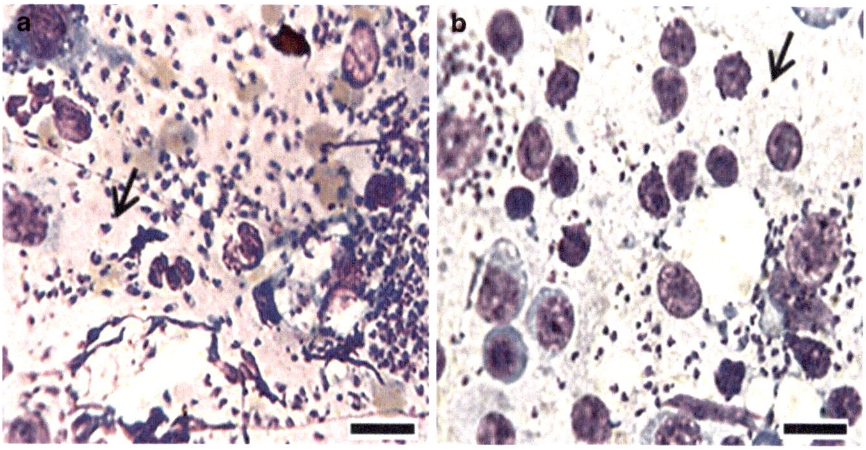

Fig. 13.5 Diagnosis of *Leishmania* sp. in a dog by cytological observation of Giemsa-stained smears. (**a**) *L. infantum* amastigotes—example with black arrow—from a skin lesion smear; (**b**) popliteal lymph node aspirate. Magnification 1000×. Bar: 20 μm

Cultivation is used for the isolation and identification of parasite species. This method is useful for some research purposes but, because it is laborious and time-consuming, is hardly applied in clinical diagnosis.

Xenodiagnosis is specific and sensitive but is also laborious and expensive. It consists in allowing laboratory-raised sand flies to feed on a dog suspected of leishmaniosis that has been sedated and constrained in a cage. Sand flies are examined after their blood meal to check for promastigote presence in the gut (Miró et al. 2011).

Serological tests are an essential part of the diagnosis of leishmaniosis. Considering the high percentage of infected dogs that present no clinical symptoms, anti-*Leishmania* antibody detection can be taken as a first indication of infection, since high antibody titers can be predictive of the eventual development of clinical disease (Reis et al. 2006). On the other hand, low antibody titers are not necessarily indicative of the disease, due to possible cross-reactivity reactions with American trypanosomes, or *Ehrlichia* sp. There are qualitative serological methods, such as immunochromatographic strips, and quantitative methods, such as immunofluorescence antibody test—IFAT—and enzyme-linked immunosorbent assay—ELISA—each one with pros and cons with respect to their clinical usefulness (Bevilacqua and Alves 2004; Sanchez et al. 2004; De Almeida Curi et al. 2006; Solano-Gallego et al. 2011). Immunochromatographic strips are based on a recombinant form of rK39 antigen, which shows high specificity for *L. infantum*. However, different studies around the world show variable results with respect to sensitivity, which probably reflects antigen polymorphism connected to geographical origin (Laurenti et al. 2014). A qualitative ELISA based on a *L. infantum* extract has similar advantages with respect to fastness and better indexes of specificity and sensitivity (Marcondes et al. 2011). Fast serological methods are useful for epidemiological surveys and as a first indication of infection in the clinic that needs to be corroborated by other diagnostic methods. Among quantitative methods, IFAT is the most widely used, although cross-reactivity with antigens of *Trypanosoma cruzi* and other pathogens decreases its specificity (OIE 2014). Quantitative ELISAs are progressively becoming the serological methods of choice for Old and New World *Leishmania* diagnosis, due to their enhanced sensitivity and specificity with respect to IFAT. A recently modified direct agglutination test—DAT—has been validated as a sensitive and specific tool of particular usefulness for large-scale epidemiological and ecological studies (OIE 2014).

Molecular diagnosis is mainly carried out by polymerase chain reaction—PCR—that allows amplifying parasite DNA from different types of samples, including tissues, blood, body fluids, or even histopathological specimens. Assays that target kinetoplastid DNA are the most sensitive for direct detection in infected tissues and can evidence the presence of parasite DNA in minimal quantities. In the clinic, it allows to assess the persistence of the parasite after treatment and also to diagnose carriers among animals destined to reproduction or blood donation. Seminested or nested PCRs display increased sensitivity, and quantitative PCRs allow quantifying parasite loads. In recent years, the loop-mediated isothermal amplification assay—LAMP—has emerged as an efficient tool in the field of leishmaniosis diagnosis. Compared to conventional PCR, LAMP is a faster, more sensitive, and cost-effective method (Mori and Notomi 2009; Takagi et al. 2009; Adams et al. 2010; Verma et al. 2017). The diagnostic efficacy of this type of detection depends on the natural course of the disease and is maximal immediately after infection (Solano-Gallego et al. 2009, 2011). It is important to note that, for the diagnosis of individual patients, the information obtained by PCR should not be separated from the data obtained after clinical, pathological, and serological evaluation.

13.2.2 Epidemiology

Leishmania sp. parasites are amply distributed around the world, extending over tropical, subtropical, and temperate regions, between latitudes 45° N and 32° S (Dawit 2013). They are transmitted to mammals by sand flies of the genera *Phlebotomus* and *Lutzomyia* in the Old and New World, respectively. As with other vector-borne pathogens, as a general rule, the geographical distribution of the vector/s determines the geographical distribution of the parasite. Thus, factors that limit the vector should be the same as those that guide the distribution of the disease (World Health Organization 2010). Sand flies reproduce on decomposing solid organic matter. They exhibit crepuscular and night activity. *Phlebotomus* populations peak from the end of spring to the end of fall, while *Lutzomyia* sand flies are active all year round. Spatial analysis demonstrated that vegetation and hydrography may be related to sand fly distribution and infected dogs (Menezes et al. 2015). Climate, especially high temperature and increased humidity, also has an impact on the geographical distribution of sand flies, their density, their activity, and their reproductive periods (Margonari et al. 2006; Ready 2008; Sangiorgi et al. 2012).

Reservoir mammalian hosts include dogs, humans, and, to a lesser degree, cats and horses (Table 13.1). Sporadic cases have been reported in cattle, sheep, and goats. Black rats and hares are proven *Leishmania* reservoirs, contributing to the maintenance of the parasite in certain areas. Foxes have been extensively studied due to their similarities to dogs. They present high prevalence of *Leishmania* infections; however, their infectiveness to sand flies, and thus their epidemiological relevance, has so far not been demonstrated (Millán et al. 2014; Mol et al. 2015).

Less than 50% of *L. infantum*-seropositive dogs develop severe disease with clear visceral and/or cutaneous signs, while the rest can be equally infectious for sand flies but remain unnoticed (Molina et al. 1994; Oliva et al. 2004; Michalsky et al. 2007). Studies on risk factors associated with breed, gender, and age are few and not always consensual (França-Silva et al. 2003; Barroso et al. 2015).

Although most leishmaniosis cases occur in sand fly-endemic regions, occasional case reports in vector-free regions are associated with journeys of dogs with their owners to endemic areas or the import of infected animals (De Souza et al. 2013; Silva et al. 2015). In addition to vectorial transmission, atypical transmission routes include transfused blood products and bite wounds (Naucke et al. 2016). Also, infected males can develop genital lesions, and parasites can reach the semen and be sexually transmitted to susceptible females. Transmission from females to males, on the other hand, has not been described so far. Vertical transmission through the passage of infected macrophages of the mother into the placenta and the fetus has also been shown (Naucke and Lorentz 2012). The high prevalence of canine leishmaniosis in foxhound dogs in the USA, especially in regions where no vectors are present, illustrates the epidemiological importance of vertical and sexual transmission (Turchetti et al. 2014).

13.3 Clinical Effects, Prevention, and Treatment

13.3.1 Clinical Effects

According to their clinical manifestations, leishmaniosis patients can be classified in three evolutionary stages: (1) asymptomatic, clinically healthy with confirmed positive diagnosis; (2) oligosymptomatic, with scarce symptomatology corresponding to initial stages of infection; and (3) symptomatic, with active and chronic infections. This simple and schematic classification might suggest a progressive development of the disease; however, there are animals that can stay in the same stage during the whole course of infection. An additional category corresponding to resistant animals, i.e., those that will never contract the disease, should also be included in this scheme (Oliva et al. 2004; OIE 2014). Another model of classification contemplates not only clinical aspects but also serological and pathological data and is thus a better reflection of the situation of a particular individual. In this scheme, a patient can display slight, moderate, severe, or very severe disease (Solano-Gallego et al. 2009, 2011). Leishmaniosis provokes multiorgan damage, with a great variety of clinical manifestations (Fig. 13.6) that can be divided into cutaneous and systemic signs (Sanchez et al. 2004).

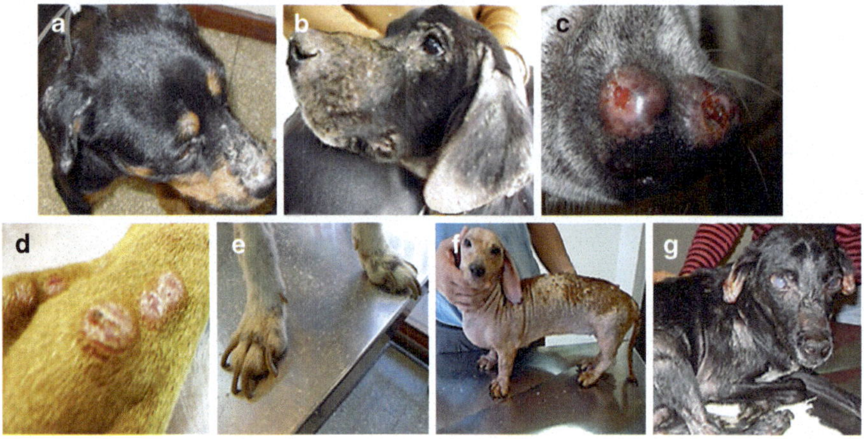

Fig. 13.6 Clinical signs of canine leishmaniosis. (**a–c**) Dermatitis and lesions found in upper head, snout, ears, and eyelids, or (**d**) limbs. (**e**) Onicogrifosis. (**f**) Exfoliative dermatitis. (**g**) Keratitis and cutaneous wounds

Cutaneous signs are generally the most evident and are often the main reason for veterinary consultation. They consist of exfoliative dermatitis without itching, progressive alopecia—especially in limbs and around the eyes and ears—as well as ulcerous lesions and wounds with scabs in ears, snout, and bone eminences. Multiple or single nodes are frequently observed in different parts of the thorax and head, in addition to onicogrifosis—an exaggerated increase in nail length (Fig. 13.6).

Systemic signs include generalized lymphadenomegaly—with ganglia increased three to five times in size—loss of weight, cachexia, hepatomegaly, splenomegaly, occasional arthritis or synovitis, and an array of ocular disorders, such as keratoconjunctivitis and eye inflammation—uveitis and blepharitis. Systemic disorders are reflected in the clinical examination. Affected animals appear thin, lethargic, feverish, anemic, generally without evident itching, and often with claudication of hind limbs. Epistaxis or hemorrhages of the digestive tract are frequently observed in severe stages of disease. In addition, pulmonary disorders, hepatic alteration, and nephrotic syndrome can also be found. These alterations influence diagnostic biochemical parameters and evidence the profound impact of this disease over different systems and organs, especially on renal function. Renal disorders can be the only clinical manifestation of canine leishmaniosis and can progress from slight proteinuria to nephrotic syndrome, with chronic renal insufficiency as the main cause of mortality. Occasionally, other clinical signs can be observed, such as neurological disorders, including ataxia, paralysis, or convulsions; myositis, evidenced by the atrophy of chewing muscles or other muscular groups; bone lesions with osteopenia foci and resorption; and synovitis.

In cats, polymorphic cutaneous signs are frequent, including localized nodular, ulcerative, crusty, or papular lesions or generalized dermatitis, alopecia, and scaling. Systemic disease, with the involvement of the liver, spleen, lymph nodes, and kidney, has been less frequently described (Pennisi et al. 2015). In horses, the clinical forms observed consist of nodular or ulcerated cutaneous lesions, occasionally disseminating, without visceralization. Spontaneous regression of the lesions was also reported (Koehler et al. 2002; Rolao et al. 2005).

13.3.2 Prevention

The choice of control measures for canine leishmaniosis has so far been controversial. In several regions of the world, *Leishmania*-seropositive dogs are euthanized, in an attempt to control the disease; however, this measure is upsetting for veterinarians and dog owners. However, since it has not been possible to counteract the advance of the disease by this extreme measure in different socioeconomic backgrounds, the World Health Organization is no longer considering it as a first choice (World Health Organization 2010). A frequent obstacle in the monitoring of canine populations in endemic areas has been the lack of fast and reliable diagnostic methods, to precisely discriminate infected from noninfected animals. This has led to the unnecessary sacrifice of false-positive animals. In addition, massive dog euthanasia campaigns are sometimes carried out in impoverished sand fly-endemic areas, where the fast reintroduction of puppies into households turns the measure completely useless.

It is important to note that the advance of this disease in regions such as South America is directly connected with the increasing environmental degradation of tropical and subtropical areas, as well as with explosive urbanization, generally under conditions of poverty and bad infrastructure.

Sand fly control measures need to be based on the population dynamics of these insects. Their presence and activity can be circumscribed to a few months in certain temperate regions or be found all year round with little fluctuation, in tropical and subtropical regions. Importantly, it has been shown that *Lutzomyia longipalpis* is the most synanthropic species of its genus in the American continent, which is evidenced by its notable adaptation to human environments.

Although of lesser epidemiological relevance, the possibility of non-vectorial transmission needs to be taken into account in areas where sand flies are absent or under relatively efficient control. Introduction of positive animals in areas where no vector has been identified needs to be restricted, especially in the case of blood transfusion donors or if animals are destined for reproduction (da Silva et al. 2009).

Although leishmaniosis is a zoonosis in which dogs are the main reservoirs in domestic environments, the contact with a sick animal does not guarantee contagion, since the presence of the vector is needed for the parasite to complete its cycle and generate infectious forms. In endemic regions, it is important to keep susceptible animals, such as dogs, cats, and horses, in closed spaces from evening to dawn during warm months, when sand flies are active. Sand fly control by ambient hygienic measures and the use of chemical substances, as well as physical barriers, such as nets—also impregnated with insecticides—are highly necessary to diminish the contact between vector and animals. Due to their small size, sand flies can get through untreated nets, unless the mesh size is much reduced. However, as poor fliers, they can be stopped by wind as that provided by a house fan. Homes and dog houses should be sprayed with insecticides. A number of efficacious formulations against sand flies—mainly piretroids—are commercially available. They are usually delivered as sprays, collars, and pour-on devices (Dantas-Torres et al. 2012; Otranto and Dantas-Torres 2013).

Development of vaccines against leishmaniosis is an active field of research. Currently, there are three commercially available subunit vaccine formulations, all of which use regular saponin or QuilA saponin as adjuvant and showed partial but considerably high protective efficacy: (1) Leish-Tec®, based on A2 recombinant protein; (2) Leishmune®, based on a *L. donovani* glycoprotein fraction fused to a fucose/mannose ligand; and (3) CaniLeish®, based on excreted/secreted proteins of *L. infantum*. No significant differences were observed in dogs from Brazilian *Leishmania*-endemic areas vaccinated with Leishmune® or Leish-Tec®, with respect to clinical signs, parasitism, IgG seropositivity, or infectiousness (Fernandes et al. 2014). Application of these vaccines in seronegative dogs older than 4 months is carried out in three doses, 21 days apart, and should be repeated annually. A new vaccine, Letifend®, has been recently approved by the European Agency of Medicaments. The formulation, with a protection efficacy of 90%, includes a recombinant form of the MON-1 antigen; peptides containing epitopes of the ribosomal proteins LiP2a, LiP2b, and LiP0; and the histone H2A, as well as CpG-ODN as adjuvant (Reguera et al. 2016).

13.3.3 Treatment

Several pharmaceutical products are under use, and, in addition, alternative products are also under investigation. Treatment leads to clinical improvement, though it might not eliminate the parasite. Currently, the most accepted formulations and protocols include:

1. Pentavalent antimonials, mainly meglumine antimoniate—Glucantime®—and sodium stibogluconate, Pentostam®, 50 mg/kg or 75–100 mg/kg/day, subcutaneous (sc) for 4 weeks.
2. Amphotericin B, as a second choice since it is more toxic than Glucantime®, 0.1–1 mg/kg/day, intravenous (iv). A liposome-encapsulated formulation that reduces toxicity and improves bioavailability is also commercialized.
3. Miltefosine, originally developed as an antineoplasic drug for humans, but with clear anti-leishmanial effects. A veterinarian product with low toxicity and oral administration is available. The recommended daily dose is of 2 mg/kg, during 4–6 weeks, and the cycle can be repeated according to the clinical response.
4. Allopurinol, with excellent clinical results but limited curative efficacy. It is recommended in combination with Glucantime® or Miltefosine, in doses of

15–30 mg/kg/day, orally. It needs to be administered for periods of 6–12 months, according to the clinical response. However, long treatments can lead to the formation of urinary xanthine stones. Thus, periodic assessment of these potential complications is needed, and low-purine diets are recommended during prolonged treatments.

5. Other drugs that have proved efficacious against leishmaniosis include paromomycin (aminosidine), pentamidine, spiramycin, metronidazole, quinolones, and ketoconazole.

Treatment efficacy can be improved by co-administration of immunomodulators such as levamisole, interferon gamma, domperidona, and others. During treatment, clinical evaluation of renal and hepatic damage, anemia, and cutaneous lesions should also be addressed. In addition, antibody titers and the evolution of the proteinogram should be monitored. As mentioned in Sect. 13.2.1, leishmaniosis generally provokes hyperglobulinemia that reverts when treatment has been successful. After reversion, a successive alteration of the proteinogram should be interpreted as a precocious sign of relapse, which suggests reinitiating treatment.

To take a decision on the treatment to be applied, as well as to give a prognosis, the severity of the alterations produced by the disease needs to be taken into account. Most importantly, the degree of renal failure is a determinant of the success of therapy and, consequently, of the life expectancy of the patient.

References

Adams ER, Schoone GJ, Ageed AF, et al. Development of a reverse transcriptase loop-mediated isothermal amplification (LAMP) assay for the sensitive detection of Leishmania parasites in clinical samples. Am J Trop Med Hyg. 2010;82:591–6. https://doi.org/10.4269/ajtmh.2010.09-0369.

Akopyants NS, Kimblin N, Secundino N, et al. Demonstration of genetic exchange during cyclical development of Leishmania in the sand fly vector. Science. 2009;324:265–8. https://doi.org/10.1126/science.1169464.

Alcolea PJ, Alonso A, Gomez MJ, et al. Temperature increase prevails over acidification in gene expression modulation of amastigote differentiation in *Leishmania infantum*. BMC Genomics. 2010;11:31. https://doi.org/10.1186/1471-2164-11-31.

Baneth G, Koutinas AF, Solano-Gallego L, et al. Canine leishmaniosis—new concepts and insights on an expanding zoonosis: part one. Trends Parasitol. 2008;24:324–30. https://doi.org/10.1016/j.pt.2008.04.001.

Barroso PA, Nevot MC, Hoyos CL, et al. Genetic and clinical characterization of canine leishmaniasis caused by *Leishmania* (Leishmania) *infantum* in northeastern Argentina. Acta Trop. 2015;150:218–23. https://doi.org/10.1016/j.actatropica.2015.08.007.

Bates PA. Transmission of Leishmania metacyclic promastigotes by phlebotomine sand flies. Int J Parasitol. 2007;37:1097–106. https://doi.org/10.1016/j.ijpara.2007.04.003.

Bevilacqua PD, Alves WA. Reflexões sobre a qualidade do diagnóstico da leishmaniose visceral canina em inquéritos epidemiológicos: o caso da epidemia de Belo Horizonte, Minas Gerais, Brasil, 1993-1997. Cad Saude Publica. 2004;20:259–65. https://doi.org/10.1590/S0102-311X2004000100043.

Bonfante-Garrido R, Melendez E, Torres R, et al. Enzootic equine cutaneous leishmaniasis in Venezuela. Trans R Soc Trop Med Hyg. 1981;75:471.
Brinkmann V, Reichard U, Goosmann C, et al. Neutrophil extracellular traps kill bacteria Brinkmann science 2004.Pdf. Science. 2004;303:1532–5. https://doi.org/10.1126/science.1092385.
Calvo-Álvarez E, Álvarez-Velilla R, Jiménez M, et al. First evidence of intraclonal genetic exchange in trypanosomatids using two *Leishmania infantum* fluorescent transgenic clones. PLoS Negl Trop Dis. 2014;10(5):e0004741. https://doi.org/10.1371/journal.pntd.0003075.
Chagas AC, Oliveira F, Debrabant A, et al. Lundep, a sand fly salivary endonuclease increases leishmania parasite survival in neutrophils and inhibits XIIa contact activation in human plasma. PLoS Pathog. 2014;10(2):e1003923. https://doi.org/10.1371/journal.ppat.1003923.
Cowell R, Tyler R, Meinkoth J, DeNicola D. Diagnostic cytology and hematology of the dog and cat. Can Vet J. 2000;41:330.
Dantas-Torres F, Solano-Gallego L, Baneth G, et al. Canine leishmaniosis in the old and new worlds: unveiled similarities and differences. Trends Parasitol. 2012;28:531–8. https://doi.org/10.1016/j.pt.2012.08.007.
Dawit G. A review on biology, epidemiology and public health significance of Leishmaniasis. J Bacteriol Parasitol. 2013;4:2–7. https://doi.org/10.4172/2155-9597.1000166.
De Almeida Curi NH, Miranda I, Talamoni SA. Serologic evidence of Leishmania infection in free-ranging wild and domestic canids around a Brazilian National Park. Mem Inst Oswaldo Cruz. 2006;101:99–101. https://doi.org/10.1590/S0074-02762006000100019.
De Souza RF, dos Santos YL, de Souza Vasconcellos R, et al. Recombinant Leishmania (Leishmania) infantum Ecto-nucleoside triphosphate Diphosphohydrolase NTPDase-2 as a new antigen in canine visceral leishmaniasis diagnosis. Acta Trop. 2013;125:60–6. https://doi.org/10.1016/j.actatropica.2012.09.011.
Desjeux P. The increase in risk factors for leishmaniasis worldwide. Trans R Soc Trop Med Hyg. 2001;95:239–43. https://doi.org/10.1016/S0035-9203(01)90223-8.
Dobson DE, Kamhawi S, Lawyer P, et al. Leishmania major survival in selective Phlebotomus papatasi sand fly vector requires a specific SCG-encoded lipophosphoglycan galactosylation pattern. PLoS Pathog. 2010;6(11):e1001185. https://doi.org/10.1371/journal.ppat.1001185.
Dostálová A, Volf P. Leishmania development in sand flies: parasite-vector interactions overview. Parasit Vectors. 2012;5:276. https://doi.org/10.1186/1756-3305-5-276.
Fatoux-Ardore M, Peysselon F, Weiss A, et al. Large-scale investigation of leishmania interaction networks with host extracellular matrix by surface plasmon resonance imaging. Infect Immun. 2014;82:594–606. https://doi.org/10.1128/IAI.01146-13.
Fernandes CB, Junior JTM, De Jesus C, et al. Comparison of two commercial vaccines against visceral leishmaniasis in dogs from endemic areas: IgG, and subclasses, parasitism, and parasite transmission by xenodiagnosis. Vaccine. 2014;32:1287–95. https://doi.org/10.1016/j.vaccine.2013.12.046.
França-Silva JC, Da Costa RT, Siqueira AM, et al. Epidemiology of canine visceral leishmaniosis in the endemic area of Montes Claros municipality, Minas Gerais state, Brazil. Vet Parasitol. 2003;111:161–73. https://doi.org/10.1016/S0304-4017(02)00351-5.
Gallego Berenguer J. Manual de parasitología: morfología y biología de los parásitos de interés sanitario. Barcelona: Edicions Universitat Barcelona; 2014.
Gossage SM, Rogers ME, Bates PA. Two separate growth phases during the development of Leishmania in sand flies: implications for understanding the life cycle. Int J Parasitol. 2003;33:1027–34. https://doi.org/10.1016/S0020-7519(03)00142-5.
Gramiccia M. Recent advances in leishmaniosis in pet animals: epidemiology, diagnostics and anti-vectorial prophylaxis. Vet Parasitol. 2011;181:23–30. https://doi.org/10.1016/j.vetpar.2011.04.019.
Gramiccia M, Gradoni L. The current status of zoonotic leishmaniases and approaches to disease control. Int J Parasitol. 2005;35:1169–80.
Guimarães-Costa AB, DeSouza-Vieira TS, Paletta-Silva R, et al. 3′-nucleotidase/nuclease activity allows Leishmania parasites to escape killing by neutrophil extracellular traps. Infect Immun. 2014;82:1732–40. https://doi.org/10.1128/IAI.01232-13.

Gupta G, Oghumu S, Satoskar AR. Chapter five—mechanisms of immune evasion in Leishmaniasis. In: Advances in applied microbiology. 2013. p. 155–84.

Holzmuller P, Bras-Gonçalves R, Lemesre J-L. Phenotypical characteristics, biochemical pathways, molecular targets and putative role of nitric oxide-mediated programmed cell death in Leishmania. Parasitology. 2006;132(Suppl):S19–32. https://doi.org/10.1017/S0031182006000837.

Kaye P, Scott P. Leishmaniasis: complexity at the host-pathogen interface. Nat Rev Microbiol. 2011;9:604–15. https://doi.org/10.1038/nrmicro2608.

Kelly PH, Bahr SM, Serafim TD, et al. The gut microbiome of the vector *Lutzomyia longipalpis* is essential for survival of *Leishmania infantum*. MBio. 2017;8:e01121-16. https://doi.org/10.1128/mBio.01121-16.

Killick Kendrick R, Rioux JA. Intravectorial cycle of Leishmania in sand flies. Ann Parasitol Hum Comp. 1991;66(Suppl 1):71–4.

Koehler K, Stechele M, Hetzel U, et al. Cutaneous leishmaniosis in a horse in southern Germany caused by *Leishmania infantum*. Vet Parasitol. 2002;109:9–17. https://doi.org/10.1016/S0304-4017(02)00246-7.

Kulkarni MM, Jones EA, McMaster WR, McGwire BS. Fibronectin binding and proteolytic degradation by Leishmania and effects on macrophage activation. Infect Immun. 2008;76:1738–47. https://doi.org/10.1128/IAI.01274-07.

Laskay T, Van Zandbergen G, Solbach W. Neutrophil granulocytes—Trojan horses for Leishmania major and other intracellular microbes? Trends Microbiol. 2003;11:210–4.

Laurenti MD, de Santana Leandro MV, Tomokane TY, et al. Comparative evaluation of the DPP?? CVL rapid test for canine serodiagnosis in area of visceral leishmaniasis. Vet Parasitol. 2014;205:444–50. https://doi.org/10.1016/j.vetpar.2014.09.002.

Lira R, Rosales-Encina JL, Argüello C. Leishmania mexicana: binding of promastigotes to type I collagen. Exp Parasitol. 1997;85:149–57. https://doi.org/10.1006/expr.1996.4127.

Marcondes M, Biondo AW, Gomes AAD, et al. Validation of a *Leishmania infantum* ELISA rapid test for serological diagnosis of Leishmania chagasi in dogs. Vet Parasitol. 2011;175:15–9. https://doi.org/10.1016/j.vetpar.2010.09.036.

Margonari C, Freitas CR, Ribeiro RC, et al. Epidemiology of visceral leishmaniasis through spatial analysis, in Belo Horizonte municipality, state of Minas Gerais, Brazil. Mem Inst Oswaldo Cruz. 2006;101:31–8. https://doi.org/10.1590/S0074-02762006000100007.

Menezes JA, Ferreira E, De C, Andrade-Filho JD, et al. An integrated approach using spatial analysis to study the risk factors for Leishmaniasis in area of recent transmission. Biomed Res Int. 2015;2015:621854. https://doi.org/10.1155/2015/621854.

de Menezes JPB, Koushik A, Das S, et al. Leishmania infection inhibits macrophage motility by altering F-actin dynamics and the expression of adhesion complex proteins. Cell Microbiol. 2016:1–9. https://doi.org/10.1111/cmi.12668.

Michalsky ÉM, Rocha MF, da Rocha Lima ACVM, et al. Infectivity of seropositive dogs, showing different clinical forms of leishmaniasis, to Lutzomyia Longipalpis phlebotomine sand flies. Vet Parasitol. 2007;147:67–76. https://doi.org/10.1016/j.vetpar.2007.03.004.

Millán J, Ferroglio E, Solano-Gallego L. Role of wildlife in the epidemiology of Leishmania infantum infection in Europe. Parasitol Res. 2014;113:2005–14.

Miró G, Gálvez R, Fraile C, et al. Infectivity to *Phlebotomus perniciosus* of dogs naturally parasitized with *Leishmania infantum* after different treatments. Parasit Vectors. 2011;4:52. https://doi.org/10.1186/1756-3305-4-52.

Mol JPS, Soave SA, Turchetti AP, et al. Transmissibility of *Leishmania infantum* from maned wolves (*Chrysocyon brachyurus*) and bush dogs (*Speothos venaticus*) to *Lutzomyia longipalpis*. Vet Parasitol. 2015;212:86–91. https://doi.org/10.1016/j.vetpar.2015.08.024.

Molina R, Amela C, Nieto J, et al. Infectivity of dogs naturally infected with *Leishmania infantum* to colonized *Phlebotomus perniciosus*. Trans R Soc Trop Med Hyg. 1994;88:491–3. https://doi.org/10.1016/0035-9203(94)90446-4.

Mori Y, Notomi T. Loop-mediated isothermal amplification (LAMP): a rapid, accurate, and cost-effective diagnostic method for infectious diseases. J Infect Chemother. 2009;15:62–9.

Müller N, Welle M, Lobsiger L, et al. Occurrence of *Leishmania* sp. in cutaneous lesions of horses in Central Europe. Vet Parasitol. 2009;166:346–51. https://doi.org/10.1016/j.vetpar.2009.09.001.
Naucke TJ, Lorentz S. First report of venereal and vertical transmission of canine leishmaniosis from naturally infected dogs in Germany. Parasit Vectors. 2012;5:67. https://doi.org/10.1186/1756-3305-5-67.
Naucke TJ, Amelung S, Lorentz S. First report of transmission of canine leishmaniosis through bite wounds from a naturally infected dog in Germany. Parasit Vectors. 2016;9:67. https://doi.org/10.1186/1756-3305-5-67.
OIE. Leishmaniosis. In: Terrestial manual; 2014. p. 1–12.
Oliva G, Foglia Manzillo V, Pagano A. Evoluzione dei protocolli terapeutici in corso di leishmaniosi canina. In: Parassitologia; 2004. p. 231–34.
Otranto D, Dantas-Torres F. The prevention of canine leishmaniasis and its impact on public health. Trends Parasitol. 2013;29:339–45. https://doi.org/10.1016/j.pt.2013.05.003.
Pennisi MG. Leishmaniosis of companion animals in Europe: an update. Vet Parasitol. 2015;208(1-2):35–47. https://doi.org/10.1016/j.vetpar.2014.12.023.
Pennisi M-G, Cardoso L, Baneth G, et al. Leishvet update and recommendations on feline leishmaniosis. Parasit Vectors. 2015;8:302. https://doi.org/10.1186/s13071-015-0909-z.
Pimenta PF, Modi GB, Pereira ST, et al. A novel role for the peritrophic matrix in protecting Leishmania from the hydrolytic activities of the sand fly midgut. Parasitology. 1997;115((Pt 4)):359–69. https://doi.org/10.1017/S0031182097001510.
Podinovskaia M, Descoteaux A. Leishmania and the macrophage: a multifaceted interaction. Future Microbiol. 2015;10:111–29. https://doi.org/10.2217/fmb.14.103.
Ramalho-Ortigao M, Saraiva EM, Traub-Csekö YM. Sand fly-Leishmania interactions: long relationships are not necessarily easy. Open Parasitol J. 2010;4:195–204. https://doi.org/10.2174/1874421401004010195.
Ready PD. Leishmaniasis emergence and climate change. Rev Sci Tech. 2008;27:399–412.
Reguera RM, Morán M, Pérez-Pertejo Y, et al. Current status on prevention and treatment of canine leishmaniasis. Vet Parasitol. 2016;227:98–114. https://doi.org/10.1016/j.vetpar.2016.07.011.
Reis AB, Teixeira-Carvalho A, Vale AM, et al. Isotype patterns of immunoglobulins: hallmarks for clinical status and tissue parasite density in brazilian dogs naturally infected by Leishmania (Leishmania) chagasi. Vet Immunol Immunopathol. 2006;112:102–16. https://doi.org/10.1016/j.vetimm.2006.02.001.
Reis AB, Giunchetti RC, Carrillo E, et al. Immunity to Leishmania and the rational search for vaccines against canine leishmaniasis. Trends Parasitol. 2010;26:341–9.
Reuss SM, Dunbar MD, Calderwood Mays MB, et al. Autochthonous Leishmania siamensis in horse, Florida, USA. Emerg Infect Dis. 2012;18:1545–7. https://doi.org/10.3201/eid1809.120184.
Ritter U, Frischknecht F, van Zandbergen G. Are neutrophils important host cells for Leishmania parasites? Trends Parasitol. 2009;25:505–10. https://doi.org/10.1016/j.pt.2009.08.003.
Rogers ME, Bates PA. Leishmania manipulation of sand fly feeding behavior results in enhanced transmission. PLoS Pathog. 2007;3:e91. https://doi.org/10.1371/journal.ppat.0030091.
Rogers ME, Chance ML, Bates PA. The role of promastigote secretory gel in the origin and transmission of the infective stage of Leishmania mexicana by the sand fly *Lutzomyia longipalpis*. Parasitology. 2002;124:495–507. https://doi.org/10.1017/S0031182002001439.
Rogers ME, Corware K, Müller I, Bates PA. Leishmania infantum proteophosphoglycans regurgitated by the bite of its natural sand fly vector, *Lutzomyia longipalpis*, promote parasite establishment in mouse skin and skin-distant tissues. Microbes Infect. 2010;12:875–9. https://doi.org/10.1016/j.micinf.2010.05.014.
Rolao N, Martins MJ, Joao A, Campino L. Equine infection with Leishmania in Portugal. Parasite. 2005;12:183–6.
Sadlova J, Yeo M, Seblova V, et al. Visualisation of leishmania donovani fluorescent hybrids during early stage development in the sand fly vector. PLoS One. 2011;6(5):e19851. https://doi.org/10.1371/journal.pone.0019851.

Sanchez MA, Diaz NL, Zerpa O, et al. Organ-specific immunity in canine visceral leishmaniasis: analysis of symptomatic and asymptomatic dogs naturally infected with Leishmania chagasi. Am J Trop Med Hyg. 2004;70:618–24.

Sangiorgi B, Miranda DN, Oliveira DF, et al. Natural breeding places for phlebotomine sand flies (Diptera: Psychodidae) in a semiarid region of Bahia state Brazil. J Trop Med. 2012;2012:124068. https://doi.org/10.1155/2012/124068.

Schlein Y, Jacobson RL, Messer G. Leishmania infections damage the feeding mechanism of the sand fly vector and implement parasite transmission by bite. Proc Natl Acad Sci U S A. 1992;89:9944–8. https://doi.org/10.1073/pnas.89.20.9944.

Schwartz E, Hatz C, Blum J. New world cutaneous leishmaniasis in travellers. Lancet Infect Dis. 2006;6:342–9.

Secundino N, Kimblin N, Peters NC, et al. Proteophosphoglycan confers resistance of Leishmania major to midgut digestive enzymes induced by blood feeding in vector sand flies. Cell Microbiol. 2010;12:906–18. https://doi.org/10.1111/j.1462-5822.2010.01439.x.

da Silva SM, Ribeiro VM, Ribeiro RR, et al. First report of vertical transmission of Leishmania (Leishmania) infantum in a naturally infected bitch from Brazil. Vet Parasitol. 2009;166:159–62. https://doi.org/10.1016/j.vetpar.2009.08.011.

Silva DA, Madeira Mde F, Figueiredo FB. Geographical expansion of canine visceral leishmaniasis in Rio de Janeiro state, Brazil. Rev Inst Med Trop Sao Paulo. 2015;57:435–8.

Solano-Gallego L, Koutinas A, Miró G, et al. Directions for the diagnosis, clinical staging, treatment and prevention of canine leishmaniosis. Vet Parasitol. 2009;165:1–18.

Solano-Gallego L, Miró G, Koutinas A, et al. LeishVet guidelines for the practical management of canine leishmaniosis. Parasit Vectors. 2011;4:86. https://doi.org/10.1186/1756-3305-4-86.

Stierhof Y-D, Bates PA, Jacobson RL, et al. Filamentous proteophosphoglycan secreted by Leishmania promastigotes forms gel-like three-dimensional networks that obstruct the digestive tract of infected sand fly vectors. Eur J Cell Biol. 1999;78:675–89. https://doi.org/10.1016/S0171-9335(99)80036-3.

Tafuri WL, Santos RDL, Arantes RME, et al. An alternative immunohistochemical method for detecting Leishmania amastigotes in paraffin-embedded canine tissues. J Immunol Methods. 2004;292:17–23. https://doi.org/10.1016/j.jim.2004.05.009.

Takagi H, Itoh M, Islam MZ, et al. Sensitive, specific, and rapid detection of Leishmania donovani DNA by loop-mediated isothermal amplification. Am J Trop Med Hyg. 2009;81:578–82. https://doi.org/10.4269/ajtmh.2009.09-0145.

Turchetti AP, Souza TD, Paixão TA, Santos RL. Sexual and vertical transmission of visceral leishmaniasis. J Infect Dev Ctries. 2014;8:403–7.

Verma S, Singh R, Sharma V, et al. Development of a rapid loop-mediated isothermal amplification assay for diagnosis and assessment of cure of Leishmania infection. BMC Infect Dis. 2017;17:223. https://doi.org/10.1186/s12879-017-2318-8.

Volf P, Hajmova M, Sadlova J, Votypka J. Blocked stomodeal valve of the insect vector: similar mechanism of transmission in two trypanosomatid models. Int J Parasitol. 2004;34:1221–7. https://doi.org/10.1016/j.ijpara.2004.07.010.

World Health Organization. Control of the leishmaniases. World Health Organ Tech Rep Ser. 2010;(949):xii–xiii, 1-186.

Trichomonas

14

Esther Collántes-Fernández, Marcelo C. Fort, Luis M. Ortega-Mora, and Gereon Schares

Abstract

The most widely known trichomonad in veterinary medicine is *Tritrichomonas foetus*. It is the etiologic agent of bovine tritrichomonosis, a sexually transmitted disease in extensively managed herds throughout many geographic regions worldwide. The same trichomonad species is also regarded as the causative agent of chronic diarrhea in the domestic cat, although more recent studies observed molecular differences between bovine- and feline-derived *T. foetus*. Trichomonosis in cats has a worldwide distribution and is mainly present among cats from high-density housing environments. Other trichomonads are found as inhabitants of the gastrointestinal tract in birds, such as *Trichomonas gallinae*. Particularly, Columbiformes, Falconiformes, Strigiformes, and wild Passeriformes can be severely affected by avian trichomonads. Diagnosis of trichomonosis is often complicated by the fragility of the parasite. To ensure valid test results, it is essential to collect and handle specimens in the right way prior to analysis. Cultivation tests, the specific amplification of parasites, or a combination of both test methods is the most efficient and most commonly used way to diagnose trichomonosis in animals. Bovine tritrichomonosis is mainly controlled by the identification and withdrawal of infected animals from bovine herds. The control of feline and avian trichomonosis relies mainly on preventive measures.

E. Collántes-Fernández • L.M. Ortega-Mora
Faculty of Veterinary Sciences, University Complutense of Madrid, Madrid, Spain
e-mail: esthercf@vet.ucm.es; luisucm@vet.ucm.es

M.C. Fort
Veterinary Public Health Group, National Institute of Agricultural Technology (INTA), Anguil, La Pampa, Argentina
e-mail: fort.marcelo@inta.gob.ar

G. Schares (✉)
Friedrich-Loeffler-Institut, Federal Research Institute for Animal Health, Institute of Epidemiology, Greifswald-Insel Riems, Germany
e-mail: gereon.schares@fli.de

M. Florin-Christensen, L. Schnittger (eds.), *Parasitic Protozoa of Farm Animals and Pets*, https://doi.org/10.1007/978-3-319-70132-5_14

14.1 Morphology, Life Cycle, and Host-Pathogen Interactions

In veterinary medicine the most widely known trichomonad is *Tritrichomonas foetus*. It is located in the urogenital tract of cattle and considered the etiologic agent of bovine tritrichomonosis, a sexually transmitted disease throughout many geographic regions worldwide (Bondurant 2005; Ondrak 2016). The same trichomonad species was initially described in 1999 and finally confirmed in 2003 to be the causative agent of chronic diarrhea in the domestic cat (Gookin et al. 1999; Levy et al. 2003). Trichomonosis in cats has a worldwide distribution and is mainly present among cats from high-density housing environments such as catteries, shelters, or breeding facilities (Yao and Köster 2015). Other trichomonads in animals, for example, *Tritrichomonas suis* in swine—presumably genetically identical to *T. foetus* of bovine origin—are commensals and rarely involved in disease. *T. suis* was once thought to cause atrophic rhinitis in pigs, further studies were not able to establish a causal relationship, and *T. suis* is now considered a harmless nasal and gastrointestinal commensal in swine (BonDurant and Honigberg 1994). Isolates of *T. suis* that reside in the stomach, caecum, and nasal cavity of pigs are not of clinical significance to their porcine hosts (Fitzgerald et al. 1958; Hibler et al. 1960; Pakandl 1994; Mostegl et al. 2011).

Trichomonads are occasionally observed in the feces from dogs with diarrhea (Gookin et al. 2005). The parasite was also detected by culture in the feces of 17.2% of puppies from French breeding kennels, which indicates that *T. foetus* may be a common parasite in dogs (Grellet et al. 2010). However, molecular identity of the enteric trichomonads observed in these dogs was not investigated, and the relevance for dogs is not clear. More studies are required to determine the prevalence and clinical significance of *T. foetus* infection in dogs, since the finding could be also attributed to opportunistic overgrowth of the commensal, *Pentatrichomonas hominis*.

In human medicine, the most studied and relevant is *Trichomonas vaginalis* that affects over 150 million people worldwide and is the most common non-viral sexually transmitted disease (Van der Pol 2007).

Other trichomonads are found as inhabitants of the gastrointestinal tract in birds such as *Tetratrichomonas gallinarum* and *Trichomonas gallinae*. Particularly, Columbiformes, Falconiformes, Strigiformes, and wild Passeriformes can be severely affected by avian trichomonads, whereas the majority of infections in Galliformes and Anatiformes are subclinical although severe infections are occasionally reported (Amin et al. 2014).

Other examples of trichomonads found as inhabitants of the gastrointestinal tract are *Trichomonas muris* of mice and *Pentatrichomonas hominis* of a variety of vertebrate species (BonDurant and Honigberg 1994).

14.1.1 Morphology

Trichomonads are taxonomically framed in the Parabasalia class and Trichomonadida order. This order includes protists with a parabasal apparatus and three to five anterior kinetosomes and one posterior kinetosome. They usually bear flagella and have a conspicuous pelta-axostyle complex, and the recurrent flagella are often associated with a lamellar undulating membrane underlain by a striated costal fiber (Adl et al. 2005). The number of free flagella characterizes each genus of the family Trichomonadidae. Thus, the genus *Tritrichomonas* is characterized by having three free flagella, whereas the genera *Tetratrichomonas* and *Pentatrichomonas* possess four and five flagella, respectively. Among the various species of trichomonads thus far identified, only a number of them are regarded as pathogens (BonDurant and Honigberg 1994).

14.1.1.1 *Tritrichomonas foetus*

Bovine *T. foetus* isolated from the urogenital tract of cattle and feline isolates found in the gastrointestinal tract of the domestic cat are morphologically indistinguishable. However, there appears to be no association between *T. foetus* infection in cats and reported exposure to cattle (Gookin et al. 2004). There is an ongoing debate whether *T. foetus* from cattle and cats should be placed into separate species. A molecular separation of feline and bovine isolates of *T. foetus* based on a number of gene loci—summarized by Yao and Köster (2015)—seems to be possible, but on the transcriptomic level, a separation remains difficult (Reinmann et al. 2012; Slapeta et al. 2010, 2012; Sun et al. 2012; Morin-Adeline et al. 2014, 2015b). There is evidence that *T. foetus* isolates from cats and cattle show differences in pH tolerance (Sect. 14.2.2), and *T. foetus* from cats are able to survive a passage through the alimentary tract of slugs (Morin-Adeline et al. 2015a; Van der Saag et al. 2011). Thus, some authors believe that feline *T. foetus* represents a different species, and a change in name—to *T. blagburni*—has been proposed (Walden et al. 2013). It has been hypothesized that feline *T. foetus* has extended its host range into the bovine reproductive tract (Morin-Adeline et al. 2015a). The picture becomes even more complicated because *T. foetus* isolated from cattle seems to be morphologically and genetically identical to *Tritrichomonas suis*, that is, a commensal observed in the nasal cavity, stomach, cecum, and colon of the domestic pig (Felleisen 1998; Hampl et al. 2001; Tachezy et al. 2002; Reinmann et al. 2012; Slapeta et al. 2012; Sun et al. 2012). Consequently, it was assumed that *T. suis* and *T. foetus* belong to the same species (Tachezy et al. 2002; Lun et al. 2005; Frey and Müller 2012; Yao and Köster 2015). However, more recent epidemiological studies suggest that cross-species transmission from pigs to cattle on the same farm—e.g., by exposure to *T. foetus*-contaminated pig feces—is unlikely to occur (Mueller et al. 2015).

T. foetus has a trophozoite stage, with a pyriform or ovoid appearance, and a size ranging from 8 to 18 μm in length and 4 to 9 μm in width (BonDurant and Honigberg 1994) (Fig. 14.1). The trophozoite has several structures with locomotor function such as flagella and the undulating membrane. The flagella originate from the basal bodies or kinetosomes—located in the apical pole of the cell. Three of the flagella are of similar length to each other and are directed forward, while the fourth flagellum, called the recurrent flagellum, is directed toward the posterior part of the body, associated with it by an undulating membrane, and continues as a free flagellum beyond the posterior end of the undulating membrane (Taylor et al. 1994; Benchimol 2004).

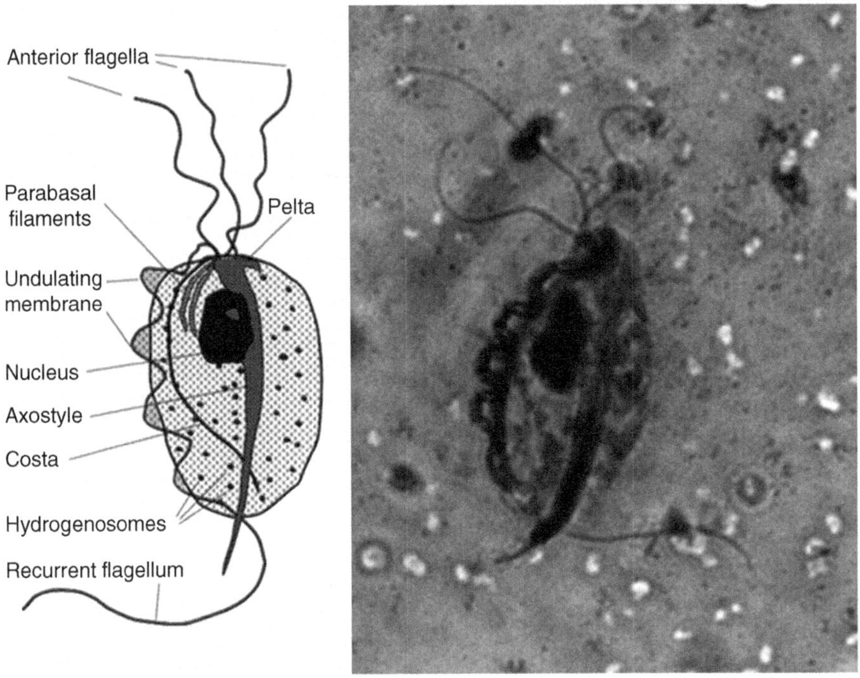

Fig. 14.1 *Tritrichomonas foetus* trophozoites (size, 8–18 × 4–9 μm): schematic and microscopic representation

The internal organelles of *T. foetus* are similar to those of other trichomonads. The cytoplasm contains a series of support elements, including the pelta-axostyle complex, the costal fiber bordering the recurrent flagellum, and the parabasal apparatus. These elements together with the flagellum make up the cytoskeleton (Benchimol 2005). The axostyle originates in the same area of the birth of the flagellum—surrounded at this point by a chromatin ring—and is directed toward the back of the parasite, making prominence at the posterior end. The pelta is a semilunar structure, very little developed, located in the anterior part of the axostyle. Both structures form the pelta-axostyle complex—composed of groups of connected microtubules—forming a kind of slit that houses the nucleus and parabasal bodies (Benchimol 2004, 2005). In *T. foetus*, the axostyle has two functions; it serves as a support organelle and participates in the processes of cell division, allowing the constriction of the nuclei during the karyokinesis (Ribeiro et al. 2002). The costa is a rigid structure that sits on the inner margin of the undulating membrane and serves as a support. The parabasal filaments are also filaments striped perpendicularly, and their mission seems to support the parabasal body—i.e., the Golgi complex. The parabasal filaments and the parabasal body constitute the parabasal apparatus, located in the anterior part of the cell (Benchimol 2004, 2005). *T. foetus* has a simple anterior nucleus and hydrogenosomes, which appear as electro-dense corpuscles that act as functional substitutes for mitochondria. Other cellular components that can be observed in the cytoplasm are free ribosomes, polysomes, glycogen granules, vesicles, and vacuoles related to processes of endocytosis, digestion, and transport (Benchimol 2004, 2005). Under unfavorable conditions, such as a low concentration of nutrients, the presence of certain drugs such as griseofulvin, or abrupt changes in temperature, trophozoites internalize the flagellum and acquire a form of pseudocyst, which is not surrounded by a manifest cell wall (Pereira-Neves and Benchimol 2009).

As regards nutrition, trichomonads lack a cytostome; they are able to capture food through the cell surface by means of pinocytosis and phagocytosis, with the resulting formation of food vacuoles of different size. Like other trichomonads that inhabit body cavities, *T. foetus* feeds mainly on bacteria, whose proliferation depends on the environment conditions where the parasite is based (Petrin et al. 1998). From the metabolic point of view, *T. foetus* is unable to de novo synthesize purine and pyrimidine nucleotides, as well as complex phosphoglycerides or cholesterol. The parasite obtains its energy through the anaerobic catabolism of carbohydrates, although the trichomonads are able to survive in the presence of oxygen. As already noted, trichomonads lack mitochondria but possess hydrogenosomes that produce molecular hydrogen in anaerobiosis and reduce oxygen. In this way, the parasite manages to keep the pH of its environment close to neutrality favoring its own development (Kleydman et al. 2004).

The reproduction of *T. foetus*—like that of all trichomonads—is asexual (Petrin et al. 1998). The parasite divides by longitudinal binary fission in which the nuclear membrane persists—a type of mitosis referred to as cryptopleuromitosis. In addition, when compared to the trophozoite form, pseudocysts present a different mitosis model, since they first divide the nuclei without dividing their cytoplasm, leading to the formation of multinucleated polymastigotes that persist if the cells are maintained under conditions of stress. When the environmental conditions are again favorable, flagella are externalized, and the new flagellated trophozoites emerge from the multinucleated cells (Pereira-Neves and Benchimol 2009).

14.1.1.2 *Trichomonas gallinae* and *Tetratrichomonas gallinarum*

T. gallinae is the only trichomonad species with a clear pathogenic potential for birds (BonDurant and Honigberg 1994; Amin et al. 2014). *T. gallinarum* is commonly found in the large intestine of gallinaceous and anseriform birds, yet its role in causing disease either in naturally infected chickens and turkeys or via experimental infection is under discussion (Amin et al. 2014). *T. gallinae* trophozoites have an ovoidal to pyriform shape with a size of about 7–11 μm. They are provided with four free anterior flagella and a fifth recurrent one, which does not become free at the posterior pole as it extends for only two-thirds of the body length (Tasca and De Carli 2003; Mehlhorn et al. 2009). Trophozoites of *T. gallinarum* appear mostly pear shaped and range in size from 6 to 15 μm (Clark et al. 2003). They also have four free anterior flagella and a fifth recurrent one, which becomes free at the posterior pole. Another difference to *T. gallinae* is the occurrence of a sphere of lacunes of the endoplasmic reticulum surrounding in a regular distance the nucleus with its typical perinuclear membranes. Furthermore, the food vacuoles appear to be very large (Mehlhorn et al. 2009).

14.1.2 Life Cycle

14.1.2.1 *Tritrichomonas foetus* in Cattle

Bovine *T. foetus* is located in the genital tract of its natural hosts, *Bos taurus taurus* and *Bos taurus indicus* cattle (Skirrow and BonDurant 1988; Bondurant 2005; Sager et al. 2007). The preferred location of the parasite in the bull is the preputial cavity—concentrating mainly in the penile mucosa and adjacent areas of the posterior preputial mucosa—specifically on the surface of the stratified squamous epithelium of the penis and the proximal foreskin in the fornix area (Clark et al. 1974; Parsonson et al. 1974; Parker et al. 1999). This epithelium undergoes numerous folds—resulting in a greater development of crypts—where *T. foetus* can develop properly by providing a suitable microenvironment for facultative or microaerophilic anaerobic microorganisms (Rhyan et al. 1999). Infection may persist for the life of the bull, in spite of the presence of a measurable humoral immune response in the preputial cavity (Rhyan et al. 1999; Campero et al. 1990; Flower et al. 1983).

In the female, once the infection has occurred, the parasite colonizes the surface of the entire genital system—vagina, cervix, endometrium, and oviduct—in a period of 2 weeks (Parsonson et al. 1976). As observed in natural infections, the parasite is preferentially concentrated in the folds of the cervix (BonDurant 1997). The infection is self-limiting, and the parasite disappears simultaneously from all areas of the genital tract after a period of at least 90–95 days (Parsonson et al. 1976; Rae et al. 2004; Bondurant 2005). In experimental infections of nonpregnant heifers, *T. foetus* infection is typically cleared from the uterus and vagina between weeks 6 and 12 following infection (Parsonson et al. 1976; Anderson et al. 1996; BonDurant et al. 1993; Skirrow and BonDurant 1990a). A very small proportion of cows in infected herds—a fraction less than 1%—have been shown to remain infected throughout pregnancy and into the following breeding season. Fortunately, such carrier cows are rare (Bondurant 2005).

Under natural conditions, *T. foetus* is transmitted directly from an infected animal to a healthy animal, almost exclusively through natural mating (Bondurant 2005). The bulls become infected during the mating of infected cows, remaining asymptomatic carriers (Fig. 14.2). Very rarely, however, the parasite can be transmitted by other routes, for example, mechanically during the practice of artificial insemination or vaginal examination, if contaminated material is used—e.g., using the same glass rod or insemination pipette for different cows or not properly disinfected specula (Murname 1959; Goodger and Skirrow 1986). Mechanical transmission seems to be possible through a healthy bull—i.e., from an infected cow to a receptive cow—if the time between two services does not exceed 20 min (Clark et al. 1977; Goodger and Skirrow 1986; Bondurant 2005; Ondrak 2016).

T. foetus was shown to be able to survive in cryopreserved semen and may be present in semen if it is contaminated with preputial fluid during manual collection

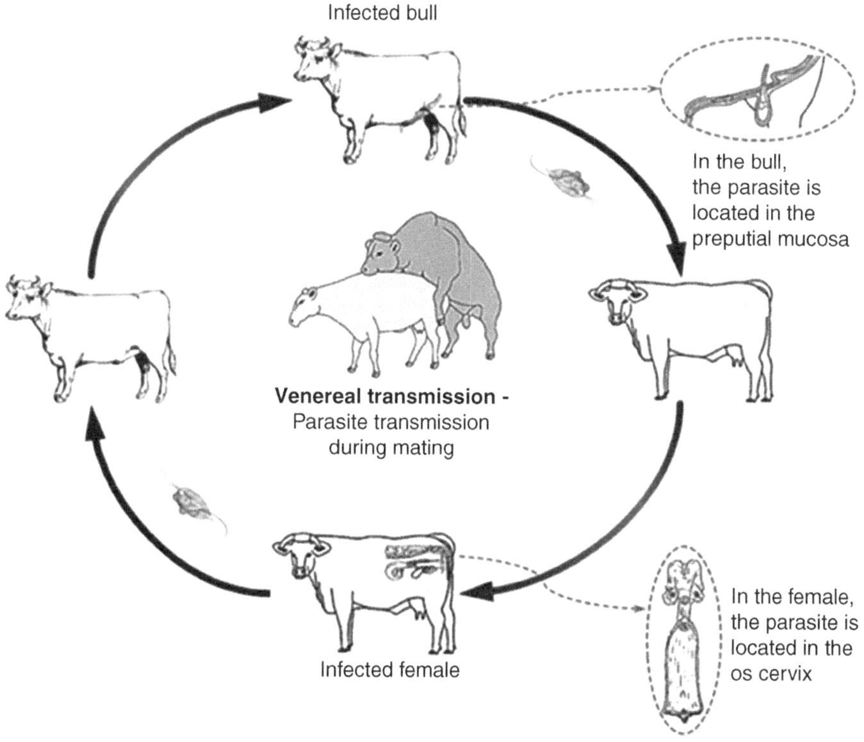

Fig. 14.2 Life cycle of bovine *Tritrichomonas foetus* (David Arranz Solís from SALUVET-UCM is acknowledged for providing this graph)

(Blackshaw and Beattie 1955). Given the resistance of this parasite in fresh, pure, or diluted semen, refrigerated and even cryopreserved, there is the possibility of transmission through artificial insemination with contaminated semen (Bondurant 2005).

Because bulls tend to mount each other, feces in the preputial cavity is commonly found. This fecal material may contain non-*T. foetus* trichomonads, such as *Pentatrichomonas hominis* and any number of *Tetratrichomonas* species that have been shown to be nonpathogenic (Taylor et al. 1994; Campero et al. 2003; Hayes et al. 2003). The opportunity for transmission of *T. foetus* between males is regarded as very limited.

14.1.2.2 *Tritrichomonas foetus* in Cats

Feline *T. foetus* appears to be host adapted, i.e., adapted to the intestinal tract of cats. After experimental orogastric infection of kittens, feline *T. foetus* has been demonstrated to colonize the lumen of the ileum, caecum, colon, and rectum 203 days after infection (Gookin et al. 2001). In naturally infected cats, massive numbers of trichomonads can be observed at the surface of the colonic epithelium and within the colonic crypts (Yaeger and Gookin 2005). Although the presence of *T. foetus* in the uterus of a cat with pyometra has been described, it has been speculated that the

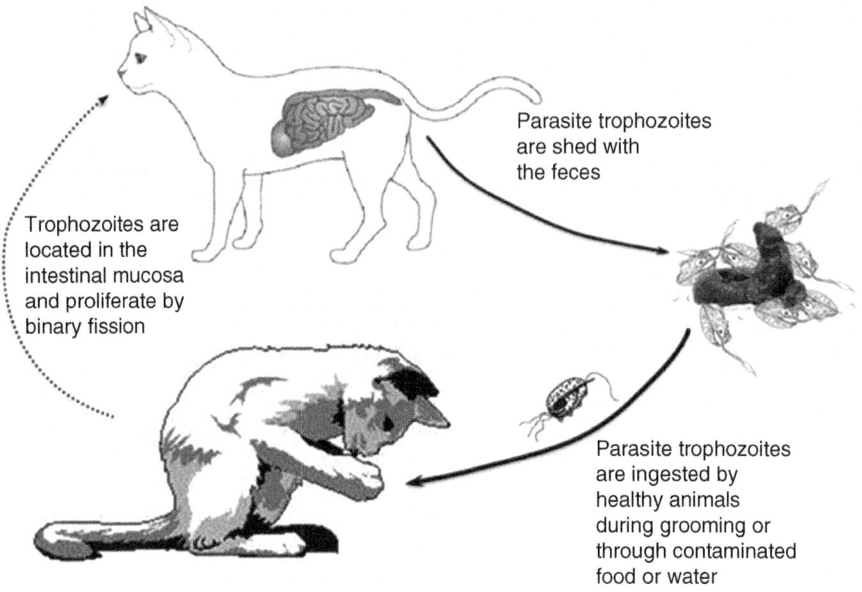

Fig. 14.3 Life cycle of feline *Tritrichomonas foetus* (David Arranz Solís from SALUVET-UCM is acknowledged for providing this graph)

parasite could have accidentally accessed the genital area through contact with contaminated feces (Dahlgren et al. 2007). However, colonization of the reproductive tract in both male and female cats from breeding grounds with a high prevalence of feline tritrichomonosis has not been observed (Gray et al. 2010). Feline *T. foetus* infection occurs by direct fecal-oral transmission. Infected cats are shedding trophozoites with their feces, and transmission occurs when two or more cats share the same litter box (Fig. 14.3). Trophozoites would adhere to the hair of the animals and could be ingested during grooming (Gookin et al. 2004; Tolbert and Gookin 2009). The viability of the parasite in the environment is limited though it can withstand several days at room temperature facilitating its transmission (Hale et al. 2009). *T. foetus*-contaminated food and less likely water may be also a relevant route for transmission. Further, shedding of viable *T. foetus* has been demonstrated in some slug species, which were fed cat food spiked with trophozoites of a feline *T. foetus* isolate, suggesting that invertebrates like slugs could play a role as mechanical vectors (Van der Saag et al. 2011).

14.1.2.3 *Trichomonas gallinae* and *Tetratrichomonas gallinarum*

The rock pigeon—*Columba livia*—was regarded as the primary host of *T. gallinae* and has been considered responsible for the worldwide distribution of this protozoal infection (Stabler 1954; Harmon et al. 1987). Other species within the Columbiformes, Falconiformes, Strigiformes, and, most recently, different Passeriformes have been recognized as potential hosts (Forrester and Foster 2008; Robinson et al. 2010). However, only a few natural occurrences of trichomonosis have been reported in gallinaceous birds like turkeys and chickens (Levine and Brandly 1939). The preferred site for *T. gallinae* is the upper digestive tract including the mouth, pharynx, esophagus, and crop, with the parasite rarely found posterior to the proventriculus (Cauthen 1936). Transmission by direct contact seems to be the most efficient route to establish an infection—e.g., via the crop milk from infected parent birds to the nestlings during feeding (Stabler 1954). In adult pigeons, the infection can occur during courtship while raptors can be infected from prey animals carrying the parasite. The infection of turkeys and chickens happens mainly via drinking water contaminated by pigeons (BonDurant and Honigberg 1994). *Trichomonas gallinae* is unable to form true cysts, even though cyst-like stages—pseudocysts—have been reported (Tasca and De Carli 2003; Mehlhorn et al. 2009). These pseudocysts may provide another route of transmission and an environmentally resistant stage during unfavorable conditions.

Tetratrichomonas gallinarum locates in the intestinal tract of different poultry species including chickens, turkeys, guinea fowl, quails, ducks, and geese and can be transmitted via consumption of contaminated food (BonDurant and Honigberg 1994). Pseudocysts of *T. gallinarum* have been reported in vivo and in vitro possibly protecting the parasite during fecal-oral transmission (Mehlhorn et al. 2009).

14.1.3 Host-Pathogen Interactions

14.1.3.1 *T. foetus* Infection in Cattle

Once infected, the male acts as an asymptomatic carrier throughout his life (Clark et al. 1974; Parsonson et al. 1974; Parker et al. 1999). Minor histological changes are observed with increased accumulations of neutrophils followed by an infiltrate of lymphocytes and plasma cells penetrating into the intraepithelial area and coalescing in the subepithelium to form lymphoid nodules (Rhyan et al. 1999; Bondurant 2005).

In the female, 2 weeks after infection, *T. foetus* may have colonized the different parts of the genital tract (Parsonson et al. 1976). The preferred location is in the cervix and cervicovaginal mucus, but the number of parasites varies throughout the estrus cycle, being higher in the days prior to estrus. The establishment of the parasite in the genital tract of the female does not seem to interfere with the fertilization nor with the early development of the embryo (Bielanski et al. 2004). In heifers experimentally infected with *T. foetus*, conceptus deaths peaked at 50–70 days of gestation (Parsonson et al. 1976). Occasional abortions of fetuses older than 4-month gestational age are reported, but typically losses occur 2 months earlier (Bondurant 2005). The infection in the female is usually self-limiting, disappearing between 2 and 4 months after the loss of the conceptus. The immunity that develops is not permanent and usually lasts for about 6 months; after 6 months, the female is again susceptible to infection. Carrier cows—these are cows that remain infected for at least 10 months—seem to fail to develop a protective immune reaction against *T. foetus*. Notable lesions in the maternal endometrium and fetal envelopes have been described only at the time of fetal loss (Parsonson et al. 1976).

The mechanisms of pathogenic actions that underlie the loss of the embryo or fetus are not known with accuracy and may include (1) the direct mechanical action of the parasite, (2) the adverse effects of enzymes secreted by the parasite, and (3) the alteration of the intrauterine environment mainly by antiparasitic inflammatory reactions of an infected dam (reviewed by Bondurant (2005) and Campero and Cobo (2006)). The increase in the number of microorganisms in the female genital tract occurs slowly and probably does not produce any relevant damage until this number exceeds a certain threshold. This fact would explain the long period of time between infection and loss of the conceptus.

14.1.3.2 *T. foetus* Infection in Cats

Few studies have examined the interaction of feline *T. foetus* with intestinal epithelium (recently reviewed by Yao and Köster (2015) and Tolbert and Gookin (2016)). Recent studies examining *T. foetus* infection in a co-culture model with monolayers of porcine intestinal epithelial cells suggest that adhesion to the intestinal epithelium occurs by means of specific receptor-ligand interactions (Tolbert et al. 2013). Pathogenesis of *T. foetus* on the intestinal epithelial cells has been suggested to be both contact-dependent and contact-independent. In the former, a cytopathic effect is mainly exerted via apoptosis induced by cell-associated proteases, whereas extracellular proteases are the major players in contact-independent cytotoxicity. Extracellular proteases may also play a role in evading complement killing.

14.1.3.3 *T. gallinae* and *T. gallinarum* Infections in Birds

The severity of the disease depends on the susceptibility of the infected birds together with the pathogenic potential of the incriminated strain and the stage of infection (Cooper and Petty 1988; Cole and Friend 1999). The severity of pathologic lesions of *T. gallinae* in the upper digestive tract varies from a mild inflammation of the mucosa to caseous areas that block the esophageal lumen (Stabler 1954). Some virulent strains are able to create diphtheritic membranes—associated with fibrinous lesions in internal organs such as the liver, lungs, and peritoneum—resulting in high mortality (Narcisi et al. 1991). Strains of moderate virulence are often associated with caseous abscesses in the upper digestive tract and oropharyngeal region, whereas no appreciable lesions are produced by avirulent strains (Cole and Friend 1999). In vitro, *T. gallinae* proliferation has been associated with a disintegration of the cell monolayer, and genetically different *T. gallinae* isolates caused diverse magnitudes of cytopathic effects on different cell lines (Amin et al. 2012a). However, little is known concerning the mechanism by which *T. gallinae* causes pathological changes in its hosts. Proteolytic proteins secreted by the parasite have been identified as contributing to the detachment of a cell monolayer (Amin et al. 2012b).

Various studies investigated the pathogenicity of *T. gallinarum* either in naturally infected chickens and turkeys or via experimental infection, with contradicting outcomes as reviewed by Amin et al. (2014). Recent studies have shown that in vitro, *T. gallinarum* has no destructive effect on cells and, in vivo, did neither produce clinical signs nor macroscopic or microscopic lesions in turkeys and specified pathogen-free chickens (Amin et al. 2011).

14.2 Clinical Effects and Diagnosis

14.2.1 Clinical Effects

14.2.1.1 Cattle

The clinical effects produced by the disease occur only in female cattle, causing early abortion and temporary infertility (reviewed by BonDurant (1997, 2005, 2007), Yule et al. (1989a), Rae and Crews (2006)). In males, *T. foetus* infection is asymptomatic and affects neither semen quality nor sexual behavior, but bulls can shed the organism indefinitely (Parsonson et al. 1974; Rhyan et al. 1999).

The parasite multiplication causes inflammation of the endometrium, cervical, and vaginal mucous membranes in cows or heifers following the infection at breeding (Parsonson et al. 1976; Rhyan et al. 1988; Anderson et al. 1996). Consequently, signs of mild vaginitis, cervicitis, or endometritis, such as mucopurulent vaginal discharge, may be observed, although generally there are no overt signs. Conception apparently proceeds normally, but almost all conceptuses are lost at some time early in gestation with early fetal death and resorption but also abortion—with a peak loss at 70–90 days (Parsonson et al. 1976; Bielanski et al. 2004). Infection can result in fetal maceration and pyometra. The consequence is infertility (Parsonson et al. 1976; BonDurant 1985; Ball et al. 1987; Anderson et al. 1996). Abortions of fetuses typically occur around 2 months of gestational age. Abortions of fetuses older than 4 months of gestational age due to trichomonosis have been occasionally reported. If the affected cow undergoes early fetal loss, it may cycle regularly without showing any signs but a prolonged interestrous interval (BonDurant 1985). Pyometra occurs in less than 5% of infected cows and is followed—as the *corpus luteum* of pregnancy is maintained—by a large purulent response (Rhyan et al. 1988); pyometra is probably a result of bacterial contamination that occurs at the time of fetal loss, when the cervix is likely to relax sufficiently to admit contamination from outside the environment (Rhyan et al. 1995a). Cows that are infected with *T. foetus* typically clear the infection within a few months, i.e., after three cycles (Parsonson et al. 1976). Immunity, however, is not permanent, and the cow will be subject to reinfection and embryonic death in subsequent breeding periods, and, as mentioned earlier, some infected cows may carry infections into the next breeding season (Skirrow 1987; Mancebo et al. 1995).

In an infected herd, bovine tritrichomonosis is associated with lowered fertility. The usual signs in the herd include return to estrus 1–3 months after breeding. At pregnancy exam time, a number of early pregnancies and open cows are observed. The period of infertility may last for another 2–6 months as a result of the infection. Other clinical features of the disease in the herd include many services per conception, poor pregnancy rates, long calving intervals, and calf crop reduction. In addition, the calving season is spread out causing batches of calves of different ages with a wide variation in weaning weights (Clark et al. 1983a; McCool et al. 1988; Rae 1989; Collantes-Fernandez et al. 2014).

14.2.1.2 Cats

T. foetus is recognized as an important cause of diarrhea in domestic cats. Typical clinical signs in natural infections are chronic or intermittent large bowel diarrhea, which can vary from subclinical to intractable (reviewed by Gookin et al. (1999), Gookin et al. (2001), Foster et al. (2004), Manning (2010), Yao and Köster (2015)). The feces are described as yellow-green in color, gassy, and malodorous with typical signs of colitis including fresh blood, mucus, fecal incontinence, tenesmus, and flatulence. The consistency of the feces can vary from liquid to semi-formed or cow pat (Stockdale et al. 2009). Severe cases may be accompanied by marked inflammation of the anal region, fecal incontinence, and rectal prolapse (Gookin et al. 1999; Foster et al. 2004; Tolbert and Gookin 2009; Bell et al. 2010). In addition, some infected cats have been reported showing systemic signs including anorexia, depression, vomiting, and weight loss (Stockdale et al. 2009). Mortality is extremely rare and only reported in kittens and is presumably caused by endotoxic shock because of deep lesions in the colonic mucosa (Holliday et al. 2009). The majority of infected cats maintain good body condition and appetite without signs of systemic illness (Gookin et al. 1999, 2001; Tolbert and Gookin 2009). The long-term prognosis for *T. foetus*-infected cats is usually good and most will eventually overcome the infection. Remission could take between 4 months and 3 years, with irregular episodes of diarrhea of variable length (Gookin et al. 1999, 2004). No abnormalities are routinely noted on hematology and serum biochemistry profile of some cats, though they remain infected and continue shedding the organism despite clinical improvement, i.e. these cats represent asymptomatic carriers. (Manning 2010). Some positive cats were also infected with other pathogens, like *Cryptosporidium* spp., *Giardia* spp., coccidian, or feline immunodeficiency virus (Gookin et al. 1999; Stockdale et al. 2009); concurrent infections may contribute and increase susceptibility and vulnerability to intestinal disease in infected cats.

14.2.1.3 Other Animals

In birds the two trichomonad species *T. gallinarum* and *T. gallinae* are commonly found (reviewed by Amin et al. (2014)). *T. gallinarum* parasitizes the large intestine of gallinaceous and anseriform birds. *T. gallinarum* induces usually a latent infection in the absence of clinical signs and lesions, and it is not clear whether *T. gallinarum* should be regarded a primary pathogen (Amin et al. 2011; Friedhoff et al. 1991). However, the presence of *T. gallinarum* may aggravate primary diseases—e.g., caused by *Histomonas meleagridis*—and coinfections have been observed (Grabensteiner and Hess 2006). *T. gallinae* is of veterinary and economic importance, as it causes avian trichomonosis, a disease with important medical and commercial implications, which is known as *pigeon canker*, *canker*, *roup*, or *Gelber Knopf* (Amin et al. 2014). *T. gallinae* is located in the upper digestive tract of pigeons, causing lesions (BonDurant and Honigberg 1994). The disease is characterized by greenish fluid and caseous lesions—whitish-yellowish fibrinous material—on the oropharyngeal membranes that can block the lumen of the esophagus impairing drinking and feeding. Clinical signs associated with avian trichomonosis are loss of appetite, vomiting, ruffled feathers, diarrhea, dysphagia, dyspnea, weight loss, increased thirst, inability to stand or to maintain balance, and a pendulous crop

(Narcisi et al. 1991). Death may occur within 3 weeks of infection. Infected birds can also remain asymptomatic due to the infection with avirulent strains of trichomonads or a lower susceptibility as seen in older birds. Avian trichomonosis may also affect domestic fowl; in earlier studies severe outbreaks have been recorded in chickens and turkeys, but we are not aware on recent cases (Hawn 1937).

14.2.2 Diagnosis

14.2.2.1 Diagnostic Techniques

In cattle, the prescribed test for international trade is the identification of *T. foetus* by culture or PCR—World Organization for Animal Health (OIE), OIE Terrestrial Manual. The OIE Terrestrial Manual provides protocols for sampling, sample transportation, transport medium, culture media, culture conditions, and how to read out the culture test. In addition, the OIE Terrestrial Manual also provides recommendations for PCR analyses, which can be applied in combination either with or after culture as an ancillary test or—more often—direct as the primary test to examine bovine samples—i.e., preputial material, uterine or vaginal secretions, or abomasal content of aborted fetuses. Protocols to diagnose bovine tritrichomonosis have been described in detail previously (Sager et al. 2007). To diagnose *T. foetus* infection in cattle, PCR tests may have a higher or at least the same sensitivity as culture tests but have several advantages, because parasites in the sample do not need to be viable and PCR results are rapidly available in contrast to results of culture tests. On the other hand, culture tests are superior to PCR because of their relative easiness (Yao 2013).

In cats, both cultivation and PCR tests are regarded as optimal methods for a sensitive and specific detection of *T. foetus* in fecal samples, although currently PCR is regarded as the gold standard assay for diagnosis of feline *T. foetus* infection since detection is independent from parasite viability (Gookin et al. 2002, 2004; Manning 2010; Yao and Köster 2015). Under optimized conditions, PCR is the method of choice when samples have to be shipped—e.g., from practitioner to a veterinary laboratory. Similar to *T. foetus* in cattle, the success of cultivation tests is largely dependent on the viability of *T. foetus* in the sample (Hale et al. 2009).

In birds infected by *T. gallinae*, an immediate diagnosis by direct microscopy of material collected via swabbing the oral cavity during clinical examination or necropsy is possible. Also, *T. gallinarum* can be observed in fecal material collected from birds—e.g., by swabbing cloacae. However, although the direct detection by light microscopy is fast and inexpensive, it is regarded as insensitive, and low numbers of parasites may not be detected (Amin et al. 2014). Also in birds the use of cultivation for the detection of trichomonads is clearly superior in sensitivity as compared to direct microscopy (Cooper and Petty 1988; Bunbury et al. 2005).

14.2.2.2 Direct Microscopic Examination

In fresh samples it is possible to diagnose the infection with trichomonads by a direct light microscopical examination. Optimal is a 200 to 400-fold magnification. It is advantageous to pre-warm slides, to retain motility of trichomonads. A drop of physiological saline is added to the slide, mixed with a nearly equal volume of material collected, and mounted with a coverslip.

It is possible to apply conventional light microscopy—either using a conventional up-light or an inverted microscope—, phase-contrast microscopy, or dark-field microscopy. In phase-contrast microscopy, it is easier to see flagella. In dark-field microscopy, trichomonads appear as small rolling luminescent footballs. In conventional light microscopy, trichomonads are identified by their characteristic movement, which is described as rolling and jerky. They are *flashing*, due to their rolling movements. The presence of multiple anterior flagella and the characteristic *refractile* undulating membrane can be observed. However, one disadvantage of direct microscopy is that at the resolution of a conventional laboratory microscope, the exact number of anterior flagella cannot be determined.

The specificity of direct microscopy is very limited. The identification of the trichomonad species observed by this method is not possible, and confirmatory PCR analyses are necessary. For inexperienced examiners it might be difficult to differentiate trichomonads from other intestinal parasites—e.g., *Giardia* spp. in cat feces (Yao and Köster 2015). Also the intestinal commensal trichomonad *Pentatrichomonas hominis* might be misinterpreted as *T. foetus*.

Another disadvantage of direct microscopy is its low sensitivity. A sample is only positive, if it contains a sufficient number of parasites per milliliter specimen. For example, in cats, the fecal examination by direct microscopy is reported to have a diagnostic sensitivity of only about 14%, and also in bovine tritrichomonosis, direct examination is estimated to be 25% less sensitive than culture diagnosis (Gookin et al. 2004; Sager et al. 2007).

The advantage of direct microscopy as diagnostic tool is its speed and the low cost of examination. This is the reason why direct microscopy is often used by practitioners for the examination of *T. foetus* in cats and *T. gallinae* infection in birds (Forrester and Foster 2008; Amin et al. 2014).

Staining of trichomonads is possible using a number of stains, including Giemsa, silver, iron hematoxylin, malachite green, methylene blue, Papanicolaou, and acridine orange (Amin et al. 2011). A fast and inexpensive staining protocol—Giemsa or Diff-Quick and iodine—has been reported (Lun and Gajadhar 1999). Single parasites might be easier to inspect; however, the chance to find small numbers of parasites in a sample might decrease because parasites can no longer be identified by their characteristic movement or *flashing* essential for parasite identification in low concentrated samples. In tissues, the use of hematoxylin-eosin (HE) and periodic acid-Schiff (PAS) stains was proven to be advantageous for identification of the flagellates, especially in organs that contained only a few protozoal cells (Amin et al. 2011, 2014). Immunohistochemical techniques and in situ hybridization have been successfully applied to demonstrate trichomonads in histological sections of cat or bird tissues, respectively (Rhyan et al. 1995b; Yaeger and Gookin 2005; Liebhart et al. 2006; Mostegl et al. 2012). Immunohistochemical detection using monoclonal antibodies against *T. foetus* has been shown to be a valuable tool (Hodgson et al. 1990). A protocol for immunohistochemical detection using the monoclonal antibody Mab 34.7C4.4 is provided in the OIE Terrestrial Manual (www.oie.int/international-standard-setting/terrestrial-manual). Staining is also applied to confirm positive cultures by morphological criteria.

14.2.2.3 Culture

In vitro culture can be performed by incubating the samples at 25–37 °C in a growth medium. If parasites are present, their numbers will multiply in the culture over time, increasing the likelihood of their detection. A large number of culture systems for trichomonads and especially *T. foetus* have been developed and published. Presumably the first culture system for an axenic cultivation of *T. foetus* isolated from an aborted fetus—i.e., cultivation without bacteria or other living organisms—was reported by a German microbiologist (Witte 1933). From that time on, numerous reports of further cultivation protocols have been published.

Until now diagnostic culturing is of outmost importance for sensitive diagnosis of bovine tritrichomonosis. Also in the diagnosis of tritrichomonosis of cats, cultivation has been widely used for epidemiological studies or diagnostic purposes (Tables 14.2 and 14.3).

In bovine tritrichomonosis cultivation became an important diagnostic tool, because parasite numbers in bovine samples—e.g., preputial smegma or cervico-vaginal mucus—are usually too low to be detected by direct microscopy and a multiplication of parasites after a few days of cultivation increases the chance to find infected bulls.

The number of organisms in preputial secretions has been estimated and ranges from less than 200/mL up to more than 80,000/mL (Skirrow and BonDurant 1988). In bovine tritrichomonosis, diagnostic sensitivity of a single culture test on infected bulls has been estimated to range between 70 and 100% (Skirrow et al. 1985; Schönmann et al. 1994; Parker et al. 1999, 2003a, b). In a large field study, including 2832 mature bulls from 124 beef herds in Argentina, Bayesian estimation revealed a diagnostic sensitivity and specificity of 72.0% (59–87%) and 95.4% (94–96%), respectively (Perez et al. 2006). A repeated testing of bulls—e.g., three times with intervals of several days—has been shown to increase the diagnostic sensitivity of the cell culture test close to 100%. Of 29 samples collected from 5 experimentally infected bulls with resting periods of 2–4 days between samplings, 24 (83%) were determined as positive (Mukhufhi et al. 2003). In another study, consecutive testing over a period of more than 7 months resulted in the determination of an infection rate of 100% in 15 bulls (Clark et al. 1971). For bulls from herds in which *T. foetus* is endemic, two to four tests per bull may be required to ensure that the bull is not infected (Parker et al. 1999). A sexual rest of bulls for a minimum of about 1–2 weeks prior to sampling increases sensitivity (Yule et al. 1989a).

Also for the analysis of females, i.e., after sampling of cervico-vaginal mucus, sensitivity of cultivation is superior to direct microscopic examination (Simmons and Laws 1957; Skirrow and BonDurant 1988). In female cattle diagnostic sensitivity of culture tests has been reported in the range of 56 and 95% (Kimsey et al. 1980; Goodger and Skirrow 1986; Skirrow and BonDurant 1988; Parsonson et al. 1976). The infection in females is usually cleared within 3 months, and it is often difficult to isolate organisms from female cattle in the late stage of their infection.

In cats, the culture method is reported to have a detection limit of about 2×10^2 trophozoites and a diagnostic sensitivity from 26.4 to 58.8% (Hale et al. 2009; Gookin et al. 2004).

To achieve optimal test sensitivity, it is essential to retain as long as possible viability of trichomonads after sampling. The number of viable organisms decreases progressively after sampling (Todorovic and McNutt 1967; Tedesco et al. 1979; Reece et al. 1983; Skirrow et al. 1985; Kittel et al. 1998; Bryan et al. 1999; Parker et al. 1999). Immediate cultivation is ideal but rarely possible. A 1-day delay is estimated to cause a loss of diagnostic sensitivity of 10% (Sager et al. 2007). Sampling the parasite into transportation media providing nutrients has been shown to be essential for the survival of trichomonads, especially if time between sampling and starting cultivation is exceeding 2 days (Kimsey et al. 1980; Hale et al. 2009). An earlier study showed that physiological saline with 5% fetal serum or lactate Ringer's solution was effective (Kimsey et al. 1980). A thyoglycolate transport medium was also shown to be suitable; however, sensitivity of the subsequent cell culture test was slightly lower than after transport using InPouch TF medium (BioMed Diagnostics, White City, OR, USA). Today the medium used for later cultivation is often also used as transportation medium—i.e., InPouch TF or Diamond's medium (Bryan et al. 1999).

An alternative is the direct sampling into a commercially available transport and culture kit—InPouch TF—containing a selective medium, a medium optimized for *T. foetus,* and a medium repressing the growth of the contaminating bacterial flora. This commercial transport and culture kit is recommended not only for sampling in cattle but also for sampling in cats or birds (Thomas et al. 1990; BonDurant 1997; Gookin et al. 2004; Hale et al. 2009; Yao and Köster 2015).

According to the OIE Terrestrial Manual (www.oie.int/international-standard-setting/terrestrial-manual), bovine samples—after being added to transport media—should be protected from exposure to daylight and extremes of temperature, which should remain above 5 °C and below 38 °C (Bryan et al. 1999). For cat fecal samples, a storage for 1 to 24 h at room temperature (23–25 °C) was superior to a 4 °C storage for the same period of time as shown in experiments performed with fecal samples spiked with different *T. foetus* concentrations (2×10^2–2×10^4 *T. foetus* per gram of feces) (Hale et al. 2009).

Several culture media have been found suitable for the cultivation of trichomonads. Overviews on media have been provided in the OIE Terrestrial Manual, www.oie.int/international-standard-setting/terrestrial-manual/, and in several reviews (Skirrow and BonDurant 1988; Sager et al. 2007).

Currently, the most widely used system is InPouch TF, a commercial transport and cultivation kit (Yao 2013). As noncommercial medium the so-called Diamond's medium is widely used, also in epidemiological studies (Tables 14.2 and 14.3). Modified Diamond's medium is a trypticase-yeast extract-maltose medium which in most studies was used modified by the addition of heat-inactivated serum—e.g., of 5% heat-inactivated horse or lamb serum (Diamond 1957; Skirrow and BonDurant 1988; Sager et al. 2007). Both the use of modified Diamond's medium and the

InPouch TF kit are recommended for diagnosis of bovine tritrichomonosis by the OIE Terrestrial Manual (www.oie.int/international-standard-setting/terrestrial-manual, accessed 22. Febr. 2017).

The InPouch TF kit seems superior to Diamond's medium to detect *T. foetus* infection in bulls (Schönmann et al. 1994; Appell et al. 1993; Mendoza-Ibarra et al. 2012; Yao 2013). As regards the cultivating of cat samples, in one study, the InPouch TF kit was found to be superior to modified Diamond's medium (Gookin et al. 2004); however, in a recent study, comparison of both systems revealed a higher sensitivity when modified Diamond's medium was used—ATCC medium 719 (Hale et al. 2009). A retrospective analysis of data revealed no statistical significant differences between cultivation with modified Diamond's medium and InPouch TF.

A modified Plastridge medium containing antibiotics and antifungal agents as well as heat-inactivated bovine serum was recommended for initial cultivation of trichomonads and can be applied combined with modified Diamond's medium for subsequent procedures—e.g., sub-cultivation (Reece et al. 1983; Skirrow and BonDurant 1988; Sager et al. 2007). In studies conducted in Argentina, a commercially available modified Plastridge medium has been applied (Mardones et al. 2008).

The preparation of modified Diamond's medium and test vials, as well as sample processing and reading the results of the culture test, is described in the OIE Terrestrial Manual for bovine samples; many of the recommendations also apply for processing cat and avian samples (www.oie.int/international-standard-setting/terrestrial-manual, accessed 22. Febr. 2017).

Samples collected via preputial scraping—e.g., vigorously by brush, insemination pipette, or aspiration, usually about 0.5–1 mL—can be inoculated directly on top of the medium of a test tube, into the transportation medium, or into the upper chamber of the InPouch TF kit. In contrast, samples collected by preputial washing need to be centrifuged and the supernatant discarded in order to reduce volume. Reading the cell culture tests is performed by microscopic detection of the trichomonads. To increase specificity of the culture test, it is recommended to confirm observed parasites by PCR (Campero et al. 2003).

For cats, voided feces sampled directly from the litter box, rectal swabs obtained from rectal mucous membranes, or feces collected by manual extraction with the aid of fecal loops or by a colon flush technique can be the starting point for trichomonads cultivation (Yao and Köster 2015; Manning 2010; Tolbert and Gookin 2009).

In case of *T. gallinae*-infected birds, a cotton-tipped applicator moistened with sterile saline is used to swab the oral cavity, and swabs are added to InPouch TF culture devices or other commercial or noncommercial culture media (Forrester and Foster 2008; Rogers et al. 2016; Girard et al. 2014; Krone et al. 2005)).

Microscopic detection of culture-growing organisms can be done by light microscopy, on a wet mount slide prepared directly from the culture or through the plastic wall of the InPouch TF kit using a plastic clip provided by the supplier. The motile organisms may be seen under a standard microscope using a 200-fold or higher magnification. An inverted microscope may be useful for examining culture flasks containing culture medium. Culture media should be inspected by microscopic examination at regular daily intervals—from day 1 to day 7 after inoculation (Bryan et al. 1999; Lun et al. 2000). During the first 4–72 h of culturing, there might be an initial increase of parasite numbers, subsequently followed by a decrease. Organisms may be identified on the basis of characteristic morphological features (OIE Terrestrial Manual, www.oie.int/international-standard-setting/terrestrial-manual, accessed 22. Febr. 2017).

It has been shown for cattle and cats that without test confirmation—e.g., by using specific PCR or sometimes by using staining or electron microscopy—the diagnostic specificity of the culture method remains well under 100%. Therefore, a subsequent PCR analysis of culture-positive samples has been recommended to avoid false-positive findings (Parker et al. 2001; Campero et al. 2003; Ceplecha et al. 2013). Intestinal trichomonads were observed in virgin bull samples submitted for confirmation of InPouch TF culture diagnosis or culture diagnosis using Sutherland medium (BonDurant et al. 1999; Michi et al. 2016). The medium of InPouch TF-Feline is thought to be highly specific to *T. foetus*, and the morphologically similar flagellates *P. hominis* and *Giardia* spp. should not survive longer in this medium than 24 h (Gookin et al. 2003). However, the InPouch TF-Feline medium seems to be not entirely selective as *P. hominis* could be successfully cultivated and identified after 3 days following inoculation of InPouch TF-Feline medium using cat feces (Ceplecha et al. 2013).

14.2.2.4 DNA Detection

DNA detection has become one of the most important methods for the diagnosis of infections with trichomonads in cattle and cats. The major advantage of DNA detection by PCR is that it is independent of parasite viability and contaminating microbes that may inhibit trichomonad cultivation. Correspondingly, sensitivity of PCR is often reported to be higher than cultivation and direct microscopical examination (recently reviewed by Yao (2013)). However, PCR analysis has also a number of disadvantages since due to its sensitivity it is prone to carry-over and cross-contamination. In addition, samples may contain inhibitory components that may reduce the sensitivity of PCR or even disable amplification. Each lab should validate the entire diagnostic process—including DNA extraction and PCR amplification—prior to carry out PCR detection as laboratory-specific conditions, equipment, or consumables may have an impact on the outcome of the diagnostic process (Hoorfar et al. 2004; Conraths and Schares 2006).

Preputial material for DNA extraction is generally collected by sheath scraping combined with aspiration or by sheath washing. Material obtained by scraping may contain blood or fecal contaminations from outside the preputium, and material collected by preputial washing may contain urine. Blood, fecal components, and urine may act as PCR inhibitors, and, therefore, it is necessary to minimize or avoid such contaminations. Analytical sensitivities of PCR protocols are high and usually sufficient to detect the DNA of a single parasite (Table 14.1). In field samples, however, analytical sensitivity might be much lower and has been reported to be around 100 organisms per sample (Mukhufhi et al. 2003). In an assessment of diagnostic sensitivity carried out with spiked cat feces, ten organisms per 200 mg of feces were

Table 14.1 Veterinary relevant diagnostic PCRs to detect trichomonads

Target	Type of PCR	Name of primer	Primer, 5′–3′	Name of probe (type of probe)	Probe (type of probe)	Reported specificity	Reported sensitivity	Remarks	Reference
ITS1/5.8S rDNA/ ITS2	End-point	TFR1, TFR2	TGC TTC AGT TCA GCG GGT CTT CC, CGG TAG GTG AAC CTG CCG TTG G	NA	NA	Amplifies *T. foetus*, *T. suis*, *T. mobilensis*, *T. vaginalis*, *T. gallinae*, *T. tenax*, *P. hominis* (Felleisen et al. 1998) . No amplification of bacterial DNA orpurified bovine genomic DNA	One or a few protozoa	Also referred to as pan-trichomonad PCR	Felleisen (1997)
ITS1/5.8S rDNA/ ITS2	End-point	TFR3, TFR4	CGG GTC TTC CTA TAT GAG ACA GAA CC, CCT GCC GTT GGA TCA GTT TCG TTA A	NA	NA	Amplifies *T. foetus*, *T. suis*, *T. mobilensis*	One or a few protozoa	Often referred to as *T. foetus*-specific PCR	Felleisen et al. (1998)
18S rDNA, ITS1, 5.8S rDNA	End-point	TF211A, TF211B	CCT GCC GTT GGA TCA GTT TCG TTA, GCG CAA TGT GCA TTC AAA GAT TCG	NA	NA	Does not amplify *Mycoplasma bovigenitalium*, *Ureaplasma diversum*, or bovine genomic DNA	1 pg *T. foetus* DNA	Reported to produce few unspecific DNA bands	Nickel et al. (2002)

(continued)

Table 14.1 (continued)

Target	Type of PCR	Name of primer	Primer, 5′–3′′	Name of probe (type of probe)	Probe (type of probe)	Reported specificity	Reported sensitivity	Remarks	Reference
ITS1-5.8S rDNA-ITS2	End-point	Tricho-F/ Tricho-R	CGG TAG GTG AAC CTG CCG TT (truncated TRF2, (Felleisen 1997)), TGC TTC AGT TCA GCG GGT CT (truncated TRF1 (Felleisen 1997))	NA	NA	Amplifies *T. foetus*, *T. suis*, *T. mobilensis* based on in silico analyses; amplified *Pentatrichomonas hominis*	NA	Used in human samples and in a study on cats (Profizi et al. 2013)	Jongwutiwes et al. (2000) and Duboucher et al. (2006)
18S rDNA, ITS1 and 5.8S rDNA	End-point	Forward, reverse	GTA GGT GAA CCT GCC GTT G (5'FAM labeled), ATG CAA CGT TCT TCA TCG TG	NA	NA	Amplifies *T. foetus* but also trichomonad DNAfrom a variety of genera; *T. foetus* (157 bp), *Tetratrichomonas* spp. (170–175 bp), *Pentatrichomonas hominis* (142 bp)	Accurate typing is possible from both the 1.0 and 0.1 pgtemplates	Using diagnostic size variants from within the internal transcribed spacer 1 (ITS1) region. Incorporation of a fluorescently labeled primer enables high sensitivity and rapid assessment of results for species identification	Grahn et al. (2005)

18S rDNA, ITS1 and 5.8S rDNA	End-point	Forward, reverse 5.8S primer	Forward primer (Grahn et al. 2005); TTC AGT TCA GCG GGT CTT C	NA	NA	Amplified *T. foetus* (367 bp), *Tetratrichomonas* sp. (379 bp), *Pentatrichomonas* sp. (333 bp), *T. gallinae* (364 bp), and *T. vaginalis* (363 bp)	0.1 pg	Analysis in a 2% agarose gel and by using fluorescent-labeled primers and 6% polyacrylamide gels; disadvantage: too much template makes typing difficult or impossible; advantage: low costs	Frey et al. (2009)
ITS1/5.8S rDNA/ ITS2	End-point nested	TFR3, TFR4 (external); TFITS-F, TFITS-R (internal)	TFR3, TFR4, primer sequences published (Felleisen et al. 1998); CTG CCG TTG GAT CAG TTT CG, GCA ATG TGC ATT CAA AGA TCG	NA	NA	*T. foetus*-specific	Sensitivity in PBS: 1 organism, 70% ; 10 organisms, 90%; 100 organisms, 100%; sensitivity in 200 mg of feces: 10 organisms, 90%; 100 organisms, 100%	Single-tube nested PCR	Gookin et al. (2002)
ITS1/5.8S rDNA/ ITS2	Real-time	TFR3, TFR 4	TFR3, TFR4, primer sequences published (Felleisen et al. 1998)	NA	NA	*T. foetus*-specific		SYBR® qPCR	Mueller et al. (2015)

(continued)

Table 14.1 (continued)

Target	Type of PCR	Name of primer	Primer, 5′–3′′	Name of probe (type of probe)	Probe (type of probe)	Reported specificity	Reported sensitivity	Remarks	Reference
5.8S rDNA	Real-time	T.foeForward (TFF2), T.foeReverse (TFR2)	GCG GCT GGA TTA GCT TTC TTT, GGC GCG CAA TGT GCA T	T.foeProbe (5'FAM/3'MGB-NFQ)	ACA AGT TCG ATC TTT G	Amplifies *T. foetus*, *T. suis*, *T. mobilensis*	3 fg DNA, 0.1–1 cells per assay	5' Taq nuclease assay using a 3' minor groove binder-DNA probe; no need for post-amplification processing	McMillen and Lew (2006)
SSU rDNA	End-point nested	External: 16Sl, 16Sr; nternal: TN3, TN4	External published by Cepicka et al. (2005): TAC TTG GTT GAT CCT GCC, TCACCTACCGTTACCTTG; internal: ATA GGA CTG CAA AGC CGA GA, TGA TTT CAC CGA GTC ATC CA	NA	NA	Amplifies *Trichomonas* sp.	NA	NA	Robinson et al. (2010)
SSU rDNA	End-point	Tgf, Tgr	GCA ATT GTT TCT CCA GAA GTG, GAT GGC TCT CTT TGA GCT TG	NA	NA	Amplifies *T. gallinarum*	One protozoon per assay	Cross-reactions with *T. gallinae*. No cross-reactions were also observed with samples from other protozoa (*Toxoplasma gondii*, *Eimeria tenella*, *Cryptosporidium* spp., *E. invadens*, and *E. ranarum*)	Grabensteiner and Hess (2006)

Not reported	End-point + southernblot by probe	TF1, TF2	CAT TAT CCC AAA TGG TAT AAC, GTC ATT AAG TAC ATA AAT TC	Probe for Southern blot	CAT CAT TAA TGC CTT TTG ATG GAT CAG GCA ACC ATT TAT A	Amplifies *T. foetus*	Ten or occasionally fewer protozoa	Southern blot necessary to identify specific band. A 400 bp product from bovine genomic DNA is amplified. Multiple amplification products from DNA from a related organism, *T. vaginalis*; Southern blot is negative for *T. vaginalis*	Ho et al. (1994)

NA not applicable

detected in 90% of nested PCR tests, and 100 organisms per 200 mg of feces were detected in 100% of nested PCR tests (Gookin et al. 2002).

It has been shown that also in PCR diagnosis the likelihood to detect *T. foetus* decreases with time between sampling and analysis. In one study it was reported that diagnostic sensitivity in PCR detection declined from 90% when samples were stored for 6 h to 31% when they were stored for 5 days (Mukhufhi et al. 2003). It has been hypothesized that hydrolases secreted by trichomonads are responsible for this effect (Thomford et al. 1996; Sager et al. 2007). Adding the DNA-stabilizing agent guanidinium thiocyanate—GuSCN in a final concentration of 200 mmol/L—or a commercial lysis buffer for sample collection known to preserve DNA for several months at room temperature to the transport medium did not improve results (Mukhufhi et al. 2003; Mendoza-Ibarra et al. 2012). Fecal samples from cats should be submitted within 24 h after sampling at room temperature or 4 °C.

To prevent amplification of carry-over contaminants, protocols that incorporate dUTP instead of dTTP as a nucleotide and allow to apply the Uracil-DNA Glycosylase (UDG) system have been reported (Longo et al. 1990; Felleisen et al. 1998). Contaminating amplicons carried over from previous PCRs can be removed by this system before a PCR amplification. Possible disadvantages of the UDG system are that it might become necessary to optimize the reaction mixture used for PCR—e.g., the $MgCl_2$ concentration—and a standard PCR buffer may not work (Felleisen et al. 1998).

It is essential to monitor the presence of a potential inhibitor in each individual sample since preputial material, cervico-vaginal mucus, and fecal samples contain PCR-inhibiting components. There are different possibilities available. The TFR3/TFR4 PCR protocol includes an artificial internal control DNA carrying TFR3 and TFR4 sequences—based on pBluescript KS+ DNA—and generates control amplicons, via PCR amplification and composite primers (Table 14.1) (Felleisen et al. 1998; Sager et al. 2007). Internal controls of unrelated DNAs have been integrated into a 5′ nuclease assay—real-time PCR assays with TaqMan probes. In this type of assay it is possible to incorporate an unrelated DNA into the sample prior to DNA purification, which is later amplified in a multiplex assay along with the parasite DNA but detected by a probe carrying a fluorophore different than that of the parasite-specific probe. One commercially available *T. foetus* real-time PCR includes such a control system (VetMAX™-Gold Trich Detection Kit, Life Technologies). This principle of inhibition control is becoming more and more popular, also for in-house assay.

There are many ways, by which DNA from preputial smegma, cervico-vaginal mucus, fecal samples, oropharyngeal swabs, culture material, and others types of samples can be extracted. Often commercial kits but also in-house methods have been applied. Also non-purified but heat-treated samples were used with success (McMillen and Lew 2006). However, the use of unpurified DNA is prone to inhibition and is not generally recommended. It has been shown that inhibiting components could be successfully removed from preputial smegma by 5% Chelex®-100 and 0.05% agar solution (Chen and Li 2001). However, in another study, the use of Chelex®-100 caused significantly lower detection rates (Mendoza-Ibarra et al. 2012).

The majority of published diagnostic PCRs for *T. foetus* are targeting rRNA-coding genes (rDNA) and their flanking regions (Table 14.1). These regions include the 18S rRNA gene, the internal transcribed spacer (ITS)1 region, the 5.8S-rRNA gene, the ITS2 region, and the 28S rRNA gene. One of the first diagnostic PCRs established—the TFR3/TFR4 PCR—is widely used. The TFR3 primer targets the 3' end of the 18S rRNA gene and the TFR4 primer the 5′ end of the 28S rRNA gene (Felleisen et al. 1998). Although this PCR assay is often referred to as being specific for *T. foetus*, also DNA of *T. mobilensis*—an intestinal parasite of squirrel monkey—or *T. suis* is amplified (Table 14.1).

The rRNA gene sequences have been widely used for phylogenetic studies in Parabasalia to which trichomonads belong. Other genes coding for cysteine proteinases—CP1, CP2, and CP4–CP9—and cytosolic malate dehydrogenase 1 (MDH1) have been used to differentiate *T. foetus* isolates from cattle and cat or to characterize new strains of *T. foetus* (Kleina et al. 2004; Cepicka et al. 2005, 2006; Gaspar da Silva et al. 2007; Kolisko et al. 2008; Sun et al. 2012; Slapeta et al. 2012; Casteriano et al. 2016). In *T. gallinae* further genes were used to define lineages (recently reviewed in Amin et al. (2014)).

Because 18S rRNA genes show limited differences between trichomonads, end-point assays have been applied using primers capable to amplify DNA of several trichomonad species simultaneously (Felleisen 1997). In these PCRs, species diagnosis was achieved in a second step, either by PCR-RFLP, by determination of the precise size of amplification products, or by single-strand conformation polymorphism (SSCP) (Hayes et al. 2003; Huby-Chilton et al. 2009).

The TFR3/TFR4 PCR protocol has been modified by using the TFR3/TFR4 primer pair for external amplification followed by an internal newly designed primer pair in a single-tube nested PCR (Gookin et al. 2002). The TFR3/TFR4 PCR protocol has been also modified into a SYBR®-based real-time PCR assay (Mueller et al. 2015).

A 5' nuclease assay—i.e., a real-time PCR applying a TaqMan probe—based on rRNA gene sequences has been established to detect *T. foetus*, *T. suis*, and *T. mobilensis* (McMillen and Lew 2006). In this assay a 57 bp region of the 5.8S rRNA gene region is amplified (Table 14.1). As mentioned earlier in this section, a commercialized 5′′ nuclease assay is available (VetMAX™-Gold Trich Detection Kit) which has been used in epidemiological studies on *T. foetus* of cattle in Southern Africa (Casteriano et al. 2016).

14.2.2.5 Serological Techniques

Serological and other antibody detection tests have been established. However, they are of no importance for the diagnosis of trichomonosis. These tests include:

- **Agglutination and hemolytic tests.** In *T. foetus*-infected cows, antibodies appear in the cervicovaginal mucus about 6 weeks after infection and persist for several months (Pierce 1947). A mucus agglutination test detected 32% (57 of 178) of cows in naturally infected herds, and no cows from clean herds tested positive (Pierce 1949). The mucus agglutination test was shown to be specific as no cross-reactions with *Campylobacter fetus* or *Brucella abortus* has been observed. However, the reliability of the test was strongly influenced by the type of mucus. The mucus agglutination test was regarded as herd test (Pierce 1949).

- A serum agglutination tests, similar to the mucus agglutination test, did not show results that correlated well with the *T. foetus* infection status of cows (Kerr 1944).
- A hemolytic assay has been established which showed in female cattle a diagnostic specificity of 96% and a diagnostic sensitivity of 94% (BonDurant et al. 1996). However, only 43% of chronically infected bulls tested positive when this test was applied (BonDurant et al. 1996).
- **Indirect ELISAs to detect parasite-specific antibodies.** An indirect ELISA (iELISA) based on the TF1.17 surface antigen of *T. foetus* showed promising results when tested with cervico-vaginal mucus of heifers (Ikeda et al. 1995). TF1.17 surface antigen-specific IgG1, IgA, and IgM antibodies in the smegma of bulls naturally infected with *T. foetus*—as determined and measured by ELISA—were observed concurrently with *T. foetus*-positive smegma cultures (Rhyan et al. 1999); to the best of our knowledge, this approach was not further elaborated. In vaccination studies, IgG1, IgG2, IgA, and IgE responses were monitored in the preputial secretions or in sera of bulls by using a whole *T. foetus* cell antigen preparation for ELISA (Cobo et al. 2009).
- An iELISA coated with whole *T. foetus* parasites and fixed with ethanol was used to determine an isotype-specific antibody response in the reproductive tract secretions and sera of *T. foetus*-infected heifers (Skirrow and BonDurant 1990a). In cervical and vaginal secretions, parasite-specific IgA and IgG1 antibodies predominated 7–12 weeks after infection, while in serum, parasite-specific IgG1 and IgG2 antibodies were detected. Interestingly, elevated antibody levels were observed after reinfection using this iELISA (Skirrow and BonDurant 1990a). A similar iELISA with immobilized whole *T. foetus* parasites was used to monitor the *T. foetus*-antibody response in immunized heifers naturally challenged by being bred with a naturally infected bull (Cobo et al. 2002). The presented serological tests have not been used or validated for routine diagnostic purposes.

An iELISA has been also used under experimental conditions to detect antibodies against *T. gallinarum* and *T. gallinae* in poultry (Amin et al. 2011). For this iELISA, the plates were coated with *T. gallinarum* parasites in carbonate buffer per well.

14.2.2.6 Intradermal Test

Analogous to the tuberculin test, a Tricin test has been developed to identify *T. foetus*-infected cattle or herds (Kerr 1944). The antigen used for an intradermal Tricin test has been prepared by fixation of cultured *T. foetus* parasites using trichloroacetic acid. Skin reactions—i.e., an increase in skin thickness—were read 30–60 min after the application of the antigen. This test was regarded to represent a herd test (Kerr 1944).

14.2.2.7 Antigen Detection

Attempts to develop a sensitive and specific diagnostic antigen test for detecting *T. foetus* antigen in cervico-vaginal mucus have failed (Yule et al. 1989b). However, immunohistochemical techniques have been successfully applied to demonstrate trichomonads in histological sections (Rhyan et al. 1995b; Yaeger and Gookin 2005). For immunohistochemical detection, monoclonal antibodies developed to characterize *T. foetus* antigens revealed to be valuable tools (Hodgson et al. 1990). A protocol for immunohistochemical detection by using the monoclonal antibody Mab 34.7C4.4 is provided in the OIE Terrestrial Manual (www.oie.int/international-standard-setting/terrestrial-manual, accessed, 22. Febr. 2017).

14.2.2.8 Diagnosis in Different Hosts

Diagnosis in Cattle

A tentative diagnosis of trichomonosis as a cause of reproductive failure in a herd is based on the clinical history, signs of early abortion, repeated returns to service, or irregular estrous cycles. Confirmation of infection depends on the demonstration of the organism in placental fluid, stomach contents of the aborted fetus, uterine washings, pyometra discharge, vaginal mucus, or preputial smegma. In infected herds, the most reliable material for diagnosis is preputial scrapings (Kittel et al. 1998; Mukhufhi et al. 2003; Parker et al. 1999; Schönmann et al. 1994).

Bulls are the main reservoir for the parasite. Control programs focus on identifying and culling infected bulls and nonpregnant cows carrying the parasite. Prevention of transmission of the disease through culling practices relies on the ability to identify infected animals accurately.

Advances in cell culture and polymerase chain reaction (PCR) have increased the ability to detect the disease in bulls. However, the collection of an adequate sample is of outmost importance, as this step affects sensitivity, specificity, and repeatability. The low repeatability observed with most sample collection techniques can cause false-negative results. The most efficient method of sampling vaginal and preputial secretions is insertion of an insemination/infusion pipette inside the vaginal fornix or preputial cavity and performing short strokes while concurrently aspirating secretions (Cobo et al. 2007). Vagina or preputial cavity can be washed with PBS to recover more organisms, although this usually dilutes the sample. Alternatively, preputial secretions can be collected by scraping with a plastic or metal brush, with no significant differences in culture sensitivity compared to using a pipette (Tedesco et al. 1979; Parker et al. 1999).

Diagnosis in Bulls

Routine herd diagnosis is optimally performed on bulls and not on females, because bulls remain permanently infected while in most female cattle infection is only transient. In addition, sampling of bulls reduces costs as preputial samples have to be taken only from a smaller number of bulls. Diagnosis in the bull involves collecting, transporting, and culturing the sample in special growth media and tentatively identifying the organism by microscopic examination. Finally a confirmation of detection using PCR has to be carried out to confirm a *T. foetus* infection (Fig. 14.4). If samples

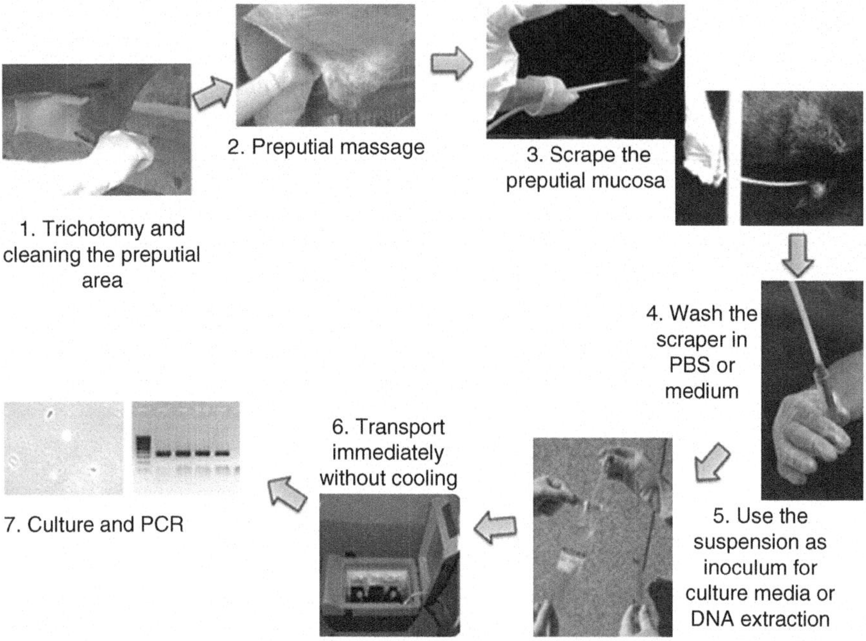

Fig. 14.4 Collection of preputial smegma samples and diagnostic methods for bovine tritrichomonosis

are processed by PCR, the use of pooled direct preputial samples is possible. However, this strategy required repeated sampling to optimize sensitivity (Garcia Guerra et al. 2013). Optimally, all bulls belonging to the same herd should be sampled. In those herds in which periodic sanitary control is carried out without previous history of venereal diseases and with good pregnancy rates, at least two consecutive samplings of all bulls should be performed, with a minimum of 1 or 2 weeks time between them (Skirrow et al. 1985; Kimsey et al. 1980). Consecutive testing is necessary due to the low diagnostic sensitivity of an individual test. If both samplings revealed negative results, the herd can be considered tritrichomonosis free.

Some recommendations have to be taken into account at the time of sampling:

- It is generally recommended to allow a sexual rest of about at least 1–2 weeks before taking a preputial sample to allow a repopulation of microorganisms in the preputial cavity and increase diagnostic sensitivity.
- It is important to have adequate facilities to perform preputial sampling particularly when sampling a large number of bulls to avoid accidents and injuries.
- The bulls should remain in a pen, without access to water approximately 12 h prior to sampling. This will prevent urination during sampling.
- The external preputial area must be cleaned with disposable paper towels without soap or disinfectants, and—if necessary—preputial hairs should be trimmed to prevent contamination.

The presence of *T. foetus* seems to be confined to the preputial cavity, and *T. foetus* is localized in the preputial secretions and does not invade the epithelium of the penis or prepuce (Parsonson et al. 1974). *T. foetus* could not be cultured from the epididymis, ampulla, seminal vesicle, pelvic urethra, or testis. Macroscopic and microscopic examination of the genital tracts of infected bulls did not reveal any lesions populated by *T. foetus* (Parsonson et al. 1974).

A number of techniques for collecting preputial samples from bulls have been described (INTA, 2014, Técnicas de muestreo para el diagnóstico de enfermedades venéreas en bovinos, https://www.youtube.com/watch?v=G-1StrWaHKA, accessed 21. Febr. 2017; Navajo Technical College, 2012, Veterinary Technicians perform Trich testing, https://www.youtube.com/watch?v=lp8fpDVDOCE, accessed 21. Febr. 2017). In all these protocols, it is important to avoid contaminations, as this may introduce intestinal protozoa or PCR-inhibiting contaminants. Samples can be collected from bulls by scraping the preputial and penile mucosa with an artificial insemination pipette or scraper, by preputial lavage, or by washing the artificial vagina after semen collection (Cobo et al. 2007; Tedesco et al. 1979; Parker et al. 1999). The latter technique is not recommended as its sensitivity may be lower (Ostrowsky et al. 1974; Parker et al. 1999).

The collection of the preputial smegma can be carried out by using different devices:

- A *Cassou* insemination pipette (Cassou straws, IMV Technologies, L'Aigle, France) covered by a plastic sheath. Inside the preputial cavity, the pipette is pulled forward through the plastic sheath to expose the tip and moved back and forward in short strokes adjacent to the glans penis, especially near the fornix, while aspirating and massaging the glands penis to release greater amounts of smegma into the pipette. A new plastic sheath is used for each bull (Cobo et al. 2007).
- A sterile, dry, plastic *insemination/*infusion pipette of 43 cm length with a 10–15 mL syringe attached to one end is placed into the preputial fornix. The pipette is scraped vigorously across the preputial epithelium without aspiration, and then negative pressure is applied with the syringe to collect preputial smegma. The negative pressure is released before removing the pipette from the sheath, to avoid aspiration of urine or other contaminants. After removing the pipette from the sheath, the sample is placed immediately into a transport or culture medium. A new syringe and pipette are used for each bull (Rodning et al. 2008).
- A plastic or metal *scraping instrument* of approximately 70 cm in length, having at the anterior end a grooved surface of approximately 10 cm in length by 0.8 cm in diameter, which is used to perform scraping of the preputial mucosa. The scraper is introduced into the preputial mucosa by performing 20–30 movements back and forward. It is then carefully withdrawn from the preputial cavity, avoiding contamination with the external part of the foreskin (Terzolo et al. 1992). The aspiration method for the collection of preputial secretions for cultural examination continues to be used although a specially designed scraping instrument was

reported to possess advantages (Bartlett et al. 1947; Clark et al. 1971; Stuka and Katai 1969).

- A *gauze sponge*, which to our knowledge is rarely used. In this procedure, the penis is extended by electrostimulation with a rectal probe. Once extended, a 16-ply gauze sponge is used to wipe around the glans and down the penile shaft and exposed preputial mucosa two to three times. Recently, it has been demonstrated that sponge sampling of *T. foetus* from bulls is a valid method (Dewell et al. 2016). The method facilitates easier collection once the penis is extended and is potentially safer for the veterinarian and the bull. Additionally, the gauze sponge method might be slightly more sensitive than the pipette method (Dewell et al. 2016).

After collection the material has to be inoculated in corresponding transport or culture media and should arrive at the laboratory within 24–36 h; during transportation, the sample should be protected from exposure to daylight and extremes of temperature.

Diagnosis in Heifers and Cows

Initially, the uterus was regarded as the definitive and persistent site of infection while the vagina was considered to be a relatively unreliable source of *T. foetus* for diagnosis (Bartlett 1947). Subsequently, however, the persistence of *T. foetus* in the vagina and/or cervix for periods up to 95 days post-infection was shown, and it was established that samples collected from vagina and the uterine cervix allow a reliable and accurate diagnosis (Parsonson et al. 1976). More recent studies of naturally infected cows reconfirmed that the cervix belongs to the preferred site of parasite location (Skirrow and BonDurant 1990b). The numbers of parasites present in the cervical-vaginal mucus fluctuate during the estrus cycle, and the largest numbers are seen a few days before estrus.

The time a cow remains infected was significantly longer for cows experiencing their first, mean 20.3 weeks, than for those experiencing either their second, mean 9.8 weeks, or their third period of infection, mean 11 weeks. However, the rate of isolation of *T. foetus* from samples of vaginal mucus collected from cows remained similar—mean 83.5%—irrespective of the period of infection or whether the cows showed normal fertility, infertility, or abortion (Clark et al. 1986).

The efficiency of diagnosis in cows increased with temporal proximity between the initial infection and the time of sampling (Clark et al. 1986). It is important to note that the detection of positive females is of value for the initial detection of the infection in a herd. However, it is not useful for a subsequent disease control because cows usually clear their infection and generally become immune, at least for the actual breeding season (BonDurant 1997; Fitzgerald 1986). Sampling should be performed when females with conception failures are observed or at the time of rectal palpation in nonpregnant females. Culture sensitivity is lower in cows than in bulls. However, it is important to point out that the optimal period for sampling is near the end of the service period (Skirrow and BonDurant 1990b; Terzolo et al. 1992).

Uterine and vaginal secretions can be collected in cows that have aborted, in those that have not been pregnant, or heifers. To perform the extraction of cervico-vaginal mucus, a sterile, dry, plastic 43 cm insemination pipette with a

10 mL syringe attached to one end or a Cassou insemination pipette (44 cm long × 0.64 cm outer diameter × 0.32 cm inner diameter; Cassou straws, IMV Technologies, L'Aigle, France) can be used. Opening the lips of vulva without having to fix the cervix rectally, the pipette is inserted in a dorsal-cranial direction into the vagina. After the pipette has been inserted, slight anteroposterior and circular movements are executed, performing aspiration simultaneously. The vacuum generated is usually sufficient to extract the cervical-vaginal mucus, which will vary in quantity and consistency depending on the moment of the estrous cycle in which it is extracted. In a low percentage of animals, the volume of mucus extracted may be scarce. In this case it is possible to introduce 5 mL of phosphate-buffered saline solution by performing a wash with subsequent extraction.

The use of a "screwhead scraper rod" for collecting of samples from the cervico-vaginal mucosa proved to be a practical method and calls for further comparative evaluation with other standard methods in use (Abbitt and Ball 1978). Apart from other extraction techniques, also the Bartlett glass pipette procedure, which is rarely used because it is complicated and the equipment often inaccessible, has been described (Hammond and Bartlett 1943).

Fetal Diagnosis

When abortion occurs, *T. fetus* can be isolated from placental fluids or cotyledons. However, the high degree of contamination of this material limits its use. Isolations can be made from samples taken from the fetal mouth. Swabbing the mucosa of the tongue and roof of the mouth has been recommended (Case and Keefer 1938). Nevertheless, the place where *T. foetus* is most consistently isolated is the abomasal fluid. The sample can be taken with sterile syringe and needle. Once the abomasum content has been extracted, it can be sent to the laboratory in the same syringe or can be inoculated in transport or cultured medium.

Diagnosis in Cats

The diagnosis of feline trichomonosis has been reviewed comprehensively (Tolbert and Gookin 2009; Manning 2010; Yao and Köster 2015). *T. foetus* infection is suspected in cats with recent—less than 6 months lasting—clinical signs of chronic large bowel diarrhea, in young, purebred cats, from densely housed origin. Routine coprological methods, like flotation-sedimentation or sodium acetate-acetic acid-formalin concentration (SAF), destroy or fix trichomonads, respectively, and fixation causes the loss of their characteristic movement which makes them hard to be recognized.

Currently, the preferred diagnostic methods for feline trichomonosis include visualization of the organism in direct smears or culture or *T. foetus* DNA detection by PCR (Gookin et al. 2001, 2002, 2004; Levy et al. 2003; Foster et al. 2004). Histopathological examinations of colon, cecum, and ileum samples are not routinely used but can be helpful to diagnose a feline tritrichomonosis.

It is important to mention that parasite shedding in feces is erratic throughout the course of infection being occasionally so low that it cannot be detected by diagnostic techniques. Consequently, it is advisable to resample cats showing clinical signs that have been tested negative or to increase diagnostic sensitivity by using more than one test method—e.g., culture and PCR (Gookin et al. 2002). Cats should not receive any antibiotics within several days prior to or at the time of testing.

Samples consist of fresh voided feces taken directly from the litter box, rectal swabs, and manual collection with the aid of fecal loops or by a colon flush technique (revised in Manning (2010), Yao and Köster (2015), Tolbert and Gookin (2009)). Samples collected with a *fecal loop* or by the *colon flush* technique are preferable. The technique of colon flush is demonstrated in a video clip the North Carolina State University, College of Veterinary Medicine website (http://www.youtube.com/watch?v=JMfZ9M80V8E, accessed 22. Febr. 2017). Freshly voided or diarrheic feces are considered ideal for testing whereas samples obtained from normal or dry stools are believed to be less suitable (Tolbert and Gookin 2009).

A diagnostic approach based on direct fecal smears, which may reveal motile trophozoites, can be employed during examination at pet clinics. Samples are suspended in saline and examined immediately under a cover slip at 200 to 400-fold magnification using a light or, preferably, a phase-contrast microscope (Fig. 14.5). Although direct smears represent a cheap, quick, and readily available technique, the sensitivity of microscopic examination of a direct fecal smear is low. The diagnostic sensitivity of a direct smear using samples from naturally infected cats has been shown to be only 14% in one study, but sensitivity can be increased by analyzing multiple fecal smears (Gookin et al. 2004). Other challenge associated with this diagnostic procedure is the skill of the practitioner in identifying motile trichomonads. *T. foetus* cannot precisely be distinguished by microscopical examination from *P. hominis* and is often misdiagnosed as *Giardia* spp. (Tolbert and Gookin 2009; Manning 2010; Yao and Köster 2015). A video clip demonstrating the classic jerky motility has been provided by the North Carolina State University, College of Veterinary Medicine website (https://www.youtube.com/watch?v=aF06jlbcF8E; accessed 22. Febr. 2017). The presence of *Giardia* spp. can be confirmed by a fecal enzyme-linked immunosorbent assay for *Giardia*-specific antigen. Differentiation

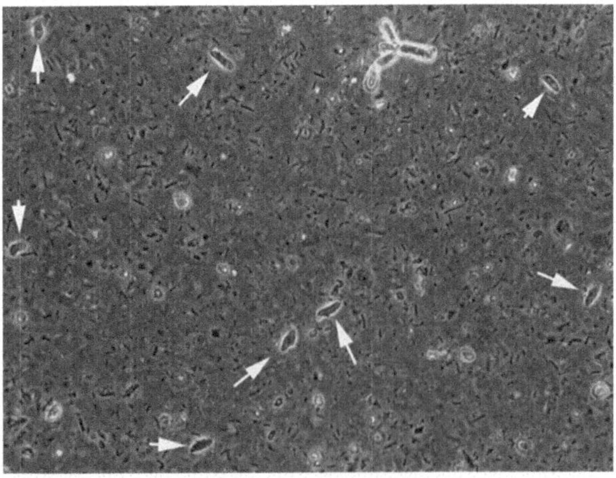

Fig. 14.5 Microscopic detection of *Tritrichomonas foetus* trophozoites (*arrows*) in feline feces (200× magnification)

of *Giardia* spp. and *T. foetus* is crucial to avoid a useless initiation of tritrichomonosis therapy using potentially neurotoxic ronidazole.

In vitro culture of *T. foetus* can be started by incubating the sample feces in a suitable growth medium (Gookin et al. 2001, 2003). Diagnostic kits—for example, the InPouch TF-Feline test kit—can be an attractive option for the clinical practice due to the commercial availability, a shelf life of a year, and the simplicity of use. Sample submission requirement for in vitro culture diagnostic includes the collection of fresh feces—roughly the size of a lima bean—and the use of transport-growth media, InPouch TF or modified Diamond's media. It is recommended to dilute the fecal samples in PBS or physiological saline—to about 1 to 20 v/v—to create a homogenous mixture with a semiliquid consistency to be used as the inoculum for the culture medium, reducing the bacteria burden in the sample (Arranz-Solis et al. 2016). Furthermore, it is crucial for a reliable diagnosis of feline trichomonosis to keep the time delay from sample collection to processing as short as possible, given the environmental fragility of *T. foetus* trophozoites (Gookin et al. 2004). In order to maximize the diagnostic sensitivity, cat feces should be inoculated into the culture within a 6-h period after voiding and to be submitted at room temperature—23 to 25 °C—to a specialized diagnostic laboratory within 24 h (Hale et al. 2009). False-negative findings can be due to refrigeration or delayed processing of feces.

After culture initiation, samples are then incubated at room temperature or at 37 °C; incubation at room temperature made more robust and long-lived cultures (Gookin et al. 2003). Cultures are examined microscopically periodically up to 7 days—normally every 24–48 h—on a wet mount slide prepared directly from the culture or through the plastic wall of the InPouch TF kit. Most of the microscopically positive culture samples are detected 72 h post-incubation. It has to be mentioned that the diagnosis by culture is more difficult for cats than for bovines due to the nature of the feces sample. As feces are directly cultured in enriched media, bacterial and/or fungal contamination is more likely to arise. Bacterial contamination severely impairs the culture of *T. foetus* and is able to reduce the sensitivity of the culture method. Incubation at 37 °C shows a quicker positive result but also more bacterial overgrowth potentially inhibiting the growth of *T. foetus*. If the commercial InPouch TF-Feline test kit is used, an accumulation of gas within the pouches due to an overgrowth of fecal flora is a commonly observed problem. Moreover, *T. foetus* and related parasites are hard to distinguish with light microscopy because of the similar morphology. Culture-positive samples require confirmation by PCR because also other trichomonads like *P. hominis* are able to grow (Ceplecha et al. 2013).

Molecular diagnosis is becoming more widely available than culturing the organism. Under optimized conditions, a direct diagnosis by PCR—i.e., without a prior cell culture test—offers a highly specific and a sensitive diagnostic alternative also able to detect nonviable parasites. Moreover, PCR is the method of choice to confirm positive results by culture. It is important to monitor a potential PCR inhibition in each individual sample. The choice of an appropriate DNA extraction technique greatly influences the reliability and sensitivity of PCR (Stauffer et al. 2008; Hale et al. 2009). Several methods have been successful employed for feline trichomonosis diagnosis, such as a modified procedure using the commercial QIAamp DNA stool minikit (QIAGEN, Hilden, Germany), Boom's method, or ZR Fecal DNA kit ZR (Zymo Research, Orange, CA, USA), which was able to detect 10 *T. foetus* organisms per 100 mg feces in 100% of PCR reactions (Stauffer et al. 2008). Diagnostic *T. foetus* PCR using primers TFR3/TFR4 and a single-tube nested PCR using primers TFITS-F/TFITS-R in combination with primers TFR3/TFR4 are widely used for feline trichomonosis diagnosis (Table 14.1) (Felleisen et al. 1998). PCR amplification of DNA extracted from feces samples appears to be more sensitive than the InPouch TF culture, and a high level of agreement has been described between the culture and PCR detection (Gookin et al. 2002; Hosein et al. 2013; Arranz-Solis et al. 2016).

Finally, colonic histopathology can be used as a diagnostic tool. Colon, cecum, and ileum samples are collected during necropsy, surgery, or endoscopy. On histologic examination, large numbers of teardrop- to crescent-shaped trichomonads can be found associated with the colonic mucosal surface and less commonly within colonic crypt lumens. Histologic features include mild to moderate lymphoplasmacytic colitis with crypt micro-abscesses, increased mitotic activity, loss of goblet cells, and attenuation of superficial colonic mucosa (Yaeger and Gookin 2005). The probability of diagnosing a *T. foetus* infection based on histopathology increases with the number of submitted samples—a minimum of six colon samples is required to have a larger than or equal to 95% confidence of detecting *T. foetus* in at least one sample (Yaeger and Gookin 2005). Recently, a fluorescence in situ hybridization assay (FISH) and a chromogenic in situ hybridization (CISH) technique have been described to allow the localization and identification of *T. foetus* in formalin-fixed and paraffin-embedded samples, respectively (Gookin et al. 2010b).

Diagnosis in Other Animals

The trichomonad *T. gallinae* of birds is of veterinary importance, while *Tetratrichomonas gallinarum* is a commensal that can become important under certain circumstances—e.g., concurrent infections with other pathogens (recently reviewed in Amin et al. (2014)). Diagnostic techniques described for *T. foetus* can be also applied to the diagnosis of bird trichomonosis, including direct microscopy, cultivation, and PCR detection.

For direct microscopy, sample material can be collected by swabbing the oral cavity, *T. gallinae*, or cloacae, *T. gallinarum*, and is mounted either diluted or undiluted on glass slides. Glass slides need to be examined immediately to observe motile protozoa (Forrester and Foster 2008; Amin et al. 2014). As mentioned, staining of trichomonads is possible and may facilitate detection (Amin et al. 2011). Similar to bovine and feline trichomonosis, the sensitivity of direct microscopy is low, and situations in which the parasite load is low may likely result in false-negative results.

The use of samples to initiate cultures allows the amplification of the number of parasites making this procedure more sensitive than the direct microscopic examination of samples. Correspondingly, in a study of pigeons, only about half of the samples positive in culture revealed a positive result in wet mount microscopy (Bunbury et al. 2005). In case of *T. gallinae*-infected birds, material collected via swabbing from the oral cavity of the animal is added to a InPouch TF culture device or other commercial or noncommercial culture media (Bunbury et al. 2007; Forrester and Foster 2008; Rogers et al. 2016; Girard et al. 2014; Krone et al. 2005). For cultivation, specimens are usually incubated at 37 °C for several days and examined microscopically every 24 h. Although birds have a higher body temperature, 37 °C seems to be the optimal temperature for the cultivation of *T. gallinarum* (De Carli and Tasca 2002; De Carli et al. 1996; Amin et al. 2010).

For PCR, a number of primer pairs have been reported that target the ITS1-5.8SrDNA-ITS2 genomic region but which are not species-specific and able to detect virtually all trichomonad species. Most commonly used are primers TFR1 and TFR2 (Felleisen 1997). A nested PCR, targeting the 18S rRNA gene, ITS1, and 5.8S rRNA gene, was established but was not species-specific, amplifying also *T. gallinae* (Frey et al. 2009; Grahn et al. 2005). Another nested PCR has been developed for the amplification of the ITS1-5.8S rDNA-ITS2 genomic region of trichomonads used to detect *T. gallinae*-specific DNA from esophageal lesions of finches sampled during an epidemic of finch mortality (Robinson et al. 2010; Cepicka et al. 2005). The analytic specificity of this PCR has not been reported. Amplification primers specifically designed for the identification of *T. gallinarum* yielded also a PCR product of specific DNA of *T. gallinae* of identical size (Grabensteiner and Hess 2006).

An indirect ELISA has only been used under experimental conditions to detect antibodies against *T. gallinarum* and *T. gallinae* in poultry (Amin et al. 2011). Because birds are often latent carriers of trichomonads, there is hope that serological analyses in bird populations might be able to identify carrier birds and may help to better understand persistence and spread of these parasites (Amin et al. 2014).

14.3 Epidemiology

14.3.1 Epidemiology of Tritrichomonosis in Cattle

Bovine tritrichomonosis is a major problem worldwide, affecting a large proportion of herds in North and South America, in parts of Europe, Africa, Asia, and Australia (de Oliveira et al. 2015; Yao 2013; Perez et al. 2006; Mendoza-Ibarra et al. 2012, 2013; Madoroba et al. 2011; Yang et al. 2012; Guven et al. 2013; McCool et al. 1988). Bovine tritrichomonosis is considered endemic in herds managed under extensive conditions and using natural service for breeding (Bondurant 2005). The economic impact of tritrichomonosis infection has to be regarded severe in particular regions of the world. The calf crop in affected beef and dairy herds can be reduced up to 50% in beef operations (Rae 1989). Economic losses are variable and depend on the percentage of bulls infected and the susceptibility of the cows in the herd. Further losses, in addition to calf crop losses, include an extended breeding season due to an increased number of repeat breeders. Due to later calving, the growing periods for calves might be shortened, and there are batches of calves of different ages with a wide variation in weaning weights. Summarizing these losses, it was estimated that tritrichomonosis in a herd caused a 35% decrease in economic return per cow in an infected herd (Rae 1989). Economic losses in a case of bovine tritrichomonosis in a large Californian dairy herd were calculated at 665 US$ per infected cow (Goodger and Skirrow 1986).

The number of tritrichomonosis reports has drastically reduced in regions or production systems in which artificial insemination is the predominant mode of breeding and in which comingling of herds is avoided—for example, in the European dairy industry. A recent survey conducted in Switzerland involving 1362 preputial samples from bulls and 60 abomasal fluid samples of aborted fetuses from beef and dairy herds revealed no *T. foetus*-positive finding (Bernasconi et al. 2014). However, also in areas assumed to be largely free of bovine tritrichomonosis, a reestablishment of the infection in herds, especially in beef herds farmed under extensive, pastoral systems, is possible as shown by findings from Spain (Mendoza-Ibarra et al. 2012; Mendoza-Ibarra et al. 2013). Knowledge on risk factors is important for the implementation of effective measures to control tritrichomonosis. Results from risk factor studies were used to model effects of vaccination against tritrichomonosis in beef herds. A number of potential herd-level risk factors were assessed, including "no. of cows," "no. of young bulls," "trichomonad testing yes/no," "no. of trichomonad tests," "shared grazing," "previous diagnosis of trichomonosis," or "duration of breeding season" (Villarroel et al. 2004).

T. foetus isolated from cattle and cats and *T. suis* from pigs are genetically very similar or in case of *T. foetus* from cattle and *T. suis* even indistinguishable. Nevertheless, there is no evidence that there are links between life cycles of *T. foetus* in cattle and cats or between life cycles of feline or bovine *T. foetus* and porcine *T. suis*. Most likely these parasites have evolved separately, and despite their genetic similarity, *T. foetus* of bovine and feline origin and porcine *T. suis* show biological traits which differ considerably.

As stated in Sect. 14.1.2, tritrichomonosis is an almost exclusively venereal transmitted disease in cattle and affects predominantly adult animals (Sager et al. 2007; Ondrak 2016). *T. foetus* is transmitted during coitus, mainly from an infected bull to an uninfected dam or vice versa (Ondrak 2016). Single mating with an infected bull may result in a 95% infection rate among susceptible heifers (Parsonson et al. 1976), but in general a transmission rate of 30–70% is assumed (Bondurant 2005).

A mechanical transmission either by uninfected bulls or by contaminated equipment or cryopreserved semen seems to be possible, but the relative importance of these routes of transmission is minor (Ondrak 2016; Clark et al. 1977; Murname 1959; Goodger and Skirrow 1986; Blackshaw and Beattie 1955; Clark et al. 1971; Skirrow and BonDurant 1988).

The following risk factors for tritrichomonosis in individual animals have been identified:

- **Carrier state and age as a risk factor of infection**: Infection in bulls is reported to persist for more than 3 years and may persist for life (Rhyan et al. 1999; Campero et al. 1990; Flower et al. 1983; Bondurant 2005). Several studies established that the likelihood of bulls being infected seems to increase with age (Skirrow et al. 1985; McCool et al. 1988; BonDurant et al. 1990; Rae et al. 1999, 2004; Mendoza-Ibarra et al. 2012) (Table 14.1). In one experiment using bulls—3 to 6 years of age—all bulls more than 4 years old became infected after three to six services while only one of two young bulls, 2–3 years of age, became infected after nine services (Clark et al. 1974). Bulls less than 4 years of age are rarely carriers of *T. foetus* (BonDurant 1985; Perez et al. 1992; Ondrak 2016; Kimsey et al. 1980; Skirrow et al. 1985; BonDurant et al. 1990). The reason for this finding is not completely understood. Old bulls may have had a higher number of sexual contacts than young bulls. An often mentioned other possible reason is the more pronounced invaginations in the penile and preputial epithelium of older bulls—i.e., the crypts of these epithelia are becoming deeper and increase in number with the age of the bull (BonDurant 1985; Skirrow et al. 1985; McCool et al. 1988; Perez et al. 1992; Ondrak 2016). However, the hypothesis that anatomical changes are the cause for older bulls found to be infected more often was recently questioned because no age-related statistically significant differences were observed in the surface architecture of the penile and preputial epithelium of bulls (Ondrak 2016; Strickland et al. 2014).

  In female cattle, age seems not to be associated with the likelihood of infection; however, there is evidence that repeated exposure induces resistance to infection (Simmons and Laws 1957; Clark et al. 1986; Skirrow and BonDurant 1990b). While mature bulls seem to remain infected for life, most cows are able to clear the infection after a few months—usually after 1 to 3 months—rarely longer (Parsonson et al. 1974, 1976; Skirrow and BonDurant 1990b; Bondurant 2005). Several studies of infected cows indicate that the os cervix is the preferred site of multiplication and persistence (Skirrow and BonDurant 1990b). Initial multiplication of *T. foetus* after infection seems to be followed by a decline of parasite numbers until next estrus (Bartlett and Hammond 1945). The numbers of parasites present in the cervicovaginal mucus seem to fluctuate during the estrus cycle, and the largest numbers are seen a few days before estrus (Hammond and Bartlett 1945). In two infected heifers that have been followed over time after experimental infection, *T. foetus* was not always observed in vaginal mucus by microscopic examinations or culture isolation suggesting fluctuations in vaginal parasite concentration; however, only in one heifer, there was a coincidence between the detection of *T. foetus* and the time of estrus (Simmons and Laws 1957). Not all cows are able to clear infection. Carrier dams have been reported—e.g., two chronically infected dams were observed in one Australian study 16 and 22 months after initial infection (Alexander 1953). In a Californian dairy herd, infected cows were found positive 9 weeks after pregnancy or 63 days after parturition (Skirrow 1987; Goodger and Skirrow 1986). In a more recent study from Argentina, several heifers remained infected up to 300 days after breeding which underlines the importance of these carrier state heifers for persistence of infection in affected herds (Mancebo et al. 1995).

- **Herd management practices:** The risk of bulls of being tested *T. foetus* positive increased when the number of bulls used per unit was higher than 10 or the bull-to-cow ratio per unit was lower than 1 to 25. The higher number of bulls and lower bull-to-cow ratios are typical management practices in large herds to increase conception rates (Rae et al. 2004). It was hypothesized that by these practices the number of potential sexual contacts per bull and the probability for an individual bull to become positive are increased (Rae et al. 2004).
- **Breed disposition:** The possibility of a breed predisposition was discussed based on study results suggesting a higher prevalence of infection in particular breeds (BonDurant et al. 1990; Perez et al. 1992; Rae et al. 1999, 2004; Skirrow et al. 1985; Abbitt and Meyerholz 1979). A number of epidemiological studies addressed this question (Table 14.2). It had been hypothesized that in *B. taurus indicus* bulls, due to their longer preputial length, the likelihood of being infected might be higher (BonDurant et al. 1990). Statistically significant differences in prevalence were observed when *B. taurus taurus* and *B. taurus indicus* or *B. taurus taurus*/*B. taurus indicus* crosses were compared; highest prevalences of infection were observed in *B. taurus taurus* (BonDurant et al. 1990). Another study in Costa Rica also found a strong association between the risk of positive findings and the *Bos taurus taurus* breed—as compared to *Bos taurus indicus* pure or hybrid breeds (Perez et al. 1992); an additional study also showed an increased risk in Angus, Charolais, Hereford, or Simmental breeds relative to *B. taurus indicus* (Rae et al. 2004) (Table 14.2). However, findings suggesting breed disposition should be interpreted with care; studies might have been biased by uneven study designs or by not paying enough attention to the differences in the way herds of particular breeds were operated (Ondrak 2016). It was hypothesized that prevalence differences could be due to the increased number of matings accomplished by *B. taurus taurus* bulls as compared to *B. taurus indicus* bulls in the same period of time, thus increasing the risk of exposure to infection (Perez et al. 1992). However, there are also some studies that did not observe differences in prevalence of infection between *B. taurus taurus* and *B. taurus indicus* (Dennett et al. 1974).

  Nevertheless, a breed disposition and genetically based predisposition should not be completely ruled out. Further studies are needed to elucidate the reason why some of the infected dams are becoming carrier of *T. foetus*—being still infected after pregnancy—while others immediately eliminate infection during the first 2 to 3 months after loss of the conceptus (Alexander 1953; Goodger and Skirrow 1986; Skirrow 1987; Mancebo et al. 1995). It is possible that like in other

Table 14.2 Risk factor studies in bovine tritrichomonosis

Country	Region	Type of herd or animal	No. of herds examined/no. of herds positive (%)	No. of animals examined/no. of animals positive (%)	Diagnosis	Herd-level risk factors	Individual-level risk factors	Type of study	Reference
Argentina	Province Buenos Aires	Beef	42/173 (24)	NA	C (modified Plastridge medium, Tricoazul), all bulls per herd, three sequential tests	Rearing herds vs. full-cycle herds, $p = 0.052$; pregnancy rate in cows ≤ 90%, $p = 0.005$[a]; shared livestock with others, $p = 0.003$[a]; rotation of bulls, $p<0.05$; abortion, $p<0.05$; *T. foetus* reported in previous year, $p = 0.001$[a]	NA	Case-control	Mardones et al. (2008)
Argentina	Province La Pampa	Beef	194/3766 (5.15)	309/29178 (1.06)	C with modified Diamond's medium (MDM), all non-virgin bulls, twice a year	Seasonal effect: highest no. of positive findings in February (pre-breeding season), no statistics provided; spatial clustering in south of La Pampa, $p = 0.008$; coinfection, high-risk cluster cells for bovine genital campylobacteriosis were also high-risk cells for bovine tritrichomonosis, $p = 0.0014$	NA	Cross-sectional	Molina et al. (2013)

(continued)

Table 14.2 (continued)

Country	Region	Type of herd or animal	No. of herds examined/ no. of herds positive (%)	No. of animals examined/no. of animals positive (%)	Diagnosis	Herd-level risk factors	Individual-level risk factors	Type of study	Reference
Australia	Victoria River District	NA	27/41 (65.5)	81/1008 (8.0)	C (modified Plastridge medium)	NA	Tritrichomonosis infection rates varied significantly with age ($p<0.0001$), that is, increasing with age ($p<0.05$); no evidence of an increased likelihood of coinfection with *Campylobacter fetus*	Cross-sectional	McCool et al. (1988)
Costa Rica	NA	Dairy, mainly	10/63 (15.9)	14/225 (6.2)	C (InPouch TF)	NA	Bull in service: no vs. yes, $p = 0.02^a$; age > 3 years, $p = 0.02^a$; breed, *B. taurus taurus* vs. *B. taurus indicus*, $p = 0.02^a$	Cross -sectional (results of pilot study not included)	Perez et al. (1992)

Spain	Asturias de la Montana	Beef	27/65 (41.5)	33/103 (32)	C (InPouch TF or thyoglycollate transport medium (TFTM) with modified Diamond medium (MDM))+PCR (Felleisen et al. 1998)	Increased number in repeat breeder cows, $p = 0.007$	Age >3 years, $p = 0.04$	Cross-sectional	Mendoza-Ibarra et al. (2012)
Spain	Asturiana de los Valles	Beef	12/229 (5.2)	13/327 (4.0)	C (InPouch TF)+PCR (Felleisen et al. 1998)	No statistical significant predictors	Age >3 years, $p<0.001$ (univariable)	Cross-sectional	Mendoza-Ibarra et al. (2013)
USA	Florida	Beef	17 (40.4)	119/1984 (6)	C	Herd size ≥500, $p = 0.004^{b}$; bull-to-cow ratio = 1:<25, $p = 0.039^{b}$	Age ≥5 years, $p = 0.022^{a}$; breed, Angus, Charolais, Hereford, Simmental vs. *B. taurus indicus*; $p<0.031^{a}$; herd management, number of bulls per group ≥10, $p = 0.002^{a}$, bull-to-cow ratio <1:25, $p = 0.03^{a}$; no knowledge of farmer on tritrichomonosis, $p = 0.003$; geographical area, South Florida vs. North Florida, $p = 0.001^{a}$	Cross-sectional	Rae et al. (2004)

(continued)

Table 14.2 (continued)

Country	Region	Type of herd or animal	No. of herds examined/ no. of herds positive (%)	No. of animals examined/no. of animals positive (%)	Diagnosis	Herd-level risk factors	Individual-level risk factors	Type of study	Reference
USA	Idaho	Beef	65/159 (40.9)	NA	dM or C (InPouch TF, modified Diamond medium (MDM))	Total cattle grazed on FS (US Forest Service) allotment >844, $p<0.05$; commingling on BLM (Bureau of Lands Management) allotment, FS (US Forest Service) allotment, or on any public land allotment, $p<0.05$	NA	Case-control	Gay et al. (1996)
USA	Wyoming	Beef	8/303+8 (2.6)	NA	C or PCR (three consecutive cell cultures or one PCR by an accredited diagnostic laboratory); herd status based on findings in the past 3 years	Allotments neighboring a positive herd(s), $p = 0.0003$; allotment type, open/public vs. private, $p = 0.003$; mingling with neighboring herd(s), $p = 0.026$	NA	Cross-sectional	Jin et al. (2014)

USA	California	Beef	9/57 (15.8)	30/729 (4.1)	C, modified Diamond medium (MDM)	NA	Age >3 years, $p<0.025$; breed (*B. taurus taurus* vs. *B. taurus indicus* (*Bi*) or *Bi*-hybrids), $p<0.001$)	Cross-sectional	BonDurant et al. (1990)
USA	Texas	Mainly non-virgin bulls, tested prior to interstate or intrastate commerce	NA	NA/NA; 1154 positive results/31202 tests (3.7)	C + real-time PCR	Spatial cluster in southeastern Texas identified ($p<0.001$)	Proportion of positive findings was highest in August (5.5%), no statistics provided	Cross-sectional	Szonyi et al. (2012)

NA not applicable, *C* culture
[a]Statistically significant in multivariate logistic regression

diseases there are genetic determinants influencing a predisposition to acquire infection and to develop immunity against the pathogen.

- **Other individual-level risk factors:** One study observed that the probability of positive findings is lower in bulls in service—i.e., in sexually active bulls (Perez et al. 1992) (Table 14.2). This is in accord with previous findings, which suggested that a depletion of the preputial *T. foetus* population might occur because of intense sexual activity. It also supports recommendations of a sexual rest of at least 1 to 2 weeks before sampling bulls in order to improve the likelihood of accurately identifying *T. foetus*-positive bulls (Clark et al. 1983a; BonDurant 1985; Yule et al. 1989a; Ondrak 2016). Variations in sexual activity between different seasons may also explain fluctuating differences in the proportion of positive findings in bulls during a year (Molina et al. 2013; Szonyi et al. 2012).

In addition, there are a number of herd-level risk factor studies—case-control and cross-sectional studies—that have been carried out to elucidate in more detail management practices and other factors increasing the risk of herds to acquire tritrichomonosis (Table 14.2) (Mardones et al. 2008; Molina et al. 2013; McCool et al. 1988; Perez et al. 1992; Mendoza-Ibarra et al. 2012, 2013; Rae et al. 2004; Gay et al. 1996; Jin et al. 2014; BonDurant et al. 1990; Szonyi et al. 2012). Many of the identified explanatory variables are related to predictors or risk factors in favor to increase the likelihood of venereal transmission of *T. foetus* in herds and between herds:

- **Transmission between herds:** Risk factors were identified which characterized the likelihood to acquire tritrichomonosis from other herds. "Allotments neighboring a positive herd(s)"; "allotment type, open/public vs. private"; and "mingling with neighboring herd(s)" were recently identified as risk factors in *T. foetus*-positive beef herds in Wyoming, USA (Jin et al. 2014). These findings confirm previous observations in beef herds from Idaho, which identified "commingling on BLM—Bureau of Lands Management—allotment; FS, US Forest Service, allotment; or on any public land allotment" as an important predictor of positive herds (Gay et al. 1996). Also in Argentina "shared livestock with others" was identified as a significant risk factor for herds testing positive, that is, having positive bulls (Mardones et al. 2008). In summary, these findings suggest that any type of mingling herds and also fence-line contact with other herds represent important risk factors and should be avoided by farmers (Jin et al. 2014).
- **Transmission within herds:** Other predictors characterized the likelihood of transmission within a herd. Obviously, herd size plays a role. A study on beef herds in Idaho, USA, identified "total cattle grazed on FS—US Forest Service—allotment more than 844" as a risk factor, and a study from Florida, USA, observed that herds with more than or equal to 500 cows had a higher risk of being positive (Gay et al. 1996; Rae et al. 2004). The latter study also identified particular herd management practices increasing the individual risk of bulls to test positive; therefore, most likely, herd size is a confounder explained by man-

agement practices in large herds in favor for *T. foetus* transmission, like "no. of bulls used per unit larger than or equal to 10" or "bull-to-cow ratio per unit smaller than 1 to 25" as discussed earlier (Rae et al. 2004). The management practice "rotation of bulls within a herd" has been identified in an Argentinian study to most likely favor to perpetuate transmission of infection within a herd (Mardones et al. 2008). The same may apply to an observation that among "rearing beef herds"—i.e., herds that rear cattle until the weight of 150–250 kg—the prevalence of positive herds was higher as compared to "full-cycle herds," herds with breeding, rearing, and fattening. This observation was explained by hypothesizing a higher proportion of reproductively active animals resulting in an increased spreading of disease in "rearing beef herds" compared to "full-cycle herds" (Mardones et al. 2008). The risk factors related to a potential perpetuation/acceleration of transmission inside herds cannot be easily used to give recommendations with respect to better management practices, because these risk factors are either unspecific or difficult to change.

Herd-level predictors, possibly related to effects of tritrichomonosis include infertility, early fetal death, and rarely, abortion. In epidemiological studies "pregnancy rate in cows higher than or equal to 90%" and reporting "abortion" were found associated with *T. foetus*-positive beef herds in Argentina (Mardones et al. 2008). An "increased number in repeat breeder cows" was identified as a predictor for positive herds in extensively managed beef cattle in Spain (Mendoza-Ibarra et al. 2012).

Additionally, in a few studies, spatial clustering of *T. foetus*-positive herds was observed, and in one of these studies, it was also observed that high-risk clusters for *T. foetus* correlated to high-risk clusters for bovine genital campylobacteriosis (Molina et al. 2013; Szonyi et al. 2012). The observed spatial clustering remained unexplained in these studies.

14.3.2 Epidemiology of Tritrichomonosis in Cats

Although the typical clinical sign in natural infection is chronic or intermittent diarrhea, for many cats, no diarrhea was reported in 6 months preceding diagnosis (Xenoulis et al. 2013; Kuehner et al. 2011). Chronic *T. foetus*-associated diarrhea in most cats is likely to resolve spontaneously within 2 years of onset, and chronic infection with *T. foetus*—without clinical signs—after resolution of diarrhea appears to be common (Foster et al. 2004). In experimentally infected kittens, *T. foetus* infections was long lasting and that, in later phases of the infection, the presence of *T. foetus* in feces was not always associated with clinical signs—such as abnormal consistency of feces (Gookin et al. 2001). Thus, in addition to disease-infected cats, cats with asymptomatic infection are able to transmit infection to other cats.

As mentioned in Sect. 14.1.1, feline *T. foetus* tolerates a broader pH range than *T. foetus* from cattle (Morin-Adeline et al. 2015a). Like for *T. foetus* from cattle, no encysted stages are known for *T. foetus* from cats. However, the formation of pseudocysts, which may support survival in the environment, has been reported for *T. foetus* (Pereira-Neves and Benchimol 2009). In addition, it has been shown that feline *T. foetus* is able to survive outside of its host for at least 30 min on dry cat food and 180 min in drinking water or urine (Rosypal et al. 2012). Other studies showed that *T. foetus* can survive in cat feces for several hours or even days or up to at least 5 days in wet cat food (Hale et al. 2009; Van der Saag et al. 2011). Even the survival of a passage through the alimentary tract of slugs (*Limax maximus*, *Limacus flavus*) was demonstrated, suggesting that slugs could transmit *T. foetus* over short distance (Van der Saag et al. 2011). The predominant mode of infection is the fecal-oral route, and most likely a close contact of cats favors the spread of transmission.

Several epidemiological studies are reported, providing data on putative risk factors for *T. foetus* infection in individual cats or catteries, cat shelters, and breeding centers (Table 14.3). Most of the studies were small-scale studies, including only low numbers of individual cats and catteries, shelters, or breeding centers. Some of the studies were restricted to cats from shows or pedigree cats. Some studies were restricted to diarrheic cats or cats with a history of chronic diarrhea. Therefore, studies listed in Table 14.3 were stratified into (1) studies including only healthy or healthy and diseased cats and (2) studies including almost only diseased cats—i.e., cats with diarrhea or a history of diarrhea. In addition to studies listed in Table 14.3, a large number of case reports and small-scale studies are available—recently reviewed by Yao and Köster (2015)—which are not mentioned in the following because these studies provided no statistical evidence on putative risk factors. As indicated in Table 14.3, some studies were included after performing own statistical analyses based on data extracted from the reports.

Individual risk factors for tritrichomonosis in cats include:

- **Diarrhea, abnormal fecal consistency, and history of diarrhea.** In a number of studies including only healthy or healthy and diseased cats, it was observed that positivity for *T. foetus* was associated with a "previous history of diarrhea," "history of diarrhea in the past 6 months," "presence of chronic diarrhea," and "abnormal fecal consistency" (Table 14.3) (Kuehner et al. 2011; Tysnes et al. 2011; Doi et al. 2012). These findings confirm that *T. foetus* is an important cause of diarrhea in cats.

  The likelihood of a cat to test *T. foetus* positive was also increasing if there was a "history of another cat in the house with diarrhea in the past 6 months" or a "history of diarrhea in cattery in the past 6 months," which suggested that individual positive cats were only part of a larger problem in a cattery, cat shelter, or breeding center (Kuehner et al. 2011; Hosein et al. 2013).

- **Age.** With a few exceptions, it is now accepted that *T. foetus* associated with large bowel diarrhea is mainly a disease of young cats. Several studies revealed that cats about 1 year old or younger are more often *T. foetus* positive than elder cats (Table 14.3). These observations were made in studies including only healthy or healthy and diseased cats as well as in studies including almost only diseased cats (Profizi et al. 2013; Kuehner et al. 2011; Queen et al. 2012; Galián

Table 14.3 Risk factor studies in feline tritrichomonosis

Sampled animals	Country	Region	Type of animal	No. of catteries, shelters, or breeding centers examined/ no. of catteries positive (%)	No. of animals examined/ no. of animals positive (%)	Diagnosis	Cattery-level risk factors	Individual cat-level risk factors	Type of study	Remarks	Reference
Only healthy or healthy and diseased cats											
	Canada	Ontario	Cats sampled at a cat clinic (n=140), cat shows (n=55), and a humane society (n=46)	NA	14/241 (5.8)	C (InPouch TF), PCR for confirmation of culture positives		Attendance to cat shows, $p<0.05$ (23.6% from the show cats were positive); history of another cat in the house with diarrhea in the past 6 months, $p<0.01$; fed a raw food diet, $p<0.01$; >5 cats per house, $p<0.01$; purebred vs. mixed breed, $p<0.01$	Cross-sectional	No significant association between the presence of *T. foetus* and diarrhea at the time of sampling or having a history of diarrhea in the past 6 months; "attendance to cat shows" was the only significant variable in bivariable models	Hosein et al. (2013)

(continued)

Table 14.3 (continued)

Sampled animals	Country	Region	Type of animal	No. of catteries, shelters, or breeding centers examined/ no. of catteries positive (%)	No. of animals examined/ no. of animals positive (%)	Diagnosis	Cattery-level risk factors	Individual cat-level risk factors	Type of study	Remarks	Reference
	France	NA	140 cats cattery-housed pedigree cats, participating in international cat shows	18/117 (15.9)	20/140 (14.3)	C (InPouch TF) + Tricho-F/ Tricho-R-PCR (TRICHO-F and antisense primer TRICHO-R targeting the ITS1-5.8S rDNA-ITS2 region as previously described (Jongwutiwes et al. 2000); (Duboucher et al. 2006) + Seq (amplicon, cloned)		Age < 1 year, $p = 0.057$	Cross-sectional	No significant associations: size of cattery, type of food, vicinity of dogs	Profizi et al. (2013)

	Germany	NA	230 purebred cats	23/124 (15.7)	36/230 (18.5); C-positive: 29/230, PCR-positive: 28/230	C + PCR (Grahn et al. 2005) + Seq (Representative amplicons)		Abnormal fecal consistency, $p<0.001$; history of diarrhea in the past 6 months, $p = 0.027$; age ≤1 year old, $p<0.034$; Norwegian Forest (NFO) cats had a significantly higher prevalence of *T. foetus* infection than other breeds, $p< 0.001$; history of diarrhea in cattery in the past 6 months ($p = 0.01$)	Cross-sectional	Any	Kuehner et al. (2011)
	Japan	Hokkaido, Saitama	147 samples, also from cats with chronic diarrhea	NA	13/147 (8.8)	Cultivation (98 samples) + PCR (all samples)	NA	Presence of chronic diarrhea, $p<0.0035$	Cross-sectional	No significant differences with respect to age, breed, and whether cats were maintained indoors/outdoors between infected and uninfected cats (Fisher exact test)	Doi et al. (2012)

(continued)

Table 14.3 (continued)

Sampled animals	Country	Region	Type of animal	No. of catteries, shelters, or breeding centers examined/ no. of catteries positive (%)	No. of animals examined/ no. of animals positive (%)	Diagnosis	Cattery-level risk factors	Individual cat-level risk factors	Type of study	Remarks	Reference
	Norway	NA	52 clinically healthy cats participating in 3 cat shows	NA	11/21 (21)	Direct microscopy (*n*=39) + C (n=39) + nPCR (*n*=52), Seq of TFR3/TFR4 amplicons	NA	Own previous history of diarrhea, $p = 0.1$ (indicated as trend)	Cross-sectional, pilot study	Four samples positive for *Giardia* spp., one TF-*Cryptosporidium* spp. coinfection	Tysnes et al. (2011)
	USA	NA	173 cats with and without clinical signs of tritrichomonosis; 32 purebred (including 2 purebred-cross cats), 143 mixed breed cats	NA	17/173 (9.8)	C + confirmation by TFR3/ TFR4-PCR + Seq (amplicons)	NA	No statistical analysis provided in reference. Retrospective statistical analysis of presented data shows that purebred cats have a higher risk than mixed bred cats, Fisher exact test, $p<0.001$	Cross-sectional, pilot study	All positive cats had diarrhea. Positive cats were between 6 weeks and 12 yrs old, Negative cats 4 weeks to 13 yrs old. Positive cats often (53%) had concurrent infections (*Giardia* spp., *Cryptosporidium* spp., Coccidia, FIP)	Stockdale et al. (2009)

	USA	NA	117 cats from 89 catteries at an international cat show	28/89 (31.5)	36/117 (30.8); In detail, fecal smear (5/36), fecal culture in modified Diamond's medium (9/36), fecal culture in InPouch TF (20/36), or nPCR (34/36)	Fecal smear examination, C in modified Diamond's medium, C in InPouch TF medium, or nPCR amplification (Gookin et al. 2002)	Loose stools or diarrhea in any cats within the past 6 months, *P*<0.052; coinfection with coccidia, *P*<0.059; square feet of facility available per cat was low, *P*<0.056	NA	Cross-sectional	An association between *T. foetus* and *Giardia* spp. infection was not significant (*p* = 0.075). There were no differences in age or sex between uninfected cats and cats having *T. foetus* infection. The proximity of the cattery to within 0.5 miles of agricultural species, type of diet fed to cattery cats (commercial or home cooked), and source of water were not associated with risk of infection	Gookin et al. (2004)
	USA	California	Diarrheic (n=219) and control cats (n=54)	NA	10/223 (4.5)	C and/or PCR	NA	Age: mean age of positive cats 0.7 ± 0.3 years, mean age of negative cats 5.4 ± 5.0 years, *p*<0.001	Case-control	Any	Queen et al. (2012)
Almost only diseased cats											

(continued)

Table 14.3 (continued)

Sampled animals	Country	Region	Type of animal	No. of catteries, shelters, or breeding centers examined/ no. of catteries positive (%)	No. of animals examined/ no. of animals positive (%)	Diagnosis	Cattery-level risk factors	Individual cat-level risk factors	Type of study	Remarks	Reference
	Europe	NA	Cats submitted to a private veterinary laboratory due to diarrhea	NA	166/1840 (9.02)	TFR3/ TFR4-PCR	NA	No statistical analysis provided in publication. Results of retrospective analysis: age ≤1 years (65/628 positive), 2–15 years (33/755 positive), statistically significant difference, chi-square, $p<0.001$	Cross-sectional	Any	Galián et al. (2011a, 2011b)
	Germany and Austria	NA	31 cats (6 weeks–14 years old; 30 had diarrhea)	NA	6/31 (19.4%)	nPCR+Seq (Amplicon)	NA	Purebred cats, $p = 0.0153$	Cross-sectional, pilot study	Coinfection with *Giardia* (ELISA) likely in four of the *T. foetus*-positive cats, there was only one cat *Giardia* positive but TF negative	Steiner et al. (2007)

	Spain	NA	Cats from densely housed origins with a history of chronic diarrhea: family cats (n=15), breeding center cats (n=28), cat shelter cats (n=50); healthy cats: family cats (n=20), cat shelter cats (n=22)	3/4 (75); 50 cats were from 1 breeding center and 3 shelters	36/93 (38.7) diarrheic cats from cattery or family cats; 0/20 (0) healthy cats	C in modified Diamond's medium and/or TFR4/ TFR3-PCR	NA	Age ≤1 years, $p = 0.014$	Cross-sectional study, cats from densely housed origins with a history of chronic diarrhea and healthy cats	No difference between purebred (Persian) and mixed breed; female vs. male, $p = 0.085$	Arranz-Solis et al. (2016)
	UK	NA	111 naturally voided diarrheic feline fecal samples	NA	16/111 (14.4)	nPCR (Gookin et al. 2002)	NA	Age ≤ 1 years, $p = 0.0026$; pedigree cat, $p = 0.018$; Siamese and Bengal breed, $p = 0.0077$	Cross-sectional, pilot study	Any	Gunn-Moore et al. (2007)

(continued)

Table 14.3 (continued)

Sampled animals	Country	Region	Type of animal	No. of catteries, shelters, or breeding centers examined/no. of catteries positive (%)	No. of animals examined/no. of animals positive (%)	Diagnosis	Cattery-level risk factors	Individual cat-level risk factors	Type of study	Remarks	Reference
	UK	NA	Fecal samples from diarrheic cats submitted to a private veterinary laboratory	NA	205/1088 (18.8)	Real-time PCR (IDEXX 8-way PCR assay: 5.8S rDNA, AF339736)		Age: 6–12 months (29.4%) vs. >12 months (15.2%), $p<0.001$; pedigree cats (37.8%) vs. non-pedigree cats (6.0%), $p<0.001$; *T. foetus*-positive findings tended to be found in combinations with *Coronavirus*, *Clostridium perfringens*, and *Giardia* sp	Cross-sectional	<6-month-old cats had lower *T. foetus* prevalence than 6–12-month-old cats	Paris et al. (2014)

NA not applicable, *C* cultivation, *nPCR* nested PCR, *Seq* sequencing et al. 2011a, b; Arranz-Solis et al. 2016; Gunn-Moore et al. 2007; Paris et al. 2014). These epidemiological findings confirm the results of follow-up studies of cats with *T. foetus* infections, which clearly showed that clinical signs spontaneously resolve within 2 years of onset and that it is difficult to detect *T. foetus* in cats after several weeks of infection (Gookin et al. 2001; Foster et al. 2004). Meta-analysis on available data confirmed this view (Yao and Köster 2015). The reasons why especially young cats are affected by *T. foetus* diarrhea are not sufficiently understood. Possible explanations are a more intense contact of kittens to their mothers or to larger numbers of other cats in breeding centers, private households, or cat shelters, which may favor transmission of *T. foetus* and may also influence the dose by which kittens become infected. Immunity against feline *T. foetus* is poorly understood; recently the possibility that parasite-specific secretory IgA mediates immunity has been discussed (Yao and Köster 2015). Histological examinations in infected cats revealed influx of lymphocytes, plasma cells, and neutrophils into the subepithelial lamina propria (Yaeger and Gookin 2005). The inflammatory response might not only be involved in pathogenesis but may also be involved in the control of infection (Tolbert and Gookin 2016).

- **Breed.** A number of studies identified an increased risk of infection in purebred, or pedigree cats, or particular breeds (Hosein et al. 2013; Stockdale et al. 2009; Steiner et al. 2007; Gunn-Moore et al. 2007; Paris et al. 2014). These observations remain largely unexplained. Most likely the conditions under which purebred or pedigree cats are reared are responsible for the increased risk. Purebred cats from breeding centers represent densely housed populations, and due to the frequent, close, and direct contact, the risk of infection might be higher than in mixed breed cats reared in less dense populations (Yao and Köster 2015; Arranz-Solis et al. 2016).
- **Other risk factors.** One other risk factor identified was "attendance to cat shows," which might be a confounding predictor; the predictor could be related to the increased risk of purebred or pedigree cats (Hosein et al. 2013). There is only a single study report on food-related predictors. In a Canadian study, "fed a raw food diet" has been identified as a risk factor (Hosein et al. 2013).
- **Other concurrent infections and impaired immune system.** In experimental *T. foetus* infection with cats which shed naturally *Cryptosporidium parvum* oocysts and Cryptosporidium non-infected cats those with *C. parvum* infection had an earlier onset, more severe diarrhea, and increased number of trichomonads on direct fecal examination, as compared to non-infected cats (Gookin et al. 2001). In epidemiological studies none of the concurrent infections examined—*Giardia* sp., *Cryptosporidium* sp., coccidia, *Clostridium perfringens*, feline infectious peritonitis, and *Coronavirus*—revealed a statistical association either to infection or disease on the individual animal level (Table 14.3). Nevertheless, it has been discussed whether *T. foetus* alone can cause clinical signs without an impaired or immature immune system, concurrent infection, or other factors resulting in alterations in the colonic microflora (Manning 2010).

In catteries, cat shelters, or breeding centers, *T. foetus* may cause outbreaks of persistent large bowel disease (Holliday et al. 2009). There is only a single study on risk factors associated with occurrence of trichomonosis in catteries (Table 14.3). Similar to the findings in individual cats, the analysis on the cattery level also identified an association between *T. foetus* and abnormal fecal consistency or diarrhea; "loose stools or diarrhea in any cats within the past 6 months" was significantly associated with *T. foetus*-positive findings (Gookin et al. 2004). A second putative risk factor identified in this study was related to the cat population density in catteries; if the "square feet of facility available per cat was low," this was associated with positive findings (Gookin et al. 2004). The observation that "coinfection with coccidia" was associated with *T. foetus*-positive findings may suggest that there are common risk factors in favor of mixed infections of coccidian parasites with *T. foetus* (Gookin et al. 2004). In epidemiological studies none of the concurrent infections examined revealed a statistical association either to infection or disease on the individual animal level.

14.3.3 Epidemiology of *T. gallinae*

As outlined in Sect. 14.2.1, *T. gallinae* affects a large number of avian species where it mainly parasitizes the oropharyngeal membranes—sinuses, mouth, throat, and esophagus—causing a disease characterized by greenish fluid and caseous lesions, of whitish-yellowish fibrinous material, on the oropharyngeal membranes (reviewed by Amin et al. (2010)).

T. gallinae infection seems not to be host species-specific. However, studies on parasite diversity suggest that there are subtypes more commonly found in certain bird species (Gerhold et al. 2008; Grabensteiner et al. 2010; Sansano-Maestre et al. 2009). *T. gallinae* is most common among domestic pigeons and wild doves, and these species may represent a reservoir also for other bird species. Accordingly, rock pigeons—*Columba liva*—were regarded as source for the worldwide distribution of *T. gallinae* (Stabler 1954; Harmon et al. 1987). Most pigeons harbor this protozoan but rarely show clinical disease (Stabler 1954). Hawks, falcons, and owls may become infected, most likely via predation of other infected birds (Rogers et al. 2016). In recent years, severe outbreaks of avian trichomonosis caused by *T. gallinae* have been recorded initially in wild finches, later in Passeriformes, canaries, and psittacines in Europe and North America since 2005 (reviewed by Amin et al. (2010)). Trichomonosis is also reported in several other bird species, including corvids in California, USA (Anderson et al. 2009). *T. gallinae* has a low tenacity in the environment and is regarded as unable to survive a gastric passage, and droppings of birds are regarded as free of *T. gallinae* (Stabler 1954). *T. gallinae* has no cyst stage. However, as mentioned for *T. foetus* (Sect. 14.1.1), the formation of pseudocysts has been reported (Tasca and De Carli 2003). The importance of pseudocysts in the transmission of *T. gallinae*—e.g., via contaminated drinking water—is not sufficiently understood. Survival times in tap water and in carcasses are regarded as limited; a survival time of 8–48 h in carcasses has been reported (Erwin et al. 2000). The following important facts about transmission of *T. gallinae* need to be taken into account:

- **Transmission by crop milk or direct contact:** Prevalences in pigeons can be very high, and pigeons are regarded as endemically infected, often without clinical signs. Feeding on infectious crop milk is regarded as an important route of infection for nestlings (Stabler 1954). In other bird species, feeding nestlings by regurgitation might also be an important route of transmission. Close contact—e.g., during courtship—is also regarded as a way by which the infection is spread between adult birds (Stabler 1954). In Southeastern USA, trichomonosis was diagnosed more often in the spring and summer months than in autumn and winter months, which might be related to the times of courtship and feeding nestlings (Gerhold et al. 2007).
- **Transmission by drinking water:** It is believed that infection of turkeys and chickens is caused by *T. gallinae*-contaminated drinking water (Stabler 1954). Drinking water is also a likely source of infection for other bird species.
- **Effect of weather events on outbreaks of avian trichomonosis:** *T. gallinae*-associated finch mortality usually peaked in summer and autumn, but a correlation with climatic events has not been observed (Neimanis et al. 2010; Robinson et al. 2010; Lawson et al. 2011). A coincidence of the emergence of avian trichomonosis with high temperature and low rainfall has been reported (Simpson and Molenaar 2006; Bunbury et al. 2007). It has been hypothesized that as a consequence of the dry and hot weather, numbers and volumes of water sources decline, which may favor, due to higher densities of birds aggregating at limited water sources, the transmission of *T. gallinae* (Bunbury et al. 2007).
- **Transmission by predation:** Pigeons and doves may also serve as a source of infection for raptors (Boal et al. 1998; Krone et al. 2005). Although *T. gallinae* is a labile protozoan which does not remain viable for a prolonged period and is rapidly killed by desiccation, survival time seems to be long enough to be transmitted also to the nestlings of raptors (Krone et al. 2005).

14.3.4 Epidemiology of *T. gallinarum*

In contrast to *T. gallinae*, *T. gallinarum* is mainly found as a commensal in the intestine of many domestic bird species, including chicken, turkey, guinea fowl, duck, and goose (Friedhoff 1982; BonDurant and Honigberg 1994). As a parasite of the intestine, *T. gallinarum* is transmitted via consumption of contaminated food or drinking water. Cloacal as well as oral experimental infection of chickens and turkeys is possible.

14.4 Prevention

14.4.1 Prevention in Cattle

Due to the widespread use of artificial insemination, bovine tritrichomonosis has almost disappeared in dairy cattle industries, like in Western European countries, in the USA, or in Canada. However, the disease is still present in many areas of the world were cattle are raised on pastures and natural mating is used. Prevention and control measures are based on the distinctive epidemiologic features of bovine tritrichomonosis, a sexually transmitted disease where infected bulls are asymptomatic carriers and represent a permanent source of infection while in heifers and cows infection is typically transient (Clark et al. 1971, 1974; Parsonson et al. 1974; Skirrow and BonDurant 1990b).

Bovine tritrichomonosis belongs to the OIE-listed diseases (http://www.oie.int/en/animal-health-in-the-world/oie-listed-diseases-2017; accessed 22. Febr. 2017). *T. foetus* as a causative agent of this venereal transmitted disease may be present in semen if the semen has been contaminated with preputial fluid during manual collection (Bondurant 2005). Recommendations for the importation of cattle and bulls for breeding can be found in the *Terrestrial Animal Health Code*. In particular, emphasis has been placed in the measures applied to bull semen donor health status in order to avoid dissemination and transmission of the disease (http://www.oie.int/index.php?id=169&L=0&htmfile=chapitre_trichomonosis.htm; accessed 22. Febr. 2017).

In dairy herds and in some beef herds, artificial insemination is a very useful measure to reduce and eliminate infection. For bulls destined for artificial insemination, quarantine and periodic testing are required. Also, in semen trade, it is of great value to know the country of origin, the reproductive history of the bull, and the tests performed by the artificial insemination center (Eaglesome and Garcia 1997). In some areas of the world, as in the EU, specific regulations are applied to bovine semen trade and to regulate the sanitary conditions of the collection center and the animals. Specifically, bulls selected for entry into artificial insemination should be tested in quarantine before admission to the center, and regular testing of animals in service is included as basic measures to avoid the presence and dissemination of bovine tritrichomonosis (European Directive 88/407/EEC of 14th June 1988 and European Directive 2003/43/EC of 26th May 2003).

In other countries, different policies have been established to control the infection. In the USA, state regulations have endeavored to control the endemic disease as to minimize economic losses by testing bulls. Only the importation of *T. foetus*-free bulls is permitted for reproductive purposes while positive bulls are culled (Yao 2013). In the province of La Pampa, Argentina, with a bovine census close to four million heads, more than 80% of bulls are tested twice each year, and positive bulls were culled. By this measure, a significant decrease in the number of infected herds and animals has been achieved in this region in the last 10 years (Fort et al. 2016). Recently, an online tool has been developed in the USA—Trich CONSULT; www.trichconsult.org, accessed 22. Febr. 2017—that uses a series of questions to assess the *T. foetus* status and management practices of a herd. Based on the responses to the questions, this page provides feedback to users allowing them to evaluate the importance of implementing suggested control strategies (Ondrak 2016).

At present, a consensus exists concerning the most relevant measures as to how to prevent and control bovine tritrichomonosis (reviewed in Ball et al. (1987), McCool et al. (1988), Bondurant (2005), Rae and Crews (2006), Campero and Gottstein (2007), Yao (2013), Ondrak (2016)).

In herds where it is not possible to introduce artificial insemination and where natural mating is the normal practice, as is the case in many regions where extensive beef cattle are raised, the following measures to prevent the entrance of the infection are recommended.

- **Quarantine and testing of replacement bulls.** Replacement should be done by virgin animals or bulls acquired only from disease-free herds with records of excellence in reproductive performance. All the bulls must be tested during quarantine before entrance in the herd. If the bull comes from a tritrichomonosis-free herd, the analysis of two samples of preputial smegma with 1–2 weeks of interval during the quarantine is recommended. Three or more samples must be analyzed if bulls are provided from an area where tritrichomonosis is known to be endemic (Campero and Gottstein 2007; Yao 2013; Ondrak 2016). The measure includes the prohibition of the use of communal, shared, or rented bulls, unless their herd of origin and individual health status are known.
- **Avoid communal pastures and keep fences in good conditions.** These measures will help to control some of the two most important risk factors influencing disease transmission (Gay et al. 1996; Mardones et al. 2008; Jin et al. 2014).
- **No introduction of cows or heifers of unknown health status during the breeding season.** Similar to bulls, heifers and cows must be acquired only from disease-free herds with records of excellence in reproductive performance. In Argentina several heifers were found infected up to 300 days after breeding; such heifers may represent carrier cows able to introduce infection into naïve herds (Mancebo et al. 1995).

In addition to the preventive measures outlined in Sect. 14.3, the following control measures are recommended to reduce the impact and eliminate the disease in case that bovine tritrichomonosis has been diagnosed in a herd or the herd is located in a tritrichomonosis endemic area:

- **Testing of bulls before the breeding season and culling of infected bulls.** Efforts to control the disease focus on using diagnostic tests with a high sensitivity, low cost, and time efficacy. It is recommended to sample the animal twice or three times before the breeding season and every time new bulls are introduced into herds (Bondurant 2005; Campero and Gottstein 2007; Yao 2013). Once new cases are not detected, annual testing is recommended to verify the non-infected status of the herd. However, testing and culling policies alone, although effective in improving reproductive efficiency, do not allow the elimination of the disease since other putative risk factors associated with the disease are usually present in the management of beef herds (Yao 2013; Collantes-Fernandez et al. 2014).

- **Average age reduction of bulls and replacement of all bulls after four breeding seasons.** Replacement with negative-tested young bulls reduces the prevalence since 3-year-old bulls seem not to be as susceptible as older bulls in natural service (Clark et al. 1971; Christensen et al. 1977). A strong association of infection rate with age has been reported in several studies (Rae et al. 1999; Mendoza-Ibarra et al. 2012).
- **Pregnancy examination.** It is mandatory to know the reproductive performance and the magnitude of the infertility problem in the herd. All open and aborted cows should be culled or segregated in high-risk sub-herds or groups (Campero and Gottstein 2007).
- **Use of commercial vaccines.** In the presence of a high prevalence of infection in an area, vaccination of all heifers and cows is a good measure to improve reproductive efficiency especially when risk factors associated with infection cannot be avoided—e.g., the use of communal pastures. Commercially available vaccines in the Americas offer an improvement in reproductive efficacy (Kvasnicka et al. 1989, 1992).
- **Segregation of the herd in low- and high-risk sub-herds or groups.** In the low-risk sub-herds, only dams that have recently calved are pregnant for more than 5 months, and virgin females must be included. These must be serviced preferably by virgin bulls or by bulls with two negative examinations and coming from negative herds. In order to follow the effect of the infection, the same group of females should be serviced with the same male until the disease is controlled (Campero and Gottstein 2007).
- **Limiting of breeding season.** The breeding season should be restricted to less than 4 months to reduce the duration of the possible transmission period as much as possible. In addition, with a shortened breeding season, it becomes easier to monitor reproductive performance.

With respect to vaccination against bovine tritrichomonosis, the main objective is to eliminate a *T. foetus* infection from the reproductive tract before fetal loss occurs without necessarily avoiding parasite colonization of the epithelium during the first 40 days post-infection (BonDurant et al. 1993; Bondurant 2005). The mucosal immune response in the genital area seems to be the main protective mechanism which is characterized by a local response with the presence of IgA and IgG1 and a mild systemic response characterized by the presence of IgG2 and IgG1 (Skirrow and BonDurant 1990a; Anderson et al. 1996; Corbeil et al. 2008). As a rule, cows immunized with *T. foetus* have a humoral response pattern similar to that described for natural infections (Herr et al. 1991; BonDurant et al. 1993). However, genital IgA levels appear to depend on the type of antigen, adjuvant, and route of administration employed.

As to our knowledge, commercial vaccines against bovine tritrichomonosis exist only in the Americas but not in Europe. One available inactivated vaccine is prepared from whole organisms and can be acquired in a monovalent formulation—Trich Guard®—but also in a polyvalent formulation combined with *Campylobacter foetus* subspecies venerealis and *Leptospira*, Trich Guard V5-L®. In addition, in Argentina, an alternative inactivated vaccine—Tricovac, Tandil Biological Laboratory—containing an oily adjuvant with a concentration of 5 × 10^7 trophozoites of *T. foetus* is available.

In several studies on heifers using whole-parasite-based vaccines, a reduction in the number of infected females, a shorter time of genital infection, and a higher percentage of pregnant females in comparison with control animals have been reported (Kvasnicka et al. 1989, 1992; Herr et al. 1991; Gault et al. 1995; Anderson et al. 1996; Cobo et al. 2002, 2004). In addition, a lower number of services, 1.44 vs. 1.73 in non-vaccinated animals, $p = 0.16$; higher percentage of pregnant animals at the first service, 66.7 vs. 33.3% pregnancies, $p < 0.05$; and a reduction of embryo/fetal losses of 56.4% were observed (Kvasnicka et al. 1992; Hudson et al. 1993).

Subunit vaccines have also been developed (Clark et al. 1983b; BonDurant et al. 1993; Corbeil et al. 1998). A trial with an experimental vaccine formulated with membrane antigens of *T. foetus* was conducted; cows were immunized with this vaccine and subsequently challenged by the vaginal route (Campero et al. 1999). The vaccine used was able to generate a specific humoral immune response and shorten the period of infection in the vaccinated animals compared to the controls. A greater efficacy of a *T. foetus* membrane vaccine compared to a whole cell vaccine was observed; the animals had a shorter duration of infection, a higher pregnancy rate, and a lower rate of fetal mortality (Cobo et al. 2002). Additional work has been done to identify *T. foetus* surface antigens such as TF1.17 and TF190 (Voyich et al. 2001). The application of the former in experimentally challenged heifers evidenced a rapid shortening of infection and a specific IgA production in genital secretions (Anderson et al. 1996; Corbeil et al. 1998).

Vaccine-induced immunity to *T. foetus* has not been studied in depth in bulls and is only mentioned in some earlier studies (Clark et al. 1983b, 1984; Soto and Parma 1989; Campero et al. 1990; Herr et al. 1991). In bulls older than 5 years, the whole cell vaccine lacked a preventive or curative effect (Clark et al. 1983b). Therefore, commercial Trich Guard® vaccine is not indicated for bulls (Herr et al. 1991; BonDurant 1997).

14.4.2 Prevention in Other Animals

Prevention of feline trichomonosis is based on interrupting the fecal-oral route of transmission, particularly in catteries, shelters, and other dense cat populations. Since the viability of the parasite in the environment is limited, strict cleaning and disinfection measures are sensible measures to be implemented (Hale et al. 2009). Additionally, contamination of food and water by *T. foetus* is also possible; consequently, regular replacement of water and food and cleaning and disinfection of watering troughs and food bowls should be recommended. Due to the suggested role of some invertebrates to function as mechanical vectors, the avoidance of their presence in the litter box area and food and drinking areas may help to prevent infection transmission (Van der Saag et al. 2011). Finally, limiting contact between infected and non-infected cats will help to interrupt transmission of *T. foetus* in the population.

Measures to prevent *T. gallinae* outbreaks in wild as well as in captive birds are focused on actions to reduce sources of infection. One of the major aims would be to prevent attracting birds to feeding places. If feeding is necessary, such places should fulfill minimum requirements with regard to sanitary conditions, like regular changing of food and disinfection as suggested in a recent review (Amin et al. 2014). Since *T. gallinae* seems to require a wet environment to persist, drying of buildings and housing facilities following washing will support to control trichomonad infections. The prevention of the infection of prey birds, like pigeons nesting near urban areas, is necessary to prevent infection of predator birds. Due to the loss of habitat, predators have replaced their traditional prey mainly by urban pigeons (Boal et al. 1998; Höfle et al. 2000; Estes and Mannan 2003; Krone et al. 2005).

14.5 Treatment

14.5.1 In Cattle

In the past various imidazoles were used to treat bulls, but none was both safe and effective, and drug resistant strains were found (reviewed in Bondurant (2005), Rae and Crews (2006)). Specifically, ipronidazole is probably the most effective drug, but due to its low pH, it frequently causes sterile abscesses at the injection sites, and resistances have also been observed (Skirrow et al. 1985). A systemic treatment using drugs like metronidazole or dimetridazole produces adverse side effects and resistant populations of trichomonads (Campero et al. 1987). Currently, there is no approved treatment for cattle infected with *T. foetus* because of concerns regarding toxic residues in meat (BonDurant 1997).

14.5.2 In Cats

Therapies traditionally used for treatment of protozoa are not successful for feline trichomonosis (reviewed in Manning (2010), Yao and Köster (2015)). Currently, ronidazole has been the most effective drug used to date and is recommended at 20–30 mg/kg orally once daily for 14 days (Gookin et al. 2006). Relapse of diarrhea is common, but cats can continue to carry the organism, and resistant strains of *T. foetus* to ronidazole have also been documented (Gookin et al. 2010a). In addition, neurological toxicity in cats treated with ronidazole in the range of 30–50 mg/kg has been reported (Rosado et al. 2007). It is therefore important that owners are informed of the potential side effects. Ronidazole is not registered for veterinary use, and informed consent is necessary prior to its use in cats, and it should only be prescribed in confirmed cases.

14.5.3 In Other Animals

The drugs of choice for the treatment of avian trichomonosis are nitroimidazoles (metronidazole, dimetridazole, ronidazole, and carnidazole) (reviewed by Amin et al. (2014)). However, subtherapeutic dosing and prophylactic use of these drugs against trichomonosis have resulted in emergence of resistant strains of *T. gallinae* (Franssen and Lumeij 1992; Munoz et al. 1998). In wild birds, treatment is not a practical approach and generally not considered an option due to the way of application (Cole and Friend 1999). These drugs can be only used in non-food-producing birds by veterinary prescription.

References

Abbitt B, Ball L. Diagnosis of Trichomoniasis in pregnant cows by culture of cervical-vaginal mucus. Theriogenology. 1978;9(3):267–70. https://doi.org/10.1016/0093-691x(78)90034-1.

Abbitt B, Meyerholz GW. *Trichomonas fetus* infection of range bulls in South Florida. Vet Med Small Anim Clin. 1979;74(9):1339–42.

Adl SM, Simpson AG, Farmer MA, Andersen RA, Anderson OR, Barta JR, Bowser SS, Brugerolle G, Fensome RA, Fredericq S, James TY, Karpov S, Kugrens P, Krug J, Lane CE, Lewis LA, Lodge J, Lynn DH, Mann DG, McCourt RM, Mendoza L, Moestrup O, Mozley-Standridge SE, Nerad TA, Shearer CA, Smirnov AV, Spiegel FW, Taylor MF. The new higher level classification of eukaryotes with emphasis on the taxonomy of protists. J Eukaryot Microbiol. 2005;52(5):399–451. https://doi.org/10.1111/j.1550-7408.2005.00053.x.

Alexander GI. An outbreak of bovine trichomoniasis in Queensland and its control. Aust Vet J. 1953;29:61–6.

Amin A, Neubauer C, Liebhart D, Grabensteiner E, Hess M. Axenization and optimization of in vitro growth of clonal cultures of *Tetratrichomonas gallinarum* and *Trichomonas gallinae*. Exp Parasitol. 2010;124(2):202–8. https://doi.org/10.1016/j.exppara.2009.09.014.

Amin A, Liebhart D, Weissenbock H, Hess M. Experimental infection of turkeys and chickens with a clonal strain of *Tetratrichomonas gallinarum* induces a latent infection in the absence of clinical signs and lesions. J Comp Pathol. 2011;144(1):55–62. https://doi.org/10.1016/j.jcpa.2010.06.002.

Amin A, Bilic I, Berger E, Hess M. *Trichomonas gallinae*, in comparison to *Tetratrichomonas gallinarum*, induces distinctive cytopathogenic effects in tissue cultures. Vet Parasitol. 2012a;186(3-4):196–206. https://doi.org/10.1016/j.vetpar.2011.11.037.

Amin A, Nobauer K, Patzl M, Berger E, Hess M, Bilic I. Cysteine peptidases, secreted by *Trichomonas gallinae*, are involved in the cytopathogenic effects on a permanent chicken liver cell culture. PLoS One. 2012b;7(5):e37417. https://doi.org/10.1371/journal.pone.0037417.

Amin A, Bilic I, Liebhart D, Hess M. Trichomonads in birds—a review. Parasitology. 2014;141(6):733–47. https://doi.org/10.1017/S0031182013002096.

Anderson ML, BonDurant RH, Corbeil RR, Corbeil LB. Immune and inflammatory responses to reproductive tract infection with *Tritrichomonas foetus* in immunized and control heifers. J Parasitol. 1996;82(4):594–600.

Anderson NL, Grahn RA, Van Hoosear K, Bondurant RH. Studies of trichomonad protozoa in free ranging songbirds: prevalence of *Trichomonas gallinae* in house finches (*Carpodacus mexicanus*) and corvids and a novel trichomonad in mockingbirds (*Mimus polyglottos*). Vet Parasitol. 2009;161(3-4):178–86. https://doi.org/10.1016/j.vetpar.2009.01.023.

Appell LH, Mickelsen WD, Thomas MW, Harmon WM. A comparison of techniques used for the diagnosis of *Tritrichomonas-fetus* infections in beef bulls. Agri-Practice. 1993;14(2):30–4.

Arranz-Solis D, Pedraza-Diaz S, Miro G, Rojo-Montejo S, Hernandez L, Ortega-Mora LM, Collantes-Fernandez E. *Tritrichomonas foetus* infection in cats with diarrhea from densely housed origins. Vet Parasitol. 2016;221:118–22. https://doi.org/10.1016/j.vetpar.2016.03.019.

Ball L, Dargatz DA, Cheney JM, Mortimer RG. Control of venereal disease in infected herds. Vet Clin North Am Food Anim Pract. 1987;3(3):561–74.

Bartlett DE. Trichomonas foetus infection and bovine reproduction. Am J Vet Res. 1947;8(29):343–52.

Bartlett DE, Hammond DM. Pattern of fluctuations in number of *Trichomonas foetus* occurring in the bovine vagina during initial infections. 2. Application in diagnosis. Am J Vet Res. 1945;6(19):91–5.

Bartlett DE, Hasson EV, Teeter KG. Occurrence of *Trichomonas foetus* in preputial samples from infected bulls. J Am Vet Med Assoc. 1947;110(839):114–20.

Bell ET, Gowan RA, Lingard AE, McCoy RJ, Slapeta J, Malik R. Naturally occurring *Tritrichomonas foetus* infections in Australian cats: 38 cases. J Feline Med Surg. 2010;12(12):889–98. https://doi.org/10.1016/j.jfms.2010.06.003.

Benchimol M. Trichomonads under microscopy. Microsc Microanal. 2004;10(5):528–50. https://doi.org/10.1017/S1431927604040905.

Benchimol M. New ultrastructural observations on the skeletal matrix of *Tritrichomonas foetus*. Parasitol Res. 2005;97(5):408–16. https://doi.org/10.1007/s00436-005-1480-x.

Bernasconi C, Bodmer M, Doherr MG, Janett F, Thomann A, Spycher C, Iten C, Hentrich B, Gottstein B, Muller N, Frey CF. *Tritrichomonas foetus*: prevalence study in naturally mating bulls in Switzerland. Vet Parasitol. 2014;200(3-4):289–94. https://doi.org/10.1016/j.vetpar.2013.12.029.

Bielanski A, Ghazi DF, Phipps-Toodd B. Observations on the fertilization and development of preimplantation bovine embryos in vitro in the presence of *Tritrichomonas foetus*. Theriogenology. 2004;61(5):821–9.

Blackshaw AW, Beattie HER. The preservation of *Trichomonas foetus* at -79°C. Aust Vet J. 1955;31:214–6.

Boal CW, Mannan RW, Hudelson KS. Trichomoniasis in Cooper's hawks from Arizona. J Wildl Dis. 1998;34(3):590–3. https://doi.org/10.7589/0090-3558-34.3.590.

BonDurant RH. Diagnosis, treatment, and control of bovine trichomoniasis. Comp Cont Educ Pract. 1985;7(3):S179–88.

BonDurant RH. Pathogenesis, diagnosis, and management of trichomoniasis in cattle. Vet Clin North Am Food Anim Pract. 1997;13(2):345–61.

Bondurant RH. Venereal diseases of cattle: natural history, diagnosis, and the role of vaccines in their control. Vet Clin North Am Food Anim Pract. 2005;21(2):383–408. https://doi.org/10.1016/j.cvfa.2005.03.002.

BonDurant RH. Selected diseases and conditions associated with bovine conceptus loss in the first trimester. Theriogenology. 2007;68(3):461–73. https://doi.org/10.1016/j.theriogenology.2007.04.022.

BonDurant RH, Honigberg BM. Trichomonads of veterinary importance. In: Kreier JP, editor. Parasitic protozoa. New York, NY: Academic; 1994. p. 111–88.

BonDurant RH, Anderson ML, Blanchard P, Hird D, Danaye-Elmi C, Palmer C, Sischo WM, Suther D, Utterback W, Weigler BJ. Prevalence of trichomoniasis among California beef herds. J Am Vet Med Assoc. 1990;196(10):1590–3.

BonDurant RH, Corbeil RR, Corbeil LB. Immunization of virgin cows with surface antigen TF1.17 of *Tritrichomonas foetus*. Infect Immun. 1993;61(4):1385–94.

BonDurant RH, van Hoosear KA, Corbeil LB, Bernoco D. Serological response to in vitro-shed antigen(s) of *Tritrichomonas foetus* in cattle. Clin Diagn Lab Immunol. 1996;3(4):432–7.

BonDurant RH, Gajadhar A, Campero CM. Preliminary characterization of a *Tritrichomonas foetus*-like protozoan isolated from preputial smegma of virgin bulls. The Bovine. Practitioner. 1999;33:124–7.

Bryan LA, Campbell JR, Gajadhar AA. Effects of temperature on the survival of *Tritrichomonas foetus* in transport, Diamond's and InPouch TF media. Vet Rec. 1999;144(9):227–32.

Bunbury N, Bell D, Jones C, Greenwood A, Hunter P. Comparison of the InPouch TF culture system and wet-mount microscopy for diagnosis of *Trichomonas gallinae* infections in the pink pigeon *Columba mayeri*. J Clin Microbiol. 2005;43(2):1005–6. https://doi.org/10.1128/JCM.43.2.1005-1006.2005.

Bunbury N, Jones CG, Greenwood AG, Bell DJ. *Trichomonas gallinae* in Mauritian columbids: implications for an endangered endemic. J Wildl Dis. 2007;43(3):399–407. https://doi.org/10.7589/0090-3558-43.3.399.

Campero CM, Cobo ER. *Tritrichomonas foetus*: patogénesis de la mortalidad embrionaria/fetal, caracterización de antígenos vacunales y respuesta inmune inducida. Rev Med Vet. 2006;87:47–56.

Campero C, Gottstein B. Control measures, tritrichomoniasis. In: Ortega-Mora LM, Gottstein B, Conraths FJ, Buxton D, editors. Protozoal abortion in farm ruminants. Wallingford, Oxfordshire: CAB International; 2007. p. 290–301.

Campero CM, Ballabene NC, Cipolla AC, Zamora AS. Dual infection of bulls with campylobacteriosis and trichomoniasis: treatment with dimetridazole chlorhydrate. Aust Vet J. 1987;64(10):320–1.

Campero CM, Hirst RG, Ladds PW, Vaughan JA, Emery DL, Watson DL. Measurement of antibody in serum and genital fluids of bulls by ELISA after vaccination and challenge with *Tritrichomonas foetus*. Aust Vet J. 1990;67(5):175–8.

Campero CM, Rossetti O, Medina D, Bretschneider G, Roppel M. Inmunización en vaquillonas nediante vacuna de membrana de *Tritrichomonas foetus*. Vet Argent. 1999;154:250–62.

Campero CM, Rodriguez Dubra C, Bolondi A, Cacciato C, Cobo E, Perez S, Odeon A, Cipolla A, BonDurant RH. Two-step (culture and PCR) diagnostic approach for differentiation of non-*T. foetus* trichomonads from genitalia of virgin beef bulls in Argentina. Vet Parasitol. 2003;112(3):167–75.

Case CH, Keefer WO. Some ways to detect and prevent the spread of trichomoniasis in cattle. J Am Vet Med Ass. 1938;93:239–40.

Casteriano A, Molini U, Kandjumbwa K, Khaiseb S, Frey CF, Slapeta J. Novel genotype of *Tritrichomonas foetus* from cattle in Southern Africa. Parasitology. 2016:1–6. https://doi.org/10.1017/S003118201600158X.

Cauthen G. Studies on *Trichomonas columbae*, a flagellate parasitic in pigeons and doves. Am J Hyg. 1936;23:132–42.

Cepicka I, Kutisova K, Tachezy J, Kulda J, Flegr J. Cryptic species within the *Tetratrichomonas gallinarum* species complex revealed by molecular polymorphism. Vet Parasitol. 2005;128(1-2):11–21. https://doi.org/10.1016/j.vetpar.2004.11.003.

Cepicka I, Hampl V, Kulda J, Flegr J. New evolutionary lineages, unexpected diversity, and host specificity in the parabasalid genus Tetratrichomonas. Mol Phylogenet Evol. 2006;39(2):542–51. https://doi.org/10.1016/j.ympev.2006.01.005.

Ceplecha V, Svoboda M, Cepicka I, Husnik R, Horackova K, Svobodova V. InPouch TF-Feline medium is not specific for *Tritrichomonas foetus*. Vet Parasitol. 2013;196(3-4):503–5. https://doi.org/10.1016/j.vetpar.2013.04.015.

Chen XG, Li J. Increasing the sensitivity of PCR detection in bovine preputial smegma spiked with *Tritrichomonas foetus* by the addition of agar and resin. Parasitol Res. 2001;87(7):556–8.

Christensen HR, Clark BL, Parsonson IM. Incidence of *Tritrichomonas foetus* in young replacement bulls following introduction into an infected herd. Aust Vet J. 1977;53(3):132–4.

Clark BL, White MB, Banfield JC. Diagnosis of *Trichomonas foetus* infection in bulls. Aust Vet J. 1971;47(5):181–3.

Clark BL, Parsonson IM, Dufty JH. Letter: infection of bulls with *Tritrichomonas foetus* through mating with infected heifers. Aust Vet J. 1974;50(4):180.

Clark BL, Dufty JH, Parsonson IM. Studies on the transmission of *Tritrichomonas foetus*. Aust Vet J. 1977;53(4):170–2.

Clark BL, Dufty JH, Parsonson IM. The effect of *Tritrichomonas foetus* infection on calving rates in beef cattle. Aust Vet J. 1983a;60(3):71–4.

Clark BL, Dufty JH, Parsonson IM. Immunisation of bulls against trichomoniasis. Aust Vet J. 1983b;60(6):178–9.

Clark BL, Emery DL, Dufty JH. Therapeutic immunisation of bulls with the membranes and glycoproteins of *Tritrichomonas foetus* var. Brisbane. Aust Vet J. 1984;61(2):65–6.

Clark BL, Dufty JH, Parsonson IM. The frequency of infertility and abortion in cows infected with *Tritrichomonas foetus* var. brisbane. Aust Vet J. 1986;63(1):31–2.

Clark S, De Gussem K, Barnes J. Flagellated protozoan infections in turkeys. World Poultry Turkey Special. 2003;5:20–3.

Cobo ER, Cano D, Rossetti O, Campero CM. Heifers immunized with whole-cell and membrane vaccines against *Tritrichomonas foetus* and naturally challenged with an infected bull. Vet Parasitol. 2002;109(3-4):169–84.

Cobo ER, Favetto PH, Lane VM, Friend A, VanHooser K, Mitchell J, BonDurant RH. Sensitivity and specificity of culture and PCR of smegma samples of bulls experimentally infected with *Tritrichomonas foetus*. Theriogenology. 2007;68(6):853–60. https://doi.org/10.1016/j.theriogenology.2007.06.019.

Cobo ER, Morsella C, Cano D, Cipolla A, Campero CM. Immunization in heifers with dual vaccines containing *Tritrichomonas foetus* and *Campylobacter fetus* antigens using systemic and mucosal routes. Theriogenology. 2004;62(8):1367–82. https://doi.org/10.1016/j.theriogenology.2003.12.034.

Cobo ER, Corbeil LB, Gershwin LJ, BonDurant RH. Preputial cellular and antibody responses of bulls vaccinated and/or challenged with *Tritrichomonas foetus*. Vaccine. 2009;28(2):361–70. https://doi.org/10.1016/j.vaccine.2009.10.039.

Cole R, Friend M. Trichomoniasis. In: Friend M, Franson JC, editors. Trichomoniasis. Washington, DC: USGS-National Wildlife Health Center; 1999. p. 201–6.

Collantes-Fernandez E, Mendoza-Ibarra JA, Pedraza-Diaz S, Rojo-Montejo S, Navarro-Lozano V, Sanchez-Sanchez R, Ruiz-Santa-Quiteria JA, Ortega-Mora LM, Osoro K. Efficacy of a control program for bovine trichomonosis based on testing and culling infected bulls in beef cattle managed under mountain pastoral systems of Northern Spain. Vet J. 2014;200(1):140–5. https://doi.org/10.1016/j.tvjl.2014.02.003.

Conraths FJ, Schares G. Validation of molecular-diagnostic techniques in the parasitological laboratory. Vet Parasitol. 2006;136(2):91–8.

Cooper JE, Petty SJ. Trichomoniasis in free-living goshawks (*Accipiter gentilis gentilis*) from Great Britain. J Wildl Dis. 1988;24(1):80–7. https://doi.org/10.7589/0090-3558-24.1.80.

Corbeil LB, Anderson ML, Corbeil RR, Eddow JM, BonDurant RH. Female reproductive tract immunity in bovine trichomoniasis. Am J Reprod Immunol. 1998;39(3):189–98.

Corbeil LB, Campero CM, Van Hoosear K, Bondurant RH. Detection of trichomonad species in the reproductive tracts of breeding and virgin bulls. Vet Parasitol. 2008;154(3-4):226–32. https://doi.org/10.1016/j.vetpar.2008.03.014.

Dahlgren SS, Gjerde B, Pettersen HY. First record of natural *Tritrichomonas foetus* infection of the feline uterus. J Small Anim Pract. 2007;48(11):654–7. https://doi.org/10.1111/j.1748-5827.2007.00405.x.

De Carli GA, Tasca T. *Trichomonas gallinae*: a possible contact-dependent mechanism in the hemolytic activity. Vet Parasitol. 2002;106(4):277–83.

De Carli GA, da Silva AC, Wendorff A, Rott M. Lysis of erythrocytes by *Trichomonas gallinae*. Avian Dis. 1996;40:228–30.

de Oliveira JM, da Silva GM, Batista Filho AF, de Melo Borges J, de Oliveira PR, Brandespim DF, Mota RA, Pinheiro JW Jr. Prevalence and risk factors associated with bovine genital campylobacteriosis and bovine trichomonosis in the state of Pernambuco, Brazil. Trop Anim Health Prod. 2015;47(3):549–55. https://doi.org/10.1007/s11250-015-0761-3.

Dennett DP, Reece RL, Barasa JO, Johnson RH. Observations on the incidence and distribution of serotypes of *Tritrichomonas foetus* in beef cattle in north-eastern Australia. Aust Vet J. 1974;50(10):427–31.

Dewell GA, Phillips PE, Dohlman TM, Harmon KM, Gauger PC. Validation of a gauze sponge sampling methodology to detect *Tritrichomonas foetus* by real-time PCR. J Vet Diagn Invest. 2016;28(5):595–8. https://doi.org/10.1177/1040638716653637.

Diamond LS. The establishment of various trichomonads of animals and man in axenic cultures. J Parasitol. 1957;43(4):488–90.

Doi J, Hirota J, Morita A, Fukushima K, Kamijyo H, Ohta H, Yamasaki M, Takahashi T, Katakura K, Oku Y. Intestinal *Tritrichomonas suis* (=*T. foetus*) infection in Japanese cats. J Vet Med Sci. 2012;74(4):413–7.

Duboucher C, Caby S, Dufernez F, Chabe M, Gantois N, Delgado-Viscogliosi P, Billy C, Barre E, Torabi E, Capron M, Pierce RJ, Dei-Cas E, Viscogliosi E. Molecular identification of *Tritrichomonas foetus*-like organisms as coinfecting agents of human *Pneumocystis pneumonia*. J Clin Microbiol. 2006;44(3):1165–8. https://doi.org/10.1128/JCM.44.3.1165-1168.2006.

Eaglesome MD, Garcia MM. Disease risks to animal health from artificial insemination with bovine semen. Rev Sci Tech. 1997;16(1):215–25.

Erwin KG, Kloss C, Lyles J, Felderhoff J, Fedynich AM, Henke SE, Roberson JA. Survival of *Trichomonas gallinae* in white-winged dove carcasses. J Wildl Dis. 2000;36(3):551–4. https://doi.org/10.7589/0090-3558-36.3.551.

Estes WA, Mannan RW. Feeding behavior of Cooper's Hawks at urban and rural nests in southeastern Arizona. Condor. 2003;105(1):107–16. https://doi.org/10.1650/0010-5422(2003)105[107:Fbocha]2.0.Co;2.

Felleisen RS. Comparative sequence analysis of 5.8S rRNA genes and internal transcribed spacer (ITS) regions of trichomonadid protozoa. Parasitology. 1997;115(Pt 2):111–9.

Felleisen RS. Comparative genetic analysis of tritrichomonadid protozoa by the random amplified polymorphic DNA technique. Parasitol Res. 1998;84(2):153–6.

Felleisen RS, Lambelet N, Bachmann P, Nicolet J, Muller N, Gottstein B. Detection of *Tritrichomonas foetus* by PCR and DNA enzyme immunoassay based on rRNA gene unit sequences. J Clin Microbiol. 1998;36(2):513–9.

Fitzgerald PR. Bovine trichomoniasis. Vet Clin North Am Food Anim Pract. 1986;2:277–82.

Fitzgerald PR, Johnson AE, Hammond DM, Thorne JL, Hibler CP. Experimental infection of young pigs following intranasal inoculation with nasal, gastric, or cecal trichomonads from swine or with *Trichomonas foetus*. J Parasitol. 1958;44(6):597–602.

Flower PJ, Ladds PW, Thomas AD, Watson DL. An immunopathologic study of the bovine prepuce. Vet Pathol. 1983;20(2):189–202.

Forrester DJ, Foster GW. Trichomonosis. In: Atkinson CT, Thomas NJ, Hunter DB, editors. Parasitic diseases of wild birds. Ames, IA: Wiley-Blackwell; 2008. p. 120–53.

Fort M, Dubié D, Sago A, Goyeneche P, Fernández E, Collantes-Fernández E, Sánchez Sánchez R, Moreno Gonzalo J, Ortega-Mora LM. Evaluation of the performance of bovine trichomonosis control program in La Pampa-Argentina. Paper presented at the XXI Inter Comgress ANEMBE, Santiago de Compostela, Spain; 2016.

Foster DM, Gookin JL, Poore MF, Stebbins ME, Levy MG. Outcome of cats with diarrhea and *Tritrichomonas foetus* infection. J Am Vet Med Assoc. 2004;225(6):888–92.

Franssen FF, Lumeij JT. In vitro nitroimidazole resistance of *Trichomonas gallinae* and successful therapy with an increased dosage of ronidazole in racing pigeons (*Columba livia domestica*). J Vet Pharmacol Ther. 1992;15(4):409–15.

Frey CF, Müller N. Tritrichomonas—systematics of an enigmatic genus. Mol Cell Probes. 2012;26(3):132–6.

Frey CF, Schild M, Hemphill A, Stunzi P, Muller N, Gottstein B, Burgener IA. Intestinal *Tritrichomonas foetus* infection in cats in Switzerland detected by in vitro cultivation and PCR. Parasitol Res. 2009;104(4):783–8. https://doi.org/10.1007/s00436-008-1255-2.

Friedhoff KT. Pathogene, intestinale Flagellaten bei Tauben, Sittichen und Papageien. Collegium Veterinarium. 1982;63:329–34.

Friedhoff KT, Kuhnigk C, Muller I. Experimental infections in chickens with *Chilomastix gallinarum*, *Tetratrichomonas gallinarum*, and *Tritrichomonas eberthi*. Parasitol Res. 1991;77(4):329–34.

Galián M, Gentil M, Heusinger A, Müller E. *Tritrichomonas foetus* as a cause of diarrhoea in cats. Tierärztliche Umschau Parasiten-Spezial. 2011a;1:5–8.
Galián M, Heusinger A, Gentil M, Müller E. *Tritrichomonas foetus* en el gato. Argos Informativo Veterinario. 2011b;134:44–5.
Garcia Guerra A, Hill JE, Waldner CL, Campbell J, Hendrick S. Sensitivity of a real-time polymerase chain reaction for *Tritrichomonas fetus* in direct individual and pooled preputial samples. Theriogenology. 2013;80(9):1097–103. https://doi.org/10.1016/j.theriogenology.2013.08.011.
Gaspar da Silva D, Barton E, Bunbury N, Lunness P, Bell DJ, Tyler KM. Molecular identity and heterogeneity of trichomonad parasites in a closed avian population. Infect Genet Evol. 2007;7(4):433–40. https://doi.org/10.1016/j.meegid.2007.01.002.
Gault RA, Kvasnicka WG, Hanks D, Hanks M, Hall MR. Specific antibodies in serum and vaginal mucus of heifers inoculated with a vaccine containing *Tritrichomonas foetus*. Am J Vet Res. 1995;56(4):454–9.
Gay JM, Ebel ED, Kearley WP. Commingled grazing as a risk factor for trichomonosis in beef herds. J Am Vet Med Assoc. 1996;209(3):643–6.
Gerhold RW, Tate CM, Gibbs SE, Mead DG, Allison AB, Fischer JR. Necropsy findings and arbovirus surveillance in mourning doves from the southeastern United States. J Wildl Dis. 2007;43(1):129–35. https://doi.org/10.7589/0090-3558-43.1.129.
Gerhold RW, Yabsley MJ, Smith AJ, Ostergaard E, Mannan W, Cann JD, Fischer JR. Molecular characterization of the *Trichomonas gallinae* morphologic complex in the United States. J Parasitol. 2008;94(6):1335–41. https://doi.org/10.1645/GE-1585.1.
Girard YA, Rogers KH, Gerhold R, Land KM, Lenaghan SC, Woods LW, Haberkern N, Hopper M, Cann JD, Johnson CK. *Trichomonas stableri* n. sp., an agent of trichomonosis in Pacific Coast band-tailed pigeons (*Patagioenas fasciata monilis*). Int J Parasitol Parasites Wildl. 2014;3(1):32–40. https://doi.org/10.1016/j.ijppaw.2013.12.002.
Goodger WJ, Skirrow SZ. Epidemiologic and economic analyses of an unusually long epizootic of trichomoniasis in a large California dairy herd. J Am Vet Med Assoc. 1986;189(7):772–6.
Gookin JL, Breitschwerdt EB, Levy MG, Gager RB, Benrud JG. Diarrhea associated with trichomonosis in cats. J Am Vet Med Assoc. 1999;215(10):1450–4.
Gookin JL, Levy MG, Law JM, Papich MG, Poore MF, Breitschwerdt EB. Experimental infection of cats with *Tritrichomonas foetus*. Am J Vet Res. 2001;62(11):1690–7.
Gookin JL, Birkenheuer AJ, Breitschwerdt EB, Levy MG. Single-tube nested PCR for detection of *Tritrichomonas foetus* in feline feces. J Clin Microbiol. 2002;40(11):4126–30.
Gookin JL, Foster DM, Poore MF, Stebbins ME, Levy MG. Use of a commercially available culture system for diagnosis of *Tritrichomonas foetus* infection in cats. J Am Vet Med Assoc. 2003;222(10):1376–9.
Gookin JL, Stebbins ME, Hunt E, Burlone K, Fulton M, Hochel R, Talaat M, Poore M, Levy MG. Prevalence of and risk factors for feline *Tritrichomonas foetus* and Giardia infection. J Clin Microbiol. 2004;42(6):2707–10. https://doi.org/10.1128/JCM.42.6.2707-2710.2004.
Gookin JL, Birkenheuer AJ, St John V, Spector M, Levy MG. Molecular characterization of trichomonads from feces of dogs with diarrhea. J Parasitol. 2005;91(4):939–43. https://doi.org/10.1645/GE-474R.1.
Gookin JL, Copple CN, Papich MG, Poore MF, Stauffer SH, Birkenheuer AJ, Twedt DC, Levy MG. Efficacy of ronidazole for treatment of feline *Tritrichomonas foetus* infection. J Vet Int Med. 2006;20(3):536–43.
Gookin JL, Stauffer SH, Dybas D, Cannon DH. Documentation of in vivo and in vitro aerobic resistance of feline *Tritrichomonas foetus* isolates to ronidazole. J Vet Int Med. 2010a;24(4):1003–7. https://doi.org/10.1111/j.1939-1676.2010.0534.x.
Gookin JL, Stone MR, Yaeger MJ, Meyerholz DK, Moisan P. Fluorescence in situ hybridization for identification of *Tritrichomonas foetus* in formalin-fixed and paraffin-embedded histological specimens of intestinal trichomoniasis. Vet Parasitol. 2010b;172(1-2):139–43. https://doi.org/10.1016/j.vetpar.2010.04.014.
Grabensteiner E, Hess M. PCR for the identification and differentiation of *Histomonas meleagridis*, *Tetratrichomonas gallinarum* and *Blastocystis* spp. Vet Parasitol. 2006;142(3-4):223–30. https://doi.org/10.1016/j.vetpar.2006.07.011.

Grabensteiner E, Bilic I, Kolbe T, Hess M. Molecular analysis of clonal trichomonad isolates indicate the existence of heterogenic species present in different birds and within the same host. Vet Parasitol. 2010;172(1-2):53–64. https://doi.org/10.1016/j.vetpar.2010.04.015.

Grahn RA, BonDurant RH, van Hoosear KA, Walker RL, Lyons LA. An improved molecular assay for *Tritrichomonas foetus*. Vet Parasitol. 2005;127(1):33–41. https://doi.org/10.1016/j.vetpar.2004.08.018.

Gray SG, Hunter SA, Stone MR, Gookin JL. Assessment of reproductive tract disease in cats at risk for *Tritrichomonas foetus* infection. Am J Vet Res. 2010;71(1):76–81. https://doi.org/10.2460/ajvr.71.1.76.

Grellet A, Bickel T, Polack B, Boogaerts C, Casseleux G, Biourge V, Grandjean D. Prevalence of *Tritrichomonas foetus* in puppies from French breeding kennels. J Vet Int Med. 2010;24:1572.

Gunn-Moore DA, McCann TM, Reed N, Simpson KE, Tennant B. Prevalence of *Tritrichomonas foetus* infection in cats with diarrhoea in the UK. J Feline Med Surg. 2007;9(3):214–8. https://doi.org/10.1016/j.jfms.2007.01.003.

Guven E, Bastem Z, Avcioglu H, Erdem H. Molecular determination of *Tritrichomonas* spp. in aborted bovine foetuses in Eastern Anatolian Region of Turkey. Vet Parasitol. 2013;196(3-4):278–82. https://doi.org/10.1016/j.vetpar.2013.03.031.

Hale S, Norris JM, Slapeta J. Prolonged resilience of *Tritrichomonas foetus* in cat faeces at ambient temperature. Vet Parasitol. 2009;166(1-2):60–5. https://doi.org/10.1016/j.vetpar.2009.07.032.

Hammond DM, Bartlett DE. The distribution of Tritrichomonas foetus in the preputial cavity of infected bulls. Am J Vet Res. 1943;4:143–9.

Hammond DM, Bartlett DE. Pattern of fluctuations in numbers of *Trichomonas foetus* occurring in the bovine vagina during initial infections. 1. Correlation with time of exposure and with subsequent estrual cycles. Am J Vet Res. 1945;6(19):84–90.

Hampl V, Pavlicek A, Flegr J. Construction and bootstrap analysis of DNA fingerprinting-based phylogenetic trees with the freeware program FreeTree: application to trichomonad parasites. Int J Syst Evol Microbiol. 2001;51(Pt 3):731–5. https://doi.org/10.1099/00207713-51-3-731.

Harmon WM, Clark WA, Hawbecker AC, Stafford M. *Trichomonas gallinae* in columbiform birds from the Galapagos Islands. J Wildl Dis. 1987;23(3):492–4.

Hawn C. Trichomoniasis of turkeys. J Infect Dis. 1937;61:184–97.

Hayes DC, Anderson RR, Walker RL. Identification of trichomonadid protozoa from the bovine preputial cavity by polymerase chain reaction and restriction fragment length polymorphism typing. J Vet Diagn Invest. 2003;15(4):390–4.

Herr S, Ribeiro LM, Claassen E, Myburgh JG. A reduction in the duration of infection with *Tritrichomonas foetus* following vaccination in heifers and the failure to demonstrate a curative effect in infected bulls. Onderstepoort J Vet Res. 1991;58(1):41–5.

Hibler CP, Hammond DM, Caskey FH, Johnson AE, Fitzgerald PR. The morphology and incidence of the trichomonads of swine, *Tritrichomonas-suis* (Gruby and Delafond), *Tritrichomonas-rotunda*, nsp and *Trichomonas-buttreyi*, nsp. J Protozool. 1960;7(2):159–71. https://doi.org/10.1111/j.1550-7408.1960.tb00725.x.

Ho MS, Conrad PA, Conrad PJ, LeFebvre RB, Perez E, BonDurant RH. Detection of bovine trichomoniasis with a specific DNA probe and PCR amplification system. J Clin Microbiol. 1994;32(1):98–104.

Hodgson JL, Jones DW, Widders PR, Corbeil LB. Characterization of *Tritrichomonas foetus* antigens by use of monoclonal antibodies. Infect Immun. 1990;58(9):3078–83.

Höfle U, Blanco J, Palma L, Melo P. Trichomoniasis in Bonelli's eagle nestlings in south-west Portugal. In: Redig PT, Cooper JE, Remple TD, editors. Raptor Biomedicine III. Minneapolis, MN: University of Minnesota Press; 2000. p. 45–51.

Holliday M, Deni D, Gunn-Moore DA. *Tritrichomonas foetus* infection in cats with diarrhoea in a rescue colony in Italy. J Feline Med Surg. 2009;11(2):131–4. https://doi.org/10.1016/j.jfms.2008.06.004.

Hoorfar J, Wolffs P, Radstrom P. Diagnostic PCR: validation and sample preparation are two sides of the same coin. APMIS. 2004;112(11-12):808–14.

Hosein A, Kruth SA, Pearl DL, Richardson D, Maggs JC, Peach HA, Peregrine AS. Isolation of *Tritrichomonas foetus* from cats sampled at a cat clinic, cat shows and a humane society in southern Ontario. J Feline Med Surg. 2013;15(8):706–11. https://doi.org/10.1177/1098612X13475617.

Huby-Chilton F, Scandrett BW, Chilton NB, Gajadhar AA. Detection and identification of Tetratrichomonas in a preputial wash from a bull by PCR and SSCP. Vet Parasitol. 2009;166(3-4):199–204. https://doi.org/10.1016/j.vetpar.2009.08.026.

Hudson DB, Ball L, Cheney JM, Mortimer RG, Bowen RA, Marsh DJ, Peetz RH. Development and testing of a bovine trichomoniasis vaccine. Theriogenology. 1993;39(4):929–35.

Ikeda JS, BonDurant RH, Corbeil LB. Bovine vaginal antibody responses to immunoaffinity-purified surface antigen of *Tritrichomonas foetus*. J Clin Microbiol. 1995;33(5): 1158–63.

Jin Y, Schumaker B, Logan J, Yao C. Risk factors associated with bovine trichomoniasis in beef cattle identified by a questionnaire. J Med Microbiol. 2014;63(Pt 6):896–902. https://doi.org/10.1099/jmm.0.074971-0.

Jongwutiwes S, Silachamroon U, Putaporntip C. *Pentatrichomonas hominis* in empyema thoracis. Trans R Soc Trop Med Hyg. 2000;94(2):185–6.

Kerr WR. The intradermal test in bovine trichomoniasis. Vet Rec. 1944;56:303–5.

Kimsey PB, Darien BJ, Kendrick JW, Franti CE. Bovine trichomoniasis—diagnosis and treatment. J Am Vet Med Assoc. 1980;177(7):616–9.

Kittel DR, Campero C, Van Hoosear KA, Rhyan JC, BonDurant RH. Comparison of diagnostic methods for detection of active infection with *Tritrichomonas foetus* in beef heifers. J Am Vet Med Assoc. 1998;213(4):519–22.

Kleina P, Bettim-Bandinelli J, Bonatto SL, Benchimol M, Bogo MR. Molecular phylogeny of Trichomonadidae family inferred from ITS-1, 5.8S rRNA and ITS-2 sequences. Int J Parasitol. 2004;34(8):963–70. https://doi.org/10.1016/j.ijpara.2004.04.004.

Kleydman Y, Yarlett N, Gorrell TE. Production of ammonia by *Tritrichomonas foetus* and *Trichomonas vaginalis*. Microbiology. 2004;150(Pt 5):1139–45. https://doi.org/10.1099/mic.0.26939-0.

Kolisko M, Cepicka I, Hampl V, Leigh J, Roger AJ, Kulda J, Simpson AG, Flegr J. Molecular phylogeny of diplomonads and enteromonads based on SSU rRNA, alpha-tubulin and HSP90 genes: implications for the evolutionary history of the double karyomastigont of diplomonads. BMC Evol Biol. 2008;8:205.

Krone O, Altenkamp R, Kenntner N. Prevalence of *Trichomonas gallinae* in northern goshawks from the Berlin area of northeastern Germany. J Wildl Dis. 2005;41(2):304–9. https://doi.org/10.7589/0090-3558-41.2.304.

Kuehner KA, Marks SL, Kass PH, Sauter-Louis C, Grahn RA, Barutzki D, Hartmann K. *Tritrichomonas foetus* infection in purebred cats in Germany: prevalence of clinical signs and the role of co-infection with other enteroparasites. J Feline Med Surg. 2011;13(4):251–8. https://doi.org/10.1016/j.jfms.2010.12.002.

Kvasnicka WG, Taylor RE, Huang JC, Hanks D, Tronstad RJ, Bosomworth A, Hall MR. Investigations of the incidence of bovine trichomoniasis in Nevada and of the efficacy of immunizing cattle with vaccines containing *Tritrichomonas foetus*. Theriogenology. 1989;31(5):963–71.

Kvasnicka WG, Hanks D, Huang JC, Hall MR, Sandblom D, Chu HJ, Chavez L, Acree WM. Clinical evaluation of the efficacy of inoculating cattle with a vaccine containing *Tritrichomonas foetus*. Am J Vet Res. 1992;53(11):2023–7.

Lawson B, Robinson RA, Neimanis A, Handeland K, Isomursu M, Agren EO, Hamnes IS, Tyler KM, Chantrey J, Hughes LA, Pennycott TW, Simpson VR, John SK, Peck KM, Toms MP, Bennett M, Kirkwood JK, Cunningham AA. Evidence of spread of the emerging infectious disease, finch trichomonosis, by migrating birds. EcoHealth. 2011;8(2):143–53. https://doi.org/10.1007/s10393-011-0696-8.

Levine D, Brandly A. A pathogenic trichomonas from the upper digestive tract of chickens. J Am Vet Med Assoc. 1939;95:77–8.

Levy MG, Gookin JL, Poore M, Birkenheuer AJ, Dykstra MJ, Litaker RW. *Tritrichomonas foetus* and not *Pentatrichomonas hominis* is the etiologic agent of feline trichomonal diarrhea. J Parasitol. 2003;89(1):99–104. https://doi.org/10.1645/0022-3395(2003)089[0099:TFANPH]2.0.CO;2.

Liebhart D, Weissenbock H, Hess M. In-situ hybridization for the detection and identification of *Histomonas meleagridis* in tissues. J Comp Pathol. 2006;135(4):237–42. https://doi.org/10.1016/j.jcpa.2006.08.002.

Longo MC, Berninger MS, Hartley JL. Use of uracil DNA glycosylase to control carry-over contamination in polymerase chain reactions. Gene. 1990;93:125–8.

Lun ZR, Gajadhar AA. A simple and rapid method for staining *Tritrichomonas foetus* and *Trichomonas vaginalis*. J Vet Diagn Invest. 1999;11(5):471–4.

Lun Z, Parker S, Gajadhar AA. Comparison of growth rates of *Tritrichomonas foetus* isolates from various geographic regions using three different culture media. Vet Parasitol. 2000;89(3):199–208.

Lun ZR, Chen XG, Zhu XQ, Li XR, Xie MQ. Are *Tritrichomonas foetus* and *Tritrichomonas suis* synonyms? Trends Parasitol. 2005;21(3):122–5. https://doi.org/10.1016/j.pt.2004.12.001.

Madoroba E, Gelaw A, Hlokwe T, Mnisi M. Prevalence of *Campylobacter foetus* and *Trichomonas foetus* among cattle from Southern Africa. Afr J Biotechnol. 2011;10(50):10311–4.

Mancebo OA, Russo AM, Carabajal LL, Monzon CM. Persistence of *Tritrichomonas foetus* in naturally infected cows and heifers in Argentina. Vet Parasitol. 1995;59(1):7–11.

Manning K. Update on the diagnosis and management of *Tritrichomonas foetus* infections in cats. Top Companion Anim Med. 2010;25(3):145–8. https://doi.org/10.1053/j.tcam.2010.08.001.

Mardones FO, Perez AM, Martinez A, Carpenter TE. Risk factors associated with *Tritrichomonas foetus* infection in beef herds in the Province of Buenos Aires, Argentina. Vet Parasitol. 2008;153(3-4):231–7. https://doi.org/10.1016/j.vetpar.2008.01.038.

McCool CJ, Townsend MP, Wolfe SG, Simpson MA, Olm TC, Jayawardhana GA, Carney JV. Prevalence of bovine venereal disease in the Victoria River District of the Northern Territory: likely economic effects and practicable control measures. Aust Vet J. 1988;65(5):153–6.

McMillen L, Lew AE. Improved detection of *Tritrichomonas foetus* in bovine diagnostic specimens using a novel probe-based real time PCR assay. Vet Parasitol. 2006;141(3-4):204–15. https://doi.org/10.1016/j.vetpar.2006.06.012.

Mehlhorn H, Al-Quraishy S, Aziza A, Hess M. Fine structure of the bird parasites *Trichomonas gallinae* and *Tetratrichomonas gallinarum* from cultures. Parasitol Res. 2009;105(3):751–6. https://doi.org/10.1007/s00436-009-1451-8.

Mendoza-Ibarra JA, Pedraza-Diaz S, Garcia-Pena FJ, Rojo-Montejo S, Ruiz-Santa-Quiteria JA, San Miguel-Ibanez E, Navarro-Lozano V, Ortega-Mora LM, Osoro K, Collantes-Fernandez E. High prevalence of *Tritrichomonas foetus* infection in Asturiana de la Montana beef cattle kept in extensive conditions in northern Spain. Vet J. 2012;193(1):146–51. https://doi.org/10.1016/j.tvjl.2011.09.020.

Mendoza-Ibarra JA, Ortega-Mora LM, Pedraza-Diaz S, Rojo-Montejo S, Ruiz-Santa-Quiteria JA, Garcia-Pena FJ, Navarro-Lozano V, Cuevas-Martin Mdel C, Osoro K, Collantes-Fernandez E. Differences in the prevalence of *Tritrichomonas foetus* infection in beef cattle farmed under extensive conditions in northern Spain. Vet J. 2013;196(3):547–9. https://doi.org/10.1016/j.tvjl.2012.10.026.

Michi AN, Favetto PH, Kastelic J, Cobo ER. A review of sexually transmitted bovine trichomoniasis and campylobacteriosis affecting cattle reproductive health. Theriogenology. 2016;85(5):781–91. https://doi.org/10.1016/j.theriogenology.2015.10.037.

Molina L, Perea J, Meglia G, Angon E, Garcia A. Spatial and temporal epidemiology of bovine trichomoniasis and bovine genital campylobacteriosis in La Pampa province (Argentina). Prev Vet Med. 2013;110(3-4):388–94. https://doi.org/10.1016/j.prevetmed.2013.02.019.

Morin-Adeline V, Lomas R, O'Meally D, Stack C, Conesa A, Slapeta J. Comparative transcriptomics reveals striking similarities between the bovine and feline isolates of *Tritrichomonas foetus*: consequences for in silico drug-target identification. BMC Genomics. 2014;15:955. https://doi.org/10.1186/1471-2164-15-955.

Morin-Adeline V, Fraser ST, Stack C, Slapeta J. Host origin determines pH tolerance of *Tritrichomonas foetus* isolates from the feline gastrointestinal and bovine urogenital tracts. Exp Parasitol. 2015a;157:68–77. https://doi.org/10.1016/j.exppara.2015.06.017.

Morin-Adeline V, Mueller K, Conesa A, Slapeta J. Comparative RNA-seq analysis of the *Tritrichomonas foetus* PIG30/1 isolate from pigs reveals close association with *Tritrichomonas foetus* BP-4 isolate 'bovine genotype'. Vet Parasitol. 2015b;212(3-4):111–7. https://doi.org/10.1016/j.vetpar.2015.08.012.

Mostegl MM, Richter B, Nedorost N, Maderner A, Dinhopl N, Weissenbock H. Investigations on the prevalence and potential pathogenicity of intestinal trichomonads in pigs using in situ hybridization. Vet Parasitol. 2011;178(1-2):58–63. https://doi.org/10.1016/j.vetpar.2010.12.022.

Mostegl MM, Wetscher A, Richter B, Nedorost N, Dinhopl N, Weissenbock H. Detection of *Tritrichomonas foetus* and *Pentatrichomonas hominis* in intestinal tissue specimens of cats by chromogenic in situ hybridization. Vet Parasitol. 2012;183(3-4):209–14. https://doi.org/10.1016/j.vetpar.2011.07.050.

Mueller K, Morin-Adeline V, Gilchrist K, Brown G, Slapeta J. High prevalence of *Tritrichomonas foetus* 'bovine genotype' in faecal samples from domestic pigs at a farm where bovine trichomonosis has not been reported for over 30 years. Vet Parasitol. 2015;212(3-4):105–10. https://doi.org/10.1016/j.vetpar.2015.08.010.

Mukhufhi N, Irons PC, Michel A, Peta F. Evaluation of a PCR test for the diagnosis of *Tritrichomonas foetus* infection in bulls: effects of sample collection method, storage and transport medium on the test. Theriogenology. 2003;60(7):1269–78.

Munoz E, Castella J, Gutierrez JF. In vivo and in vitro sensitivity of *Trichomonas gallinae* to some nitroimidazole drugs. Vet Parasitol. 1998;78(4):239–46.

Murname D. Field and laboratory observations on trichomoniasis of dairy cattle in Victoria. Austr Vet J. 1959;35:80–3.

Narcisi EM, Sevoian M, Honigberg BM. Pathologic changes in pigeons infected with a virulent *Trichomonas gallinae* strain (Eiberg). Avian Dis. 1991;35(1):55–61.

Neimanis AS, Handeland K, Isomursu M, Agren E, Mattsson R, Hamnes IS, Bergsjo B, Hirvela-Koski V. First report of epizootic trichomoniasis in wild finches (family Fringillidae) in southern Fennoscandia. Avian Dis. 2010;54(1):136–41. https://doi.org/10.1637/8952-060509-Case.1.

Nickel DD, Olson ME, Schultz GA. An improved polymerase chain reaction assay for the detection of *Tritrichomonas foetus* in cattle. Can Vet J. 2002;43(3):213–6.

Ondrak JD. *Tritrichomonas foetus* prevention and control in cattle. Vet Clin North Am Food Anim Pract. 2016;32(2):411–23. https://doi.org/10.1016/j.cvfa.2016.01.010.

Ostrowsky JE, Frene AJ, Rodriguez Dubra C, Rutter B. Obtención de muestras prepuciales para el diagnóstico de *Tritrichomonas foetus* porraspado de mucosas. Rev Med Vet. 1974;55:525–8.

Pakandl M. The prevalence of intestinal protozoa in wild and domestic pigs. Vet Med (Praha). 1994;39(7):377–80.

Paris JK, Wills S, Balzer HJ, Shaw DJ, Gunn-Moore DA. Enteropathogen co-infection in UK cats with diarrhoea. BMC Vet Res. 2014;10:13. https://doi.org/10.1186/1746-6148-10-13.

Parker S, Campbell J, Ribble C, Gajadhar A. Comparison of two sampling tools for diagnosis of *Tritrichomonas foetus* in bulls and clinical interpretation of culture results. J Am Vet Med Assoc. 1999;215(2):231–5.

Parker S, Lun ZR, Gajadhar A. Application of a PCR assay to enhance the detection and identification of *Tritrichomonas foetus* in cultured preputial samples. J Vet Diagn Invest. 2001;13(6):508–13.

Parker S, Campbell J, Gajadhar A. Comparison of the diagnostic sensitivity of a commercially available culture kit and a diagnostic culture test using Diamond's media for diagnosing *Tritrichomonas foetus* in bulls. J Vet Diagn Invest. 2003a;15(5):460–5.

Parker S, Campbell J, Ribble C, Gajadhar A. Sample collection factors affect the sensitivity of the diagnostic test for *Tritrichomonas foetus* in bulls. Can J Vet Res. 2003b;67(2):138–41.

Parsonson IM, Clark BL, Dufty J. The pathogenesis of *Tritrichomonas foetus* infection in the bull. Aust Vet J. 1974;50(10):421–3.

Parsonson IM, Clark BL, Dufty JH. Early pathogenesis and pathology of *Tritrichomonas foetus* infection in virgin heifers. J Comp Pathol. 1976;86(1):59–66.

Pereira-Neves A, Benchimol M. *Tritrichomonas foetus*: budding from multinucleated pseudocysts. Protist. 2009;160(4):536–51. https://doi.org/10.1016/j.protis.2009.05.001.

Perez E, Conrad PA, Hird D, Ortuno A, Chacon J, Bondurant R, Noordhuizen J. Prevalence and risk-factors for *Trichomonas-fetus* infection in cattle in northeastern Costa-Rica. Prev Vet Med. 1992;14(3-4):155–65. https://doi.org/10.1016/0167-5877(92)90013-6.

Perez A, Cobo E, Martinez A, Campero C, Spath E. Bayesian estimation of *Tritrichomonas foetus* diagnostic test sensitivity and specificity in range beef bulls. Vet Parasitol. 2006;142(1-2):159–62. https://doi.org/10.1016/j.vetpar.2006.06.021.

Petrin D, Delgaty K, Bhatt R, Garber G. Clinical and microbiological aspects of *Trichomonas vaginalis*. Clin Microbiol Rev. 1998;11(2):300–17.

Pierce AE. The demonstration of an agglutinin to *Trichomonas foetus* in the vaginal discharge of infected heifers. J Comp Pathol Ther. 1947;57(2):84–97.

Pierce AE. The mucus agglutination test for the diagnosis of bovine trichomoniasis. Vet Rec. 1949;61:347–9.

Profizi C, Cian A, Meloni D, Hugonnard M, Lambert V, Groud K, Gagnon AC, Viscogliosi E, Zenner L. Prevalence of *Tritrichomonas foetus* infections in French catteries. Vet Parasitol. 2013;196(1-2):50–5. https://doi.org/10.1016/j.vetpar.2013.01.021.

Queen EV, Marks SL, Farver TB. Prevalence of selected bacterial and parasitic agents in feces from diarrheic and healthy control cats from northern California. J Vet Int Med. 2012;26(1):54–60. https://doi.org/10.1111/j.1939-1676.2011.00843.x.

Rae DO. Impact of trichomoniasis on the cow-calf producer's profitability. J Am Vet Med Assoc. 1989;194(6):771–5.

Rae DO, Crews JE. *Tritrichomonas foetus*. Vet Clin North Am Food Anim Pract. 2006;22(3):595–611. https://doi.org/10.1016/j.cvfa.2006.07.001.

Rae DO, Chenoweth PJ, Genho PC, McIntosh AD, Crosby CE, Moore SA. Prevalence of *Tritrichomonas fetus* in a bull population and effect on production in a large cow-calf enterprise. J Am Vet Med Assoc. 1999;214(7):1051–5.

Rae DO, Crews JE, Greiner EC, Donovan GA. Epidemiology of *Tritrichomonas foetus* in beef bull populations in Florida. Theriogenology. 2004;61(4):605–18.

Reece RL, Dennett DP, Johnson RH. Some observations on cultural and transport conditions for *Tritrichomonas foetus* var. brisbane. Aust Vet J. 1983;60(2):62–3.

Reinmann K, Muller N, Kuhnert P, Campero CM, Leitsch D, Hess M, Henning K, Fort M, Muller J, Gottstein B, Frey CF. *Tritrichomonas foetus* isolates from cats and cattle show minor genetic differences in unrelated loci ITS-2 and EF-1alpha. Vet Parasitol. 2012;185(2-4):138–44. https://doi.org/10.1016/j.vetpar.2011.09.032.

Rhyan JC, Stackhouse LL, Quinn WJ. Fetal and placental lesions in bovine abortion due to *Tritrichomonas foetus*. Vet Pathol. 1988;25(5):350–5.

Rhyan JC, Blanchard PC, Kvasnicka WG, Hall MR, Hanks D. Tissue-invasive *Tritrichomonas foetus* in four aborted bovine fetuses. J Vet Diagn Invest. 1995a;7(3):409–12.

Rhyan JC, Wilson KL, Burgess DE, Stackhouse LL, Quinn WJ. Immunohistochemical detection of *Tritrichomonas foetus* in formalin-fixed, paraffin-embedded sections of bovine placenta and fetal lung. J Vet Diagn Invest. 1995b;7(1):98–101.

Rhyan JC, Wilson KL, Wagner B, Anderson ML, BonDurant RH, Burgess DE, Mutwiri GK, Corbeil LB. Demonstration of *Tritrichomonas foetus* in the external genitalia and of specific antibodies in preputial secretions of naturally infected bulls. Vet Pathol. 1999;36(5):406–11.

Ribeiro KC, Mariante RM, Coutinho LL, Benchimol M. Nucleus behavior during the closed mitosis of *Tritrichomonas foetus*. Biol Cell. 2002;94(4-5):289–301.

Robinson RA, Lawson B, Toms MP, Peck KM, Kirkwood JK, Chantrey J, Clatworthy IR, Evans AD, Hughes LA, Hutchinson OC, John SK, Pennycott TW, Perkins MW, Rowley PS, Simpson VR, Tyler KM, Cunningham AA. Emerging infectious disease leads to rapid population declines of common British birds. PLoS One. 2010;5(8):e12215. https://doi.org/10.1371/journal.pone.0012215.

Rodning SP, Wolfe DF, Carson RL, Wright JC, Stockdale HD, Pacoli ME, Busby HC, Rowe SE. Prevalence of *Tritrichomonas foetus* in several subpopulations of Alabama beef bulls. Theriogenology. 2008;69(2):212–7. https://doi.org/10.1016/j.theriogenology.2007.09.014.

Rogers KH, Girard YA, Woods L, Johnson CK. Avian trichomonosis in spotted owls (*Strix occidentalis*): Indication of opportunistic spillover from prey. Int J Parasitol Parasites Wildl. 2016;5(3):305–11. https://doi.org/10.1016/j.ijppaw.2016.10.002.

Rosado TW, Specht A, Marks SL. Neurotoxicosis in 4 cats receiving ronidazole. J Vet Int Med. 2007;21(2):328–31.

Rosypal AC, Ripley A, Stockdale Walden HD, Blagburn BL, Grant DC, Lindsay DS. Survival of a feline isolate of *Tritrichomonas foetus* in water, cat urine, cat food and cat litter. Vet Parasitol. 2012;185(2-4):279–81. https://doi.org/10.1016/j.vetpar.2011.11.003.

Sager H, Ferre I, Henning K, Ortega-Mora LM. Tritrichomoniasis. In: Ortega-Mora LM, Gottstein B, Conraths FJ, Buxton D, editors. Protozoal abortion in farm ruminants. Wallingford, Oxfordshire: CAB International; 2007. p. 232–62.

Sansano-Maestre J, Garijo-Toledo MM, Gomez-Munoz MT. Prevalence and genotyping of *Trichomonas gallinae* in pigeons and birds of prey. Avian Pathol. 2009;38(3):201–7. https://doi.org/10.1080/03079450902912135.

Schönmann MJ, BonDurant RH, Gardner IA, Van Hoosear K, Baltzer W, Kachulis C. Comparison of sampling and culture methods for the diagnosis of *Tritrichomonas foetus* infection in bulls. Vet Rec. 1994;134(24):620–2.

Simmons GC, Laws L. Observations on bovine tritrichomoniasis. Aust Vet J. 1957;33:249–53.

Simpson V, Molenaar F. Increase in trichomonosis in finches. Vet Rec. 2006;159(18):606.

Skirrow S. Identification of trichomonad-carrier cows. J Am Vet Med Assoc. 1987;191(5):553–4.

Skirrow SZ, BonDurant RH. Bovine trichomoniasis. Vet Bull. 1988;58:591–603.

Skirrow SZ, BonDurant RH. Immunoglobulin isotype of specific antibodies in reproductive tract secretions and sera in *Tritrichomonas foetus*-infected heifers. Am J Vet Res. 1990a;51(4):645–53.

Skirrow SZ, BonDurant RH. Induced *Tritrichomonas foetus* infection in beef heifers. J Am Vet Med Assoc. 1990b;196(6):885–9.

Skirrow S, BonDurant R, Farley J, Correa J. Efficacy of ipronidazole against trichomoniasis in beef bulls. J Am Vet Med Assoc. 1985;187(4):405–7.

Slapeta J, Craig S, McDonell D, Emery D. *Tritrichomonas foetus* from domestic cats and cattle are genetically distinct. Exp Parasitol. 2010;126(2):209–13. https://doi.org/10.1016/j.exppara.2010.04.024.

Slapeta J, Müller N, Stack CM, Walker G, Lew-Tabor A, Tachezy J, Frey CF. Comparative analysis of *Tritrichomonas foetus* (Riedmuller, 1928) cat genotype, *T. foetus* (Riedmuller, 1928) cattle genotype and *Tritrichomonas suis* (Davaine, 1875) at 10 DNA loci. Int J Parasitol. 2012;42(13–14):1143–9. https://doi.org/10.1016/j.ijpara.2012.10.004.

Soto P, Parma AE. The immune response in cattle infected with *Tritrichomonas foetus*. Vet Parasitol. 1989;33(3-4):343–8.

Stabler RM. *Trichomonas gallinae*: a review. Exp Parasitol. 1954;3(4):368–402.

Stauffer SH, Birkenheuer AJ, Levy MG, Marr H, Gookin JL. Evaluation of four DNA extraction methods for the detection of *Tritrichomonas foetus* in feline stool specimens by polymerase chain reaction. J Vet Diagn Invest. 2008;20(5):639–41.

Steiner JM, Xenoulis PG, Read SA, Suchodolski JS, Globokar M, Huisinga E, Thucre S. Identification of *Tritrichomonas foetus* DNA in feces from cats with diarrhea from Germany and Austria. J Vet Int Med. 2007;21(3):649.

Stockdale HD, Givens MD, Dykstra CC, Blagburn BL. *Tritrichomonas foetus* infections in surveyed pet cats. Vet Parasitol. 2009;160(1-2):13–7. https://doi.org/10.1016/j.vetpar.2008.10.091.

Strickland L, Edmondson M, Maxwell H, et al. Surface architectural anatomy of the penile and preputial epithelium of bulls. Clin Therio. 2014;6:445–51.

Stuka P, Katai P. Rapid demonstration of bull trichomoniasis in unstained smear preparations from preputial scrapings. Acta Vet Hung. 1969;19:385–9.

Sun Z, Stack C, Slapeta J. Sequence differences in the diagnostic region of the cysteine protease 8 gene of *Tritrichomonas foetus* parasites of cats and cattle. Vet Parasitol. 2012;186(3-4):445–9. https://doi.org/10.1016/j.vetpar.2011.12.001.

Szonyi B, Srinath I, Schwartz A, Clavijo A, Ivanek R. Spatio-temporal epidemiology of *Tritrichomonas foetus* infection in Texas bulls based on state-wide diagnostic laboratory data. Vet Parasitol. 2012;186(3-4):450–5. https://doi.org/10.1016/j.vetpar.2011.11.075.

Tachezy J, Tachezy R, Hampl V, Sedinova M, Vanacova S, Vrlik M, Van Ranst M, Flegr J, Kuldaa J. Cattle pathogen *Tritrichomonas foetus* (Riedmuller, 1928) and pig commensal *Tritrichomonas suis* (Gruby & Delafond, 1843) belong to the same species. J Eukaryot Microbiol. 2002;49(2):154–63.

Tasca T, De Carli GA. Scanning electron microscopy study of *Trichomonas gallinae*. Vet Parasitol. 2003;118(1-2):37–42.

Taylor MA, Marshall RN, Stack M. Morphological differentiation of *Tritrichomonas foetus* from other protozoa of the bovine reproductive tract. Br Vet J. 1994;150(1):73–80. https://doi.org/10.1016/S0007-1935(05)80098-3.

Tedesco LF, Errico F, Del Baglivi LP. Diagnosis of *Tritrichomonas foetus* infection in bulls using two sampling methods and a transport medium. Aust Vet J. 1979;55(7):322–4.

Terzolo HR, Argento E, Catena MC, Cipolla AL, Martinez AH, Tejada G, Villa C, Betancor R, Campero CM, Cordeviola JM, Pasini MI. Procedimientos de laboratorio para el diagnóstico de la Campylobacteriosis y Tricomoniasis genital bovina. In: Comisión Científica Permanente de Enfermedades Venéreas de los Bovinos. Balcarce, Argentina: INTA; 1992. p. 1–33.

Thomas MW, Harmon WM, White C. An improved method for the detection of *Tritrichomonas-fetus* Infection by culture in bulls. Agric Pract. 1990;11(1):13–7.

Thomford JW, Talbot JA, Ikeda JS, Corbeil LB. Characterization of extracellular proteinases of *Tritrichomonas foetus*. J Parasitol. 1996;82(1):112–7.

Todorovic R, McNutt SH. Diagnosis of *Trichomonas foetus* infection in bulls. Am J Vet Res. 1967;28(126):1581–90.

Tolbert MK, Gookin J. Tritrichomonas foetus: a new agent of feline diarrhea. Compend Contin Educ Vet. 2009;31(8):374–81. 390; quiz 381.

Tolbert MK, Gookin JL. Mechanisms of *Tritrichomonas foetus* pathogenicity in cats with insights from venereal trichomonosis. J Vet Int Med. 2016;30(2):516–26. https://doi.org/10.1111/jvim.13920.

Tolbert MK, Stauffer SH, Gookin JL. Feline *Tritrichomonas foetus* adhere to intestinal epithelium by receptor-ligand-dependent mechanisms. Vet Parasitol. 2013;192(1-3):75–82. https://doi.org/10.1016/j.vetpar.2012.10.019.

Tysnes K, Gjerde B, Nodtvedt A, Skancke E. A cross-sectional study of *Tritrichomonas foetus* infection among healthy cats at shows in Norway. Acta Vet Scand. 2011;53:39. https://doi.org/10.1186/1751-0147-53-39.

Van der Pol B. *Trichomonas vaginalis* infection: the most prevalent nonviral sexually transmitted infection receives the least public health attention. Clin Infect Dis. 2007;44(1):23–5. https://doi.org/10.1086/509934.

Van der Saag M, McDonell D, Slapeta J. Cat genotype *Tritrichomonas foetus* survives passage through the alimentary tract of two common slug species. Vet Parasitol. 2011;177(3-4):262–6. https://doi.org/10.1016/j.vetpar.2010.11.054.

Villarroel A, Carpenter TE, BonDurant RH. Development of a simulation model to evaluate the effect of vaccination against *Tritrichomonas foetus* on reproductive efficiency in beef herds. Am J Vet Res. 2004;65(6):770–5.

Voyich JM, Ansotegui R, Swenson C, Bailey J, Burgess DE. Antibody responses of cattle immunized with the Tf190 adhesin of *Tritrichomonas foetus*. Clin Diagn Lab Immunol. 2001;8(6):1120–5. https://doi.org/10.1128/CDLI.8.6.1120-1125.2001.

Walden HS, Dykstra C, Dillon A, Rodning S, Givens D, Bird R, Newton J, Lindsay D. A new species of Tritrichomonas (Sarcomastigophora: Trichomonida) from the domestic cat (*Felis catus*). Parasitol Res. 2013;112(6):2227–35. https://doi.org/10.1007/s00436-013-3381-8.

Witte J. Bakterienfreie Züchtung von Trichomonaden und dem Uterus des Rindes in einfachen Nährböden. Zentralblatt für Bakteriologie, Parasitenkunde, Infektionskrankheiten und Hygiene. 1933;128:188–95.

Xenoulis PG, Lopinski DJ, Read SA, Suchodolski JS, Steiner JM. Intestinal *Tritrichomonas foetus* infection in cats: a retrospective study of 104 cases. J Feline Med Surg. 2013;15(12):1098–103. https://doi.org/10.1177/1098612X13495024.

Yaeger MJ, Gookin JL. Histologic features associated with *Tritrichomonas foetus*-induced colitis in domestic cats. Vet Pathol. 2005;42(6):797–804. https://doi.org/10.1354/vp.42-6-797.

Yang N, Cui X, Qian W, Yu S, Liu Q. Survey of nine abortifacient infectious agents in aborted bovine fetuses from dairy farms in Beijing, China, by PCR. Acta Vet Hung. 2012;60(1):83–92. https://doi.org/10.1556/AVet.2012.007.

Yao C. Diagnosis of *Tritrichomonas foetus*-infected bulls, an ultimate approach to eradicate bovine trichomoniasis in US cattle? J Med Microbiol. 2013;62(Pt 1):1–9. https://doi.org/10.1099/jmm.0.047365-0.

Yao C, Köster LS. *Tritrichomonas foetus* infection, a cause of chronic diarrhea in the domestic cat. Vet Res. 2015;46:35. https://doi.org/10.1186/s13567-015-0169-0.

Yule A, Skirrow SZ, Bondurant RH. Bovine trichomoniasis. Parasitol Today. 1989a;5(12):373–7.

Yule A, Skirrow SZ, Staats J, Bondurant RH. Development and preliminary assessment of a polyclonal antibody-based enzyme immunoassay for the detection of *Tritrichomonas foetus* antigen in breeding cattle. Vet Parasitol. 1989b;31(2):115–23.

Amoebae

15

Julia Walochnik

Abstract

The term *amoebae* refers to protozoan organisms that are able to change their shape and feed by phagocytosis. Two entirely unrelated eukaryotic groups include amoeboid organisms of medical relevance, namely, the Amoebozoa and the Excavata. Most amoebae have a cosmopolitan distribution, and only few species are strict parasites. Among these are the intestinal entamoebae that cannot propagate without a host (= obligatory parasites) and survive in the environment only as cysts. However, most amoebae do not need a host and are, thus, irrespective of their phylogenetic affiliations, all subsumed under the term *free-living amoebae*. Nevertheless, when they accidentally enter a host, some of these may also cause severe disease. The most common amoebic infections in animals are the intestinal infections by enteric amoebae, of which infections with *E. histolytica* in primates and with *E. invadens* in reptiles are of highest medical relevance. Furthermore, the amoebic gill disease (AGD)—mainly caused by *Neoparamoeba perurans*—has become a significant problem in fish farms in the past few years. Infections with *Acanthamoeba* spp., *Balamuthia mandrillaris*, and *Naegleria fowleri*, are very rare and are always accidental. Generally, amoebic infections in animals can show severe and fatal progressions, mainly due to low awareness leading to delayed diagnosis and to the unavailability of sufficiently effective treatment. This chapter intends to give an overview of the amoeboid organisms known to infect animals and the corresponding diseases.

J. Walochnik
Institute of Specific Prophylaxis and Tropical Medicine, Center for Pathophysiology, Infectiology and Immunology, Medical University of Vienna, Vienna, Austria
e-mail: julia.walochnik@meduniwien.ac.at

M. Florin-Christensen, L. Schnittger (eds.), *Parasitic Protozoa of Farm Animals and Pets*, https://doi.org/10.1007/978-3-319-70132-5_15

The *amoebae* represent a most diverse group of morphologically different and entirely unrelated genera and species. All they truly have in common is that they are unicellular eukaryotes using pseudopodia for locomotion and feeding by phagocytosis. The amoebic cell typically has a granuloplasma containing the cell organelles and a hyaloplasma producing the pseudopodia. The biodiversity of amoeboid organisms is enormous, and they are generally among the most common protozoa worldwide. Most amoebae have a cosmopolitan distribution, their main habitats being freshwater and soil, but several species have adapted to extreme environments with high temperatures, high salinities or other stresses. Moreover, many species have highly resistant cysts allowing them to withstand extreme conditions for a certain period of time. Indeed amoebae can be isolated from almost any environment, including ice water and hot desert sand, and they can also infect various animals, where they can inhabit almost any organ, depending mainly on the immune system of the respective host. Some amoebae are even strictly parasitic and cannot live without a host such as entamoebae (Eukaryota-Amoebozoa) mostly causing infections of the intestinal tract. Some of these invade and lyse tissue and disseminate to other organs. However, most amoebae are primarily free-living, and their propagation is independent of the host, but upon contact, they cause severe and eventually fatal infections. The most representative taxa are *Acanthamoeba* spp. and *Balamuthia mandrillaris* (both Eukaryota-Amoebozoa) and *Naegleria fowleri*, (Eukaryota-Excavata). Clinically relevant amoebic infections in animals are rare and include *Entamoeba invadens* infections of reptiles and amoebic gill disease (AGD) of farmed fish—caused by *Neoparamoeba perurans*—that has become a significant problem in recent years. However, they may show severe progressions, and an effective and easily manageable treatment is in many cases not available.

Except the saprophilic *E. moshkovskii*, all *Entamoeba* species are anaerobic and are strict parasites. Exclusively *E. histolytica* in humans, captive primates, and occasionally in dogs; *E. nuttalli* in nonhuman primates; and *E. invadens* in captive reptiles, particularly in snakes and lizards, invade epithelia and lyse tissue after intestinal colonization eventually leading to severe bleeding and to dissemination of the parasite to other organs. The remaining enteric amoebae are noninvasive colonizers rather than pathogens. Some related genera, such as *Endolimax nana*, *Iodamoeba bütschlii,* or *Malpighamoeba mellificae*, also infect various animals (Clark and Stensvold 2015; Hooshyar et al. 2015).

Among the primarily free-living amoebozoans, mainly the genera *Neoparamoeba*, *Acanthamoeba* and *Balamuthia* are of medical relevance. *N. perurans* is the causative agent of AGD. *Acanthamoeba* spp. and *Balamuthia mandrillaris* can cause disseminating infections such as skin lesions and granulomatous amoebic encephalitis (GAE), particularly in immunocompromised vertebrate hosts (Visvesvara et al. 2010; Khan 2006). *Acanthamoeba* spp. are also the causative agents of *Acanthamoeba* keratitis (AK), a sight-threatening infection of the eye, which has very rarely be seen in animals. In humans, particularly in the industrialized countries, this infection occurs mainly in contact lens wearers (Marciano-Cabral and Cabral 2003). Besides these two genera, also the genera

Sappinia and *Vermamoeba* have been associated with disease in humans, whereas infections in animals have not yet been described (Schuster and Visvesvara 2004a). However, particularly the genera *Acanthamoeba* and *Vermamoeba* have also medical importance as hosts, vehicles, and training grounds for bacteria (Barker and Brown 1994). Several bacteria, including important human pathogens such as legionellae, have been shown to be *primed* for virulence by passage through amoebae.

The only excavate amoeba species of true medical relevance is *Naegleria fowleri* causing the usually fatal primary amoebic meningoencephalitis (PAME) in a wide range of animals, most infections having been reported from mammals. However, also *N. fowleri* is primarily free-living and causes disease if it accidentally gets access to the central nervous system (CNS) of its respective host.

Table 15.1 gives an overview of the most important amoebic taxa infecting animals, their phylogenetic affiliations and main characteristics.

Table 15.1 Amoeboid organisms as animal pathogens (medically relevant infections and main hosts in bold)

	Size in μm (T/C)	No. of nuclei in cyst	Disease/ medical relevance[a]	Animals[a]	Selected references
Excavates					
Primarily free-living					
Naegleria fowleri	12–25/7–15	1	**PAME**	Bovines, humans, nonhuman primates, reptiles, tapir	Daft et al. (2005), Frank and Bosch (1972), Lozano-Alarcón et al. (1997), Morales et al. (2006), Visvesvara et al. (2005)
Willaertia magna	50–100/18–28	1 (or more)	Gastritis?	Dogs	Steele et al. (1997)
Amoebozoans					
Parasitic					
Endolimax nana	6–15/8–10	4	Diarrhoea?	Amphibians, birds, humans, nonhuman primates, reptiles, rodents	Frank and Bosch (1997), Mc Dougald (1997), Poulsen and Stensvold (2016)
Entamoeba anatis	10–20/9–18	4	Diarrhoea, pharyngeal disease	**Ducks**, other birds	Silvanose et al. (1998)
E. bovis	5–25/4–15	1	None	**Bovines**, caprines, ovines	Al-Habsi et al. (2017)

(continued)

Table 15.1 (continued)

E. coli	25–50/20–33	8	None	**Humans**, nonhuman primates, dogs, marsupials, rodents	Verweij et al. (2003) and Clark and Stensvold (2015)
E. dispar	9–25/11–20	4	Diarrhoea?	**Humans**, nonhuman primates	Ali (2015), Verweij et al. (2003) and Clark and Stensvold (2015)
E. gallinarum	9–25/12–15	8	None	**Chickens**, ducks, geese, guinea fowl, turkeys	McDougald (1997), Tyzzer (1920)
E. gingivalis	10–20	No cyst	Paradontitis?	Cats, dogs, horses, **Humans**, nonhuman primates, pigs	Bergquist (2009) and Clark and Stensvold (2015)
E. hartmanni	5–10/3–5	4	None	**Humans**, nonhuman primates	Clark and Stensvold (2015)
E. histolytica	9–25/10–18	4	**Haemorrhagic diarrhoea, liver abscess**	**Humans**, (cats, dogs, nonhuman primates, pigs)	Alam et al. (2015), Amyx et al. (1978), Beaver et al. (1988), Clark and Stensvold (2015) and Verweij et al. (2003)
E. invadens	10–35/10–20	4	**Haemorrhagic diarrhoea, gross lesions**	**Reptiles**	Donaldson et al. (1975), Jacobson et al. (1983) and Ratcliffe and Geiman (1934)
E. moshkovskii	9–25/11–20	4	Diarrhoea?	**Sewage**, humans, nonhuman primates, reptiles	Ali (2015), Heredia et al. (2012)
E. nuttalli	9–25/11–20	4	**Diarrhoea, liver abscess**	Humans, **nonhuman primates**	Levecke et al. (2015) and Tachibana et al. (2016)
E. polecki	10–20/10–18	1	Diarrhoea	Birds, humans, nonhuman primates, **pigs, wild boars**	Solaymani-Mohammadi and Petri (2006) and Pakandl (1994)
E. ranarum	11–30/6–20	1	Diarrhoea, extraintestinal disease?	**Frogs**, toads, snakes	Poynton and Whitaker (2001) and Richter et al. (2008)
E. suis	5–25/4–17	1	Diarrhoea	Nonhuman primates, **pigs**	Matsubayashi et al. (2014)

Table 15.1 (continued)

	Size in µm (T/C)	No. of nuclei in cyst	Disease/ medical relevance[a]	Animals[a]	Selected references
Iodamoeba bütschlii	8–20/5–18	1	None	**Humans**, nonhuman primates, pigs	Verweij et al. (2003), Clark and Stensvold (2015) and Pakandl (1994)
Malpighamoeba mellificae	5–15/6–7	2	Spring time disease	**Bees**	Evans and Schwarz (2011) and Prell (1926)
Primarily free-living					
Acanthamoeba spp.	15–45/12–32	1	**Keratitis; GAE (opportunistic)**	Amphibians, birds, bovines, dogs, fishes, horses, humans, kangaroo, mice, ovines, nonhuman primates, reptiles, invertebrates	Bosch and Deichsel (1972), Culbertson et al. (1958), Dubey et al. (2005), Dykova et al. (1999), Frank and Bosch (1972), Jones et al. (1975), Kent et al. (2011), Mortazavi et al. (2010), Visvesvara et al. (2010), Valladares et al. (2014) and Walochnik et al. (1999)
Balamuthia mandrillaris	12–60/12–30	1	**GAE (mainly opportunistic)**	Dogs, horses, humans, nohuman primates, sheep	Kinde et al. (1998) and Visvesvara et al. (1993)
Neoparamoeba perurans	41–56	No cyst	**AGL**	Fish (particularly salmonids)	Munday (1986), Oldham et al. (2016) and Young et al. (2007)
Sappinia spp.	50–80/18–37	1–2	GAE (opportunistic)	Humans	Gelman et al. (2001)
Vermamoeba vermiformis	12–37/4–9	1	CNS and eye infections?	Humans	Centeno et al. (1996), Aitken et al. (1996) and Lorenzo-Morales et al. (2007a, b)

GAE granulomatous amoebic encephalitis, *PAME* primary amoebic meningoencephalitis
[a]Relevant clinical manifestations and main hosts are printed bold

15.1 Morphologies and Life Cycles

15.1.1 Amoebozoans

The amoebozoan amoebae have two life cycle stages, a feeding and dividing trophozoite and a metabolically inactive cyst stage, which enables the amoebae to survive without nutrients and withstand desiccation and heat (Figs. 15.1 and 15.2). The cyst morphology has been widely used for identification and classification, but in many cases, morphological species are not supported by molecular data.

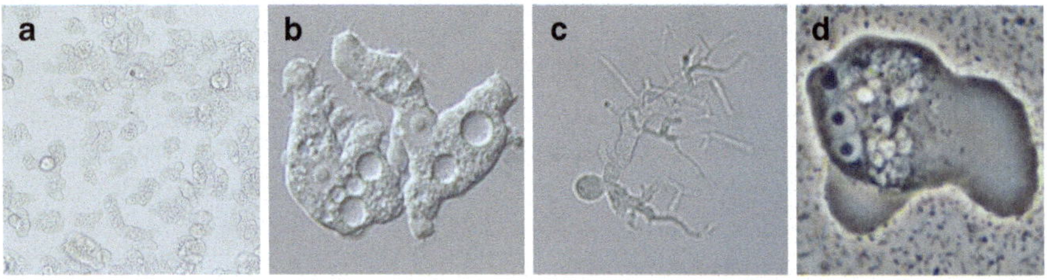

Fig. 15.1 Amoebozoans in cell culture. *Entamoeba histolytica* (×200) (**a**), *Acanthamoeba* morphological Group III (×1000) (**b**), *Balamuthia mandrillaris* (×1000) (**c**), *Sappinia diploidea* (×400) (**d**). Orig

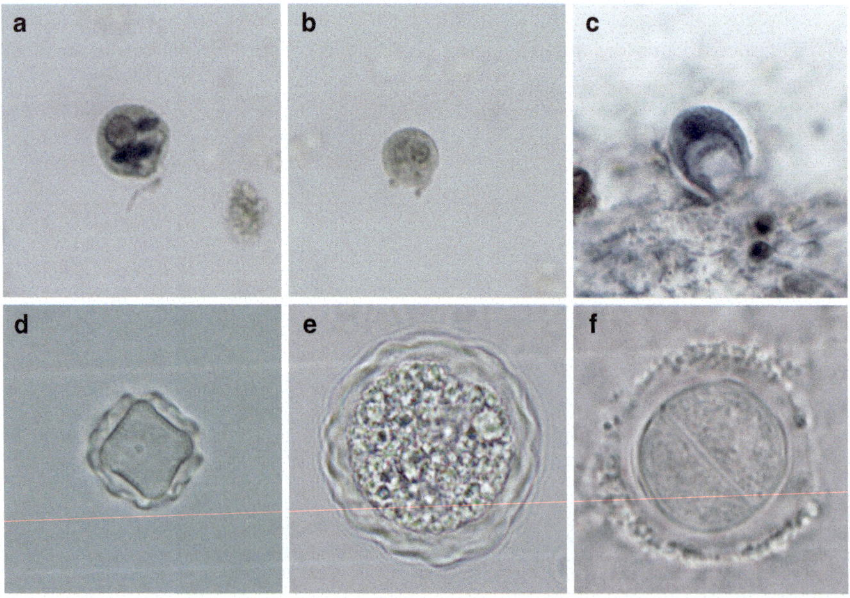

Fig. 15.2 Cysts of amoebozoans. *Entamoeba histolytica* (**a**), *Entamoeba hartmanni* (**b**), *Iodamoeba bütschlii* (**c**), *Acanthamoeba* morphological Group II (**d**), *Balamuthia mandrillaris* (**e**), *Sappinia diploidea* (**f**). All ×1000. Orig

The genus *Entamoeba* can be recognized by a distinctive nucleus—composed of a ring representing the peripheral chromatin and a dot representing the karyosome or nucleolus—allowing to divide it into species groups based on cyst morphology and the presence of one, four, or eight nuclei in mature cysts (Table 15.1). However, this genus also includes species that do not form cysts—e.g. *E. gingivalis*—where this classification scheme cannot be applied (Clark 2000). In this case, also the host species, the morphology and position of the karyosome and the existence of chromatoid bars—aggregates of ribosomes—are important characters for species discrimination. Altogether 51 species have been described, but the validity of several species is still uncertain (Hooshyar et al. 2015). The medically most important species are in alphabetical order: *E. anatis* (birds), *E. gingivalis* (humans), *E. histolytica* (humans), *E. invadens* (reptiles), *E. nuttalli* (nonhuman primates), *E. polecki* (humans, other primates and pigs) and *E. suis* (pigs). The two truly pathogenic *Entamoeba* species—*E. histolytica* and *E. invadens*—belong to separate clades within the genus. The potentially invasive *E. nuttalli* is rather closely related to *E. histolytica*. The sizes of the trophozoite stage of intestinal amoebae vary between 5 μm in *E. hartmanni*, and around 50 μm in *E. coli*, *E. histolytica*, *E. dispar*, and *E. moshkovskii* are morphological identical and cannot be distinguished by microscopy. Their cysts (Fig. 15.2a) measure between 10 and 20 μm—mostly around 12 μm—and have a pronounced chitinious cell wall, chromatoid bodies, and one to four nuclei, four displayed in the mature state. *E. hartmanni* has also four nuclei in the mature cyst, yet they measure only 3.8 μm in diameter allowing to differentiate this species (Fig. 15.2b). The trophozoites of *E. invadens* have a single large ringlike nucleus with a small central karyosome and a size of 10–40 μm. The cyst stage has four nuclei and measures around 12–14 μm in diameter. The largest species is *E. coli* with cysts up to 33 μm in diameter. The cysts of *E. polecki* range from 10 to 18 μm in diameter and have a single nucleus, whereas the cysts of *Iodamoeba bütschlii* scale 5–18 μm in diameter and are characterized by a rather large nucleolus and a large glycogen mass—vacuole—within the cell (Fig. 15.2c). While the tetra-nucleated cyst-producing species seem to exhibit little intraspecific genetic variability, *E. polecki* and *E. coli* comprise four and two genetic subtypes, respectively (Clark and Stensvold 2015, Som et al. 2000).

In all enteric amoebae, only the mature cyst is able to survive the gastric passage and represents the infective parasite stage. In multinucleated species, excystation takes place in the large intestine with one cyst typically releasing four to eight motile trophozoites. Trophozoites feed mainly on bacteria by phagocytosis and divide by binary fission, but some species can also invade the mucosa. Usually, trophozoites start to encyst when infection subsides. Encystment typically takes place in the lower parts of the large intestine, yet the triggers of cyst formation are still unknown. Most entamoebae cannot encyst outside their host, and trophozoites cannot survive for a long time in the environment. Only the cysts are resistant against harsh conditions in the environment, and the viability of these resistant stages may be weeks to months.

The genera *Neoparamoeba* and *Paramoeba* include amoebae that have short blunt digitiform subpseudopodia and are characterized by the presence of parasomes—endosymbionts of the *Perkinsela amoebae*-type—which are 5–8 μm in size and are located adjacent to the nucleus. The trophozoites of *N. perurans*—the most important causative agent of AGD—are 41–56 μm long, have numerous endocytotic vesicles, and, in contrast to *Paramoeba* spp., lack surface structures (Young et al. 2007; Oldham et al. 2016). Cysts are unknown, but pseudocysts have been observed recently (Lima et al. 2016).

The genus *Acanthamoeba* comprises three morphological groups, which are—as in the case of enteric amoebae—based on the morphologies of the cysts (Pussard and Pons 1977). The cysts are 12–32 μm in size and have a double-layered wall, with an outer ectocyst wall consisting of proteins and polysaccharides and an inner endocyst wall, mainly composed of cellulose (Fig. 15.2d) (Neff and Neff 1969). The ecto- and endocyst join at the covered ostioles, from which the amoeba emerges during excystation and which give the cysts a characteristic polygonal form, but intraspecific polymorphism is common (Chavez-Mungua et al. 2005). The cysts are usually uninuclear and have a large, dense, centrally located nucleolus. Around 25 species and 22 18S rDNA genotypes (T1-T22) have been described, yet several morphological species are not supported by molecular data (Gast et al. 1996; Stothard et al. 1998). The most common genotype in *Acanthamoeba* infections is T4, which also is the most abundant genotype. A classification into virulent and nonvirulent genotypes has not been possible so far (Booton et al. 2005; Walochnik et al. 2015). *Acanthamoeba* trophozoites are 15–45 μm in size. Their movement is sluggish, usually involving a single pseudopodium, referred to as lobopodium, and characteristic spiny surface projections, referred to as acanthopodia. The acanthapodia are important during the infection process, since they allow trophozoites to interact with the host cell surface (Omaña-Molina et al. 2004). When they attach to surfaces, they are rounded and flat, but they also have a spatially shaped floating form. Although multinucleate cells can be found, trophozoites are generally uninucleate. In contrast to entamoebae, free-living amoebozoans have mitochondria.

B. mandrillaris is the only species that constitutes this genus. The cysts of this species are more or less spherical and have a size that ranges between 12 and 30 μm corresponding to a mean diameter of 15 μm (Fig. 15.2e). They are uninucleate but frequently also two nucleoli are found. By light microscopy, cysts appear to be round or oval in shape with a wavy and irregular outer wall without pores. According to Visvesvara et al. (1993), the cyst wall has three layers, the thin and irregular outer ectocyst, the thick endocyst, and the middle amorphous fibrillar mesocyst. The trophozoites of *B. mandrillaris* are 12–60 μm long and have typically numerous tubular pseudopodia.

The cysts of *Sappinia* spp. are around 18–25 μm in diameter. The formation of bicellular cysts as a result of two copulating trophozoites, which later transform into unicellular binucleate cysts, is considered characteristic for *S. diploidea* (Fig. 15.2f). Trophozoites are 50–60 μm long and 20–30 μm wide and typically have two nuclei. They have a monopodial locomotion with a large hyaloplasma in the anterior part of the cell and a cell surface without any sub-pseudopodial projections. Moreover, *Sappinia* spp. feature so-called standing forms or standing amoeba stage.

In free-living amoebozoans, the infective and invasive stage is usually the trophozoite, which accesses the host by contact of wounds, mucosa, or corneal epithelium to contaminated water or soil. In addition, cysts are easily spread through the air and can be inhaled into the lower respiratory tract or brought upon skin lesions of the host. Under appropriate conditions, cysts transform rapidly into trophozoites, which under optimal conditions multiply every 6–8 h by binary fission. Most species prefer temperatures of around 30 °C, but many isolates grow also at elevated temperatures of up to even 45 °C (Griffin 1972). In contrast to all other amoebae, *Acanthamoeba* spp. and *B. mandrillaris* can encyst within the host tissue making treatment extremely difficult. In the environment, *Acanthamoeba* cysts maintain viability for at least 25 years (Mazur et al. 1995).

15.1.2 Heteroloboseans

All representatives of the genus *Naegleria* have three life cycle stages, a trophozoite, a cysts, and a biflagellate flagellate stage, which allows for rapid dispersal. The cysts are uninucleate and round or oval with a smooth surface and four to seven plugged pores (Fig. 15.3). The nucleus has a characteristic perinuclear ring visible in the cysts as well as in the trophozoites. In contrast to the amoebozoans, *Naegleria* spp. have discoid cristae in their mitochondria, and they lack a typical Golgi apparatus. Trophozoites are of an elongated shape, divide by binary fission, and exhibit a rapid eruptive locomotion, reaching a velocity of more than 4 times their own body length per minute. Moreover, they have cytoplasmic extensions of

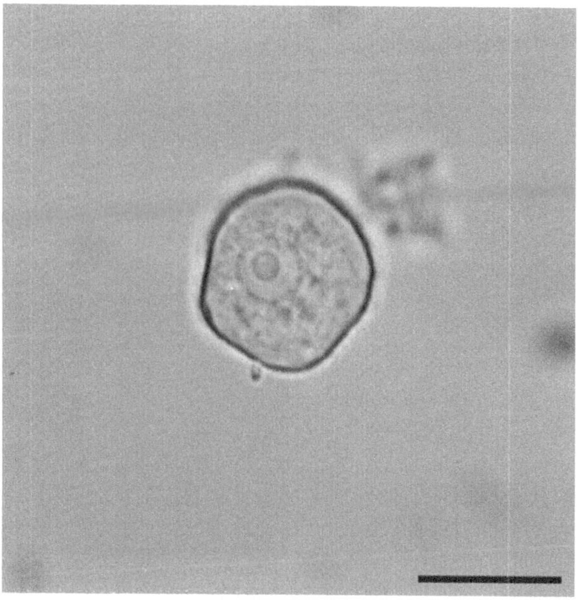

Fig. 15.3 *Naegleria fowleri* (G). Scale bar: 10 µm. Orig

the surface—food cups or *amoebastomes*—with which they ingest bacteria and also cellular debris. The genus *Naegleria* holds currently 47 described species. Although several species have been shown to be virulent after experimental infection of mice, exclusively *N. fowleri* has been described to cause disease after natural infection (De Jonckheere 2014). The cysts of *N. fowleri* range in size between 7 and 15 μm, and the trophozoites measure 12.5–25 μm. The flagellate form is pear-shaped, has a body length of 12–18 μm, and is usually biflagellate, though occasionally multiple flagella can be observed. Even though the actively invading form of *N. fowleri* is the trophozoite, all three forms of parasite stages are able to initiate an infection.

15.2 Host-Pathogen Interactions

It is important to differentiate between the simple presence of amoebal stages—particularly cysts which may simply pass through the intestinal tract—, colonization verified by the presence of amoebal trophozoites, and true infection of various organs characterized not only by multiplying amoebae but also by an immune reaction of the host (Fig. 15.4).

While many of the enteric amoebae are specific for one group of animals, in most hosts within their host range they are only colonizers and are invasive and virulent only in a few specific hosts. For example, *E. invadens* is known to potentially cause fatal enteritis in reptiles. However, while crocodiles and turtles typically act as carriers and are not susceptible to severe infections, snakes and lizards show high mortality (Kojimoto et al. 2001; Richter et al. 2008).

Concerning the facultatively parasitic free-living amoebae, various animal taxa seem to be similarly susceptible to infection. Importantly, the vast majority of healthy individuals, humans or animals, do not develop disease in spite of regular contact with free-living amoebae. Besides risky behaviour such as having contact with contaminated waters and mechanical prerequisites or such as breaks in the skin, the immune status of the respective host plays an important role for the establishment of disease. AGD typically occurs in salmonid fish farms, where crowding and other

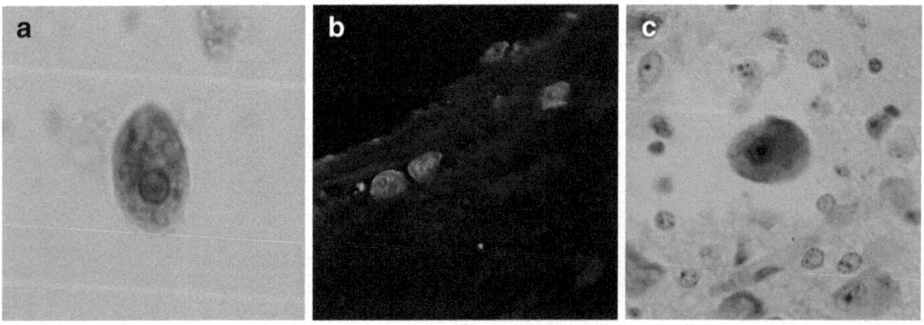

Fig. 15.4 Histology. *Entamoeba histolytica* in the gut mucosa (**a**), *Acanthamoeba* in a human cornea (**b**), *Naegleria fowleri* in the mouse brain (**c**). A+C: x1000, B: ×400. Orig

stresses make fish particularly susceptible. Although GAE has been described in a broad range of animals, most of them have been in captivity. Mice, guinea pigs, and rabbits, develop a CNS infection comparable to GAE when experimentally infected with *Acanthamoeba* spp. or *B. mandrillaris*. They are therefore used as animal models for GAE, but generally all vertebrates and even some invertebrates seem to be susceptible to the infection (Mortazavi et al. 2010; Visvesvara and Stehr-Green 1990). Naturally occurring AK seems to be extremely rare in animals but may remain unnoticed and/or undiagnosed. A single case has been described in a dog with an underlying chronical ocular surface disease that was treated with a long-term immunosuppression, but the corneas of several animals—e.g. pigs and hamster—are generally susceptible to AK and the respective hosts used as models (Alizadeh et al. 1995; He et al. 1992; Beckwith-Cohen et al. 2016). In contrast to all other free-living amoebae infections, PAME does not seem to be related to a compromised immune state. Here, the major risk factor seems to be the introduction of *N. fowleri* contaminated water into the nostrils, whereby a damaged nasal mucosa—e.g. by a common rhinovirus infection—is assumed to favour the penetration of the amoebae.

In general, all amoebae feed mainly on bacteria and other microorganisms by phagocytosis and only under certain conditions become cytolytic to the—typically much larger—cells of their respective hosts. The course of pathogenesis of an amoeba infection depends on adhesion to the host cells, contact-mediated cytolysis, and the immunoreaction of the host. The ability of amoebae to lyse cells is mainly based on lysosomal hydrolases and phospholipases. In addition, pore-forming proteins released after cell-cell contact that make pores into the host cell membrane have been described for *Entamoeba*, *Acanthamoeba*, and also for *Naegleria* (Young et al. 1982; Herbst et al. 2002; Michalek et al. 2013). *Acanthamoeba* can adhere particularly firmly to other cells and artificial surfaces by the use of their acanthopodia. Whereas *Acanthamoeba* spp. and *B. mandrillaris* cause usually granulomatous lesions, invasive entamoebae and *N. fowleri* cause typically necroses.

In general, *Entamoeba* infections gradually decline with serial infections due to the acquired immunity. In contrast, free-living amoeba infections initiate an innate immune response essentially involving macrophages, neutrophils, and the complement system which does not allow the establishment of an immunological memory (Cursons et al. 1980). In this case, the uncontrolled activation of complement seems to be partially responsible for oedema and damage to blood vessel walls that are characteristic for *Acanthamoeba* and *B. mandrillaris* infections.

15.3 Epidemiology

Most amoebae have a cosmopolitan distribution. The first amoebic pathogen described was *Entamoeba histolytica*, identified as the causative agent of chronic diarrhoea in humans by the Russian physician Fedor Aleksandrovich Lösch in 1875 (Loesch 1875). Amoeboesis in reptiles due to *Entamoeba invadens* was first recorded by Ratcliffe and Geiman in 1934. *Acanthamoeba* has been discovered as a contaminant in a fungal cell culture by the Italian tropical disease specialists Aldo

Castellani in 1930 and was at that time thought to belong to the genus *Hartmannella* (*Castellani* 1930; *Volkonsky* 1931). Their medical relevance has been established when Culbertson discovered in 1958 their ability to cause meningoencephalitis in mice and monkeys. Only a few years later, Fowler and Carter identified an amoeba—later named *Naegleria fowleri*—as the causative agent of several cases of fatal meningoencephalitis in children in southern Australia (Fowler and Carter 1965). Amoebae—later identified as *Neoparamoeba perurans*—as causative agents of AGD were first isolated in 1984 from farmed salmon off the coast of Tasmania (Munday 1986). *Balamuthia mandrillaris* was first isolated from the brain of a pregnant mandrill baboon that had died from meningoencephalitis in 1986 in the San Diego Wildlife Park (Visvesvara et al. 1990, 1993).

Although entamoebae are generally distributed worldwide, the transmission cycles of the respective species are enhanced in settings that are densely populated and/or that favour faecal contamination of water and food. In addition, the cysts remain viable for longer periods of time in moist and non-freezing environments. Entamoebae include parasites of all classes of vertebrates and some invertebrates, and most species show a relatively high host specificity. However, depending on the infective dose, the virulence of the respective strain, and the immune status of the respective host animal, a much broader group of animals is usually susceptible to infection. While the mammalian entamoebae have a preferred temperature of around 37 °C, *E. invadens*—the causal agent of amoebiasis in reptiles—thrives at low temperatures.

N. perurans is a marine cosmopolitan free-living amoeba. It causes the syndrome of AGD, which has first been reported from Tasmania, but meanwhile affects also fish farms in Europe and the Americas. Further outbreaks have been reported in Ireland, Scotland, Norway, France, Spain, Chile, and New Zealand (Oldham et al. 2016).

The ubiquitous *Acanthamoeba* spp. are among the most versatile protozoan organisms being able to feed on almost anything, to thrive under aerobic and anaerobic conditions, and to survive under extreme conditions with regard to pH, salinity and temperature. They have been isolated from fresh- and brackish waters, bottled mineral waters, dialysis machines, cooling towers of electric and nuclear power plants, heating, ventilating and air conditioning units, soil, dust, contact lens paraphernalia, ear discharge, pulmonary secretions, and stool samples (Marciano-Cabral and Cabral 2003; Visvesvara et al. 2007). Besides humans, other mammalian vertebrates seem to be particularly susceptible for cerebral infections, which are usually associated with immunodeficiency, whereas from other animals, infections of various organs have been reported.

B. mandrillaris has always been presumed to be free-living but could not be recovered for a long time from environmental samples. Principally this amoeba is a soil amoeba. Schuster et al. (2003) reported the first environmental isolation, namely, from soil of a potted plant in the home of a 3-year-old child, who had died from amoebic encephalitis in northern California. Dunnebacke et al. (2004) reported the second environmental isolation from a soil sample from an outdoor potted plant, also in California.

N. fowleri is thermophilic—37–45 °C—and is found often in very high densities in thermal waters, e.g. in poorly maintained swimming pools, hot springs, hydrotherapy pools, lakes and ponds during the warm season, and in thermally polluted waters and soil. It is resistant against low levels of under 1 mg/mL chlorine and has thus an advantage over most other protozoa in many man-made habitats. The highly motile flagellate form facilitates its dispersal. In contrast to *Acanthamoeba* spp., *Naegleria* spp. have never been isolated from seawater samples as it is more sensitive to high levels of osmolarity.

15.3.1 Farm Animals

There is limited data on amoebic infections in farm animals. *Entamoeba polecki* and *E. suis* infections are rather common in pigs—particularly in young piglets—but have long been assumed to be without clinical relevance. In recent years, this view has changed a bit, and they are now considered partly responsible for diarrhoea and haemorrhagic colitis in pigs (Matsubayashi et al. 2014). In the case of *E. polecki*, pigs play a major role as reservoir hosts for zoonotic infections in humans, and this is also assumed for *E. histolytica* (Schuster and Visvesvara 2004a, Solaymani-Mohammadi and Petri 2006). *E. bovis*, which is considered to have no clinical relevance, and which might be synonymous with *E. ovis*, can be isolated from cattle and various other ruminants (Stensvold et al. 2010). GAE, either caused by *Acanthamoeba* spp. or by *B. mandrillaris*, is known to occur naturally in horses, sheep, and bovines (Dwivedi and Singh 1965; McConnell et al. 1986; Fuentealba et al. 1992; Kinde et al. 1998). PAME caused by *N. fowleri* has been described in cows (Daft et al. 2005; Visvesvara et al. 2005).

15.3.2 Pets

Dogs can be infected by *E. histolytica*, but they have long been assumed not to function as reservoir hosts since the parasite rarely encysts (Thompson and Smith 2011). However, in a large screening in Pakistan, a high prevalence of the morphologically indistinguishable *E. histolytica/dispar/moshkovskii* group was identified in faecal samples of different dog populations—of which the highest prevalence was found in symptomatic household dogs—and authors concluded that dogs may play a significant role in the epidemiology of *E. dispar* and *E. moshkovskii* since a low prevalence of *E. histolytica* could be determined (Alam et al. 2015). In addition, several cases of cerebral *Acanthamoeba* infections and cases of GAE caused by *B. mandrillaris* have been described in dogs (Ayers et al. 1972; Pearce et al. 1985; Bauer et al. 1993; Foreman et al. 2004; Dubey et al. 2005; Finnin et al. 2007; Hodge et al. 2011).

Moreover, there has been a singular report of a dog with a gastric infection with *Willaertia magna*—a heterolobosean amoeba (Steele et al. 1997).

15.3.3 Birds

Various *Entamoeba* and related species, including *E. anatis*, *E. gallinarum*, and *Endolimax* spp., are known to occur in different groups of birds, albeit with rather low clinical relevance (McDougald 1997). *E. anatis* has been identified as causative agent of pharyngeal disease in captive bustards (Silvanose et al. 1998). An *Acanthamoeba* infection of the liver was diagnosed post-mortem in a keel-billed toucan (Visvesvara et al. 2007).

15.3.4 Reptiles

The most serious amoebal pathogen of reptiles is *E. invadens*, causing ulcerative enteritis and hepatitis associated with high morbidity and mortality, particularly in captive snakes and lizards (Donaldson et al. 1975; Jacobson et al. 1983; Ozaki et al. 2000). While *E. invadens* is considered a commensal to herbivorous reptile species in many carnivorous reptiles, it invades the mucosa causing a severe infection. The optimal growth temperature for *E. invadens* ranges between 20 and 30 °C, and an infection cannot be established at higher temperatures of 34–37 °C (Meerovitch 1961).

Bosch and Deichsel (1972) investigated gut and tissue samples of 71 reptiles for amoeba infections and isolated 114 strains pertaining to the genera *Entamoeba* and *Acanthamoeba*. Singular cases of infections with various amoeba species in various tissues of reptiles have been described (Frank and Bosch 1972, Madrigal Sesma et al. 1988; Schuster et al. 2003; Walochnik et al. 1999).

15.3.5 Fishes

In fish, AGD caused by *Neoparamoeba perurans*—and possibly other *Neoparamoeba* and *Paramoeba* species—has become a significant problem in marine fish farming of salmonids (Roubal et al. 1989, Oldham et al. 2016). In 1977, Taylor described the first systemic infection of fish by *Acanthamoeba*. Besides *Acanthamoeba*, he also isolated *Naegleria* and *Vahlkampfia* species from various organs of different species of freshwater fishes (Taylor 1977). Other genera that may become pathogenic for fish include *Cochliopodium* and *Thecamoeba*, and representatives of numerous other genera have been isolated from the gills of asymptomatic fishes (Dyková and Lom 2004).

15.3.6 Other Animals

E. histolytica—although considered to infect specifically humans—is frequently also isolated from nonhuman primates in captivity; their role as a potential reservoirs for amoebae remains unknown, but there seem to be genetic differences between strains isolated from humans and nonhuman primates (Beaver et al. 1998, Loomis et al. 1983, Takano et al. 2007, Levecke et al. 2010). *E. nuttalli* infections are rather common in captive nonhuman primates but possibly occur and cause disease also in wild monkeys and apes (Levecke et al. 2010; Tachibana et al. 2016).

Malpighamoeba mellificae is known to infect adult bees in tropical but also in temperate regions. The amoebae are transmitted by the faecal-oral route and live in the Malpighian tubules, causing mostly chronic infections. They lyse tissue and gradually damage the Malpighian tubules. Infections may remain without any symptoms but may also be associated with spring dwindling of bee colonies (Evans and Schwarz 2011; Prell 1926).

Mostly fatal infections of the CNS and other organs of various captive primates, including gorillas, a mandrill, an orangutan, and a SIV-infected rhesus macaque that were either caused by *Acanthamoeba* spp. or by *B. mandrillaris*, have been reported (Anderson et al. 1986; Visvesvara et al. 1990; Canfield et al. 1997; Rideout et al. 1997; Westmoreland et al. 2004). Moreover, one case of GAE in a captive kangaroo, most probably caused by *Acanthamoeba*, was reported from Australia (Norton and Hartley 1993). Lorenzo-Morales et al. (2007a, b) screened intestinal swab samples of wild squirrels from the Canary Islands and Morocco and could isolate various strains of *Acanthamoeba* spp.

A spontaneous PAME case due to a *N. fowleri* infection has been reported from a South American tapir in a zoo in Phoenix, Arizona, USA (Lozano-Alarcón et al. 1997).

15.4 Clinical Effects, Diagnosis, Prevention and Treatment

The two most important amoebal infections in veterinary medicine are intestinal amoebiasis in reptiles and AGD occurring mainly in Atlantic salmon. Both diseases are linked to a compromised immune status in the respective animals, caused by captivity, crowding and other stress factors, and both diseases are characterized by high mortality. Diagnostics rely on direct parasite detection in the affected tissues, whereby it is difficult for unexperienced microscopists to distinguish amoebal trophozoites from other cells—particularly macrophages—by morphological identification, which is why molecular biological methods are gaining more and more importance. Treatment of amoebal infections in animals is problematic as no highly effective and specific drugs are available.

15.4.1 Intestinal Amoebiasis

With the exception of *E. gingivalis* and *E. moshkovskii*, all entamoebae are parasites of the large intestine, where they primarily feed on bacteria, but occasionally become invasive causing commonly a necrotic colitis. Clinical signs vary between no abnormalities to anorexia, dehydration, bloody faeces, and general debility of the infected animals (Schuster and Visvesvara 2004b).

While in most animals *Entamoeba* infections have an asymptomatic and self-limiting progression, in reptiles, amoebiasis is one of the most serious diseases (Donaldson et al. 1975; Jacobson et al. 1983). Principally, meat-eating reptiles are more prone to amoebiasis than plant-eating reptiles, whereby carnivorous snakes, including boas, colubrids, elapids, vipers and crotalids, and also giant tortoises, are particularly susceptible. However, many reptiles—e.g. garter snakes, northern black racers, and box turtles—function as asymptomatic carrier hosts and represent reservoirs of infection. Transmission happens via the faecal-oral route; cysts are ingested with contaminated water or—less common—with food. The incubation period is usually around 2 weeks. First signs may be regurgitation of undigested food and a significantly increased water intake, indicating dehydration associated with hepatic and renal insufficiency. Further typical symptoms are anorexia, weight loss, and severe mucoidal or haemorrhagic diarrhoea, which finally may lead to death. Gross lesions may extend from the stomach to the cloaca; the cloacal region is typically swollen and firm. In chronic cases, the amoebae may cause multifocal hepatic abscesses containing numerous *E. invadens* trophozoites. Trophozoites or cysts are detected in wet preparations of fresh faeces, tissue impressions or tissue sections. Immediate examination of fresh, bloody faeces may reveal moving trophozoites. In recent years, antigen detection and several PCR protocols have been established as highly specific and very sensitive assays that are available as commercial diagnostic kits. Serological methods are of importance for extraintestinal amoebiasis. The intestine shows areas of ulceration that tend to coalesce and are characterized by caseous necrosis, oedema, and haemorrhage. Frequently, *E. invadens* amoebiasis is epidemic in large snake collections.

Reptile amoebiasis is treated with metronidazole, the drug of choice for the treatment of almost all infections with enteric protozoa (Denver et al. 1999, Innis et al. 2007). Strict sanitation and hygiene measures can significantly reduce the risk of infection. The joint housing of turtles and snakes should be avoided.

Depending on the animal species, several other non-pathogenic *Entamoeba* species and enteric amoebae can be detected in stool specimens (Table 15.1). Usually, trophozoites are found in diarrhoeal faecal specimens, while cysts are rather observed in formed stool.

15.4.2 Amoebic Gill Disease (AGD)

Atlantic salmon appears to be the most susceptible species to AGD. Outbreaks in Pacific salmon have also been observed but are generally minor and sporadic. However, AGD can also affect several other fish species cultured in the marine environment, including other salmonids, turbot, European sea bass, and sharpsnout sea bream (Mitchell and Rodger 2011). Several species of the genera *Neoparamoeba* and *Paramoeba* have been described as the causative agents of AGD, though some of these have been possibly misidentified (Dyková and Lom 2004; Oldham et al. 2016). The most important species is *N. perurans*, a marine amoebozoan first described in 2007 by Young and colleagues. Predisposing factors for AGD are a water temperature above 16 °C, high salinity—approaching 35‰—crowding, and poor water circulation inside the pens. The fishes typically develop symptoms around 2 months after transfer to the pens. Clinical signs are lethargy, anorexia, dyspnoea, rapid opercular movements, and increased ventilation. The infected fish tend to congregate at the water surface and often show white mucoid patches and gross lesions on their gills. Laboratory diagnosis relies on microscopic investigation of histological sections or PCR. Epithelial hyperplasia, lamellar fusion, the presence of interlamellar vesicles, and amoebae-containing parasomes, are observed in tissue sections (Oldham et al. 2016). Interestingly, fish that had clinical AGD seem to be relatively resistant to reinfection (Mitchell and Rodger 2011). The recommended treatment for AGD is repetitive freshwater baths (Findlay et al. 1995). Furthermore, early and efficient mortality removal is crucial to reduce the infection risk for cohabiting fishes. In addition, levamisole, chloramine, and chlorine dioxide have been used with some success to lower mortality within the pens.

15.4.3 Granulomatous Amoebic Encephalitis (GAE)

GAE can be caused by *Acanthamoeba* spp., *B. mandrillaris*, and *Sappinia diploidea* and is a chronic or subacute progressing infection of the brain that typically occurs in immunocompromised individuals. Nevertheless, some cases of *B. mandrillaris* GAE have been reported in apparently healthy humans. The mode of infection is usually via skin lesion contact with contaminated water or by inhalation and subsequent haematogenous spread, whereby the amoebae are assumed to disseminate rather via a perivascular route than with the blood. The incubation period can vary from several weeks to months. In many cases, the disease starts to manifest at the primary foci—either as dermatitis or as pneumonia. Within the brain, the amoebae spread centrifugally and mainly affect the cerebral lobes, the cerebellum, and the brain stem. Typical symptoms of GAE in animals are lethargy, vomiting, seizures, circling, and head tilt to either side, incapability of any movement or dragging an extremity, coma and death follows. In many cases, *post-mortem* also other organs are found to be affected.

In tissue sections, the disease presents as general inflammation with sparse granules and amoebic infiltrates. Importantly, trophozoites and cysts are detected in biopsy material, including the cerebrospinal fluid (CSF) in the case of a cerebral infection. The direct detection of the causative agent in the clinical sample is the only reliable diagnostic method of GAE. The trophozoites stain well in several staining methods, and cysts exhibit autofluorescence. Culturing on non-nutrient agar plates remains the gold standard of diagnosis, but also PCR-based techniques and immunofluorescence stainings have been established (Schuster and Visvesvara 2004b). For morphological investigation, hanging drop preparations and differential interference contrast (DIC) are generally recommended for all amoebae. However, reliable identification below the genus level requires molecular methods (Walochnik et al. 2015). Serological techniques are of limited diagnostic value—particularly for *Acanthamoeba* infections—as due to the ubiquity of this group of amoebae seropositivity is high in the normal population.

Currently, there is no established specific treatment for *Acanthamoeba* and/or *Balamuthia* infections. Importantly, cysts play a significant role for the progression of disease, as they may survive treatment within the tissue and lead to reinfection. Miltefosine is available as an investigational drug from the Centres of Disease Control (CDC) in the USA (CDC 2013).

15.4.4 Primary Amoebic Encephalitis (PAME)

Although several species of the genus *Naegleria* exhibit cytopathogenicity, only *N. fowleri* has been found to cause PAME in natural infections. The main risk factor for PAME is the introduction of contaminated water into the nostrils during activities such as swimming, diving, bathing, or nasal irrigation. PAME is an acute infection of the brain; the incubation period lasts a few days, and—if not diagnosed and treated early—it is fatal within about a week. *N. fowleri* invades the brain by penetrating the nasal mucosa, migrating along the olfactory nerve tracts and crossing the cribriform plate. The amoebae produce necroses, whereby mainly the frontal and olfactory lobes are affected, but also deeper layers of the brain may be involved. Clinically, the disease presents as diffuse meningitis and mainly peripheral encephalitis, with cerebral haemorrhages. Typical symptoms of PAME in animals are anorexia, facial paralysis, circling, ataxia, convulsions, head tilt to either side and the disability to stand without assistance. Nasal swabs or CSF reveals trophozoites, which can be demonstrated by phase contrast or in several stains. The amoebae can easily be cultured on bacteria-coated non-nutrient agar but also in different liquid media. The optimal growth temperature lies at around 37 °C. *Naegleria* can be recognized by its biflagellate stage; enflagellation is induced by mounting the plate culture with sterile saline and takes around 30–60 min. Species determination is mainly achieved by molecular biological methods. Serology provides no help in diagnosis as there is insufficient time for a humoral antibody reaction to occur, but several PCRs and a very potent immunofluorescence staining are available (Schuster and Visvesvara 2004b).

Amphotericin B is considered the drug of choice for treatment of PAME in humans and animals. Recently, also miltefosine has been used with some success (Heggie and Küpper 2017; Linam et al. 2015).

References

Aitken D, Hay J, Kinnear F, et al. Amebic keratitis in a wearer of disposable contact lenses due to a mixed *Vahlkampfia* and *Hartmannella* infection. Ophthalmology. 1996;103:485–94.

Alam MA, Maqbool A, Nazir MM, Lateef M, Khan MS, Lindsay DS. *Entamoeba* infections in different populations of dogs in an endemic area of Lahore, Pakistan. Vet Parasitol. 2015;207(3–4):216–9.

Al-Habsi K, Yang R, Ryan U, Jacobson C, Miller DW. Morphological and molecular characterization of an uninucleated cyst-producing *Entamoeba* spp. in captured Rangeland goats in Western Australia. Vet Parasitol. 2017;235:41–6.

Ali IK. Intestinal amebae. Clin Lab Med. 2015;35(2):393–422.
Alizadeh H, He YG, McCulley JP, Ma D, Stewart GL, Via M, Haehling E, Niederkorn JY. Successful immunization against *Acanthamoeba* in a pig model. Cornea. 1995;14:180–6.
Amyx HL, Asher DM, Nash TE, Gibbs CJ Jr, Gajdusek DC. Hepatic amebiasis in spider monkeys. Am J Trop Med Hyg. 1978;27(5):888–91.
Anderson MP, Oosterhuis JE, Kennedy S, Benirschke K. Pneumonia and meningoencephalitis due to amoeba in a low-land gorilla. J Zoo Anim Med. 1986;17:87–91.
Ayers KH, Billups LH, Garner FM. Acanthamoebiasis in a dog. Vet Pathol. 1972;19:221–6.
Barker J, Brown M. Trojan Horses of the microbial world: protozoa and the survival of bacterial pathogens in the environment. Microbiology. 1994;140:1253–9.
Bauer RW, Harrison LR, Watson CW, Styer EL, Chapman WL Jr. Isolation of *Acanthamoeba* sp. from a greyhound with pneumonia and granulomatous amebic encephalitis. J Vet Diagn Investig. 1993;5:386–91.
Beaver PC, Blanchard JL, Seibold HR. Invasive amebiasis in naturally infected New World and Old World monkeys with and without clinical disease. Am J Trop Med Hyg. 1988;39(4):343–52.
Beckwith-Cohen B, Gasper DJ, Bentley E, Gittelman H, Ellis AE, Snowden KF, Shock BC, Yabsley MJ, Dubielzig RR. Protozoal infections of the cornea and conjunctiva in dogs associated with chronic ocular surface disease and topical immunosuppression. Vet Ophthalmol. 2016;19(3):206–13.
Bergquist R. Parasitic infections affecting the oral cavity. Periodontology. 2009;49(1):96–105.
Booton GC, Visvesvara GS, Byers TJ, Kelly DJ, Fuerst PA. Identification and distribution of *Acanthamoeba* species genotypes associated with nonkeratitis infections. J Clin Microbiol. 2005;43(4):1689–93.
Bosch I, Deichsel G. Morphological study of pathogenic and potentially pathogenic amoebae of the types "*Entamoeba*" and "*Hartmannella-Acanthamoeba*" from reptiles. Z Parasitenkd. 1972;40(2):107–29.
Canfield PJ, Vogelnest L, Cunningham MI, Visvesvara GS. Amoebic meningoencephalitis caused by *Balamuthia mandrillaris* in an orang utan. Aust Vet J. 1997;75(2):97–100.
Castellani A. Amoeba found in culture of yeast. J Trop Med Hyg. 1930;33:160.
Centers for Disease Control and Prevention (CDC). Investigational drug available directly from CDC for the treatment of infections with free-living amebae. MMWR Morb Mortal Wkly Rep. 2013;62(33):666.
Centeno M, Rivera F, Cerva L, Tsutsumi V, Gallegos E, Calderón A, Ortiz R, Bonilla P, Ramírez E, Suárez G. *Hartmannella vermiformis* isolated from the cerebrospinal fluid of a young male patient with meningoencephalitis and bronchopneumonia. Arch Med Res. 1996;27(4):579–86.
Chávez-Munguía B, Omaña-Molina M, González-Lázaro M, González-Robles A, Bonilla P, Martínez-Palomo A. Ultrastructural study of encystation and excystation in *Acanthamoeba castellanii*. J Eukaryot Microbiol. 2005;52(2):153–8.
Clark CG. The evolution of *Entamoeba*, a cautionary tale. Res Microbiol. 2000;151(8):599–603.
Clark CG, Stensvold CR. The continuously expanding universe of entamoeba. In: Nozaki T, Bhattacharya A, editors. Amebiasis: biology and pathogenesis of *Entamoeba*. Tokyo: Springer; 2015. p. 9–25.
Culbertson CG, Smith JW, Minner JR. *Acanthamoeba*: observations on animal pathogenicity. Science. 1958;127:1506.
Cursons RT, Brown TJ, Keys EA, Moriarty KM, Till D. Immunity to pathogenic free-living amoebae: role of humoral antibody. Infect Immun. 1980;29:401–7.
Daft BM, Visvesvara GS, Read DH, Kinde H, Uzal FA, Manzer MD. Seasonal meningoencephalitis in Holstein cattle caused by *Naegleria fowleri*. J Vet Diagn Investig. 2005;17(6):605–9.
De Jonckheere JF. What do we know by now about the genus *Naegleria*? Exp Parasitol. 2014;145(Suppl):S2–9.
Denver MC, Cranfield MR, Graczyk TK, Blank P, Wisnieski A, Poole V. A review of reptilian amebiasis and current research on the diagnosis and treatment of amebiasis at the Baltimore Zoo. Proc Am Assoc Zoo Vet. 1999;1:1–15.

Donaldson M, Heyneman D, Dempster R, Garcia L. Epizootic of fatal amebiasis among exhibited snakes: epidemiologic, pathologic, and chemotherapeutic considerations. Am J Vet Res. 1975;36:807–17.
Dubey JP, Benson JE, Blakeley KT, Booton GC, Visvesvara GS. Disseminated *Acanthamoeba* sp. infection in a dog. Vet Parasitol. 2005;128(3–4):183–7.
Dunnebacke TH, Schuster FL, Yagi S, Booton GC. *Balamuthia mandrillaris* from soil samples. Microbiology. 2004;150(Pt 9):2837–42.
Dwivedi JN, Singh CM. Pulmonary lesions in an Indian buffalo associated with *Acanthamoeba* sp. Indian J Microbiol. 1965;5:31–4.
Dyková I, Lom J, Schroeder-Diedrich JM, Booton GC, Byers TJ. *Acanthamoeba* strains isolated from organs of freshwater fishes. J Parasitol. 1999;85(6):1106–13.
Dyková I, Lom J. Advances in the knowledge of amphizoic amoebae infecting fish. Folia Parasitol. 2004;51(2–3):81–97.
Evans JD, Schwarz RS. Bees brought to their knees: microbes affecting honey bee health. Trends Microbiol. 2011;19(12):614–20.
Finnin PJ, Visvesvara GS, Campbell BE, Fry DR, Gasser RB. Multifocal *Balamuthia mandrillaris* infection in a dog in Australia. Parasitol Res. 2007;100(2):423–6.
Findlay V, Helders M, Munday BL, Gurney R. Demonstration of resistance to re-infection with *Paramoeba* sp. by Atlantic Salmon, *Salmo salar* L. J Fish Dis. 1995;18:639–42.
Foreman O, Sykes J, Ball L, Yang N, De Cock H. Disseminated infection with *Balamuthia mandrillaris* in a dog. Vet Pathol. 2004;41(5):506–10.
Fowler M, Carter RF. Acute pyogenic meningitis probably due to *Acanthamoeba* sp.: a preliminary report. Br Med J. 1965;5464:740–2.
Frank W, Bosch I. Isolation of amoebae of the type "*Hartmannella-Acanthamoeba*" and "*Naegleria*" from cold-blooded vertebrates. Zeitschrift fur Parasitenkunde. 1972;40(2): 139–50.
Fuentealba IC, Wikse SE, Reed WK, Edwards JF, Visvesvara GS. Amebic meningoencephalitis in a sheep. J Am Vet Med Assoc. 1992;200:363–5.
Gast RJ, Ledee DR, Fuerst PA, Byers T. Subgenus systematics of *Acanthamoeba*: four nuclear 18S rDNA sequence types. J Eukaryot Microbiol. 1996;43:498–504.
Gelman BB, Rauf SJ, Nader R, Popov V, Borkowski J, Chaljub G, Nauta HW, Visvesvara GS. Amoebic encephalitis due to *Sappinia diploidea*. JAMA. 2001;285(19):2450–1.
Griffin JL. Temperature tolerance of pathogenic and nonpathogenic free-living amoebae. Science. 1972;178:869–70.
He YG, McCulley JP, Alizadeh H, Pidherney M, Mellon J, Ubelaker JE, Stewart GL, Silvany RE, Niederkorn JY. A pig model of *Acanthamoeba* keratitis: transmission via contaminated contact lenses. Invest Ophthalmol Vis Sci. 1992;33(1):126–33.
Heggie TW, Küpper T. Surviving *Naegleria fowleri* infections: a successful case report and novel therapeutic approach. Travel Med Infect Dis. 2017;16:49–51.
Herbst R, Ott C, Jacobs T, Marti T, Marciano-Cabral F, Leippe M. Pore-forming polypeptides of the pathogenic protozoon *Naegleria fowleri*. J Biol Chem. 2002;277(25):22353–60.
Heredia RD, Fonseca JA, López MC. *Entamoeba moshkovskii* perspectives of a new agent to be considered in the diagnosis of amebiasis. Acta Trop. 2012;123(3):139–45.
Hodge PJ, Kelers K, Gasser RB, Visvesvara GS, Martig S, Long SN. Another case of canine amoebic meningoencephalitis--the challenges of reaching a rapid diagnosis. Parasitol Res. 2011; 108(4):1069–73.
Hooshyar H, Rostamkhani P, Rezaeian M. An annotated checklist of the human and animal *Entamoeba* (Amoebida: Endamoebidae) species—a review article. Iran J Parasitol. 2015;10(2):146–56.
Innis C, Papich M, Young D. Pharmacokinetics of metronidazole in the red-eared slider turtle (*Trachemys scripta elegans*) after single intracoelomic injection. J Vet Pharmacol Ther. 2007;30(2):168–71.
Jacobson E, Clubb S, Greiner E. Amebiasis in red-footed tortoises. J Am Vet Med Assoc. 1983; 183:1192–4.

Jones DB, Visvesvara GS, Robinson NM. *Acanthamoeba polyphaga* keratitis and *Acanthamoeba* uveitis associated with fatal meningoencephalitis. Trans Ophthalmol Soc U K. 1975;95: 221–32.

Khan NA. *Acanthamoeba*: biology and increasing importance in human health. FEMS Microbiol Rev. 2006;30(4):564–95.

Kent M, Platt SR, Rech RR, Eagleson JS, Howerth EW, Shoff M, Fuerst PA, Booton G, Visvesvara GS, Schatzberg SJ. Multisystemic infection with an *Acanthamoeba* sp. in a dog. J Am Vet Med Assoc. 2011;238(11):1476–81.

Kinde H, Visvesvara GS, Barr BC, Nordhausen RW, Chiu PH. Amebic meningoencephalitis caused by *Balamuthia mandrillaris* (leptomyxid ameba) in a horse. J Vet Diagn Investig. 1998;10:378–81.

Kojimoto A, Uchida K, Horii Y, Okumura S, Yamaguch R, Tateyama S. Amebiasis in four ball pythons, *Python reginus*. J Vet Med Sci. 2001;63(12):1365–8.

Levecke B, Dreesen L, Dorny P, Verweij JJ, Vercammen F, Casaert S, Vercruysse J, Geldhof P. Molecular identification of *Entamoeba* spp. in captive nonhuman primates. J Clin Microbiol. 2010;48(8):2988–90.

Levecke B, Dorny P, Vercammen F, Visser LG, Van Esbroeck M, Vercruysse J, Verweij JJ. Transmission of *Entamoeba nuttalli* and *Trichuris trichiura* from nonhuman primates to humans. Emerg Infect Dis. 2015;21(10):1871–2.

Lima PC, Taylor RS, Cook M. Pseudocyst formation in the marine parasitic amoeba *Neoparamoeba perurans*: a short-term survival strategy to abrupt salinity variation. J Fish Dis. 2016;40(8):1109–13. 21 December (Epub).

Linam WM, Ahmed M, Cope JR, Chu C, Visvesvara GS, da Silva AJ, Qvarnstrom Y, Green J. Successful treatment of an adolescent with *Naegleria fowleri* primary amebic meningoencephalitis. Pediatrics. 2015;135(3):e744–8.

Loesch FA. Massive development of amebas in the large intestine. Am J Trop Med Hyg. 1875;24(3):383–92.

Loomis MR, Britt JO Jr, Gendron AP, Holshuh HJ, Howard EB. Hepatic and gastric amebiasis in black and white colobus monkeys. J Am Vet Med Assoc. 1983;183(11):1188–91.

Lorenzo-Morales J, López-Darias M, Martínez-Carretero E, Valladares B. Isolation of potentially pathogenic strains of *Acanthamoeba* in wild squirrels from the Canary Islands and Morocco. Exp Parasitol. 2007a;117(1):74–9.

Lorenzo-Morales J, Martínez-Carretero E, Batista N, Alvarez-Marín J, Bahaya Y, Walochnik J, Valladares B. Early diagnosis of amoebic keratitis due to a mixed infection with *Acanthamoeba* and *Hartmannella*. Parasitol Res. 2007b;102(1):167–9.

Lozano- Alarcón F, Bradley GA, Houser BS, Visvesvara GS. Primary amebic meningoencephalitis due to *Naegleria fowleri* in a South American tapir. Vet Pathol. 1997;34(3):239–43.

Madrigal Sesma MJ, Chamorro Garcia L, Guillen LP. Free-living amoebas and cold- blooded animals. J Parasitol. 1988;74:883–4.

Marciano-Cabral F, Cabral G. *Acanthamoeba* spp. as agents of disease in humans. Clin Microbiol Rev. 2003;16(2):273–307.

Matsubayashi M, Suzuta F, Terayama Y, Shimojo K, Yui T, Haritani M, Shibahara T. Ultrastructural characteristics and molecular identification of *Entamoeba suis* isolated from pigs with hemorrhagic colitis: implications for pathogenicity. Parasitol Res. 2014;113(8):3023–8.

Mazur T, Hadaś E, Iwanicka I. The duration of the cyst stage and the viability and virulence of *Acanthamoeba* isolates. Trop Med Parasitol. 1995;46(2):106–8.

McConnell EE, Garner FM, Kirk JH. Hartmannellosis in a bull. Pathol Vet. 1986;5:1–6.

McDougald LR. Protozoa: others protozoan diseases of the intestinal tract. In: Calnek BW, Barnes HJ, editors. Diseases of poultry. 10th ed. Iowa: Iowa State University Press; 1997. p. 890–9.

Meerovitch E. Infectivity and pathogenicity of polyxenic and monoxenic *Entamoeba invadens* to snakes kept at normal and high temperatures and the natural history of reptile amoebiasis. J Parasitol. 1961;47:791–4.

Mitchell SO, Rodger HD. A review of infectious gill disease in marine salmonid fish. J Fish Dis. 2011;34(6):411–32.

Michalek M, Sönnichsen FD, Wechselberger R, Dingley AJ, Hung CW, Kopp A, Wienk H, Simanski M, Herbst R, Lorenzen I, Marciano-Cabral F, Gelhaus C, Gutsmann T, Tholey A, Grötzinger J, Leippe M. 2013. Structure and function of a unique pore-forming protein from a pathogenic acanthamoeba. Nat Chem Biol. 2013;9(1):37–42.

Morales JA, Chaves AJ, Visvesvara GS, Dubey JP. *Naegleria fowleri*-associated encephalitis in a cow from Costa Rica. Vet Parasitol. 2006;139(1–3):221–3.

Mortazavi PN, Goldsworthy G, Kirk R, Khan NA. *Acanthamoeba* produces disseminated infection in locusts and traverses the locust blood-brain barrier to invade the central nervous system. BMC Microbiol. 2010;10:186.

Munday BL. Diseases of salmonids. In: Humphrey JD, Langdon JS, editors. Proceedings on the workshop of diseases of Australian fish and shellfish. Benalla, VIC: Department of Agriculture and Rural Affairs; 1986. p. 127–41.

Neff RJ, Neff RH. The biochemistry of amoebic encystment. Symp Soc Exp Biol. 1969;23:51–81.

Norton JH, Hartley WJ. Amoebic meningoencephalitis in a young grey kangaroo (*Macropus giganteus*). Aust Vet J. 1993;70(11):421.

Oldham T, Rodger H, Nowak BF. Incidence and distribution of amoebic gill disease (AGD)—an epidemiological review. Aquaculture. 2016;457:35–42.

Omaña-Molina M, Navarro-García F, González-Robles A, Serrano-Luna Jde J, Campos-Rodríguez R, Martínez-Palomo A, Tsutsumi V, Shibayama M. Induction of morphological and electrophysiological changes in hamster cornea after in vitro interaction with trophozoites of *Acanthamoeba* spp. Infect Immun. 2004;72(6):3245–51.

Ozaki K, Matsuo K, Tanaka O, Narama I. Amoebosis in the flat-shelled spider tortoise (*Acinixys planicauda*). J Comp Pathol. 2000;123(4):299–301.

Page FC. The classification of the naked amoebae (Phylum Rhizopoda). Arch Protistenkd. 1987;133:199–217.

Page FC. Nackte Rhizopoda. In: Matthes D, editor. Protozoen fauna, Band 2. Stuttgart: G. Fischer; 1991. p. 3–145.

Pakandl M. The prevalence of intestinal protozoa in wild and domestic pigs. Vet Med (Praha). 1994;39(7):377–80.

Pearce JR, Powell HS, Chandler FW, Visvesvara GS. Amebic meningoencephalitis caused by *Acanthamoeba castellanii* in a dog. J Am Vet Med Assoc. 1985;187:951–2.

Poulsen CS, Stensvold CR. Systematic review on *Endolimax nana*: a less well studied intestinal ameba. Trop Parasitol. 2016;6(1):8–29.

Poynton SL, Whitaker BR. Protozoa and metazoa infecting amphibians. In: Wright KM, Whitaker BR, editors. Amphibian medicine and captive husbandry. Malabar, FL: Krieger; 2001. p. 193–222.

Prell H. The amoeba-disease of adult bees: a little-noticed spring-time disease. Bee World. 1926;8:10–3.

Pussard M, Pons R. Morphologie de la paroi kystique et taxonomie du genre *Acanthamoeba* (Protozoa, Amoebida). Protistologica. 1977;8:557–98.

Ratcliffe HL, Geiman QM. Amoebiasis in reptiles. Science. 1934;79:324–5.

Richter B, Kübber-Heiss A, Weissenböck H. Diphtheroid colitis in a *Boa constrictor* infected with amphibian *Entamoeba* sp. Vet Parasitol. 2008;153(1–2):164–7.

Rideout BA, Gardiner CH, Stalis IH, Zuba JR, Hadfield T, Visvesvara GS. Fatal infections with *Balamuthia mandrillaris* (a free-living amoeba) in gorillas and other Old World primates. Vet Pathol. 1997;34(1):15–22.

Roubal FR, Lester RJG, Foster CK. Studies on cultured and gill-attached *Paramoeba* spp. (Gymnamoebae: Paramoebidae) and the cytopathology of paramoebic gill disease in Atlantic salmon, *Salmo salar* L. from Tasmania. J Fish Dis. 1989;12:481–93.

Schuster FL, De Jonckheere JF, Moura H, Sriram R, Garner MM, Visvesvara GS. Isolation of a thermotolerant *Paravahlkampfia* sp. from lizard intestine: biology and molecular identification. J Eukaryot Microbiol. 2003;50:373–8.

Schuster FL, Dunnebacke TH, Booton GC, Yagi S, Kohlmeier CK, Glaser C, Vugia D, Bakardjiev A, Azimi P, Maddux-Gonzalez M, Martinez AJ, Visvesvara GS. Environmental isolation of

Balamuthia mandrillaris associated with a case of amebic encephalitis. J Clin Microbiol. 2003;41:3175–80.

Schuster FL, Visvesvara GS. Free-living amoebae as opportunistic and non-opportunistic pathogens of humans and animals. Int J Parasitol. 2004a;34(9):1001–27.

Schuster FL, Visvesvara GS. Amebae and ciliated protozoa as causal agents of waterborne zoonotic disease. Vet Parasitol. 2004b;126(1-2):91–120.

Silvanose CD, Samour JH, Naldo JL, Bailey TA. Oro-pharyngeal protozoa in captive bustards: clinical and pathological considerations. Avian Pathol. 1998;27(5):526–30.

Steele KE, Visvesvera GS, Bradley GA, Lipscomb TP, Gardiner CH. Amebiasis in a dog with gastric ulcers and adenocarcinoma. J Vet Diagn Invest. 1997;9(1):91–3.

Solaymani-Mohammadi S, Petri WA Jr. Zoonotic implications of the swine-transmitted protozoal infections. Vet Parasitol. 2006;140(3–4):189–203.

Som I, Azam A, Bhattacharya A, Bhattacharya S. Inter- and intra-strain variation in the 5.8S ribosomal RNA and internal transcribed spacer sequences of *Entamoeba histolytica* and comparison with *Entamoeba dispar*, *Entamoeba moshkovskii* and *Entamoeba invadens*. Int J Parasitol. 2000;30(6):723–8.

Stensvold CR, Lebbad M, Clark CG. Genetic characterisation of uninucleated cyst-producing *Entamoeba* spp. from ruminants. Int J Parasitol. 2010;40(7):775–8.

Stothard DR, Schroeder-Dietrich JM, Awwad MH, Gast RJ, Ledee DR, Rodriguez-Zaragoza S, Dean CL, Fuerst PA, Byers T. The evolutionary history of the genus *Acanthamoeba* and the identification of eight new 18S rDNA gene sequence types. J Eukaryot Microbiol. 1998;45: 45–54.

Tachibana H, Yanagi T, Feng M, Bandara KB, Kobayashi S, Cheng X, Hirayama K, Rajapakse RP. Isolation and molecular characterization of *Entamoeba nuttalli* strains showing novel Isoenzyme patterns from wild toque macaques in Sri Lanka. J Eukaryot Microbiol. 2016;63(2): 171–80.

Takano J, Narita T, Tachibana H, Terao K, Fujimoto K. Comparison of *Entamoeba histolytica* DNA isolated from a cynomolgus monkey with human isolates. Parasitol Res. 2007;101(3):539–46.

Taylor PW. Isolation and experimental infection of fee-living amoebae in freshwater fishes. J Parasitol. 1977;63:232–7.

Thompson RC, Smith A. Zoonotic enteric protozoa. Vet Parasitol. 2011;182(1):70–8.

Tyzzer EE. Amoebae of the caeca of the common fowl and of the turkey—*Entamoeba gallinarum* sp. nov. and *Pygolimax gregariniformis* gen. et sp. nov. J Med Res. 1920;41(2):199–210.

Valladares M, Reyes-Batlle M, Mora-Peces I, Martín-Navarro CM, López-Arencibia A, Dorta-Gorrín A, Comyn-Afonso E, Martínez-Carretero E, Maciver SK, Piñero JE, Valladares B, Lorenzo-Morales J. A multisystemic *Acanthamoeba* infection in a dog in Tenerife, Canary Islands, Spain. Vet Parasitol. 2014;205(3–4):707–11.

Verweij JJ, Vermeer J, Brienen EA, Blotkamp C, Laeijendecker D, van Lieshout L, Polderman AM. *Entamoeba histolytica* infections in captive primates. Parasitol Res. 2003;90(2):100–3.

Visvesvara GS, Stehr-Green JK. Epidemiology of free-living ameba infections. J Protozool. 1990;37(4):25S–33S.

Visvesvara GS, Martinez AJ, Schuster FL, Leitch GJ, Wallace SV, Sawyer TK, Anderson M. Leptomyxid ameba, a new agent of amebic meningoencephalitis in humans and animals. J Clin Microbiol. 1990;28:2750–6.

Visvesvara GS, Schuster FL, Martinez AJ. *Balamuthia mandrillaris*, N. G., N. Sp., agent of amebic meningoencephalitis in humans and other animals. J Eukaryot Microbiol. 1993;40:504–14.

Visvesvara GS, De Jonckheere JF, Sriram R, Daft B. Isolation and molecular typing of *Naegleria fowleri* from the brain of a cow that died of primary amebic meningoencephalitis. J Clin Microbiol. 2005;43(8):4203–4.

Visvesvara GS, Booton GC, Kelley DJ, Fuerst P, Sriram R, Finkelstein A, Garner MM. In vitro culture, serologic and molecular analysis of *Acanthamoeba* isolated from the liver of a keel-billed toucan (*Ramphastos sulfuratus*). Vet Parasitol. 2007;143(1):74–8.

Visvesvara GS, Shoff ME, Sriram R, Booton GC, Crary M, Fuerst PA, Hanley CS, Garner MM. Isolation, morphologic, serologic and molecular identification of *Acanthamoeba* T4 genotype from the liver of a Temminck's tragopan (*Tragopan temminckii*). Vet Parasitol. 2010;170(3–4): 197–200.

Volkonsky M. *Hartmanella castellanii* Douglas, et classification des hartmannelles. Arch Zoo Exp Gen. 1931;72:317–39.

Walochnik J, Hassl A, Simon K, Benyr G, Aspöck H. Isolation and identification by partial sequencing of the 18S ribosomal gene of free-living amoebae from necrotic tissue of *Basiliscus plumifrons* (Sauria: Iguanidae). Parasitol Res. 1999;85(7):601–3.

Walochnik J, Scheikl U, Haller-Schober EM. Twenty years of *Acanthamoeba* diagnostics in Austria. J Eukaryot Microbiol. 2015;62(1):3–11.

Westmoreland SV, Rosen J, MacKey J, Romsey C, Xia DL, Visvesvera GS, Mansfield KG. Necrotizing meningoencephalitis and pneumonitis in a simian immunodeficiency virus-infected rhesus macaque due to *Acanthamoeba*. Vet Pathol. 2004;41(4):398–404.

Young JD, Young TM, Lu LP, Unkeless JC, Cohn ZA. Characterization of a membrane pore-forming protein from *Entamoeba histolytica*. J Exp Med. 1982;156(6):1677–90.

Young ND, Crosbie PB, Adams MB, Nowak BF, Morrison RN. *Neoparamoeba perurans* n. sp., an agent of amoebic gill disease of Atlantic salmon (*Salmo salar*). Int J Parasitol. 2007;37(13):1469–81.

Check for updates

Genomics and Genetic Manipulation of Protozoan Parasites Affecting Farm Animals

16

Carlos E. Suarez, Heba F. Alzan, and Brian M. Cooke

Abstract

In this chapter we present a brief but state-of-the-art account of the genomics and current gene manipulation methods that can be used to improve our understanding of the genetics and the biology of an arbitrary group of 17 protozoan parasites responsible for diseases that affect animals worldwide, including babesiosis, toxoplasmosis, theileriosis, cryptosporidiosis, eimeriosis, trypanosomiasis, and trichomoniasis. Complete genomes are available for all parasites discussed, except for *Besnoitia*, an apicomplexan parasite responsible for dermatitis and other disorders with high infection rates, but low mortality. Dramatic differences in genome sizes are evident among the group of parasites under study, consistent with the distinct dependency of parasitic lifestyle for each organism. In addition, linear regression analysis correlating the ratios of the number of genes per genome and genome size among all the selected protozoan parasites suggests a strong association between these two parameters, in alignment with the notion that smaller protozoan genomes are generally more compact than larger genomes. A brief description of the methods for genome manipulations, including transient and stable transfections and gene editing methods, is provided. These methods, required to understand gene

C.E. Suarez (✉)
Department of Veterinary Microbiology and Pathology, Washington State University, Pullman, WA, USA

Animal Disease Research Unit, ARS-USDA, Pullman, WA, USA
e-mail: ces@vetmed.wsu.edu

H.F. Alzan
Department of Veterinary Microbiology and Pathology, Washington State University, Pullman, WA, USA

Parasitology and Animal Diseases Department, National Research Center, Cairo, Egypt

B.M. Cooke
Department of Microbiology, Monash University, Clayton, VIC, Australia

M. Florin-Christensen, L. Schnittger (eds.), *Parasitic Protozoa of Farm Animals and Pets*, https://doi.org/10.1007/978-3-319-70132-5_16

function and for improving control measures, have been successfully developed so far in most parasites selected. Rapid progress of genomic and gene manipulation techniques will likely result in the constant emergence of novel integrated methods for the interrogation and modification of genomes, leading to our better understanding of parasite lifestyle and, ultimately, to the rational design of improved methods for the control of animal infectious diseases.

16.1 Introduction

Protozoan parasites remain as important infectious agents affecting animal and human health globally. Their often complex life cycles and relationship with their hosts present numerous challenges to our ability to develop improved methods of control. Thus, new, combined research strategies are urgently required to hasten the rate of discoveries. In this chapter, we focus on the application of genomics and genetic manipulation techniques as tools to improve our understanding of the biology of a selected group of typically neglected protozoan parasites of veterinary and medical importance (Table 16.1).

Ideally, and in addition to other measures, new and improved vaccines and novel drugs are needed to prevent or treat most of the burden of disease caused by this group of diverse protozoan pathogens. In particular, developing improved control using rational approaches requires an advanced level of understanding of the parasites' biology, their interactions with their hosts, and particularly for vaccine development and the mechanisms of protective immunity. Recent significant advances in our understanding of the biology of most protozoan parasites affecting farm animals began with the arrival of the "omics" era, including genomics, proteomics, transcriptomics, metabolomics, lipidomics, as well as other existing, or future, "omics."

The emergence of genomics, perhaps the initial "omics," and the provision of the first complete and annotated organism genomes permitted the identification of numerous species-specific genes, but it was quickly realized that genomic approaches alone were insufficient to understand gene function. The value and utility of genomic data however is greatly increased when complemented with additional approaches such as transcriptomics, proteomics, lipidomics, and metabolomics. Thus, the "omics" field is extremely dynamic, and it can be expected that progress will be accelerated as novel computer-aided strategies of data management and analysis are able to integrate the massive incoming data arriving from all of these diverse research fields. The simplified scheme shown in Fig. 16.1 depicts a model of analysis involving several different "omics" strategies. For example, genome annotation and re-annotation, an area related to genomics, usually depends on proteomic as well as transcriptomic data and so on. Additionally, distinct epigenetic markers in identical genomes can influence gene transcription and hence affect everything occurring downstream in the natural flow of information in a cell.

Metabolomics aims at integrating the flow of pathways and metabolites involved in cell function at a certain moment of the life of the cell, and lipidomics and

Table 16.1 Full genome, and "omic" studies currently performed on a selected group of protozoan parasites of veterinary importance

Organism	Genome size (Mbp)	Total gene number	"Omic" studies EST	T	P	M	References
Babesia bovis	8.18	3781	+	+	+	–	de Vries et al. (2006), Al-Khedery and Allred (2006), Lau et al. (2007), Pedroni et al. (2013), Johnson et al. (2017), Mesplet et al. (2011), Rachinsky et al. (2008) and Laughery et al. (2009)
Babesia bigemina	13.84	5136	–	+	+	–	Suarez et al. (2003), Vichido et al. (2008) and Shompole et al. (1994)
Babesia microti	6.41	3560	–	+	+	–	Silva et al. (2016a, b)
Theileria parva	8.35	4167	+	+	+	–	Savadye (1999), Bishop et al. (2005), Kishima et al. (1995), Nene et al. (2004) and Sugimoto et al. (1989a, b, 1992)
Theileria orientalis	9.01	4058	+[a]	–	+	–	Hayashida et al. (2012)
Theileria equi	11.67	5397	+[a]	–	+	–	Silva et al. (2013)
Theileria annulata	8.36	3845	+[a]	+	+	–	Oura et al. (2001, 2006) and Witschi et al. (2013)
Toxoplasma gondii	61.90–64.52	8563–8920	+	+	+	+	Manger et al. (1998), Cleary et al. (2002), Li et al. (2003), Radke et al. (2005), Xia et al. (2008) and Zhou et al. (2015, 2016)
Neospora caninum	59.10	7266	+	+	+	+	Bruno et al. (2004), Reid et al. (2012), Lee et al. (2003, 2004, 2005), Shin et al. (2005) and Elsheikha et al. (2014)
Eimeria spp.	43.67–72.24	6037–10254	+	+	+	–	Aarthi et al. (2011), Min et al. (2005), Miska et al. (2008), Schwarz et al. (2010), Jiang et al. (2005), Lal et al. (2009) and Oakes et al. (2013)
Sarcocystis spp.	124.41–130.2	7089–7174	+	+[b]	+	–	Howe (2001) and Howe et al. (2005)
Cryptosporidium spp.	8.50–9.25	3886–3981	+	+	+	+	Strong and Nelson (2000), Mauzy et al. (2012), Zhang et al. (2012), Siddiki (2013), Siddiki and Wastling (2009), Snelling et al. (2007), Hublin et al. (2013) and Ng et al. (2012)

(continued)

Table 16.1 (continued)

Organism	Genome size (Mbp)	Total gene number	"Omic" studies				References
			EST	**T**	**P**	**M**	
Trypanosoma spp.	20.93–83.51	3397–12,050	+	+	+	+	Agüero et al. (2004), Cerqueira et al. (2005), El-Sayed et al. (1995), Verdun et al. (1998), Archer et al. (2011), Greif et al. (2013), Kolev et al. (2010), Li et al. (2016), Atwood et al. (2005), dos Santos et al. (2015), Paba et al. (2004), Panigrahi et al. (2009), Urbaniak et al. (2012), Barrett et al. (2010) and Creek et al. (2013, 2015)
Leishmania spp.	29.03–35.21	8135–9101	+	+	+	+	Gentil et al. (2007), Levick et al. (1996), Cantacessi et al. (2015), Dillon et al. (2015), Saxena et al. (2007), de Jesus et al. (2014), Drummelsmith et al. (2003), Kumar et al. (2014), Menezes et al. (2013), Singh and Dubey (2016), Veras and Bezerra de Menezes (2016), Arjmand et al. (2016), Kovarova (2016), Scheltema et al. (2010) and Westrop et al. (2015)
Acanthamoeba spp.	42.02–99.59		+	–	+	–	Kong et al. (2001), Caumo et al. (2014) and Gawryluk et al. (2014)
Tritrichomonas spp.-*Trichomonas vaginalis*	176.41	99009	+	+	+	–	Huang et al. (2009, 2012), Gould et al. (2013), Woehle et al. (2014), de Miguel et al. (2010), Benchimol et al. (2017) and Oyhenart and Breccia (2014)
Besnoitia spp.	–	–	–	–	–	–	

EST, expressed sequence tag library; **T**, transcriptome; **P**, proteome; **M**, metabolome

[a]EST data is sourced from: http://piroplasmadb.org/piro/app/record/dataset/DS_6889a51dab

[b]Sourced from: http://www.ebi.ac.uk/

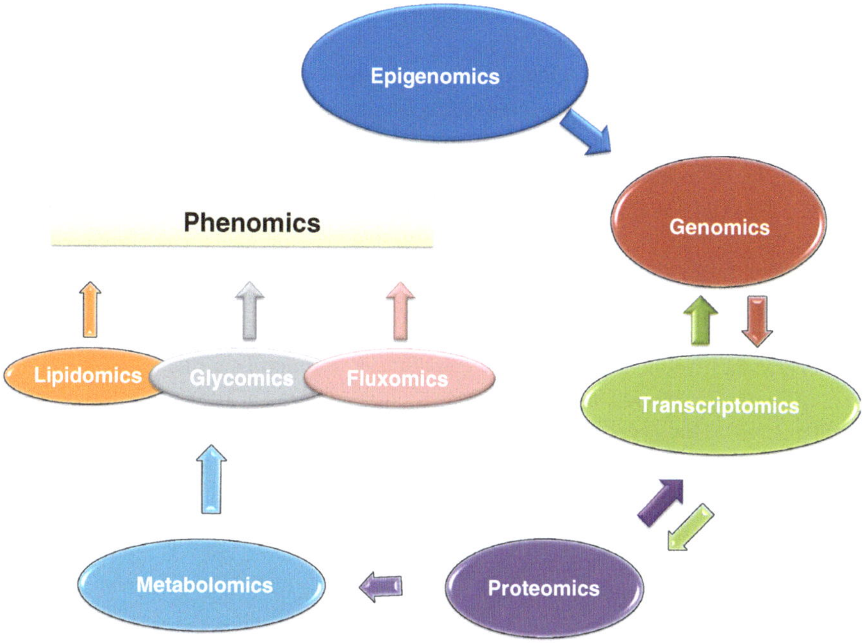

Fig. 16.1 Interrelationships among commonly used "omics" methods applied to the functional characterization of protozoan parasites

glycomics involve the study of pathways and networks of cellular lipids and sugars in biological systems, respectively, while fluxomics is aimed at determining the rates of metabolic reactions in biological systems. Finally, phenomics studies the set of physical and biochemical traits of a given organism as they respond to mutations and environmental changes. In fact, the relationships among the different "omics" are highly dynamic, and information may flow in any direction between them. In general, the combined use of these techniques used in "cross-sectional" studies can provide useful snapshots that may deliver insights into a parasite lifestyle, status, and survival strategies. In any case, comparing integrated "omics" profiles of different stages of a parasite cycle can provide useful information on the lifestyle of any unicellular organism, as is the case here for protozoa. Additionally, comparison of virulent and attenuated parasite strain/lines using "omics" approaches may also provide revealing insights into regulatory and metabolic networks and can be useful for the identification of virulence factors.

16.2 Genomics and Beyond

Genomics greatly facilitated the development of methods of genetic manipulation of these protozoan parasites, besides helping to provide a general blueprint of the biology of the organisms. Important progress has been so far achieved in genome-wide, transcriptomic, and proteomic analysis and genetic manipulation on most of the protozoan parasites listed in Table 16.1 (1–94). However, a few have received relatively little attention, as in the case of *Besnoitia* where, for example, "omics" analysis and, predictably, gene manipulations remain unavailable. Genome size varies largely among this arbitrary selected group of highly diverse protozoan parasites (Fig. 16.2, Table 16.1), differing by a factor of more than 25 times between the smallest (*B. microti*, 6.4 Mbp) and the largest (*Tritrichomonas* spp., ~176.4 Mbp). In general, there appears to be an association between the size of an organism's genome and its dependency on intracellular parasite lifestyles (Sundberg and Pulkkinen 2015) with gene reduction being generally more drastic for obligate intracellular organisms that depend almost entirely on their host for survival. Finding out the possible associations between lifestyle and genome size and the evolutionary significance of genome size differences among this diverse collection of protozoan parasites would be of great interest. However, a better understanding of parasitic lifestyle also requires comprehensive, integrative, and comparative molecular, functional, metabolic studies, as well as improved knowledge of the parasite-host relationships. Regardless,

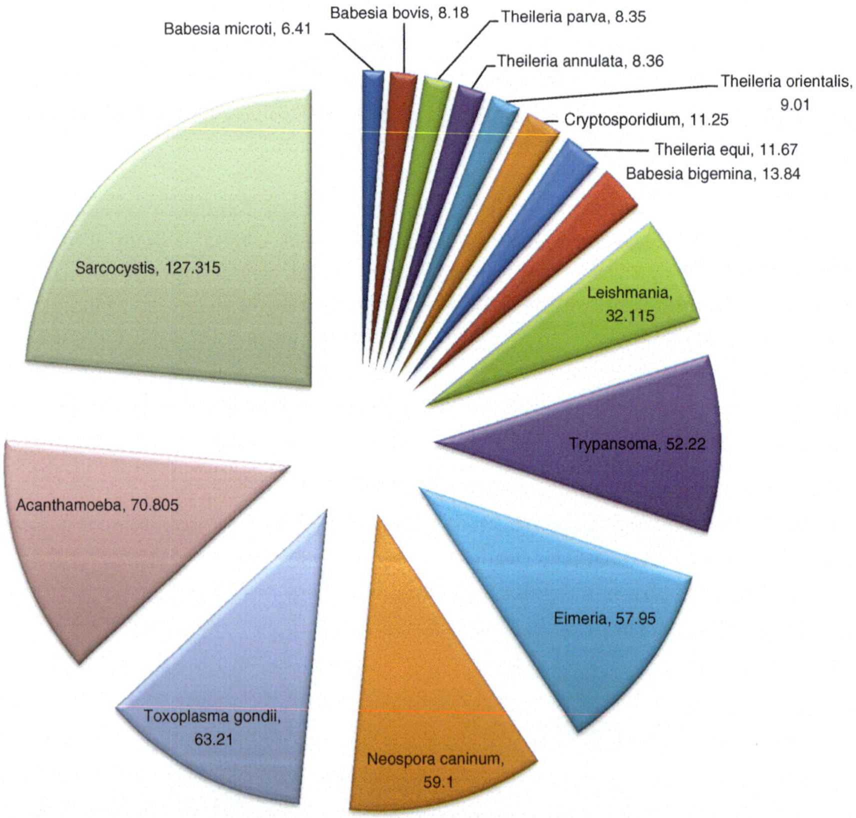

Fig. 16.2 Schematic comparison of the genome sizes of selected protozoan parasites

genomic comparative studies performed among related apicomplexans so far have resulted in a better, albeit somewhat limited, understanding of their biology (Blake 2015; Lv et al. 2015).

For instance, basic cellular and genomic research performed on *Toxoplasma gondii*, which is widely considered as a "model" apicomplexan parasite (Kim and Weiss 2004), is generally applicable to other related apicomplexans mainly because several key mechanisms, such as apical subcellular organelle formation and function, apicoplast and mitochondria function, signaling, gliding motility, intracellular molecule trafficking, cell invasion, etc., are overall well conserved among most of them (Kim and Weiss 2004; Ngô et al. 2004).

Linear regression analysis correlating the ratios of the number of genes per genome (represented as gene density) and genome size among all the protozoan parasites in Table 16.1, with the exception of *Tritrichomonas*, gives a significant linear negative correlation, with an r^2 coefficient of 0.85, suggesting a strong association between these two parameters (Fig. 16.3). This information is consistent with the notion that smaller protozoan genomes are generally more compact than larger genomes, containing more genes per megabase of DNA, likely as a result of having overall similar average gene sizes, but less repeated/redundant regions and less and/or shorter introns and noncoding intergenic regions. *Tritrichomonas* was not included in these comparisons as it has a highly atypical large and vastly

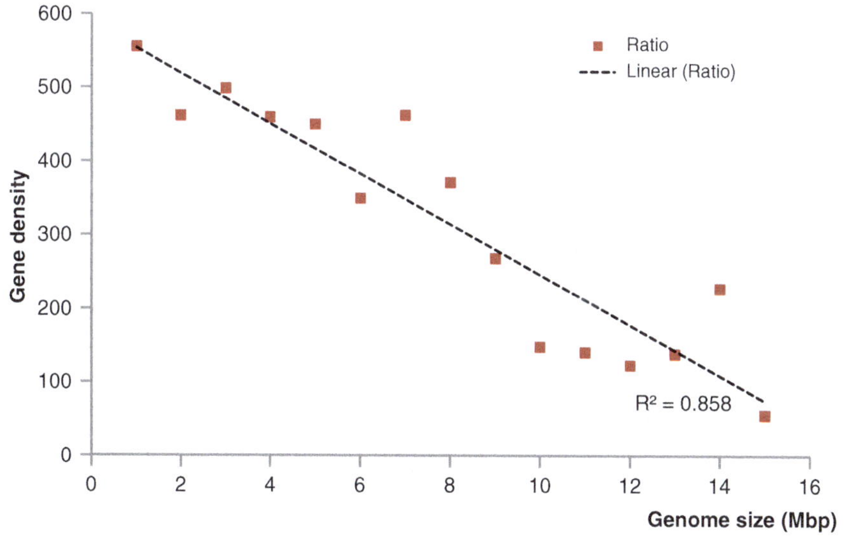

Fig. 16.3 Representation of the relationship between gene density and genome size for selected protozoan parasites. Gene density (genes/Mbp) was calculated by dividing total gene number by genome size (in Mpb) for each parasite species. The "X" axis is organized in ascending order, according to genome size: *1. Babesia microti*, *2. Babesia bovis*, *3. Theileria parva*, *4. Theileria annulata*, *5. Theileria orientalis*, *6. Cryptosporidium* spp., *7. Theileria equi*, *8. Babesia bigemina*, *9. Leishmania* spp., *10. Trypanosoma* spp., *11. Eimeria* spp., *12. Neospora caninum*, *13. Toxoplasma gondii*, *14. Acanthamoeba* spp., *15. Sarcocystis* spp

repetitive genome. As shown in Fig. 16.2 and Table 16.1, coccidian parasites, such as *Toxoplasma* and *Neospora*, have a significantly larger genome size compared to the compacted genomes of the piroplasmid (*Babesia* and *Theileria*). An exception to this pattern is *Cryptosporidium* spp., a protozoan parasite with a genome size and gene density comparable to piroplasmid parasites. Interestingly, *Cryptosporidium* parasites lack a plastid, and no genes of plastid origins were identified, suggesting the early loss of the symbiotic apicoplast by these parasites (Abrahamsen et al. 2004; Xu et al. 2004), as well as the loss of the mitochondrial genome. Genome analysis on *Cryptosporidium* also revealed extremely streamlined metabolic pathways and an absence of many cellular structures and metabolic pathways found in other apicomplexans (Bouzid et al. 2013). Remarkably, *Plasmodium* parasites have a genome size that is intermediate between these two clades (~23 Mbp). Both piroplasmid and *Plasmodium* spp. share the existence of intraerythrocytic stages, but the *Plasmodium* sporozoites inoculated by the mosquito vectors are only capable of invading liver cells, and thus these parasites, in addition to an intrahepatic stage, have many other significant life cycle differences during the arthropod life stages. Interesting differences among piroplasma include the unique ability of *Babesia* parasites to transmit via transovarial mechanisms and the ability of *Theileria*, but not *Babesia* sporozoites, to invade and transform leucocytes of the mammalian host (Lau 2009). Clearly, each of these parasites faces distinct adaptive challenges, including their different strategies for causing persistent infections, requiring a unique and specific genome composition. Other remarkable differences among coccidian and piroplasmid parasites include the ability of some coccidians, such as *Toxoplasma*, to invade multiple distinct cell types in their vertebrate hosts and to form cysts. Once more, all of these phenotypic differences may account for the unique requirements in the number and quality of genes that can sustain the distinct parasitic lifestyles with different levels of complexity occurring in each species. However, and consistent with a more conservative value in the number of genes found for each of these organisms, genome size is also related to the sizes of the noncoding and intergenic areas in their genomes, as reflected in the ratios shown in Fig. 16.3. Again, a special case is the large genome of the highly related *Trichomonas* and *Tritrichomonas* parasites. *Tritrichomonas* are members of the eukaryotic supergroup Excavata, a group of free-living organisms that may or may not have a parasitic lifestyle. Their name is derived from the existence of an "excavated" ventral feeding groove. They are anaerobic parasites that lack classical mitochondria but instead contain specialized organelles, called hydrogenosomes, which are responsible for anaerobic metabolism. Consistent with its extracellular living status, *Trichomonas vaginalis* possesses a large genome, which is largely comprised by repeats and transposable elements (Carlton et al. 2007) and is shared by other related *Tritrichomonas* parasites of veterinary importance. A draft genome sequence of *T. foetus* showed that 72% of the open reading frames (ORFs) were found to be similar to those of *Trichomonas vaginalis* (Benchimol et al. 2017). In both parasites, the superabundance of repeats resulted in a highly fragmented sequence, preventing an investigation of genome architecture. The other 28% remaining ORFs have no significant results with any other genome. The assembled genome of *T. foetus*, together with the functional annotation, is available at http://www.labinfo.lncc.br/index.php/tritrichomonas_foetus. Other study, using homology analysis, suggested that massive expansions might have occurred in the *T. foetus* genome in a similar way it was also predicted for *Trichomonas vaginalis*, while conservation assessment showed that duplications have been acquired after differentiation of the two species (Oyhenart and Breccia 2014). The authors of the former study concluded upon comparing the two genomes that gene duplications might be common among these parasitic protozoans (Oyhenart and Breccia 2014). In view of these findings, we included the genome of *T. vaginalis*, a human pathogen, together with *T. foetus* in Table 16.1. The high level of similarities among the genomes of *Trichomonas* and *Tritrichomonas* might simplify gene functional analysis using one of these organisms as a model.

Genomic and genetic manipulation studies performed on the "model apicomplexan" *Toxoplasma gondii* and *Babesia* sp. parasites, discussed below, exemplify the potential of these techniques toward improved parasite control. It is expected that the application of these approaches in other still poorly researched protozoa of veterinary importance will enhance our understanding of the biology of these parasites and their relationships with their hosts. It can be predicted that this new knowledge will translate into improved control of important yet neglected diseases with a high public health and economic impact globally. Certainly, more dramatic advances are expected to occur with the massive application of "omics" and vaccinology approaches in the near future.

16.3 Genomic Resources for Protozoan Parasites

Complete genomes of most protozoans of veterinary, medical, and zoonotic importance are currently available (Table 16.1). Furthermore, comparison between genome sequences among apicomplexan and other protozoans is now greatly facilitated using the Eukaryotic Pathogen Database Resource, EuPathDB (http://eupathdb.org/eupathdb/). This database provides access to the full genomes of *Babesia* spp. and *Theileria* spp. organized into the PiroplasmaDB (http://piroplasmadb.org/piro/showApplication.do), *Acanthamoeba* spp. genomes organized in the AmoebaDB (http://amoebadb.org/amoeba/showApplication.do), *Cryptosporidium* spp. at the CryptoDB (http://cryptodb.org/cryptodb/showApplication.do), and Coccidian genomes, including *Toxoplasma gondii*, *Neospora caninum*, *Sarcocystis neurona*, and *Eimeria* sp. at the ToxoDB (http://toxodb.org/toxo/showApplication.do). The *Trichomonas vaginalis* genome sequence is at the TrichDB (http://trichdb.org/trichdb/showApplication.do); and *Leishmania* spp. and *Trypanosoma* spp. can be found at the TriTrypDB (http://tritrypdb.org/tritrypdb/showApplication.do).

Importantly, the information on the EuPathDB is easily available and comprehensive and not limited to genome sequences. The site also provides easy access to analytical genomic tools such as Blast and available EST, microarray, RNA-seq, and proteomics data for these organisms, among other useful information.

16.4 Genetic Manipulation of Protozoan Parasites of Veterinary and Zoonotic Importance

Genetic manipulation techniques are important tools that allow access to multiple research approaches, including identification of virulence factors, subunit vaccine components, and parasite transmission factors. Importantly, genetically manipulated parasites themselves can be potentially used for vaccine development since targeted knockout of genes encoding known virulence factors might result in the production of genetically defined attenuated parasites. Another application of interest is the development of novel vaccine delivery platforms by manipulating attenuated parasites to express foreign genes coding for exogenous or stage-specific endogenous protective antigens. Also, genetically manipulated parasites used in vaccines can be easily distinguished from their wild-type counterparts, facilitating the discrimination among vaccinated and naturally infected animals. In addition to classic gene manipulation using transfection or gene editing techniques, RNA interference (RNAi) methods are also tools for gene function characterization (Meissner et al. 2007). However, *Trypanosoma* spp. and *Leishmania* spp. parasites, as well as most apicomplexan parasites (including *Babesia* and *Plasmodium*), lack the enzymes required for this pathway, and RNAi is not generally regarded so far as a useful method of gene analysis for these parasites. The mechanisms leading to the loss of the RNAi genes in these organisms, with no recognizable traces of their past presence, are unknown, although chromosomal rearrangements may have contributed to their disappearance (Kolev et al. 2011). Interestingly, the genome sequence of *T. gondii* revealed the existence of Dicer, AGO, and RdRp homologues (Braun et al. 2010) that appear to have plant/fungal (Dicer/RdRp) and metazoan (AGO) signatures. Initial reports thus suggested that *T. gondii* is the only apicomplexan with a functional RNAi pathway (Kolev et al. 2011). However, reported experimental results on the activity of this pathway were not reproducible, and the occurrence of this mechanism in this parasite has been put into doubt, highlighting the need for more research. A recent report also described the use of RNAi techniques to inhibit *B. bovis* in vitro growth upon the targeting of three distinct genes (AbouLaila et al. 2016), but the possible mechanisms involved remain uncertain given the absence of canonical RNAi genes in *B. bovis*. In other protozoan parasites, such as *Trichomonas vaginalis*, the presence of a Dicer-like gene and two Argonaute genes suggests the existence of the RNAi pathway (Carlton et al. 2007). Identification of these components raises the possibility of using RNAi technology to manipulate *T. vaginalis* gene expression.

The most widely used genetic manipulation methods include classic transfections based on the insertion of exogenous DNA using homologous recombination mechanisms (de Koning-Ward et al. 2000) and more recently CRISP/CAS9 (Lander 2016; Wright et al. 2016) and other gene editing methods such as TALENs and zinc-finger nucleases, based on programmable nucleases (Ma and Liu 2015). Such methods have been extensively used for the genetic manipulation of apicomplexan and other protozoa as will be described below.

16.4.1 Classic Transfection Methods

In transfection, DNA (or RNA) molecules may be introduced either as extrachromosomal replicating episomes or inserted into chromosomes by homologous recombination. Stably transfected lines can then be used for multiple applications including the study of gene function and by creating parasite lines that either overexpress or lack specific genes of interest. However, a limitation of reverse genetic approaches for functional gene analysis is that essential genes may be impossible to knock out, since this will result in nonviable parasites. These limitations can now be overcome at least partially by using inducible promoter strategies, including the use of tetracycline-inducible promoters. By choice of appropriate 5′ and 3′ flanking regions in the transfecting plasmid DNA endogenous chromosome, genes can be targeted and deleted. Additionally, the method can be also used to create transfected parasites that may function as vaccine delivery systems. The study of genetically transformed parasite lines can provide important clues about gene function during the parasite life cycle.

Perhaps mirroring their importance as human pathogens, *Leishmania* and other trypanosomatid parasites were first targeted for genetic transformation using transfection methods (Bellofatto and Cross 1989; Cruz and Beverley 1990; Laban and Wirth 1989; Lee and Van der Ploeg 1990; Ten Asbroek et al. 1990). The first report of the genetic modification of an apicomplexan parasite was the description of a transient transfection method for *Toxoplasma gondii* (Soldati and Boothroyd 1993). This led shortly thereafter to the development of a method for the stable transfection of this organism (Kim et al. 1993). Transient transfection is regarded as a useful approach for finding appropriate electroporation settings and to identify and test the function and efficacy of regulatory elements (promoters and termination signals) mediating gene expression and regulation. Based in part on these findings, transient and stable transfection techniques were later also applied to some species of *Plasmodium* parasites (Goonewardene et al. 1993; Van Dijk et al. 1995).

Transient transfection methods are useful for characterizing and defining promoters and other regulatory factors and later became essential components of advanced gene engineering and editing techniques, such as those based on the CRISPR/Cas9. Briefly, transient transfection techniques (Fig. 16.4) are designed to introduce and express foreign DNA, usually in the form of a plasmid, into a nucleated cell in a non-stable manner. Thus, in transient transfection, the introduced plasmid nucleic acid does not integrate into the genome of the target cells, and the transfected genes will not be replicated. After being developed in *T. gondii* (Soldati and Boothroyd 1993), transient transfection methods were applied on *Babesia bovis* (Suarez and McElwain 2008, 2010; Suarez et al. 2004, 2006, 2007), *Eimeria mitis* (Qin et al. 2014), *Sarcocystis neurona* (Gaji et al. 2006), *Theileria parva* (De Goeyse et al. 2015), and *T. annulata* (Adamson et al. 2001). This approach proved to be useful for the definition of promoters in *B. bovis* and later in *B. bigemina* (Silva et al. 2016a, b) and settled the basis for the development of stable transfections systems for *T. gondii* (Kim et al. 1993), *Sarcocystis neurona* (Gaji et al. 2006), *Acanthamoeba castellanii* (Peng et al. 2005), and *B. bovis* (Suarez and McElwain

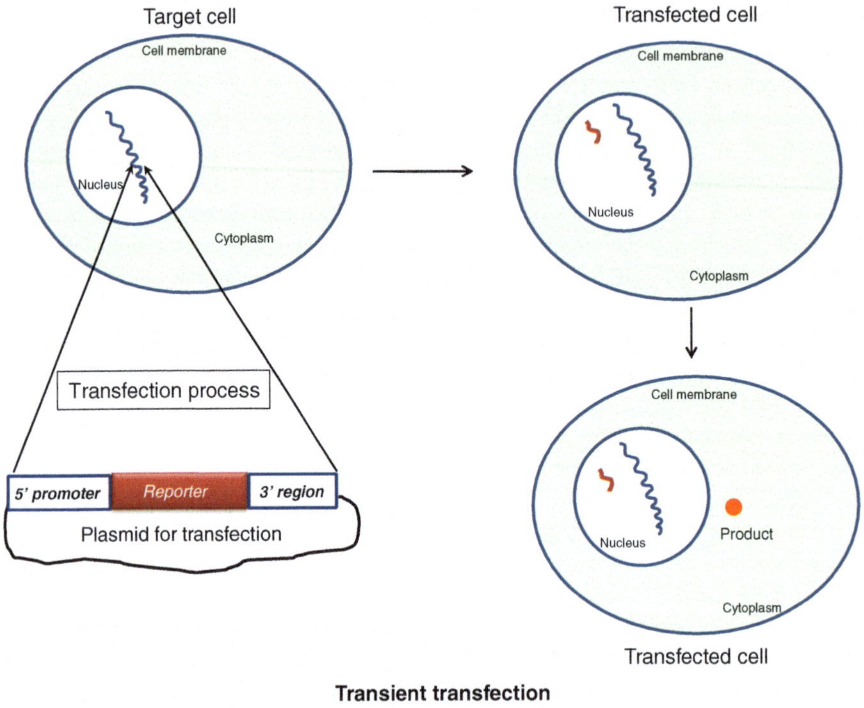

Fig. 16.4 Schematic representation of the principles and elements involved in transient transfection methods. A transfection plasmid (**a**) containing a reporter gene (*red box*) under the control of promoter and termination regions is transferred into the nucleus of a target parasite (**b**) using physical or chemical methods (transfection process). The plasmid DNA transferred into the target cell (in *red*) is not integrated stably into the genome (**c**), but it can be processed by the transcription and translation machinery of the cell to generate a product (*red dot*) (**d**) that can be quantified (e.g., by measuring luciferase activity)

2009, 2010). Transient transfection plasmids typically include a reporter gene or a gene that needs to be expressed transiently (such as required for current CRISPR/Cas9 methods), placed under the transcriptional control of a promoter and transcription and translation regions located at the 5′ and 3′ ends, respectively (Fig. 16.4). An appropriate amount of the transient transfection plasmid then is introduced into the target cells using distinct methods. These include those relying on physical treatments such as electroporation, nucleofection, biolistic delivery (gene gun), or microinjection and those relying on chemical entities, such as liposomes (Kepczynski and Róg 2016).

Physical methods create reversible "holes" in the cell membranes to insert the nucleic acids, whereas chemical methods are based on the use of transfection reagents, sometimes in the form of cationic lipids that allow membrane fusion and intracellular/intranuclear delivery of the foreign DNA into the target cells. More recently, nanoparticles and other polymers have been applied to this end. In general, the transiently transfected plasmids are designed so that they do not integrate into the genome but remain as episomes in the target cell, where the gene of interest is expressed for a limited period of time. However, promoter strength studies and comparisons using transient transfection approaches are relative and limited, and so the data should be analyzed strictly in the context in which these experiments are performed. This is so because this approach is based on measuring promoter activity by promoter regions which are cloned in transiently transfected plasmids. This experimental approach would preclude estimating the possible regulatory role and contributions of distantly located or "trans" enhancers, the potential competition for transcription factors among the native and the plasmid-cloned promoters, as well as the possible contributions to promoter activity that depends on other regulatory elements such as epigenetic factors.

The stable transfection techniques are essentially based on the ability of the parasites to insert genetic material in their DNAs using homologous recombination mechanisms, in a fashion allowing expression of the transfected genes. Again, as for transient transfection, these techniques rely either on the use of liposomes such as Lipofectamine or on the application of a controlled electrical pulse, such as in electroporation, or later in nucleofection. These procedures allow the incorporation of exogenous DNA, usually provided in the form of a circular or linearized plasmid, into the nuclear compartment of a eukaryotic cell. Therefore, basic steps involved include (1) identification of a suitable genetic marker to select for transgenic parasites ("selectable marker"); (2) preparation of a transfection plasmid vector containing, at a minimum, a selectable marker gene under the control of a suitable promoter, and 5'and 3'regions to target integration of the construct into the genome; (3) a liposome or electroporation/nucleofection protocol which does not greatly compromise the viability of the target cells; and (4) a method for selection of transfected parasites. A schematic representation of a typical stable transfection vector is shown in Fig. 16.5.

Specific integration of the transfected gene(s) at the intended site into the genome of the target parasite depends upon the operation of homologous recombination mechanisms. However, this requirement may affect the efficiency of the integration process. In fact, the efficiency of the integration process in protozoan parasites is highly variable among species, and depends heavily on the available DNA repair mechanisms operating in each cell, as found for *Toxoplasma* parasites. However, the efficiency of exogenous gene integration can also be affected by the particular DNA base composition of the target cells, as in the case of the A + T-rich genome of *Plasmodium* parasites. Interestingly, *Toxoplasma* parasites are also difficult to engineer using classic transfection technologies because they have the ability to randomly insert the foreign DNA in sites different from the targeted. This occurs, at least in part, due to the action of the NHEJ repair mechanisms based on the activity of a gene encoding the KU80 protein. This limitation has been recently addressed by preparing a genetically transformed *Toxoplasma gondii* line lacking the *KU80* gene, which makes it amenable to gene targeting using homologous recombination mechanisms (Huynh and Carruthers 2009). In contrast, transfection work performed in *Babesia bovis* suggests that this technique is efficient, at least in terms of targeting, when applied to this organism (Suarez et al. 2015). Yet, differences in the gene

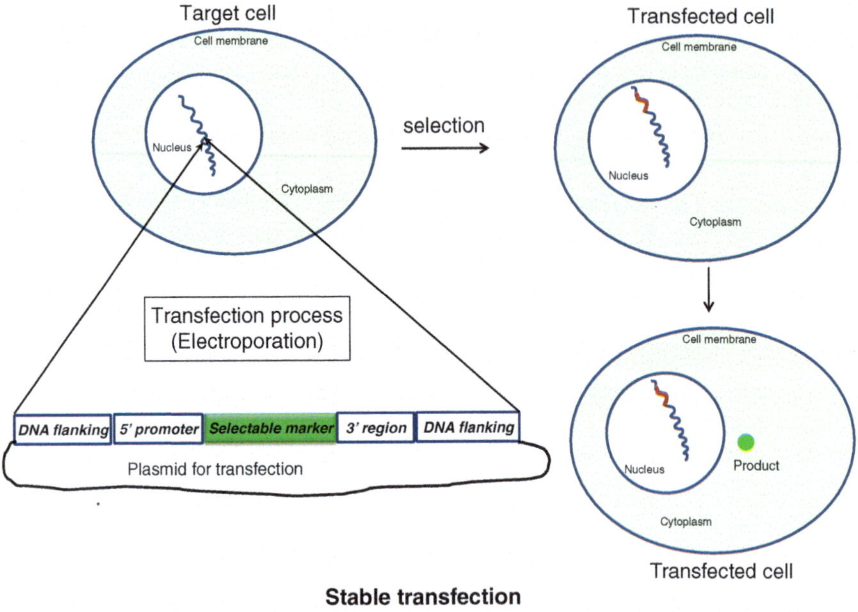

Fig. 16.5 Schematic representation of the principles and elements involved in stable transfection methods. A transfection plasmid containing a selectable drug resistance gene (*green box*) (such as bsd, dhfr, pyrimethamine, etc.) with a gene coding a fluorescent marker (such as GFP) under the control of promoter and termination regions, in addition to the 5′ and 3′ flanking regions required for homologous recombination, is transferred into the nucleus of a target parasite using physical or chemical methods. The incorporated plasmid DNA (in *red*) is integrated stably into the genome of the target cell by homologous recombination and processed by the transcription and translation machinery of the cell to generate a product (i.e., GFP-BSD) (*green dot*). Drug selection is performed to eliminate non-transfected parasites and to obtain a cell line of transfected parasites

repertoires and gene structure for other proteins involved in gene repair mechanisms among *B. bovis* and *T. gondii* together with differential regulation of their expression might also help explain the differences observed among the mechanism of gene repair operating in these two organisms. In contrast to *Plasmodium* and *Toxoplasma*, *B. bovis* appears to be quite amenable for stable transfection and, consistently, is able to specifically and efficiently integrate foreign genes. Thus, stable transfection techniques for *B. bovis* allowed highly specific KO and KO reversion experiments that are needed to study gene function (Asada et al. 2012a, b, 2015; Suarez et al. 2015). Recent progress in *Babesia* transfection technology includes the demonstration of a method for functional gene analysis by generating gene KO followed by gene function recovery (Asada et al. 2015) and the demonstration of cross-species promoter function (Silva et al. 2016a, b). This later study describes the ability of a *B. bovis* *ef-1α* promoter to function efficiently in *B. bigemina*. This observation suggests that common regulatory signals should exist, allowing the control of promoter functions among these two parasites.

16.4.2 Gene Editing Using Programmable Nucleases

Targeted genetic editing methods that allow precise modifications in a genome were more recently developed. These methods offer great potential for the manipulation of the genomes of *Toxoplasma*, *Plasmodium*, and other protozoan parasites, where transfection methods based solely on homologous recombination typically demonstrate very low efficiency. A key factor dramatically increasing efficiency of programmable nucleases is their ability to generate blunt double-strand breaks (DSB) in the target DNA of interest. The DSB results in the intervention of repairing systems of the cells, such as error-prone nonhomologous end joining (NHEJ) mechanisms, which can repair the break without the presence of donor homologous DNA. Alternatively, the breaks can be repaired by homology-directed repair (HDR) mechanisms in the presence of homologous donor double- or single-strand DNA, leading to the insertion of exogenous genetic material. The two mechanisms of DNA repair are exemplified in Fig. 16.5. The existence of these alternative pathways also suggests the possibility of using different gene manipulation strategies. Thus, introduction of simple mutations resulting in gene inactivation or disruption can be generated by double break followed by NHEJ. This repair mechanism can generate either insertions or deletions (indels) in the target gene resulting in frameshifts that disrupt the continuity of the open reading frame, usually leading to the knockout of the gene. If the objective is the insertion of foreign genes, such as reporter genes, it may then be necessary to add donor plasmid DNA containing the gene which is intended to be inserted with the addition of homologous flanking regions, to facilitate accurate targeting. In this case, the insertion of the foreign gene will likely be mediated by HDR repair mechanisms. Importantly, new discoveries on the mechanisms of DNA repair in apicomplexan parasites revealed the participation of certain proteins such as rad51 and KU80. As discussed earlier, targeted mutation of the *KU80* gene resulted in a *Toxoplasma gondii* mutant line that is more efficient for gene targeting, since it favors the KU80-indpendent HDR mechanism of repair and prevents random incorporation of transfected genes, which was commonly found in this parasite. This cell line is thus ideally suited for gene function analysis in *Toxoplasma* using homologous recombination KO approaches (Huynh and Carruthers 2009; Smolarz et al. 2014).

The specific design of gene editing experiments depends on the programming nuclease method of choice. The programmable gene editing methods currently available include the use of engineered proteins such as zinc-finger nucleases (ZFN), transcription activator-like effector nucleases (TALENs) or RNA-guided engineered nucleases (RGEN). However, despite perceived improved target specificity of TALEN methods, the RGEN methods have several advantages over the other two, including their simple design, versatility, and cost. Briefly, the ZFN attach cutting domains derived from the prokaryote *Flavobacterium okeanokoites* to proteins called zinc fingers that can be customized to recognize certain three-base-pair DNA codes. On the other hand, TALENs fuse the same cutting domains to different proteins called TAL effectors. Both ZFN and TALENs require two cutting domains in order to cleave double-stranded DNA. Excellent reviews on the use of ZFN and TALEN approaches for gene editing are available elsewhere (Ma and Liu 2015).

The most widely used RGEN method is based on the CRISPR/Cas9 system. Deeper coverage on the discovery and function of the CRISPR/Cas9 system was described elsewhere (Lander 2016; Wright et al. 2016). Briefly, this system is divided into three types based on the Cas proteins involved. Only the simpler type II system is used for gene editing and is essentially based in just a single effector Cas9 protein, although other putative effectors can now also be used. The principle of the method is illustrated in Fig. 16.6 and its applications for gene editing in Fig. 16.7. Briefly, the acronym CRISPR is derived from "clustered regularly interspaced short palindromic repeats," which together with the Cas ("CRISPR-associated" proteins) endonucleases, such as Cas9, are part of an adaptive immune system against phages of bacteria and archaebacteria (Wright et al. 2016).

This bacterial immune system provides RNA-mediated immunity against viruses and plasmids based on copying and specifically cleaving exogenous genetic materials. It was soon realized that, upon providing the necessary components to target cells, this system could be also manipulated to edit DNA in virtually any cell. Together, CRISPR and Cas9 are able to target and cut almost any DNA in vivo, and together with transfection techniques, they quickly became an important asset as efficient and specific tools for gene editing.

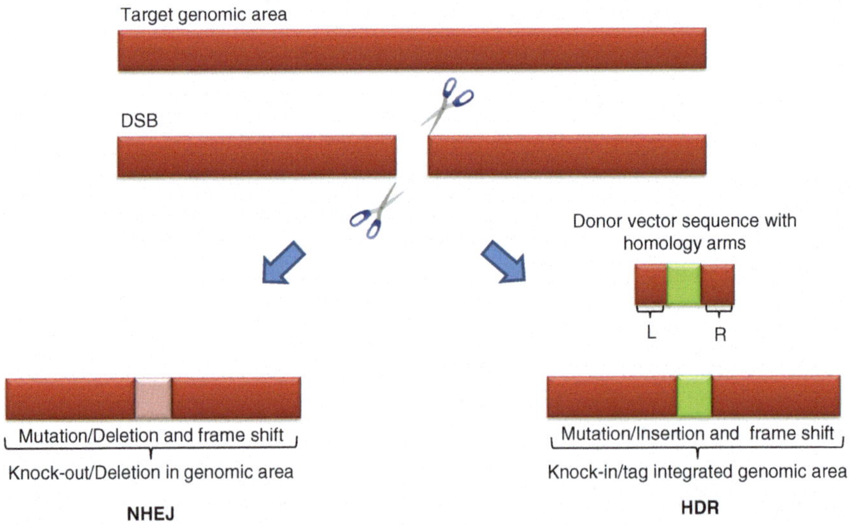

Fig. 16.6 Principles and elements involved in current gene editing methods. A targeted genome area is specifically cleaved with a double-strand break (DSB) by a nuclease and can be repaired using two different mechanisms: "nonhomologous end joining" (NHEJ) or "homology direct repair" (HDR) of the targeted cell. As a result, mutations, such as insertion/deletions (*pink box*), are caused, resulting in the inactivation of the target gene (NHEJ) or, in the presence of a donor sequence with homologous arms, the stable incorporation of new genetic material (*green box*) in the integrated genomic locus (HDR). L and R: left and right flanking homology arms

The most commonly used CRISPR/Cas9 systems are adapted from *Streptococcus pyogenes*. A CRISPR-Cas9 system specifically cleaves a DNA sequence through a two-stage recognition process, as depicted in Fig. 16.6. Initially, as more detailed below, a Cas9-sgRNA complex will be able to attach stably to a DNA sequence only if an appropriate, short (often only a few base pairs) protospacer adjacent motif (PAM) is located in close proximity (Fig. 16.7). Therefore, an important advantage of the CRISPR/Cas9 type II system is its simplicity, since only three components are required to achieve site-specific DNA recognition and cleavage. These include a CRISPR RNA (crRNA) and a trans-activating crRNA (tracrRNA) which are required in order to guide the Cas9 enzyme to its target sequence (Fig. 16.7). These two elements (crRNA and tracrRNA) are usually combined into a single synthetic guide RNA (sgRNA). According to experimental design and specific gene targeting, the sgRNAs can be designed to include the specific base sequence that matches the target gene of interest. In that way, the complex can redirect the Cas9 enzyme to almost any preferred sequence. The *S. pyogenes* Cas9 endonuclease, which should bear nuclear localization signals (NLS), preferably requires an NGG PAM (with "N" representing any nucleobase followed by two guanine or "G" nucleobases). However, NAG and NGA PAM motifs can also sometimes be recognized. The 20 bp long sequence in the guide RNA then recognizes the homologous DNA target sequence by Watson-Crick base pairing. If a complete target sequence is confirmed, allosteric activation of the Cas9's two

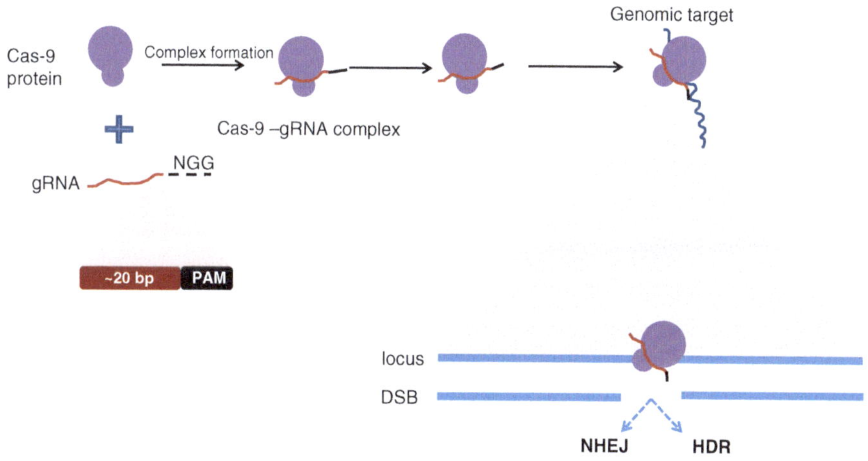

Fig. 16.7 Basic elements involved in gene editing methods based on CRISPR/Cas9. The 20 nucleotide guide RNA (gRNA) (represented in *red*) and PAM (*black box*) complexed with Cas9 is targeted to a specific sequence in the genome using Watson and Crick complementary base pairing. The complex locks into the targeted locus where it generates a double-strand break (DSB) caused by Cas9, which can be repaired using the "nonhomologous end joining" (NHEJ), or "homology direct repair" (HDR) mechanisms of the target cell. As a result, the targeted gene could be mutated or a new sequence can be integrated stably into the genome of the target cell. Gene-edited cells can later be selected using positive or negative selection procedures

nuclease domains, RuvC and HNH, will result in dual cleavage and, accordingly, a complete double-strand break in the target sequence. Clearly, the specificity of any CRISPR-Cas9 system depends heavily on the proper design of the guide RNA. This can be done sometimes using algorithms that minimize the likelihood of off-target effects. In other words, a CRISPR/Cas9 gene editing experiment requires the design of a 20 nucleotide guiding RNA (sgRNA) that can hybridize specifically with a sequence in the target gene. The sequences coding guiding RNA, the gene coding for Cas9 (including NLS), and donor DNA need to be provided to the target cells for expression in the form of plasmid DNA. Co-expression of these DNAs can be achieved following single vector or multiple vector strategies. Thus, for example, a single vector can include the genetic information necessary for the co-expression of Cas9, sgRNA, and donor DNA. This can be achieved using transient transfection of a properly engineered transfection plasmid having each gene under the transcriptional control of distinct promoters that need to be functional in the target parasite. This is now facilitated in *B. bovis* by the discovery that at least one heterologous *B. bigemina* promoter is also active in this parasite (Silva et al. 2016a, b). Alternative strategies include the delivery of in vitro transcribed sgRNA, as was the case for *T. cruzi* (Peng et al. 2015). Gene editing based on CRISPR/Cas9 has been used successfully for genetic analysis of several apicomplexan parasites of veterinary importance (Cui and Yu 2016), including *Cryptosporidium parvum* (Vinayak et al. 2015) and *Toxoplasma gondii* (Shen et al. 2014). In contrast to *T. gondii*, the low level of nonhomologous or random integration of exogenous transfected genes in *B. bovis* suggests that this parasite uses mainly HDR rather than NHEJ repair mechanisms. In addition, CRISPR/Cas9 approaches have also been used for genetic modification of trypanosomatid parasites such as *Trypanosoma cruzi*, *T. brucei*, and *Leishmania* spp. (Zhang and Matlashewski 2015). The versatility of *T. gondii* as a model apicomplexan was also employed for further development of a CRISPR/Cas9-based genome-wide genetic screen toward the identification of *T. gondii* essential genes during infection of human fibroblasts (Sidik et al. 2016). This approach allowed the description of an apicomplexan-conserved invasion factor termed claudin-like apicomplexan microneme protein (CLAMP). This novel approach has potential to be applied to other apicomplexan parasites.

Trypanosomatid parasites present the additional challenge of possessing a diploid genome; thus, deletion of an entire gene requires at least two distinct selection markers (Lander 2016). The use of these systems greatly accelerates our knowledge of the genetics of these parasites and the development of new vaccines. A website to guide the design of CRISPR tools in protozoan pathogens is a useful resource that is currently freely available (http://grna.ctegd.uga.edu/batch.html). Finally, different strategies for the selection of edited parasites, with or without the use of selectable markers, are also available (Mogollon et al. 2016).

Despite possible off-target cleavage and other potential limitations, gene editing procedures can be used for understanding gene function, generation of mutated attenuated parasites, or as a tool for the development of novel vaccines and therapeutics, thus improving the control of parasites of veterinary interest.

Conclusions

Combination of current "omics" and gene manipulation methods can improve dramatically our understanding of the genetics and the biology of protozoan parasites of veterinary and enzootic relevance. However, the rapid pace of progress of biotechnology, "omics," and other molecular and computer-aided tools will likely result in the constant emergence of novel integrated methods for the interrogation and modification of genomes, leading to a better understanding of parasite lifestyle, and, ultimately, to the rational design of improved methods for the control of animal infectious diseases.

References

Aarthi S, Raj GD, Raman M, Blake D, Subramaniam C, Tomley F. Expressed sequence tags from *Eimeria brunetti*—preliminary analysis and functional annotation. Parasitol Res. 2011;108(4):1059–62.

AbouLaila M, Yokoyama N, Igarashi I. RNA interference (rnai) for some genes from *Babesia bovis*. RJAB. 2016;2:81–92. ISSN.2356:9433.

Abrahamsen MS, Templeton TJ, Enomoto S, Abrahante JE, Zhu G, Lancto CA, et al. Complete genome sequence of the apicomplexan, *Cryptosporidium parvum*. Science. 2004;304(5669):441–5.

Adamson R, Lyons K, Sharrard M, Kinnaird J, Swan D, Graham S, et al. Transient transfection of *Theileria annulata*. Mol Biochem Parasitol. 2001;114(1):53–61.

Agüero F, Abdellah KB, Tekiel V, Sánchez DO, González A. Generation and analysis of expressed sequence tags from *Trypanosoma cruzi* trypomastigote and amastigote cDNA libraries. Mol Biochem Parasitol. 2004;136(2):221–5.

Al-Khedery B, Allred DR. Antigenic variation in Babesia bovis occurs through segmental gene conversion of the ves multigene family, within a bidirectional locus of active transcription. Mol Microbiol. 2006;59(2):402–14.

Archer SK, Inchaustegui D, Queiroz R, Clayton C. The cell cycle regulated transcriptome of *Trypanosoma brucei*. PLoS One. 2011;6(3):e18425.

Arjmand M, Madrakian A, Khalili G, Najafi A, Zamani Z, Akbari Z. Metabolomics-based study of logarithmic and stationary phases of promastigotes in *Leishmania* major by 1H NMR spectroscopy. Iran Biomed J. 2016;20(2):77.

Asada M, Tanaka M, Goto Y, Yokoyama N, Inoue N, Kawazu S-i. Stable expression of green fluorescent protein and targeted disruption of thioredoxin peroxidase-1 gene in *Babesia bovis* with the WR99210/dhfr selection system. Mol Biochem Parasitol. 2012a;181(2):162–70.

Asada M, Goto Y, Yahata K, Yokoyama N, Kawai S, Inoue N, et al. Gliding motility of *Babesia bovis* merozoites visualized by time-lapse video microscopy. PLoS One. 2012b;7(4):e35227.

Asada M, Yahata K, Hakimi H, Yokoyama N, Igarashi I, Kaneko O, et al. Transfection of *Babesia bovis* by double selection with WR99210 and blasticidin-S and its application for functional analysis of thioredoxin peroxidase-1. PLoS One. 2015;10(5):e0125993.

Atwood J, Weatherly D, Minning T, Bundy B, Cavola C, Opperdoes F, et al. The *Trypanosoma cruzi* proteome. Science. 2005;309(5733):473–6.

Barrett MP, Bakker BM, Breitling R. Metabolomic systems biology of trypanosomes. Parasitology. 2010;137(09):1285–90.

Bellofatto V, Cross GA. Expression of a bacterial gene in a trypanosomatid protozoan. Science. 1989;244(4909):1167–70.

Benchimol M, de Almeida LGP, Vasconcelos AT, et al. Draft genome sequence *of Tritrichomonas foetus* strain K. Genome Announc. 2017;5(16):e00195-17. https://doi.org/10.1128/genomeA.00195-17.

Bishop R, Shah T, Pelle R, Hoyle D, Pearson T, Haines L, et al. Analysis of the transcriptome of the protozoan *Theileria parva* using MPSS reveals that the majority of genes are transcriptionally active in the schizont stage. Nucleic Acids Res. 2005;33(17):5503–11.

Blake DP. *Eimeria* genomics: where are we now and where are we going? Vet Parasitol. 2015;212(1):68–74.

Bouzid M, Hunter PR, Chalmers RM, Tyler KM. *Cryptosporidium* pathogenicity and virulence. Clin Microbiol Rev. 2013;26(1):115–34.

Braun L, Cannella D, Ortet P, Barakat M, Sautel CF, Kieffer S, et al. A complex small RNA repertoire is generated by a plant/fungal-like machinery and effected by a metazoan-like Argonaute in the single-cell human parasite *Toxoplasma gondii*. PLoS Pathog. 2010;6(5):e1000920.

Bruno S, Duschak VG, Ledesma B, Ferella M, Andersson B, Guarnera EA, et al. Identification and characterization of serine proteinase inhibitors from *Neospora caninum*. Mol Biochem Parasitol. 2004;136(1):101–7.

Cantacessi C, Dantas-Torres F, Nolan MJ, Otranto D. The past, present, and future of *Leishmania* genomics and transcriptomics. Trends Parasitol. 2015;31(3):100–8.

Carlton JM, Hirt RP, Silva JC, Delcher AL, Schatz M, Zhao Q, et al. Draft genome sequence of the sexually transmitted pathogen *Trichomonas vaginalis*. Science. 2007;315(5809):207–12.

Caumo KS, Monteiro KM, Ott TR, Maschio VJ, Wagner G, Ferreira HB, et al. Proteomic profiling of the infective trophozoite stage of *Acanthamoeba polyphaga*. Acta Trop. 2014;140:166–72.

Cerqueira GC, DaRocha WD, Campos PC, Zouain CS, Teixeira SM. Analysis of expressed sequence tags from *Trypanosoma cruzi* amastigotes. Mem Inst Oswaldo Cruz. 2005;100(4):385–9.

Cleary MD, Singh U, Blader IJ, Brewer JL, Boothroyd JC. *Toxoplasma gondii* asexual development: identification of developmentally regulated genes and distinct patterns of gene expression. Eukaryot Cell. 2002;1(3):329–40.

Creek DJ, Nijagal B, Kim D-H, Rojas F, Matthews KR, Barrett MP. Metabolomics guides rational development of a simplified cell culture medium for drug screening against *Trypanosoma brucei*. Antimicrob Agents Chemother. 2013;57(6):2768–79.

Creek DJ, Mazet M, Achcar F, Anderson J, Kim D-H, Kamour R, et al. Probing the metabolic network in bloodstream-form *Trypanosoma brucei* using untargeted metabolomics with stable isotope labelled glucose. PLoS Pathog. 2015;11(3):e1004689.

Cruz A, Beverley SM. Gene replacement in parasitic protozoa. Nature. 1990;348(6297):171.

Cui Y, Yu L. Application of the CRISPR/Cas9 gene editing technique to research on functional genomes of parasites. Parasitol Int. 2016;65(6):641–4.

De Goeyse I, Jansen F, Madder M, Hayashida K, Berkvens D, Dobbelaere D, et al. Transfection of live, tick derived sporozoites of the protozoan Apicomplexan parasite *Theileria parva*. Vet Parasitol. 2015;208(3):238–41.

de Jesus JB, Mesquita-Rodrigues C, Cuervo P. Proteomics advances in the study of *Leishmania* parasites and leishmaniasis. Proteins and proteomics of *Leishmania* and *Trypanosoma*. Berlin: Springer; 2014. p. 323–49.

de Koning-Ward TF, Janse CJ, Waters AP. The development of genetic tools for dissecting the biology of malaria parasites. Annu Rev Microbiol. 2000;54(1):157–85.

de Miguel N, Lustig G, Twu O, Chattopadhyay A, Wohlschlegel JA, Johnson PJ. Proteome analysis of the surface of *Trichomonas vaginalis* reveals novel proteins and strain-dependent differential expression. Mol Cell Proteomics. 2010;9(7):1554–66.

de Vries E, Corton C, Harris B, Cornelissen AW, Berriman M. Expressed sequence tag (EST) analysis of the erythrocytic stages of *Babesia bovis*. Vet Parasitol. 2006;138(1):61–74.

Dillon LA, Okrah K, Hughitt VK, Suresh R, Li Y, Fernandes MC, et al. Transcriptomic profiling of gene expression and RNA processing during *Leishmania* major differentiation. Nucleic Acids Res. 2015;43:6799–813.

dos Santos ADCM, Kalume DE, Camargo R, Gómez-Mendoza DP, Correa JR, Charneau S, et al. Unveiling the *Trypanosoma cruzi* nuclear proteome. PLoS One. 2015;10(9):e0138667.

Drummelsmith J, Brochu V, Girard I, Messier N, Ouellette M. Proteome mapping of the protozoan parasite *Leishmania* and application to the study of drug targets and resistance mechanisms. Mol Cell Proteomics. 2003;2(3):146–55.

El-Sayed NM, Alarcon CM, Beck JC, Sheffield VC, Donelson JE. cDNA expressed sequence tags of *Trypanosoma brucei* rhodesiense provide new insights into the biology of the parasite. Mol Biochem Parasitol. 1995;73(1–2):75–90.

Elsheikha HM, Alkurashi M, Kong K, Zhu X-Q. Metabolic footprinting of extracellular metabolites of brain endothelium infected with *Neospora caninum in vitro*. BMC Res Notes. 2014;7(1):406.

Gaji RY, Zhang D, Breathnach CC, Vaishnava S, Striepen B, Howe DK. Molecular genetic transfection of the coccidian parasite *Sarcocystis neurona*. Mol Biochem Parasitol. 2006;150(1):1–9.

Gawryluk RM, Chisholm KA, Pinto DM, Gray MW. Compositional complexity of the mitochondrial proteome of a unicellular eukaryote (*Acanthamoeba castellanii*, supergroup Amoebozoa) rivals that of animals, fungi, and plants. J Proteome. 2014;109:400–16.

Gentil LG, Lasakosvitsch F, Silveira JFD, Santos MRMD, Barbiéri CL. Analysis and chromosomal mapping of *Leishmania* (*Leishmania*) amazonensis amastigote expressed sequence tags. Mem Inst Oswaldo Cruz. 2007;102(6):707–11.

Goonewardene R, Daily J, Kaslow D, Sullivan TJ, Duffy P, Carter R, et al. Transfection of the malaria parasite and expression of firefly luciferase. Proc Natl Acad Sci. 1993;90(11):5234–6.

Gould SB, Woehle C, Kusdian G, Landan G, Tachezy J, Zimorski V, et al. Deep sequencing of *Trichomonas vaginalis* during the early infection of vaginal epithelial cells and amoeboid transition. Int J Parasitol. 2013;43(9):707–19.

Greif G, De Leon MP, Lamolle G, Rodriguez M, Piñeyro D, Tavares-Marques LM, et al. Transcriptome analysis of the bloodstream stage from the parasite *Trypanosoma vivax*. BMC Genomics. 2013;14(1):149.

Hayashida K, Hara Y, Abe T, Yamasaki C, Toyoda A, Kosuge T, et al. Comparative genome analysis of three eukaryotic parasites with differing abilities to transform leukocytes reveals key mediators of *Theileria*-induced leukocyte transformation. MBio. 2012;3:e00204-212. https://doi.org/10.1128/mBio.00204-12. PubMed: 22951932. 2012.

Howe D. Initiation of a Sarcocystis neurona expressed sequence tag (EST) sequencing project: a preliminary report. Vet Parasitol. 2001;95(2):233–9.

Howe DK, Gaji RY, Mroz-Barrett M, Gubbels M-J, Striepen B, Stamper S. Sarcocystis neurona merozoites express a family of immunogenic surface antigens that are orthologues of the *Toxoplasma gondii* surface antigens (SAGs) and SAG-related sequences. Infect Immun. 2005;73(2):1023–33.

Huang K-Y, Chien K-Y, Lin Y-C, Hsu W-M, Fong I-K, Huang P-J, et al. A proteome reference map of *Trichomonas vaginalis*. Parasitol Res. 2009;104(4):927.

Huang K-Y, Huang P-J, F-M K, Lin R, Alderete JF, Tang P. Comparative transcriptomic and proteomic analyses of *Trichomonas vaginalis* following adherence to fibronectin. Infect Immun. 2012;80(11):3900–11.

Hublin JSN, Ryan U, Trengove R, Maker G. Metabolomic profiling of faecal extracts from *Cryptosporidium parvum* infection in experimental mouse models. PLoS One. 2013;8(10):e77803.

Huynh M-H, Carruthers VB. Tagging of endogenous genes in a *Toxoplasma gondii* strain lacking Ku80. Eukaryot Cell. 2009;8(4):530–9.

Jiang L-L, Huang B, Han H-Y, Zhao Q, Dong H, Chen Z. Comparison of the proteome of the sporutated oocysts of *Eimeria tenella* diclazuril sensitive strain with diclazuril resistant strain. Chin J Biotechnol. 2005;21(3):435–9.

Johnson WC, Taus NS, Reif KE, Bohaliga GA, Kappmeyer LS, Ueti MW. Analysis of stage specific protein expression during *Babesia bovis* development within female *Rhipicephalus microplus*. J Proteome Res. 2017. https://doi.org/10.1021/acs.jproteome.6b00947.

Kepczynski M, Róg T. Functionalized lipids and surfactants for specific applications. Biochim Biophys Acta Biomembr. 2016;1858(10):2362–79.

Kim K, Weiss LM. Toxoplasma gondii: the model apicomplexan. Int J Parasitol Parasites. 2004;34(3):423–32.

Kim K, Soldati D, Boothroyd JC. Gene replacement in *Toxoplasma gondii* with chloramphenicol acetyltransferase as selectable marker. Science. 1993;262:911.

Kishima M, Dolan T, Njamunggeh R, Nkonge C, Spooner P. Humoral immune responses to *Theileria parva* in cattle as measured by two-dimensional western blotting. Parasitol Res. 1995;81(4):334–42.

Kolev NG, Franklin JB, Carmi S, Shi H, Michaeli S, Tschudi C. The transcriptome of the human pathogen *Trypanosoma brucei* at single-nucleotide resolution. PLoS Pathog. 2010;6(9):e1001090.

Kolev NG, Tschudi C, Ullu E. RNA interference in protozoan parasites: achievements and challenges. Eukaryot Cell. 2011;10(9):1156–63.

Kong H-H, Hwang M-Y, Kim H-K, Chung D-I. Expressed sequence tags (ESTs) analysis of *Acanthamoeba healyi*. Korean J Parasitol. 2001;39(2):151–60.

Kovarova J. Unravelling metabolism of *Leishmania* by metabolomics: University of Glasgow; 2016.

Kumar A, Misra P, Sisodia B, Shasany AK, Sundar S, Dube A. Mass spectrometry-based proteomic analysis of *Leishmania donovani* soluble proteins in Indian clinical isolate. Pathog Dis. 2014;70(1):84–7.

Laban A, Wirth DF. Transfection of *Leishmania enriettii* and expression of chloramphenicol acetyltransferase gene. Proc Natl Acad Sci. 1989;86(23):9119–23.

Lal K, Bromley E, Oakes R, Prieto JH, Sanderson SJ, Kurian D, et al. Proteomic comparison of four *Eimeria tenella* life-cycle stages: Unsporulated oocyst, sporulated oocyst, sporozoite and second-generation merozoite. Proteomics. 2009;9(19):4566–76.

Lander ES. The heroes of CRISPR. Cell. 2016;164(1):18–28.

Lau AO. An overview of the *Babesia*, *Plasmodium* and *Theileria* genomes: a comparative perspective. Mol Biochem Parasitol. 2009;164(1):1–8.

Lau AO, Tibbals DL, McElwain TF. *Babesia bovis*: the development of an expression oligonucleotide microarray. Exp Parasitol. 2007;117(1):93–8.

Laughery JM, Lau AO, White SN, Howell JM, Suarez CE. *Babesia bovis*: transcriptional analysis of rRNA gene unit expression. Exp Parasitol. 2009;123(1):45–50.

Lee MG-S, Van der Ploeg LH. Homologous recombination and stable transfection in the parasitic protozoan *Trypanosoma brucei*. Science. 1990;250(4987):1583–8.

Lee EG, Kim JH, Shin YS, Shin GW, Suh MD, Kim DY, et al. Establishment of a two-dimensional electrophoresis map for *Neospora caninum* tachyzoites by proteomics. Proteomics. 2003;3(12):2339–50.

Lee E-G, Kim J-H, Shin Y-S, Shin G-W, Kim Y-H, Kim G-S, et al. Two-dimensional gel electrophoresis and immunoblot analysis of *Neospora caninum* tachyzoites. J Vet Sci. 2004;5(2):139–46.

Lee E-G, Kim J-H, Shin Y-S, Shin G-W, Kim Y-R, Palaksha K, et al. Application of proteomics for comparison of proteome of *Neospora caninum* and *Toxoplasma gondii* tachyzoites. J Chromatogr B. 2005;815(1):305–14.

Levick MP, Blackwell JM, Connor V, Coulson RM, Miles A, Smith HE, et al. An expressed sequence tag analysis of a full-length, spliced-leader cDNA library from *Leishmania* major promastigotes. Mol Biochem Parasitol. 1996;76(1–2):345–8.

Li L, Brunk BP, Kissinger JC, Pape D, Tang K, Cole RH, et al. Gene discovery in the apicomplexa as revealed by EST sequencing and assembly of a comparative gene database. Genome Res. 2003;13(3):443–54.

Li Y, Shah-Simpson S, Okrah K, Belew AT, Choi J, Caradonna KL, et al. Transcriptome remodeling in *Trypanosoma cruzi* and human cells during intracellular infection. PLoS Pathog. 2016;12(4):e1005511.

Lv Z, Wu Z, Zhang L, Ji P, Cai Y, Luo S, et al. Genome mining offers a new starting point for parasitology research. Parasitol Res. 2015;114(2):399–409.

Ma D, Liu F. Genome editing and its applications in model organisms. Genomics Proteomics Bioinformatics. 2015;13(6):336–44.

Manger ID, Hehl A, Parmley S, Sibley LD, Marra M, Hillier L, et al. Expressed sequence tag analysis of the bradyzoite stage of *Toxoplasma gondii*: identification of developmentally regulated genes. Infect Immun. 1998;66(4):1632–7.

Mauzy MJ, Enomoto S, Lancto CA, Abrahamsen MS, Rutherford MS. The *Cryptosporidium parvum* transcriptome during *in vitro* development. PLoS One. 2012;7(3):e31715.

Meissner M, Agop-Nersesian C, Sullivan WJ. Molecular tools for analysis of gene function in parasitic microorganisms. Appl Microbiol Biotechnol. 2007;75(5):963–75.

Menezes JPBD, Almeida TFD, Petersen ALOA, Guedes CES, Mota M, Lima JGB, et al. Proteomic analysis reveals differentially expressed proteins in macrophages infected with *Leishmania amazonensis* or *Leishmania major*. Microb Infect. 2013;15(8):579–91.

Mesplet M, Palmer GH, Pedroni MJ, Echaide I, Florin-Christensen M, Schnittger L, et al. Genome-wide analysis of peptidase content and expression in a virulent and attenuated *Babesia bovis* strain pair. Mol Biochem Parasitol. 2011;179(2):111–3.

Min W, Lillehoj HS, Ashwell CM, Van Tassell CP, Dalloul RA, Matukumalli LK, et al. Expressed sequence tag analysis of *Eimeria*-stimulated intestinal intraepithelial lymphocytes in chickens. Mol Biotechnol. 2005;30(2):143–9.

Miska K, Fetterer R, Rosenberg G. Analysis of transcripts from intracellular stages of *Eimeria acervulina* using expressed sequence tags. J Parasitol. 2008;94(2):462–6.

Mogollon CM, van Pul FJ, Imai T, Ramesar J, Chevalley-Maurel S, de Roo GM, et al. Rapid generation of marker-free *P. falciparum* fluorescent reporter lines using modified CRISPR/Cas9 constructs and selection protocol. PLoS One. 2016;11(12):e0168362.

Nene V, Lee D, Kang'a S, Skilton R, Shah T, de Villiers E, et al. Genes transcribed in the salivary glands of female *Rhipicephalus appendiculatus* ticks infected with *Theileria parva*. Insect Biochem Mol Biol. 2004;34(10):1117–28.

Ng JSY, Ryan U, Trengove RD, Maker GL. Development of an untargeted metabolomics method for the analysis of human faecal samples using *Cryptosporidium*-infected samples. Mol Biochem Parasitol. 2012;185(2):145–50.

Ngô HM, Yang M, Joiner KA. Are rhoptries in Apicomplexan parasites secretory granules or secretory lysosomal granules? Mol Microbiol. 2004;52(6):1531–41.

Oakes RD, Kurian D, Bromley E, Ward C, Lal K, Blake DP, et al. The rhoptry proteome of *Eimeria tenella* sporozoites. Int J Parasitol. 2013;43(2):181–8.

Oura C, Tait A, Shiels B. *Theileria annulata*: identification, by differential mRNA display, of modulated host and parasite gene expression in cell lines that are competent or attenuated for differentiation to the merozoite. Exp Parasitol. 2001;98(1):10–9.

Oura CA, McKellar S, Swan DG, Okan E, Shiels BR. Infection of bovine cells by the protozoan parasite *Theileria annulata* modulates expression of the ISGylation system. Cell Microbiol. 2006;8(2):276–88.

Oyhenart J, Breccia JD. Evidence for repeated gene duplications in *Tritrichomonas foetus* supported by EST analysis and comparison with the *Trichomonas vaginalis* genome. Vet Parasitol. 2014;206(3–4):267–76.

Paba J, Santana JM, Teixeira AR, Fontes W, Sousa MV, Ricart CA. Proteomic analysis of the human pathogen *Trypanosoma cruzi*. Proteomics. 2004;4(4):1052–9.

Panigrahi AK, Ogata Y, Zíková A, Anupama A, Dalley RA, Acestor N, et al. A comprehensive analysis of *Trypanosoma brucei* mitochondrial proteome. Proteomics. 2009;9(2):434–50.

Pedroni MJ, Sondgeroth KS, Gallego-Lopez GM, Echaide I, Lau AO. Comparative transcriptome analysis of geographically distinct virulent and attenuated *Babesia bovis* strains reveals similar gene expression changes through attenuation. BMC Genomics. 2013;14(1):763.

Peng Z, Omaruddin R, Bateman E. Stable transfection of *Acanthamoeba castellanii*. Biochim Biophys Acta. 2005;1743(1):93–100.

Peng D, Kurup SP, Yao PY, Minning TA, Tarleton RL. CRISPR-Cas9-mediated single-gene and gene family disruption in *Trypanosoma cruzi*. MBio. 2015;6(1):e02097–14.

Qin M, Liu XY, Tang XM, Suo JX, Tao GR, Suo X. Transfection of *Eimeria* mitis with yellow fluorescent protein as reporter and the endogenous development of the transgenic parasite. PLoS One. 2014;9(12):e114188.

Rachinsky A, Guerrero FD, Scoles GA. Proteomic profiling of *Rhipicephalus* (*Boophilus*) *microplus* midgut responses to infection with Babesia bovis. Vet Parasitol. 2008;152(3):294–313.

Radke JR, Behnke MS, Mackey AJ, Radke JB, Roos DS, White MW. The transcriptome of *Toxoplasma gondii*. BMC Biol. 2005;3(1):26.

Reid AJ, Vermont SJ, Cotton JA, Harris D, Hill-Cawthorne GA, Könen-Waisman S, et al. Comparative genomics of the apicomplexan parasites *Toxoplasma gondii* and *Neospora caninum*: Coccidia differing in host range and transmission strategy. PLoS Pathog. 2012;8(3):e1002567.

Savadye D. Establishment of an expressed sequence-tag database, mapping of *theileria parva* schizont cDNAs and molecular characterisation of a parasite-specific protein: University of Zimbabwe; 1999.

Saxena A, Lahav T, Holland N, Aggarwal G, Anupama A, Huang Y, et al. Analysis of the *Leishmania donovani* transcriptome reveals an ordered progression of transient and permanent changes in gene expression during differentiation. Mol Biochem Parasitol. 2007;152(1):53–65.

Scheltema RA, Decuypere S, T'kindt R, Dujardin J-C, Coombs GH, Breitling R. The potential of metabolomics for *Leishmania* research in the post-genomics era. Parasitology. 2010;137(09):1291–302.

Schwarz RS, Fetterer RH, Rosenberg GH, Miska KB. Coccidian merozoite transcriptome analysis from *Eimeria maxima* in comparison to *Eimeria tenella* and *Eimeria acervulina*. J Parasitol. 2010;96(1):49–57.

Shen B, Brown KM, Lee TD, Sibley LD. Efficient gene disruption in diverse strains of *Toxoplasma gondii* using CRISPR/CAS9. MBio. 2014;5(3):e01114-14.

Shin Y-S, Shin G-W, Kim Y-R, Lee E-Y, Yang H-H, Palaksha K, et al. Comparison of proteome and antigenic proteome between two *Neospora caninum* isolates. Vet Parasitol. 2005;134(1):41–52.

Shompole S, McElwain TF, Jasmer DP, Hines SA, Katende J, Musoke AJ, et al. Identification of *Babesia bigemina* infected erythrocyte surface antigens containing epitopes conserved among strains. Parasite Immunol. 1994;16(3):119–27.

Siddiki A. Sporozoite proteome analysis of *Cryptosporidium parvum* by one-dimensional SDS-PAGE and liquid chromatography tandem mass spectrometry. J Vet Sci. 2013;14(2):107–14.

Siddiki A, Wastling JM. Charting the proteome of *Cryptosporidium parvum* sporozoites using sequence similarity-based BLAST searching. J Vet Sci. 2009;10(3):203–10.

Sidik SM, Huet D, Ganesan SM, Huynh MH, Wang T, Nasamu AS, Thiru P, Saeij JP, Carruthers VB, Niles JC, Lourido S. A genome-wide CRISPR screen in *Toxoplasma* identifies essential Apicomplexan genes. Cell. 2016;166(6):1423–1435.e12.

Silva MG, Graça T, Suarez CE, Knowles DP. Repertoire of *Theileria equi* immunodominant antigens bound by equine antibody. Mol Biochem Parasitol. 2013;188(2):109–15.

Silva JC, Cornillot E, McCracken C, Usmani-Brown S, Dwivedi A, Ifeonu OO, et al. Genome-wide diversity and gene expression profiling of *Babesia microti* isolates identify polymorphic genes that mediate host-pathogen interactions. Sci Rep. 2016a;6:35284.

Silva MG, Knowles DP, Suarez CE. Identification of interchangeable cross-species function of elongation factor-1 alpha promoters in *Babesia bigemina* and *Babesia bovis*. Parasit Vectors. 2016b;9(1):576.

Singh S, Dubey VK. Quantitative proteome analysis of *Leishmania donovani* under Spermidine starvation. PLoS One. 2016;11(4):e0154262.

Smolarz B, Wilczyński J, Nowakowska D. DNA repair mechanisms and *Toxoplasma gondii* infection. Arch Microbiol. 2014;196(1):1–8.

Snelling WJ, Lin Q, Moore JE, Millar BC, Tosini F, Pozio E, et al. Proteomics analysis and protein expression during sporozoite excystation of *Cryptosporidium parvum* (*Coccidia*, Apicomplexa). Mol Cell Proteomics. 2007;6(2):346–55.

Soldati D, Boothroyd JC. Transient transfection and expression in the obligate intracellular parasite *Toxoplasma gondii*. Science. 1993;260:349.

Strong WB, Nelson RG. Preliminary profile of the *Cryptosporidium parvum* genome: an expressed sequence tag and genome survey sequence analysis. Mol Biochem Parasitol. 2000;107(1):1–32.

Suarez CE, McElwain TF. Transient transfection of purified *Babesia bovis* merozoites. Exp Parasitol. 2008;118(4):498–504.

Suarez CE, McElwain TF. Stable expression of a GFP-BSD fusion protein in *Babesia bovis* merozoites. Int J Parasitol. 2009;39(3):289–97.

Suarez CE, McElwain TF. Transfection systems for *Babesia bovis*: a review of methods for the transient and stable expression of exogenous genes. Vet Parasitol. 2010;167(2):205–15.

Suarez CE, Palmer GH, Florin-Christensen M, Hines SA, Hötzel I, McElwain TF. Organization, transcription, and expression of rhoptry associated protein genes in the *Babesia bigemina* rap-1 locus. Mol Biochem Parasitol. 2003;127(2):101–12.

Suarez CE, Palmer GH, LeRoith T, Florin-Christensen M, Crabb B, McElwain TF. Intergenic regions in the rhoptry associated protein-1 (rap-1) locus promote exogenous gene expression in *Babesia bovis*. Int J Parasitol Parasites. 2004;34(10):1177–84.

Suarez CE, Norimine J, Lacy P, McElwain TF. Characterization and gene expression of *Babesia bovis* elongation factor-1α. Int J Parasitol Parasites. 2006;36(8):965–73.

Suarez C, Lacy P, Laughery J, Gonzalez MG, McElwain T. Optimization of *Babesia bovis* transfection methods. Parassitologia. 2007;49:67.

Suarez CE, Johnson WC, Herndon DR, Laughery JM, Davis WC. Integration of a transfected gene into the genome of *Babesia bovis* occurs by legitimate homologous recombination mechanisms. Mol Biochem Parasitol. 2015;202(2):23–8.

Sugimoto C, Conrad PA, Mutharia L, Dolan T, Brown W, Goddeeris B, et al. Phenotypic characterization of *Theileria parva* schizonts by two-dimensional gel electrophoresis. Parasitol Res. 1989a;76(1):1–7.

Sugimoto C, Mutharia LM, Conrad PA, Dolan TT, Brown WC, Goddeeris BM, et al. Protein changes in bovine lymphoblastoid cells induced by infection with the intracellular parasite *Theileria parva*. Mol Biochem Parasitol. 1989b;37(2):159–69.

Sugimoto C, Mutharia L, Brown W, Pearson T, Dolan T, Conrad PA. Analysis of *Theileria parva* immunodominant schizont surface antigen by two-dimensional polyacrylamide gel electrophoresis and immunoblotting. Parasitol Res. 1992;78(1):82–5.

Sundberg L-R, Pulkkinen K. Genome size evolution in macroparasites. Int J Parasitol. 2015;45(5):285–8.

Ten Asbroek AL, Ouellette M, Borst P. Targeted insertion of the neomycin phosphotransferase gene into the tubulin gene cluster of *Trypanosoma brucei*. Nature. 1990;348:174–5.

Urbaniak MD, Guther MLS, Ferguson MA. Comparative SILAC proteomic analysis of *Trypanosoma brucei* bloodstream and procyclic lifecycle stages. PLoS One. 2012;7(5):e36619.

Van Dijk M, Waters A, Janse C. Stable transfection of malaria parasite blood stages. Science. 1995; 268(5215):1358.

Veras PST, Bezerra de Menezes JP. Using proteomics to understand how *Leishmania* parasites survive inside the host and establish infection. Int J Mol Sci. 2016;17(8):1270.

Verdun RE, Di Paolo N, Urmenyi TP, Rondinelli E, Frasch AC, Sanchez DO. Gene discovery through expressed sequence tag sequencing in *Trypanosoma cruzi*. Infect Immun. 1998;66(11):5393–8.

Vichido R, Falcon A, Ramos JA, Alvarez A, Figueroa JV, Norimine J, et al. Expression analysis of heat shock protein 20 and Rhoptry-associated protein 1a in sexual stages and Kinetes of *Babesia bigemina*. Ann N Y Acad Sci. 2008;1149(1):136–40.

Vinayak S, Pawlowic MC, Sateriale A, Brooks CF, Studstill CJ, Bar-Peled Y, et al. Genetic modification of the diarrhoeal pathogen *Cryptosporidium parvum*. Nature. 2015;523(7561):477–80.

Westrop GD, Williams RA, Wang L, Zhang T, Watson DG, Silva AM, et al. Metabolomic analyses of *Leishmania* reveal multiple species differences and large differences in amino acid metabolism. PLoS One. 2015;10(9):e0136891.

Witschi M, Xia D, Sanderson S, Baumgartner M, Wastling J, Dobbelaere D. Proteomic analysis of the *Theileria annulata* schizont. Int J Parasitol Parasites. 2013;43(2):173–80.

Woehle C, Kusdian G, Radine C, Graur D, Landan G, Gould SB. The parasite *Trichomonas vaginalis* expresses thousands of pseudogenes and long non-coding RNAs independently from functional neighbouring genes. BMC Genomics. 2014;15(1):906.

Wright AV, Nuñez JK, Doudna JA. Biology and applications of CRISPR systems: harnessing nature's toolbox for genome engineering. Cell. 2016;164(1):29–44.

Xia D, Sanderson SJ, Jones AR, Prieto JH, Yates JR, Bromley E, et al. The proteome of *Toxoplasma gondii*: integration with the genome provides novel insights into gene expression and annotation. Genome Biol. 2008;9(7):R116.

Xu P, Widmer G, Wang Y, Ozaki LS, Alves JM, Serrano MG, et al. The genome of *Cryptosporidium hominis*. Nature. 2004;431(7012):1107–12.

Zhang W-W, Matlashewski G. CRISPR-Cas9-mediated genome editing in *Leishmania donovani*. MBio. 2015;6(4):e00861-15.

Zhang H, Guo F, Zhou H, Zhu G. Transcriptome analysis reveals unique metabolic features in the *Cryptosporidium parvum* Oocysts associated with environmental survival and stresses. BMC Genomics. 2012;13(1):647.

Zhou C-X, Zhou D-H, Elsheikha HM, Liu G-X, Suo X, Zhu X-Q. Global metabolomic profiling of mice brains following experimental infection with the cyst-forming *Toxoplasma gondii*. PLoS One. 2015;10(10):e0139635.

Zhou C-X, Zhou D-H, Elsheikha HM, Zhao Y, Suo X, Zhu X-Q. Metabolomic profiling of mice serum during toxoplasmosis progression using liquid chromatography-mass spectrometry. Sci Rep. 2016;6:19557.

The manufacturer's authorised representative in the EU is Springer Nature Customer Service Centre GmbH, Europaplatz 3, 69115 Heidelberg, Germany. If you have any concerns regarding our products, please contact ProductSafety@springernature.com

Printed and bound by CPI Group (UK) Ltd, Croydon, CR0 4YY
07/10/2026
02229322-0001